CANADIAN RED RIVER
AND
ASSINNIBOINE AND SASKATCHEWAN
EXPEDITIONS

NARRATIVE

OF

THE CANADIAN RED RIVER

EXPLORING EXPEDITION OF 1857

AND OF THE

ASSINNIBOINE AND SASKATCHEWAN

EXPLORING EXPEDITION OF 1858

BY

HENRY YOULE HIND, M.A. F.R.G.S.

PROFESSOR OF CHEMISTRY AND GEOLOGY IN THE UNIVERSITY OF TRINITY COLLEGE, TORONTO

In Charge of the Assinniboine and Saskatchewan Expedition

TWO VOLUMES IN ONE

M. G. HURTIG LTD.
Booksellers & Publishers
EDMONTON

International Standard Book No. 0-88830-044-1

First Hurtig edition published 1971

PRINTED IN JAPAN

INTRODUCTION TO THE NEW EDITION

How the world has shrunk since 1860, the year which saw the first publication of these two stout volumes. To the long process of world shrinkage, many have contributed—each in his own way, some with a careless passion, others with a stern dedication. By reason of his contribution to an increased knowledge of a large and important area of the world, the part that was played in this process by Henry Youle Hind, the author of these volumes, though not a starring role in the drama of exploration, is yet not an insignificant one. In the company of the great, he cuts no mean figure. He brought light into the darkness surrounding a substantial part of the vast area over which the Hudson's Bay Company had ruled as the "true and absolute Lordes and Proprietors" for nearly two hundred years.

Around the middle of last century the sands began to run out for the Hudson's Bay Company as an imperial factor. Opposition to its monopoly of trade and commerce began to solidify. Its license had to be renewed in 1859. Little was known of the vast territory over which it held dominion, except by fur traders who, in their own interest, were not telling what they knew. Curiosity about the suitability of the area for general settlement was growing apace. On the petition of the Canadian Parliament, the Imperial Parliament set up in 1857, a Parliamentary select committee on the Hudson's Bay Company "to consider the state of those British possessions in America which are under the administration of the Hudson's Bay Company or over which they possess a license of trade."

In the endeavour to push back the unknown frontier, both the Imperial and the Canadian governments organized expeditions to explore and report on an area which is today included in the provinces of Manitoba, Saskatchewan and Alberta. A British army officer, Captain John Palliser, was placed in charge of the British expedition. Palliser spent three years—1857, 1858 and 1859—in western exploration. The Canadian expedition confined itself to the years 1857 and 1858. Its purpose was "to ascertain the practicability of establishing an emigrant route between Lake Superior and Selkirk Settlement, and to acquire some knowledge of the natural capabilities and resources of the Valley of Red River and the Saskatchewan."

In 1857, a party under the nominal command of George Gladman, a retired Hudson's Bay Company official, but whose moving spirit was Henry Youle Hind, a professor of chemistry and geology at the University of Toronto, devoted its energies to the exploration of the Red River. In 1858, a second party, with Hind at its head, explored the Assiniboine and Saskatchewan Rivers.

Hind prosecuted his task of exploring a simpler and a sterner world than he had previously known with zeal and energy. Like a glutton, he gorged himself on the experiences which his travels afforded to him. These volumes are the fruits of those experiences.

In writing, Hind held his pen in a firm hand. If he does not qualify as a master stylist, he did have the supreme virtue of a prose writer. He was always in command of his meaning. He did make himself clear. If he had a fault, it was the Victorian passion for piling detail upon detail, in polysyllabic phrases, until the foundation of his discourse was in danger of collapsing. He told too much, rather than too little, thus dulling the edge of his narrative with too many words.

As an explorer, he did not suffer unduly from an itch to prove preconceived notions. He let the facts stand on their own. For the most part his narrative is objective. Now and then, he enlivened it by obtruding a personal experience; as, for example, when he wrote of his visit to the home of John Gowler, a successful pioneer farmer who invited him to stay to lunch.

Mrs. Gowler laid only one plate upon the table. "Mr. Gowler turned inquiringly to his wife, saying, 'And where is my plate?' 'Oh, John! you would not think of sitting at table with gentlemen?' Mr. John seemed puzzled for a moment; his son-in-law and children were looking in silence from different corners of the room. He cast a hasty glance around, and the true feelings of independence and manly right showed themselves as he exclaimed, 'Give me a chair and a plate; am I not a gentleman, too? Is not this my house, my farm, and these my victuals? Give me a plate.' "

In recounting this experience Hind commented on "how long old associations linger in the recollections of the European portion of the population in this remote area." European settlers imported with them into the wilderness their deep-seated social values, and in their new environment, it took some time for the old notion to die that Jack might not be as good a man as his master.

Gowler's farm was situated on the banks of the Assiniboine, some nine miles from Fort Garry. He farmed about fifty acres. In 1856, he sold potatoes for sixpence a bushel. He did not grow much wheat for there was little market for it. He had grown fifty-six measured bushels to the acre. Hind wrote of his root crops: "His turnips [Swedes] were magnificent; four of them weighed 70 lbs, two weighing 39 lbs, and two others 31 lbs. . . . A portion of the potato crop was still in the ground . . . Mr. Gowler very kindly turned them up out of the soil wherever I pointed out. I counted thirteen, fourteen, and sixteen potatoes, averaging three and a half inches in diameter, at each root respectively. The potatoes were planted on the 1st June, and were ready for eating on the 16th or 18th August." Hind also spoke of Indian corn, onions of rare dimensions, melons and tobacco, which Gowler grew at Red River with marked success.

Gowler explained to him the position of the farmer in Red River in these words: "If I found it worth my while, I could enclose 50, 100, or 500 acres, and from every acre get 30 to 40 bushels of wheat, year after year. I could grow Indian corn, barley, oats, flax, hemp, hops, turnips, tobacco, anything you wish, and to any amount, but what would be the use? There are no markets. . . ."

First hand observations and discussions with men who spoke from practical experience soon convinced Hind that the prairies were capable of absorbing many thousands of settlers and of providing them with a good living. He reported enthusiastically on the possibilities of opening the West to agriculture. But there were blind spots in his vision. He wrote that "every succeeding hour's experience shows the necessity of relying on personal observation alone in all that relates to the physical aspect of the country and its immense capabilities." But he did not always follow his own prescription. He accepted the current notion of a great country extending from the Lake of the Woods to the Rocky Mountains, give to this part of British America a more than passing interest. The idea of a route across the Continent of America lying wholly within British Territory, is daily becoming more settled and defined.

The trade of China and Japan, now on the point of being opened to British enterprise, the gold wealth of British Columbia, and the FERTILE BELT forming the northern boundary of the great American desert, all give importance to the Basin of Lake Winnipeg, which increases with our contemplation of its possible and indeed probable future.

The illimitable wastes of Siberia, extending over eighty degrees of longitude, are traversed by Russian couriers in far less time than with all our appliances of steam and telegraph, we can receive "news" from China. The same postal system which there prevails can be far more easily maintained in British America, and with this vast advantage, that from the Lake of the Woods to the Rocky Mountains the route would lie through a tract of country not only remarkably fertile, but possessing rich stores of timber for fuel, lignite coal, iron, and salt—the most important elements of industry and wealth.

The chief difficulty in the way of rapid transit across the continent lies between Lake Superior and Rainy Lake. The liberality which has already been manifested by the Parliament of Canada, in voting supplies to explore and open this line of communication, will doubtless be perse-

desert in the heart of the continent, which spilled over from the American side of the international border to the Canadian side. Thus he was led to underestimate the great potential of the prairies. He estimated that there were only 11,100,000 acres of arable land on the Red, Assiniboine and Saskatchewan Rivers. But there can be no quarrel with the enthusiasm with which he recommended that the area be opened up for immediate settlement. "It is a physical reality of the highest importance," he wrote, and his words are printed in capital letters in his book, "to the interests of British North America that this continuous belt can be settled and cultivated from a few miles west of the Lake of the Woods to the passes of the Rocky Mountains, and any line of communication, whether by waggon road or railroad, passing through it, will eventually enjoy the great advantage of being fed by an agricultural population from one extremity to the other."

With equal enthusiasm, he addressed himself to the problem of how to get settlers from their old homes to new homes on the prairies. He suggested a route by land and water which would bring Fort Garry within twenty-two days of Liverpool. "It is a question of great moment to the future colony in the Basin of Lake Winnipeg," he wrote, "of vast importance to Canada, and in view of the extraordinary gold-wealth of British Columbia, [it will be remembered that gold was discovered on the Fraser River in 1857] of paramount interest to the Imperial Government, if British laws, institutions, and civilization are to stretch across the American continent, and connect the eastern with the western hemisphere. As an emigrant route, the outlay of a few thousand pounds can make it available for summer communication. An emigrant could then start from Liverpool and proceed to Quebec by steam [eleven days], from Quebec to Collingwood, Lake Huron, by rail [two days], from Collingwood to Fort William by steamer [three days], and from Fort William to Fort Garry via Arrow Lake and the Boundary Line [six days], or twenty-two days in all from Liverpool to Selkirk Settlement."

His book contains pages of sober scientific facts relating to the geology, topography, climate and soil of the areas over which he travelled. When dealing with such matters he is on his own speciality and at his best. His comments on the Indians and the Métis whom he encountered on his travels do not bear repeating, based as they are on ignorance and prejudice. It is sufficient to say here that he believed that the world was made for the man whose skin is white.

It is unfortunate that the Indians, who "were a people, beginning," a people not yet fused, when their culture was overwhelmed by an alien culture which had had a long head start—it is unfortunate that the Indians had not reached the stage in their development when they, too, could have been guilty of "the sin of authorship" (the phrase is Susanna Moodie's). How interesting it would be to know what they thought of Hind and the many other members of his race who swarmed over their territories! How good it would be to have some first hand evidence for the other side of the record!

The hardships with which Hind had to contend as he journeyed along his way were light in comparison with those suffered by some of the earlier explorers. Lack of an occasional meal, unappetizing food, the possibility of an attack by hostile Indians, dangerous rapids and angry waters, muskeg and back-breaking portages, mosquitoes and grasshoppers—these about make up the tally. Mosquitoes were a great source of annoyance. On the west side of Lumpy Hill, Hind came upon some raspberries of great size and thought he would like to sample them. But they were guarded by hordes of mosquitoes. "I offered the Cree guide," he wrote, "a piece of tobacco for a tin cup full of raspberries, he tried to win it, but after a short struggle with these terrible insects he rushed from the hillside and buried his face in the smoke of the fire we had lit in the hope of expelling them from the neighborhood of our camp."

Grasshoppers were almost as much a problem as mosquitoes. "Their principal food," wrote Hind, "is the prairie grass and the leaves of shrubs, but they will attack any substances presented to them, even such indigestible articles as leather, travelling bags, woollen garments, saddle girths, and harness. In a few minutes they ate the varnish from the leather case of a telescope I left on the ground in 1858, and so disfigured a valise that the owner who had seen it sound and untouched a few minutes before we stopped to camp, could not recognize it after it had lain ten minutes on the grass."

All writing, from the most introspective poem to the most matter-of-fact scientific report, has an autobiographical basis. If one accepts Hind's book as an expression of his personality, as a reflection of his essential essence, he must have been a pedantic, rather narrow, school-masterish, self-opinionated and aggressive sort of man—one who had a firm belief in himself. And yet, he was a man who had a ready eye for all the manifold beauties of nature. His love of nature, particularly when she was in her more tempestuous moods, keeps breaking through the surface of his narrative. Here is a typical passage from his pen to stand for many such passages in which he manifested his delight in the beauties of nature: "At 8 p.m.," he wrote of early October, 1858, "the sky was quite free from clouds; the comet shone with brilliant lustre, a flashing aurora gradually spread over the northern sky, the stars glittered like diamonds in the south, and the whole heavens assumed that aspect of silent beauty which renders night in the wilderness so impressive and sublime."

Henry Youle Hind was born in England on June 1, 1823. He came to Canada in 1846, settling in Toronto, where he was employed first as a lecturer in chemistry and mathematics at the provincial normal school. He later joined the staff of the University of Trinity. He was in his middle thirties when he visited the rivers of the West. After finishing this assignment, he went on other extensive exploring expeditions and produced other substantial volumes on his travels and findings. In 1890, he became president of a church school in Edgehill, Nova Scotia. He died on August 9, 1908, at Windsor, Nova Scotia.

No longer need the serious student or the general reader search for copies of Hind's two volumes on his western explorations in reference libraries and blow the dust from their fading covers. Here is presented a bright new edition of a minor classic. It should interest all men and women who seek to cultivate a sense of the past, that they may better understand the present, and have fewer fears of what the future may hold in store.

"One foot of Future's glass should rest on Past.
Where Hist'ry is not, Prophecy is guess."

Roy St. George Stubbs

Winnipeg, Manitoba
June 9, 1970

PREFACE.

THE objects for which the Explorations described in these volumes were undertaken, necessarily involved a more minute topographical examination than would be thought necessary in a general survey of a comparatively unknown country.

It was desirable to ascertain the practicability of establishing an emigrant route between Lake Superior and Selkirk Settlement, and to acquire some knowledge of the natural capabilities and resources of the Valley of Red River and the Saskatchewan.

The country between Lake Superior and Red River is therefore minutely delineated with reference to the object of the exploration of 1857, and the first four chapters are mainly devoted to topographical details of less interest to the general reader than the subsequent narrative. The same remark applies, though in a less degree, to the description of the country west of Red River, the object being to show its fitness, or the contrary, for settlement.

The establishment of a new Colony in the Basin of Lake Winnipeg, and the discovery of a FERTILE BELT of

vered in until the route is well established. The Governor of British Columbia sees in "means of communication" the most expeditious way of calling the inert gold wealth of that distant colony into activity, and it remains for the Imperial Government to determine how soon a postal communication shall be established across the Basin of Lake Winnipeg, and the first step taken in establishing a permanent route through British Territory, between the Atlantic and the Pacific.

London, October 1860.

KA-KA-BEKA FALLS, KAMINISTIQUIA RIVER

see p. 36, vol. 1

NARRATIVE

OF

THE CANADIAN RED RIVER

EXPLORING EXPEDITION OF 1857

AND OF THE

ASSINNIBOINE AND SASKATCHEWAN

EXPLORING EXPEDITION OF 1858

BY

HENRY YOULE HIND, M.A. F.R.G.S.

PROFESSOR OF CHEMISTRY AND GEOLOGY IN THE UNIVERSITY OF TRINITY COLLEGE, TORONTO

In Charge of the Assinniboine and Saskatchewan Expedition

VOLUME I

CONTENTS

OF

THE FIRST VOLUME.

THE

CANADIAN RED RIVER EXPLORING EXPEDITION

OF 1857.

CHAPTER I.

TORONTO TO FORT WILLIAM. — LAKE SUPERIOR.

CHAP. II.

THE KAMINISTIQUIA ROUTE.—FORT WILLIAM.—LAKE SUPERIOR TO THE HEIGHT OF LAND.

CHAP. III.

THE HEIGHT OF LAND TO RAINY LAKE.

CHAP. IV.

RAINY LAKE TO THE SOURCE OF THE WINNIPEG RIVER.

CHAP. V.

THE WINNIPEG RIVER.

CHAP. VI.

RED RIVER SETTLEMENTS.

CHAP. VII.

THE WEST AND EAST BANKS OF RED RIVER, BETWEEN FORT GARRY AND THE BOUNDARY LINE.

CHAP. VIII.

BRIEF HISTORY OF THE COLONY.—STATISTICS OF POPULATION.—ADMINISTRATION OF JUSTICE.—TRADE AND OCCUPATIONS.

CHAP. IX.

THE MISSIONS AT RED RIVER.

CHAP. X.

EDUCATION IN THE SETTLEMENT.—AGRICULTURAL INDUSTRY.

CHAP. XI.

SKETCH OF THE COUNTRY WEST OF RED RIVER.

CHAP. XII.

THE JOURNEY TO CANADA VIA ST. PAUL.

THE

ASSINNIBOINE AND SASKATCHEWAN EXPLORING EXPEDITION

OF 1858.

CHAP. XIII.

FORT GARRY TO THE BOUNDARY LINE, VIA THE ASSINNIBOINE AND LITTLE SOURIS.

CHAP. XIV.

FROM THE BOUNDARY LINE TO THE QU'APPELLE LAKES VIÂ FORT ELLICE.

CHAP. XV.

THE QU'APPELLE VALLEY.—FROM THE MISSION TO SAND HILL LAKE.

CHAP. XVIII.

FROM THE ELBOW OF THE SOUTH BRANCH OF THE SASKATCHEWAN TO THE NEPOWEWIN MISSION ON THE MAIN SASKATCHEWAN.

CHAP. XIX.

FROM THE NEPOWEWIN MISSION ACROSS THE COUNTRY TO FORT ELLICE.

CHAP. XX.

THE QU'APPELLE VALLEY.—PORT PELLY TO THE SETTLEMENTS ON RED RIVER.

CHAP. XXI.

FROM FORT À LA CORNE, DOWN THE SASKATCHEWAN, TO THE GRAND RAPID AND LAKE WINNIPEG.

LIST OF ILLUSTRATIONS

IN

THE FIRST VOLUME.

*** *The following Illustrations are from Photographs taken by Mr.* Humphrey Lloyd Hime, *Photographer to the Assinniboine and Saskatchewan Expedition, or from Sketches by Mr.* John Fleming, *Assistant Surveyor and Draughtsman.*

CHROMOXYLOGRAPHS.

WOODCUTS.

MAPS AND PLANS.

THE

CANADIAN RED RIVER EXPLORING EXPEDITION

OF 1857

THE

CANADIAN RED RIVER EXPLORING EXPEDITION

OF 1857.

INTRODUCTION.

IN July, 1857, the Canadian Government organised and despatched an expedition to examine the country between Lake Superior and the Red River of the North, with a view to determine the best route for opening a communication between that lake and the settlements on Red River. The expedition consisted of the following members: —

GEORGE GLADMAN, Director, and one Assistant.
HENRY YOULE HIND, Geologist, and one Assistant.
W. H. E. NAPIER, Engineer, and five Assistants.
S. J. DAWSON, Surveyor, and four Assistants.
One road superintendent.

The *voyageurs* were composed of twelve Iroquois Indians from Caughnawaga, near Lachine, a Scotchman from the Ottawa, a French Canadian from Collingwood, a French Canadian (Lambert) from Fort William, a half-breed, engaged on board the steamer Collingwood, where he was employed in the capacity of cook, and twelve Ojibway Indians from Fort William, thus making the number of the party forty-four persons in all, when the canoe voyage commenced.

The following extracts from Mr. Gladman's instructions

will exhibit the designs of the Canadian Government in despatching this expedition: —

"The primary object of the expedition is to make a thorough examination of the tract of country between Lake Superior and Red River, by which may be determined the best route for opening a facile communication through British territory, from that lake to the Red River Settlements, and ultimately to the great tracts of cultivable land beyond them. With this view the following suggestions are offered for your guidance, so far as you will find them practicable, and supported by the topography.

"In the first place, after being landed at Fort William, to proceed by the present Hudson's Bay canoe route — by the Kaministiquia River, Dog Lake, Lake of the Thousand Islands, &c., to Lac La Croix, and thence by Rainy Lake, Lake of the Woods, Winnipeg River to Lake Winnipeg, and up the Red River to Fort Garry.

"From Rainy Lake to Lake Winnipeg, the route as at present affords a good navigation for boats of considerable size, with the interruption however of some short portages; but from Rainy Lake eastward to Lake Superior, the route is very much interrupted, and rendered laborious, tedious and expensive, by the great number of portages, some of considerable length, which have to be encountered to avoid the falls and rapids in the ravines and creeks which this route follows.

"For the establishment of a suitable communication for the important objects aimed at, it is believed that the construction of a road throughout, from some point on Lake Superior, probably either at Fort William, or at or near the mouth of the Pigeon River to Rainy Lake, must be undertaken. To ascertain, therefore, at present, by general exploration, what the route for this road should be, whether in the vicinity of the Hudson's Bay route, or by the line of country in which lies the chain of waters from Rainy Lake to the mouth of Pigeon River; this question can obviously be only satisfactorily determined by the difficult portions of both being tested instrumentally, but in either case, as the construction of such road would be a matter of time and much expense, it is considered necessary that the portages, &c.,

of either of the routes above described should be improved, so as to be made more available and facile, and to be auxiliary to the works of the road by facilitating the transport of men, supplies, &c.

"To determine, therefore, the portages to be improved, and the best mode of doing so, and whether the present reaches of canoe or boat navigation may not be further extended by the removal of shoals or the erection of dams, will be points to which you will direct the attention of the engineering and surveying branches of your party.

"From Rainy Lake by Lake of the Woods, and Lake Winnipeg to Fort Garry, as before described, is now comparatively a good water communication, but very circuitous; and should the character of Rat River, which rises at no great distance from the Lake of the Woods, and falls into the Red River above Fort Garry, be found susceptible of its being made a boat channel, a saving probably of 150 miles in length might be effected; or on an exploration of the country through which that river flows, it may be found more desirable to construct a road along it from Red River, and should this be so, the nature of the communication between Red River and Lake Superior eventually would be about 100 miles of road from Red River to the Lake of the Woods, thence about 140 miles of water communication to the eastern end of Rainy Lake, and from that point a continuous road to Lake Superior of from 160 to 200 miles in length." *

My duties in connection with this expedition are explained in the following instructions:—

"Secretary's Office,

"Toronto, 22nd July, 1857.

"Sir,—I have the honour to inform you that his Excellency, the Administrator of the Government, has been pleased to nominate you Geologist and Naturalist to the party which is to leave this city immediately for Fort William, for the purpose, in the first instance, of examining the lines and state of the

* Report on the Exploration of the country between Lake Superior and the Red River Settlement. Printed by order of the Legislative Assembly, 1858.—Instructions and Communications, page 5.

communication thence to Fort Garry, on the Red River. It being indispensable to the satisfactory result of the expedition, as well as to the safety of the party, that one individual should be invested with the general control and management of it, Mr. Gladman has been intrusted with this authority and responsibility, for which he is considered eminently qualified, from his long residence in the territory, his acquaintance with the leading lines of communication, with the trading posts, with the tribes of Indians with whom the party will necessarily come in contact, and with the extent and nature of the supplies which can safely be calculated on as procurable in the country during the course of the expedition. By him, therefore, will be regulated and determined the movements of the party, the routes to be taken and explored, and all matters connected with the provisioning and transport of the party, the hiring and payment of the men, and all other matters of detail whatever comprised in the general conduct of the expedition.

"From the nature of your duties, it may be necessary that you should occasionally separate yourself from the party. In such cases you will state so to Mr. Gladman, who will take care that you are provided with the necessary provisions and means of transport, and with all such necessaries as you may require; and he will arrange with you as to the places and times for your reuniting yourself with the main body.

"As you will require the services of an Assistant, the appointment of an efficient one is left with you, his remuneration not to exceed 20*l*. per month. That of the Geologist, Engineer, and Surveyor is fixed at thirty shillings per day each.

"The objects to which your attention is requested are of a general character, comprising a description of the main geological features of the country you traverse, and of whatever pertains to its natural history which you may have an opportunity of observing and recording.

"In relation to its geology, you will be guided by the memorandum furnished you by Sir William Logan; giving especial attention, as far as lies in your power, to the following points:

1. The boundaries of formations.
2. The distribution of limestone.

3. The collection of fossils.
4. The occurrence of economic minerals.
5. The exact position of all facts, and the attitude of the rocks.

"The distribution of limestone should be made a constant subject of question with every one you meet.

"With reference to natural history, you will, if at the time convenient, and the object capable of transportation, collect whatever may appear to be new or of interest; and you are requested to record in a daily journal, such facts in connection with this subject as may present themselves to your notice, when not susceptible of representation by specimen or illustration.

"A general description of the whole of the country you traverse, from Fort William westward, is very desirable; and it is advisable to note, as minutely as possible, all leading features of topography, vegetation, and soil, along your line of route.

"You will proceed with the main party to Fort William, and continue with it, or with such party as may be detached from it, as much as is consistent with the efficient prosecution of your own exploration and researches. It may, of course, be occasionally necessary, as already adverted to, that you should separate from the others for a short time, for which course Mr. Gladman will afford you all requisite accommodation; but as that gentleman's instructions require him to explore not only the present canoe route of the Hudson's Bay Company, from Fort William by Dog Lake, Lake of the Thousand Islands, Lac Croix, Lake of the Woods, and Lake Winnipeg, to Fort Garry, but also in returning to examine the former North West Company's route by Pigeon River; and further to survey or examine the line of Rat River, from the Red River to its source, and the intervening country between it and the Lake of the Woods; it is not probable that there will be much necessity for your leaving the party for more than a few days at a time, which is desirable, from its limited number and the late season of the year.

"It is arranged with Mr. Gladman, that he is to send a messenger, some time hence, with despatches to the Government, explanatory of the progress made towards carrying out the objects of the expedition; and by this means you will also have an opportunity of making such *ad interim* report as you may consider desirable. You will determine the return route to be taken by you and your assistant, whether by Lake Superior or by St. Paul, as you may be led to believe will most conduce to the attainment of the object of your branch of the exploration.

"When materials for illustrating the geology and natural history of the country accumulate, so as to render their transportation an inconvenience, you will hand them over in packages, properly made up and directed, to Mr. Gladman, who will take care that they are safely lodged at some of the posts, and arrangements made for their being securely conveyed to this city.

"Your reports and communications upon the various subjects to which your attention is directed will be addressed to the Hon. Provincial Secretary; and it is presumed to be unnecessary to impress upon you the propriety and expediency of taking care that the subject of such reports, and the results of your labour, shall be only so communicated.

"I have the honour, &c.

(Signed,) "T. L. TERRILL,

"Provincial Secretary.

"H. Y. Hind, Esq.

"Professor, &c., Trinity College."

CHAPTER I.

TORONTO TO FORT WILLIAM, LAKE SUPERIOR.

Departure of the Expedition from Toronto.—Squall on Lake Superior.—Fog Phenomena.—Steamer strikes on a Rock off Michipicoten Island.—Michipicoten Harbour.—Refraction.—Thunder Bay.—Pie Island.—Thunder Cape.—The Kaministiquia.—Fort William.—Mr. McIntyre.—Old North-West Company.—Accessibility of Lake Superior.—The Sault Ste. Marie Canal.—Trade of Lake Superior.—Ship Route to the Ocean.—Canadian Canals.—Elevation of Lake Superior.—Variations in the Levels of the great Canadian Lakes.—Influence and Importance of.—Obstacles to Progress westward of Lake Superior.—Dividing Ridges.—Connection of the Valley of the Mississippi with Lake Superior.—Access to the Valley of Rainy River.

On the 23rd of July, 1857, the Red River Expedition left Toronto for Collingwood, Lake Huron, and during the afternoon of the next day embarked on board the steamer Collingwood, bound to Fort William, Lake Superior. We passed through the magnificent locks of the Sault Ste. Marie Canal at 3 P.M. on the 27th, and when entering Lake Superior were met by an inposing but threatening spectacle, which instantly arrested and fixed the attention of all. A huge cloud, dense and black at its base, seemed to lie with one extremity resting on the Gros Cap de Superior, the other on Point Iroquois, the distance between those elevations being about six miles. The form of the cloud was that of a double cone, with the bases joined together, and the apices resting upon the opposite heights. A little attention showed that the cloud was in rapid motion towards us, and as it approached masses seemed to detach themselves from the main body, and be whirled or driven in its van. For a space of five degrees, the sky beneath the cloud, and between it and the horizon of Lake Superior, was clear and blue, and as the great mass of vapour slowly rose from the lake to the height of about eight degrees, its lower edge became brilliantly tinted by the rays of the sun, which it had hitherto completely obscured; below it a shadow of the deepest purple, sharply bounded by a greenish white line, added extreme beauty and singularity to the spectacle. Its form changed rapidly, and a white line of crested waves beneath it gave warning of an approaching squall, which soon came down with great force, and compelled us to seek shelter in Whisky Bay.

As soon as the morning dawned, preparations were made for steaming out of our harbour of refuge. Fogs, so common in Lake Superior, began to appear about 9 A. M., and continued throughout the day. Fog-bows, several degrees broad, very low, and of little variety of colour, were visible whenever the sun's light succeeded in penetrating our misty screen. On looking over the side of the vessel a double halo of very brilliant colours might be seen encircling the shadow of the observer's head projected on the dark coloured waters. Every man saw "his own halo," but not that of his neighbour. Towards evening a sharp look-out was kept for land on either hand. Caribou Island was supposed to be lying to the south, and Michipicoten Island to the north of our course. The day was very cold; thermometer at 3 P.M., 42°; at 4, 40°; and at 4.30, 39°.5, which was the lowest point it reached. The waters of the lake showed also a temperature of 39°.5. Mr. Wilson, a fellow-passenger, who has resided two summers on Michipicoten Island, says that the Lake Superior summer fogs begin about 9 A.M., and disappear generally at 10 or 11 P.M., but sometimes they last for a week. They are low, and from the mountain on Michipicoten Island, at an elevation of 800 feet above the lake, they may be seen resting on its waters as far as the eye can reach.

In consequence of the variation of the compass being reported to be much affected by local attractions in this part of Lake Superior, two of our Indians were placed in the bow to look out for land at the approach of night, and in addition to the usual watch, the captain, mate, and some of the passengers were walking the deck until past 11 o'clock. An evident feeling of anxiety was common to both passengers and crew; several of the former went to their berths without taking off their clothes. The night was extremely foggy; it was impossible to see more than a few yards beyond the bow of the vessel. The lead was cast several times, with no bottom at 288 feet. At a quarter to 12 P.M. no soundings were obtained with twenty fathoms; a few minutes afterwards the lead showed forty-five feet of water; the signal was given to stop her, and then to "back water," but it was too late, a harsh grating noise, a sudden uplifting of the bow of the steamer, and a very decided shock quivering through the vessel, told that she had struck. The alarm and anxiety inseparable from such an incident followed, and it was several minutes before a rapid inspection by torch light of the ledge of rock on which we had struck showed that there was no immediate danger to be apprehended. Anchors, chains, and fuel were moved aft, but all efforts to get the vessel off were without the least effect. Her bow was five feet out of the water, her stern in thirty-six feet water; the ledge on which she struck dipped gradually to the south-east, while on either hand, and not removed from the vessel more than fifteen or twenty feet, were huge masses of rock a few feet below the surface of the water.

When morning dawned, and the mists had partially cleared away, the steamer was found to be firmly lodged upon a low rocky island of about two acres in area, lying a mile south of Michipicoten Island, and about two miles from Michipicoten harbour. A boat was despatched to Michipicoten Island to procure timber for derricks, with which it was hoped that her bows might be raised from the ledge and the vessel slipped off into deep water, a result which was fortunately attained during the afternoon by the aid of derricks, steam, and a continued rolling from side to side by the united efforts of the passengers running with measured step from one side of the vessel to the other.

Steamer Collingwood on a rock near Michipicoten Island, Lake Superior.

It was soon ascertained that the sheeting was quite sound, and neither hull or machinery had sustained any material injury from the shock of the preceding night, and the efforts made to move the vessel. Nevertheless the captain thought it would be judicious to go into Michipicoten harbour and examine her more narrowly, as well as to shift the cargo and coal into their proper places. We reached the entrance of Michipicoten harbour in safety, but had scarcely advanced more than a few hundred yards when we again grounded on a shoal, and remained firmly fixed in a new position. In the evening the Agate Islands in Michipicoten harbour were visited, and very beautiful agates found in great abundance in the trap; but it was difficult to procure good specimens, on account of the hardness of the matrix.

During the afternoon the most singular effects of mirage were continually changing the outline of a few low rocks which projected above the level of the water, some two or three miles from us; and occasionally the steep wooded hills of Michipicoten Island seemed to be filled with brilliant ever-changing little lakes, which, if they preserved their apparent form for more than a minute, were most curiously delusive. All the phenomenon which sunshine and fog are capable of producing seem to be of constant occurrence near the middle of Lake Superior; not an hour passed during the daytime without our witnessing fog-bow, halo, or mirage of very singular beauty. At 4 P.M. on the 30th we steamed out of Michipicoten harbour, and pursued a straight course to Thunder Bay.

Early on the following morning we sighted the Paps and Isle Royale. The magnificent scenery of Thunder Bay and the adjacent shores of Lake Superior were gradually revealed as the mist slowly disappeared with the advancing day. Passed Thunder Cape at 2 P.M., and anchored off Fort William about half-past 4. The scenery of Thunder Bay is of the most imposing description. Pie Island, with its round eminence Le Pâté, 850 feet above the Lake, and Thunder Cape rising boldly 1350 feet, stand on either hand as you enter the deep inlet. Mackay's mountain uplifts a broad front to the height of 1000 feet on the mainland, in the direction of Fort William. The waters of Thunder Bay are coloured by the Kaministiquia for a considerable distance from the three mouths of that river.

Early on the 1st August the expedition, with baggage and stores, were landed at Fort William. Mr. McIntyre, the officer in charge of the Fort, received us with much courtesy and hospitality, kindly placing some of his appartments at our disposal for the night. The Iroquois made themselves comfortable under the canoes we had brought with us from Collingwood, which in the first instance were procured from Lachine, and transported by rail and steamboat to near the western extremity of Lake Superior.

The present position of Lake Superior and its tributaries in relation to Montreal or the Atlantic seaboard, is wholly changed since the period when the old North-West Company, established in 1783, and amalgamated with the Hudson Bay Company in 1821, maintained large establishments at Fort William and at Fort Charlotte, on the Pigeon River, some thirty-five miles in a south-westerly direction from the mouths of the Kaministiquia. In those days of canoe transportation, merchandise was conveyed up the Ottawa, across the height of land to Lake Huron, thence by the north shore of Lake Superior to Fort William, the starting point of the long journey into the great interior valleys of Red River, the Saskatchawan, and the Mackenzie. In these days ships can sail from European or Atlantic ports, and without breaking bulk, land their cargoes at Fort William for less than one-fiftieth part of the cost involved during the period when the North-West

Company became a powerful, wealthy, and influential body.

The completion of the Sault Ste. Marie Canal*, in May, 1855, established an uninterrupted water communication for sea-going vessels between Lake Superior and the ocean. The first ship which sailed from Chicago to Liverpool was the Dean Richmond, in 1856; this craft measured 379 tons American measurement, or 266 tons according to the English method of determining the tonnage of a vessel. Since that period the number of sea-going vessels from the Upper Lake ports has been increasing with great regularity. The trade of Lake Superior is also becoming of unexpected importance. In 1859, between the 1st day of June and the 1st November, the value of the different articles which passed through the St. Mary's Canal amounted to 5,703,433 dollars, and the number of passengers to 11,622. Fifteen years since three schooners constituted the entire fleet engaged in the Lake Superior trade. The number of vessels which passed through the St. Mary's Canal in the seasons of 1858 and 1859 were respectively 443 and 847, with a tonnage 149,307 and 304,860.†

The heights and distances enumerated in the subjoined Table, show a profile of this ship route between Anticosti, in the Gulf of St. Lawrence, and Fort William, at the mouth of the Kaministiquia River, Lake Superior.‡

* The Sault Ste. Marie Canal is one mile and an eighth in length, 70 feet wide at bottom, and 100 at water-line, depth 12 feet. The average lift of the locks is 17 feet 6 inches.

† Detroit Advertiser. From official returns.

‡ See a Map of the Province of Canada, showing the connection by steam navigation of the region of the great lakes with Europe, by the route of the St. Lawrence and the great lakes, prepared for the Canadian Commissioners of the Paris Exhibition, by Thomas Keefer, C. E., Montreal, 1855.

Names.	Distance from Anticosti in miles.	Elevation above the Sea level.	Number of Locks.	Length of Locks in feet.	Breadth of Locks in feet.	Total Lockage in feet.
Anticosti						
Quebec - - - - -	410					
Montreal - - - - -	590	14				
Lachine Canal - - -	598½	14–58	5	200	45	44¾
Beauharnois do. - - -	614	58·5–141·3	9	200	45	82½
Cornwall do. - - -	662½	142·6–185·6	7	200	45	43
Farren's Point do. - -	673½	190·5–195	1	200	45	4
Rapid Plat do. - -	688	195·3–207	2	-	-	12
Pt. Iroquois Canal - -	699½	207–213	1	-	-	6
Galops do. - - -	714½	213–225	2	-	-	8
Lake Ontario - - -	766	234				
Welland Canal - - -	1016	234–564	27	150	26½	330
Lake Erie - - - - -	1041	564				
Detroit River - - -	1280	564				
Lake St. Clair						
River St. Clair						
Lake Huron - - -	1355	573				
River Ste. Marie - -	1580	573–582·5				
Sault Ste. Marie Canal -	1650	582·5–600	2	550	75	17½
Lake Superior - - -	1650	600				
Fort William - - -	1910					
Superior City - - -	2030					

With the single exception of the Sault Ste. Marie Canal, all the great public works which have been contrived and executed for the purpose of reducing the obstacles to uninterrupted navigation between the great lakes and the ocean, lie within Canadian territory, and are under the control of the Canadian Government.*

According to the results of a recent survey of the Ottawa, made with a view to connect the St. Lawrence with Lake Huron by that river, the distance from the

* The cost of the construction of these remarkable links in the chain of unbroken communication which now penetrates a distance exceeding 2000 miles into the interior of the North American continent, approaches 15,000,000 dollars, and the annual revenue has risen from 131,000 dollars, in 1850, to 369,110 dollars in 1858.

mouth of the French River to Montreal by the route surveyed, is 430·76 miles, of which 35·18 miles is already a good navigation, requiring no improvement. Of the other 78·95 miles, 29·32 will require to be canal navigation, and 49·63 miles improved, so as to connect the whole into a first class navigation for vessels drawing 12 feet of water.

The cost, exclusive of deepening the Lachine Canal and Lake St. Louis, and apart from land damages and expenses, is estimated at 12,026,351 dollars.

This route would effect a saving of distance between Chicago and Montreal, over the existing one by the Welland Canal, of 343 miles; but with an increased lockage of 15 locks, and an additional rise and fall of 169·60 feet. The lake navigation by the existing route is 1145 miles in extent, and the inland or river 134; but by the Ottawa, the former is 575 miles, and the latter 401.*

The elevation of Lake Superior above the ocean has been variously estimated by different observers. Captain Bayfield considered it to be 627 feet above the level of the sea, which altitude is adopted by the narrators of Agassiz's tour in that region, and by Messrs. Foster and Whitney, in their report on the geology of the Lake Superior Land District. Sir William Logan, in his Geological Report for 1846–7, states that its surface is 597 feet above the ocean; in Professor Hall's Geology of the 4th District, N. Y., 596 feet is its assigned elevation. Sir John Richardson assumed its level to be 641 feet above the ocean.

The altitude deduced by Mr. Keefer for the map prepared for the Canadian Commissioners at the Paris Exhi-

* Report of the Commissioner of Public Works, 1859, based on the report of T. C. Clarke, Esq., C. E. Engineer Ottawa Survey.

bition in 1855, with the advantages and information derived from the levels obtained in the construction of various railways and canals from the ocean to Lake Superior, established a difference of only three feet in excess of that obtained by Sir William Logan in 1847.

The occasional fluctuations in the level of the waters of Lake Superior certainly exceed three feet, so that an elevation of 600 feet is probably a correct estimate of the mean height of the waters of this Kitchi-gum-mi*, or "Great Lake" of the Ojibways above the ocean.

In the region about Lake Superior the years 1845–6 were unusually dry, and in 1847 the lake had reached a very low stage of water. The years 1849–50 were wet, and the level of the lake in 1851 was from three to three and a half feet above the level of 1847.†

The variations in the levels of the Great Canadian Lakes are phenomena of the utmost importance to commercial interests.‡ The supply of water to the Erie and Welland Canals is dependent upon the relative height of the waters of Lake Erie. Periods of great anxiety have occurred among mercantile men at Buffalo respecting the supply of water to the great artery which unites Lake Erie with the Hudson River. If Lake Erie should subside to the zero of comparison adopted by Dr. Houghton, the depth of

* Spelt by Longfellow Gitche-Gumee, Big Sea Water (Hiawatha).

† Report on the Geology of the Lake Superior Land District, by J. W. Foster and J. D. Whitney, U.S., Geologists.

‡ The commerce of the Lakes is increasing with marvellous rapidity. Three thousand and sixty-five steamers passed up from Lake Erie to Lakes Huron and Superior, by Detroit, in 1859, and three thousand one hundred and twenty-one passed down. The greatest number up in a single day was eighty-five—down seventy-three. Detroit statistics show that five steamers, seven propellers, four barques, seven brigs, and eighty-five schooners were more or less engaged in the Lake Superior trade during the same year. Forty vessels left during the season for European and outward ports, some of which have returned, and one has taken her second departure.

water on the mitre sill at the Black Rock Guardlock would be less than five feet, through which all the water for the supply of a canal 150 miles long would have to flow. This contingency formed the subject of a memorial to the Legislature of the State of New York in 1854.

The following tables and memoranda are abbreviated from a paper by Major Lachlan, communicated to the Canadian Institute in July 1854.*

VARIATIONS IN THE LEVEL OF LAKE ERIE.

Date.	Comparative Level.	Authorities.
1790	1st *maximum*, being 5 ft. 6 in. above lowest level.	Hall, Higgins, Whittlesey, Mather, &c.
1795	1st *minimum*.	Weld, Whittlesey.
1801	2nd *maximum*.	Higgins, Houghton, Whiting.
1810	2nd *minimum*. Reported 6 feet below 1838.	Whittlesey.
1815	3rd *maximum*, 2 feet less than 1838.	Houghton, Higgins, &c.
1820	3rd *minim.* OR ZERO OF COMPARISON.	„ „ Whittlesey.
1827–30	4th *maximum*.	Houghton, Higgins, Whiting.
1832	4th *minimum*.	
1838	5th *maximum*. 5 ft. 3 in. above zero.	American Journal of Science, Dewey, &c. &c.
1846	5th *minimum*. 2 feet above zero.	
1853	6th *maximum*.	
1859	April, 5 feet 6 inches above zero of comparison.	

Years of maximum Level of Lake Erie.	Years of minimum Level of Lake Erie.
1790	1795
1801	1810
1815	1820 zero of comparison.
1827	1832
1838	1841
1853	1846
1859 5 ft. 6 in. above zero.	

* See Canadian Journal, 1st Series. On the Periodical Rise and Fall of the Lakes, by Major Lachlan.

Changes in lake levels are important physical phenomena in the basin of Lake Winnipeg. The rise or fall of a few feet in Lakes Maintobah and Winnipego-sis determines the character of an immense area of country on the low shores of those great inland lakes. Many hundred square miles of splendid pasturage are accessible at low lake levels, which are converted into marshes or swamps during periods of high water. The shores of the Great Canadian Lakes are generally high, and are not affected by a rise or fall of a few feet. The following additional notices of extraordinary changes in lake levels, which have occurred in the St. Lawrence valley, may tend to throw some light upon similar phenomena in the valley of Lake Winnipeg during recent periods.

From 1788 to 1790, the lakes generally, and Lake Erie in particular, are stated to have been as high as in 1838 (5 ft. 3 in. above zero). Professor Hall mentions evidence of a higher level than in 1838, as recorded by ridges and submerged trees.

In 1819 and 1820, the central and lower lakes are described as being unusually low. Lake Superior in 1827 and 1828 was lower than ever before known. In 1838 Lake Erie was 5 ft. 9 in. higher during the month of August than in 1819. Much land was overflowed, and trees of 100 years' growth destroyed.

In October, 1849, the water level of Lake Ontario was at a minimum; in June, 1853, it was 4 feet 5 inches above the minimum. In the winter of 1855 it again sank to the minimum; and during the summer of 1858 the rise amounted to 4 feet 3 inches. No less than 40 inches, or 3 feet 4 inches, of this rise in the mean level was attained during the summer of 1858.

As the result of observations extended over twelve years (1846 to 1857) in the variations of the level of Lake Ontario, the following facts have been established: —

1. The mean minimum level is attained in January or February.

2. The mean maximum level in June.

3. The mean annual variation is 25 inches.

4. The maximum variation in twelve years is 4 feet 6 inches.

5. There is no periodicity observable in the fluctuation of the lakes, and recent observations tend to show that there is no flux and reflux dependent upon lunar influence.*

The Bishop of Montreal states in his journal that it is only during an extraordinary concurrence of circumstances that the whole of Lake Superior can freeze over.† He was assured that this remarkable event happened in the winter of 1843, after a calm of four days, and during intensely cold weather. No other instance is said to be on record.

The greatest supposed depth of Lake Superior is 1200 feet. Its area is about 32,000 square miles, its coast line about 1,500 miles, and it contains probably 4,000 cubic miles of water.

The barrier which opposes further progress by steam or boat navigation westward of Lake Superior follows the general direction of the north-western and western coast of that lake. Near Fond du Lac, in the territory of the United States, the dividing ridge separating the valley of Lake Superior from that of the Mississippi, is distant from the St. Louis River about 18 miles in a southerly direction, and here the elevation of the ridge is 475 feet above the waters of the lake.

* Chas. Whittlesey and C. Dewey. American Journal of Science and Arts, May 1859.

† Journal of the Bishop of Montreal during a visit to the Church Missionary Society's North West American Missions.

The dividing ridge between the Embarras River, a tributary of the St. Louis River, and Vermilion River, which flows into the valley of Rainy Lake, is about 48 miles in an air line from the north-west coast of Lake Superior. On the Pigeon River, which forms the boundary between the United States and Canada, the dividing ridge separating the St. Lawrence from the Winnipeg basin is only 28 miles in an air line from the north-west coast of the same great water level, but by the canoe route on Pigeon River, the height of land is 53 miles from the coast.

Within the territory of the United States, the country between Lake Superior and the valley of the Mississippi presents no difficulties for the construction of a railroad. The valley of the Mississippi is in direct communication with that of the Red River of the north by traveled roads, so that the approach to the valley of Lake Winnipeg from the head of Lake Superior is only a question of time, and will not involve any considerable outlay when the necessities of the country, or of commerce, render the opening of this line of communication desirable.

Kettle River, flowing into the St. Croix, a tributary of the Mississippi, issues from a small lake not 20 miles from Lake Superior, and the distance of the navigable portion of the Mississippi, adjoining Sandy Lake, is scarcely 45 miles from Fond du Lac. The Mississippi is said to be navigable for steamers of light draught from Crow Wing to beyond this point, and Crow Wing is 130 miles from St. Paul by the traveled road, and less than 120 miles in an air line from Superior City.

St. Paul and Crow Wing will soon be connected by a railway. A large portion of the heavy work on this line is completed, and if no unforeseen events occur, the connection will have been established before the publication of this narrative. This important chain of communication will be referred to in succeeding pages.

The construction of a plank road between Superior City and Crow Wing, is already in contemplation, and the route is even now occasionally traveled. It will no doubt become of great commercial importance to the region of the Upper Mississippi and its numerous tributaries; and it is not improbable that its influence may rapidly extend to other water-sheds, viz. those of Rainy River, Red River, and the Saskatchewan.

In Canadian territory there are two established routes by which access is gained from the valley of Lake Superior to that of Rainy River. The most southerly of these is the old North-west Company's frontier route by Pigeon River, the second by the Kaministiquia River, which forms the subject of the first part of this narrative.

CHAP. II.

THE KAMINISTIQUIA ROUTE. — FORT WILLIAM. — LAKE SUPERIOR TO THE HEIGHT OF LAND.

Thunder Bay.—Fort William.—M'Kay's Mountain.—The Mission of the Immaculate Conception.—The Rev. Jean Pierre Choné.—Indian Treaty. —Mass.—Current River.—Garden at Fort William.—Remains of former Industry.—The first Brigade.—Iroquois and Ojibways.—A Dance.—The River.—Scenery of Kakabeka Falls.—Valley of the Kaministiquia.— Little Dog Lake.—The Great Dog Portage.— Little Dog River.—The Great Falls on Little Dog River.—Their Beauty.—Winter Road to Dog Lake.—Summer Road to Dog Lake.—Area of Dog Lake.—Description of.—Dog River.—Character of the Country.—Prairie River.—Upper Dog River.—Prairie Portage.—Viscous Lakes.—Description of Prairie Portage. —Atmospheric Phenomena.—Scarcity of Animal Life.

THUNDER BAY, which receives the waters of the Kaministiquia*, forms a portion of the north-west expansion of Lake Superior. It is the most southerly of three large and deep land-locked bays, which characterize that part of the coast; and it is situated between the parallels 48° 15′ and 48° 35′ north latitude, and in longitude 89°, and 89° 25′ west of Greenwich. Its greatest length in a north-easterly direction is 32 miles, and its breadth from Thunder Cape to the mouth of the Kaministiquia, upon which Fort William is situated, about 14 miles.

The main entrance to the bay is between the imposing headlands of Thunder Cape, 1,350 feet above the lake level, and Pie Island, 5 miles south-west of the Cape, with

* Spelt Kaministikwoya by Sir Jno. Richardson, "the river that runs far about."

an altitude of 850 feet. The depth of water in this broad entrance exceeds 180 feet, and a measure of 60 to 120 feet is maintained in many parts of the bay.

Seven miles south-east of Thunder Cape the lake is 630 feet deep, with a muddy bottom.

Immediately opposite, and east of the three mouths of the Kaministiquia, the Welcome Islands are distant about two miles, and inside of these islands from 30 to 60 feet of water is shown on Bayfield's chart. Within half a mile of the river's mouth the water shoals rapidly, and the bar has a variable depth of $3\frac{1}{2}$ to $5\frac{1}{2}$ feet water upon it; but within 1,000 yards of the north, or main channel, 12 to 14 feet water is maintained. Land is forming fast near the mouths of the river, and large areas in advance of the increasing delta, sustain a thick growth of rushes.

Fort William, looking up the river.

At a distance of about half a mile from the exit of the northern or main channel, Fort William is situated, upon the left or north bank. Opposite to it is a large island formed by the middle channel of the Kaministiquia, which branches off from the main stream about one and a half mile from the bay. In the time of the North-west Company, this island was denuded of the trees it sustained, which consisted mainly of tamarack, for fuel and other purposes, and the greater portion is now covered with second growth. A large area south of the fort still remains denuded of wood, and forms the site of an Ojibway village, besides serving as an excellent open pasture ground for a herd of cows belonging to the Hudson's Bay Company, which swim across the river every morning, a distance of 400 feet, and return at an early hour in the afternoon to the farm yard in the vicinity of the fort.

The banks of the river here are low and flat, not exceeding ten feet in altitude In the rear of the fort, tamarack of small but dense growth prevails. The soil is a light sandy loam reposing on yellowish clay.

Two miles above the fort, and in a direction nearly south from it, the third or southern outlet separates from the main channel. The banks of the river continue to rise above the level of its waters until they attain at the Mission of the Immaculate Conception, an altitude of 18 or 20 feet. Near the Mission the Indian Reserve of about 25 square miles begins; it embraces the best and largest area of cultivable land in the valley of the Kaministiquia, and much of it being situated on the flanks of McKay's mountain range, some portions possess many advantages which do not belong to the available tracts near the shores of Thunder Bay.

The general course of the river above the Mission for a distance of nine miles is towards the south-west, by very tortuous windings. Five miles from Fort William it approaches the base of the elevated but broken table land to which McKay's Mountain forms an imposing and abrupt termination. McKay's Mountain has an elevation of 1000 feet above the lake, and is the north-eastern boundary of an irregular but extended trap range, whose south-eastern flank follows the trend of the coast as far as Pigeon River.

It is worthy of remark, that the flanks of McKay's Mountain support a heavy growth of hardwood timber (maple, &c.), and through various sources I was informed that this heavily-timbered land stretches far to the south-west, on the side and borders of the trap range. The rock formations which comprise the country between the Kaministiquia and Pigeon Rivers indicate the presence of a fertile soil on the flank of the irregular table land; the trap with which the slates are associated giving rise upon disintegration to a soil of superior character. At the Mission, a light reddish loam constitutes the soil, having a depth of six feet, and resting upon a bluish grey clay, which extends to the water's edge.

The Mission of the Immaculate Conception is under the charge of the Rev. Jean Pierre Choné, who has resided on the banks of the Kaministiquia for nine years. From that gentleman, who kindly afforded me much information respecting this valley, I obtained numerous facts of interest in relation to its adaptation for settlement. At the Mission there are already congregated from thirty to thirty-five houses, substantially built of wood; in their general arrangement and construction they are far superior to the log houses of Canadian pioneers in the forest. Many of them had gardens attached to them, a few of which were in a good state of cultivation; some small fields fenced with post and rail were in the rear of the most thriving. The river here is from 60 to 70 yards wide, its waters are very turbid, with a current not exceeding two miles an hour.

M. Choné's room, into which we were admitted, gave

us a clue to the prosperity, cleanliness, and appearance of industry which distinguished the mission. A young tame partridge was hopping about the floor when we entered. A number of books occupied a small table in one corner, the other was taken up by a turning lathe, and various articles manufactured by the curé were lying about the room. A low bed covered with a buffalo robe filled another corner, and while we were conversing an old chief, dressed in scarlet cloth, quietly entered and placed himself on a chair by the side of a small carpenter's bench, which filled the remaining angle.

Among many interesting facts with which we were furnished by the kindness of M. Choné, we learned various particulars respecting the condition of the Indians and their relation to the Government of Canada, which an inspection of the treaty confirmed. In 1850 a treaty was concluded by the Hon. W. B. Robinson on behalf of Her Majesty and the Government of the Province with the Chiefs of the Ojibway Indians, inhabiting the northern shore of Lake Superior from Batchewanaung Bay to Pigeon River, and inland to the height of land between Canada and the territories in the occupation of the Hudson's Bay Company. For the sum of £2000 currency, and an annual payment of £200, to be paid at Fort William and Michipicoten, the chiefs surrendered all their right and title to the above territory, with the exception of the following reserves made over to them for the purposes of residence and cultivation, allowance being given under certain reasonable restrictions that they shall still hunt over the territory and fish in the waters as heretofore. The number of Indians included in this treaty was 1240. The reservations made for their benefit were as follow :—

First. For Joseph Peau de Chat and his tribe ; the reserve to commence about two miles from Fort William on the right bank of the river Kaministiquia, thence westerly six miles parallel to the shores of the Lake, thence northerly five miles, thence easterly to the right bank of the said river so as not to interfere with any acquired rights of the Honourable Hudson's Bay Company.

Second. Four miles square at Gros Cap for Po-to-mi-nai and tribe; and

Third. Four miles square on Gull River, near Lake Superior, on both sides of the river, for the chief Mish-i-muck-qua.

Our Iroquois being desirous of going to mass at the Mission on Sunday, August 2nd, several of the party accompanied them, and witnessed the rather rare spectacle of a numerous and most attentive Indian congregation engaged in Christian worship. The chapel is a very spacious and well-constructed building of wood, with a semi-circular ceiling painted light blue. The walls were panelled to the height of about four feet, and altogether the interior arrangements and decorations exceeded our anticipations, and everywhere showed the industrious hand or intelligent direction of the Rev. M. Choné. The Indians forming the regular congregation were arranged in the most orderly manner; the left side of the chapel being appropriated to the men and boys, the right to the women and girls. The boys and girls were placed in front of their seniors. The men were provided with forms, the women sat upon the floor. The utmost decorum prevailed throughout the service, and the chanting of both men and women was excellent, that of the squaws being remarkably low and sweet. Few of the male portion of the congregation took their eyes from the priest or their books during the service. The squaws drew their shawls or blankets over the head and showed the utmost attention. The Curé delivered a long sermon in the Ojibway language with much energy, and seemingly with the greatest fluency. After the ordinary service of the day was over, being before requested by one of our party, he delivered an admirable sermon in French. His style, language, and manner, were of a very superior order, and the drift of his words seemed to go far in shadowing forth the philanthropic impulses which sustained him in his solitary work of love, so remote from society, comfort, and civilisation.

In the afternoon I visited the mouth of Current River, six miles from Fort William. The river reaches the Lake by a succession of sloping falls over an argillaceous rock, which in the aggregate exceed forty feet in height within half a mile from the Lake. The common chive was found occupying in abundance the cracks and fissures of the shale on the banks of the river.

I visited during the day the garden of the fort; its area is about $1\frac{3}{4}$ acres. The shallots were small, but the potatoes looked well, being at the time in flower, and Mr. McIntyre thinks that varieties may be found which will ripen well near the fort. Tomatoes do not ripen here; turnips and cabbages are very liable to be destroyed by the cut-worm or grub; currant bushes procured from the forest flourish admirably, and produce a very large berry; the red currant was just beginning to ripen. This part of the country appears to abound in currants, raspberries, strawberries and gooseberries; they were seen growing in the woods in every direction, where direct light penetrated. A patch of oats in the garden showed a most remarkable development of stalk and leaf, and the ears were beginning to show themselves. The soil of the garden was brought from the foot of the Ka-ka-beka falls in the time of the North West Company's glory.

The average period when the Kaministiquia freezes, is

from the third to the fifteenth of November, and it becomes free from ice between the twentieth and twenty-third April. The year 1857 proved an exception in many respects; the ice did not pass out of the river until the thirteenth of May, and on the first of August, the day of my visit, the waters of the river were higher than they had ever been known before at that season of the year.

Indian corn will not succeed in this settlement, early and late frosts cutting it off. Frost occurs here under the influence of the cold expanse of Lake Superior, until the end of June, and begins again towards the end of August. A few miles further up the river, west of McKay's Mountain, the late and early frosts are of rare occurrence, and it was stated that Indian corn would ripen on the flanks of McKay's Mountain.

All kinds of small grain succeed well at the Mission, and the reason why they have not been more largely cultivated is owing to the want of a mill for the purpose of converting them into flour or meal. Near the lake, at Fort William for instance, oats do not always ripen; the cold air from the lake, whose surface, thirty and fifty miles from land, showed a temperature of 39° 5′, at the close of the hottest month of the year, is sufficient to prevent many kinds of vegetables from acquiring maturity, which succeed admirably four or five miles up the river.

Fragments of limestone have been procured in the neighbourhood, but the locality could not be pointed out by any of its inhabitants. The ruins of a lime kiln, used by the North West Company, have been discovered, and it is very probable that the limestone was obtained from crystalline layers, the existence of which has been established over wide areas in Thunder Bay, by Sir William Logan, and are noticed by him as being of a "reddish white colour, and very compact, some of which would yield good material for burning." These beds of impure limestone are mentioned by Mr. Murray (Geological Survey of Canada, for 1846-7) as occurring in the lower portions of the formation occupying this valley.

It is worthy of notice that substantial records of far more extensive settlements than now exist, showing a much higher degree of civilisation and improvement, are found at or near the various posts along this route, and particularly at Fort William.

Most of these remains of former industry and art, date from the time when the North West Company occupied the country, and there is reason to believe that much valuable knowledge respecting the resources of particular localities has been forgotten, or is hidden in the memories of those who may not have the opportunity to make it known.

Mr. Keating* mentions the ruins of the old Fort de Meuron, erected by Lord Selkirk. He was also shown the remains of a winter road opened by that enterprising nobleman, from the Kaministiquia to the Grand Portage on the Pigeon River, about thirty-six miles distant. The remains of a road to White Fish Lake is also still to be seen, and, indeed, it forms a winter route for half-breeds and Indians at the present day between the lakes on the Pigeon River, and the valley of the Kaministiquia. The Canadian government have recently laid out the valley of the Kaministiquia below the Ka-ka-beka Falls into two townships, named respectively PAI-POONGE and NEE-BING.

On the 3rd August we prepared for our immediate departure, and were all ready, with the exception of the

* Narrative of an Expedition to the Source of St. Peter's River.

Iroquois Indians, by 10 A.M. The delay with them arose from an indisposition to separate and be associated in different canoes with the Ojibways we were obliged to hire; by noon, however, an arrangement was made, it being determined that one brigade of three canoes should proceed at once, the other follow on the morrow. Just before starting a large body of heathen Indians, from the camp on the opposite side of the river, came over in a number of small canoes and commenced a dance outside of the pickets of the fort. They were painted and feathered in various ways, and furnished an admirable subject for our artists. Having danced on the outside of the fort for some minutes, they entered and arranged themselves in a semicircle in the quadrangle. The medicine-man and his assistant, gaudily painted and decked with eagles' feathers, sat on the ground beating a drum, and near to them squatted some half dozen squaws, with a few children. About sixty men and boys, headed by the chief, painted and feathered similar to the medicine man, danced or jumped round the ring. Our party being collected in front of the chief, he made a short speech, which was interpreted by a half-breed attached to the expedition to the following effect:—"They were happy to see us on the soil, they were hungry and required food, and trusted to our generosity and the plenty by which we were surrounded." The pipe of peace was then lit, and handed in turns for each to take a whiff. The picture of a hand across the mouth and cheek was admirably drawn in black on the faces of the chief and medicine-man. The Ka-ki-whe-on, or insignia, consisted of eagle's feathers stuck in a strip of red cloth about four feet long, and attached to a cedar pole. The whole scene was highly ridiculous, and many of the performers were wretched looking creatures, being dreadfully affected with scrofula. Some of the men, however, possessed splendid looking figures, but the progress of civilisation will soon close the history of these wretched Indians of the Kaministiquia.

Our first brigade, consisting of two large five fathom, and one middle size canoe, containing twenty-six men in all, started from Fort William at 5 P. M., and arrived opposite McKay's Mountain at about half-past six. Half a mile above the mission we noticed a very neat house in a clearing of about ten acres in extent, the last effort of civilisation to be seen, with the exception of an occasional post of the Hudson's Bay Company, for many hundred miles. The first camp was pitched about three-quarters of a mile beyond McKay's Mountain.

Opposite this magnificent exposure of trap, the clay banks of the river are about 14 feet high, and continue to rise on one side or the other until they attain an elevation of nearly 60 feet, often, however, retiring from the present bed of the river, and giving place to an alluvial terrace, some 8 or 10 feet in altitude, and clothed with the richest profusion of grasses and twining flowering plants. The current begins to be rapid about nine miles above Fort William soon after passing Point de Meuron, the site of the fort established by Lord Selkirk before referred to, and continues so, in the ascending course of the stream, to the foot of the first demi-portage, called the *Décharges des Paresseux*, where an exposure of shale creates the rapids which occasion the portage. The fall here is 5 feet 1 inch, in a space of 924 feet. The distance of this portage from the lake, by the windings of the river, is about 22¼ miles, and the total rise probably reaches 35 feet.

The current continues rapid to the foot of the Grand Falls, and high rock exposures commence on the precipitous banks three miles below them. These gradually assume the form of mural cliffs, capped with drift, increasing in altitude until they attain at the foot of the Grand Falls, the height of about 160 feet on the left bank, while on the opposite side of the river the mountain

Décharges des Paresseux.

portage path winds round the steep hill side of a bold projecting escarpment, 91 feet in altitude, and nearly half a mile from the falls.

At our camp, seven miles below the Grand or Ka-ka-beka Falls (cleft rock) as they are termed, the level of the river was estimated to be 40 feet above Lake Superior, and the foot of the falls 16 feet higher. The Grand Falls themselves were found by leveling to have an altitude of 119·05 feet, and involved a portage of 62 chains or three quarters of a mile. They are distant from the mouth of the river by its windings about 30 miles, and in an air line 17 miles.

As the altitude of these falls has attracted the attention of several observers, the different results obtained may not be without interest:—

	Feet.
Altitude ascertained by leveling (Mr. Dawson, August, 1857)	119·05
Capt. (now Col.) Lefroy, barometrical measurement	115·00
Mr. Murray, of the Canadian Geological Survey	119·00
Major Delafield	125·00
Sir John Richardson, barometrical measurement	127·00
Lieuts. Scott and Denny*	130·00

Assuming the height of Ka-ka-beka to be 119 feet, the summit will be 175 feet above Lake Superior. This result includes the rapids at the foot of the falls. The levels were taken along the portage path, and, if the rapids be deducted, the true height of Ka-ka-beka probably does not exceed 105 feet.

The scenery of the Grand Falls is extremely beautiful. The river precipitates its yellowish-brown waters over a sharp ledge into a narrow and profound gorge. The plateau above the portage cliff, and nearly on a level with the summit of the falls, is covered with a profusion of blueberries, strawberries, raspberries, pigeon cherry, and various flowering plants, among which the bluebell was most conspicuous. On the left side of the falls a loose talus is covered with wild mint and grasses which grow luxuriantly under the spray. Beautiful rainbows of very intense colour are continually projected on this talus, when the position of the sun and the clearness of the sky are favourable. Numerous small springs trickle down a perpendicular cliff of about 12 feet in altitude at the base of the talus, whose coolness and clearness, compared with the warm, coloured waters of the river, make them a delicious beverage, the difference

* Sir Jno. Richardson's Arctic Searching Expedition.

between the temperature of the springs and river being about 20°. The right side of the cliff at the falls is perpendicular for a height of more than 100 feet, and exposes the stratification with perfect fidelity. The peculiar rounded forms into which the rock divides itself, noticed by Mr. Murray, were well marked.

The alluvial valley of the river from about three miles below the mountain portage to Fort William varies in breadth from a few hundred yards to one mile; the breadth occupied by land of a quality which might fit it for agricultural purposes extends to near the summit of the flank of a low table land, which marks the true limit of the river valley, and the average breadth of this may be double that of the strictly alluvial portion.

The low table land is thinly wooded with small pine, and the soil is poor and dry; the alluvial valley sustains elm, aspen, balsam, poplar, ash, butternut, and a very luxuriant profusion of grasses, vetches, and climbing plants; among which the wild hop, honeysuckle, and convolvulus, are the most conspicuous. The rear portion of the valley, with an admixture of the trees just named, contains birch, balsam-spruce, white and black spruce, and some heavy aspens. The underbrush embraces hazelnut, cherries of two varieties, &c.

Occasionally the flanks of the low table land approach the river, contract the valley, and give an unfavourable aspect to the country. This occurs near the Décharges des Paresseux, and at most of the heavier rapids. The area available for agricultural purposes below the Grand Falls, probably exceeds twenty thousand acres, but if the flanks of McKay's Mountain be included in the estimate, a large addition may with propriety be assumed.

The Grand Falls mark the limits of a tract of country differing in many important physical aspects from the valley of the river lower down. From black argillaceous slates of Huronian (Cambrian) age we pass to a region in which granite, gneiss, and chloritic schist prevail, and where the vegetation is often scanty and poor.

The course of the river is now almost due north to Little Dog Lake, and its flow is much broken by falls and rapids, which occasion in a distance of nineteen miles, six portages and five discharges. The names, altitudes and distances from Fort William of the falls and portages are given in a table at the end of the second volume.

In the forests which lined the banks at the different discharges, the canoe birch was frequently seen eighteen

Second Falls, Kaministiquia River.

inches in diameter; the underbrush consisted chiefly of hazel nut. Whenever the gneissoid and syenitic rocks prevailed, the valley of the river was much contracted, the timber light, and the soil shallow and full of boulders or detached masses of rock. The volume of water in the river appeared to be very small, considering its unusual height at this season of the year: an approximate measurement at one of the rapids gave a breadth of seventy with an average depth of two feet.

FALL AT THIRD PORTAGE ABOVE KA-KA-BEKA

Extensive areas covered with burnt forest trees, consisting chiefly of pine, occur in the valley of the river as far as Little Dog Lake, when the formidable barrier of the Great Dog Mountain, sustaining a heavy growth of timber, comes into view. Occasionally aspens of large dimensions may be seen from the canoe, but it is not until the plateau of the Great Dog Mountain is attained that they acquire a diameter varying from eighteen to twenty-four inches, five feet from the ground. Trees of this species, and of the above dimensions, are found in abundance on the elevated barrier which separates the region of Great

Entrance to Little Dog Lake, from the Kaministiquia River.

Dog Lake from the valley of the Kaministiquia, 350 feet below.

The Great Dog Portage rises 490 feet above the level of Little Dog Lake, and at the point of greatest elevation the ridge cannot be less than 500 feet over the same lake. The difference between the levels of Little and Great Dog Lakes is 347·81 feet, and the length of the portage between them one mile and fifty-three chains. The view from the summit of the Great Dog is very striking; Little Dog Lake lies at our feet, an unbroken forest of pines dotted with groves of aspen and birch, and in the swamp portions with tamarack, stretches in all directions from east to west, being bounded in the view by the distant undulating outline of the wooded hills which limit the valley of the Kaministiquia. A portion of the abrupt escarpment of the elevated table land in the neighbourhood of McKay's Mountain is distinctly visible in clear weather.

The base of the Great Dog Mountain consists of a gneissoid rock supporting numerous boulders and fragments of the same materials. Micaceous rock was observed in position by Mr. Keating on the east side of the portage.* A level plateau of clay then occurs for about a quarter of a mile, from which rises, at a very acute angle and to an altitude of 283 feet above Little Dog Lake, an immense bank or ridge of stratified sand, holding small water worn pebbles. The bank of sand continues to the summit of the portage, or 185 feet above the clay plateau. The portage path does not pass over the highest part of the sand ridge; east of the path it is probable that its summit is 500 feet, as before stated, above Little Dog Lake.

In an endeavour to reach the head of Little Dog River†, before it begins to make in its short course of about four or five miles a precipitous descent of 347 feet, I found that much of the soil on the flanks of the Great Dog Mountain was far superior to the average quality in the valley of the Kaministiquia. It consisted of a clay loam, with a gravelly subsoil, containing numerous pebbles and

* Expedition to the Sources of the St. Peter's River.

† Little Dog River is a continuation of the Kaministiquia, but, in accordance with the Indian custom, it is named from the lake into which it flows.

BEGINNING OF GREAT DOG PORTAGE

water-worn fragments of rock; this was particularly noticed on the flanks and surface of the lower plateau.

The upturned roots of trees in the track of a tornado, which must have occurred here some years since, afforded an excellent opportunity for examining the soil and subsoil of the lowest plateau, and the flank of the upper one. The upturned roots of large aspens, birch, and pine showed everywhere a gravelly loam containing pebbles from one to six inches in diameter. On approaching the source of Little Dog River, a black spruce swamp was found to occupy an extensive area but little above the level of the river. The clay soil in this swamp was covered to the depth of two feet with moss, which was again largely overgrown with the Labrador tea plant. Small holes in the moss, filled with clear, cool, limpid fluid, afforded a striking contrast to the heated waters of the rivers and lakes; the temperature of these shallow wells did not exceed 42°, while the water of Great Dog Lake, tested a few hours afterwards (half-past 5 P. M.), was 69°, a difference of 27°.

The Great Dog Mountain derives its name from a murderous conflict between the Sioux and Ojibways, which occurred some centuries since on or near this eminence. The figure of a dog, in commemoration of this event, is carved on the side of the mountain. It was nearly obliterated when Major Long passed through the country in 1823, and we could not discover it in 1857. The Sioux and Ojibways were at war when the French traders and missionaries first visited the head of Lake Superior, which event may be placed as early as the year 1620.*

The great falls of Little Dog River are surprisingly

* Schoolcraft. History of the Indian Tribes of the United States. Part VI.

beautiful. The difference in level between Little and Great Dog Lake is 347 feet, which is descended by the foaming torrent in six successive leaps. The course of the canoe route lies some distance to the right of the falls, hence the reason why they have not been described by former travellers in these regions. In picturesque beauty they far surpass Ka-ka-beka, and would probably take rank with the most charming and attractive falls on the continent. They have not the grandeur of the Silver Falls on the Winnipeg, nor do they approach Niagara in magnificence or sublimity, but their extraordinary height, and the broken surface they present, impart to them singular and beautiful peculiarities. The strange aspect they must possess in winter, when fringed with masses of frozen spray, would probably be unrivalled ; and in spring, when the feeding lake is from three to four feet higher than during the summer months, their augmented volume would give them an appearance of magnitude which is lost when the waters are low, in consequence of the succession of ledges of rock over which they leap being partially screened by the foliage of overhanging trees.

The shoals, rapids, and falls on the Kaministiquia will always prevent that river from being used as a mean of communication with the interior for commercial purposes. The first large area of open water on this route is Dog Lake, and with a view to reach that elevated sheet of water a road from the shores of Lake Superior in as direct a line as possible will be required.

About six miles in a north-east by east direction from Fort William the waters of Current River are seen to fall over a precipitous ledge of black argillaceous slate, of Huronian age, within a few yards of their exit into Thunder Bay.

GREAT FALLS ON LITTLE DOG RIVER

The valley of this river forms the winter route of the Indians from Thunder Bay to Great Dog Lake, and while the Great Dog Portage, by the circuitous route of the Kaministiquia, is not less than 43 miles from Fort William, Great Dog Lake is reached by the Valley of Current River, in an eighteen or twenty miles march from Thunder Bay.

In making their winter journey to Great Dog Lake, the Indians generally proceed from the Mission in the neighbourhood of Fort William to the mouth of Current River, and ascend its open and unencumbered course, reaching Dog Lake in one day from Fort William. A cursory inspection of the map will show that the direct line of route from Fort William, or rather from Point Meuron, through the forest, if a track were cleared, would save several miles.*

The height of Great Dog Lake above Lake Superior is 710 feet, and to reach it in canoes or boats by the route of the Kaministiquia involves portages, which in the aggregate amount to 325 chains, or four miles in length.

As a mean of communication between Thunder Bay and Great Dog Lake, the Indian trail up the valley of Current River, appears to be of sufficient importance to require this special notice; a bird's-eye view of the country, from the summit of the Great Dog Portage, showed no mountainous range between that point and Lake Superior apparently equal in altitude to the great barrier of Dog Lake, which exceeds 850 feet above Lake Superior. It acquires additional importance from the fact that a travelled Indian canoe route and winter road exists

* In Current River speckled trout are numerous, and its valley abounds with red and black currants, raspberries, strawberries, and gooseberries, wherever sufficient light and air for their growth obtains admittance into the forest which covers the country.

between Dog Lake and Thousand Lakes, on the west side of the height of land.*

The area of Great Dog Lake, according to Mr. Murray†, whose opportunities of examining it were considerably greater than those of the members of the Exploring Expedition, probably exceeds 200 square miles; the country surrounding it is hilly, and covered with forests in which white spruce prevails, interspersed with groves of aspens, and occasionally dotted with the Weymouth and Banksian pines; white and yellow birch are abundant, and some of them of large dimensions. The lake is bounded by bold primary rocks, and studded with innumerable islands.

The traverse of the canoe route, from the head of the Great Dog Portage to the mouth of Dog River, is about 11 miles in length, and the lake is seen to stretch far to the north of the last named point; the canoe route follows closely the direction of its longest diameter, which is nearly due north and south. The depth of water, as ascertained by occasional soundings along the line of traverse, is very considerable; in one instance, 72 feet was recorded about 200 yards from a low rocky shore,

* During the year 1858, Mr. S. Dawson was employed in examining the country between Lake Superior and Red River with a view to establish a line of communication. In his report of the operations carried on during that year he speaks favourably of a line of road between Thunder Bay and Great Dog Lake. The subjoined extract is from Mr. Dawson's report for 1858:—"To commence at Lake Superior, a land road would be required from Thunder Bay to Dog Lake, as the navigation of the Kaministiquia is utterly impracticable, except for canoes, and could only be rendered otherwise at an enormous outlay. Dog Lake is distant from Lake Superior 22½ miles, and at a higher elevation by 718 feet, a difference of level which renders a canal out of the question, notwithstanding that the supply of water in the Kaministiquia would be ample. The only way of reaching it, therefore, is by land, and the surveys have progressed so far as to show that a good line may be obtained in a distance of 28 miles."

† Report of Progress for the year 1846-7. (Geological Survey of Canada.)

and another sounding showed 90 feet half a mile from land. The waters of Dog Lake at the time of our visit (Aug. 8th) appeared to be teeming with countless millions of animalculæ visible to the unassisted eye. The water marks showed an elevation slightly exceeding three feet above the level at the time of observation.

In making the traverse I timed the voyageurs at their paddles, and found they made one stroke a second, or sixty a minute, with remarkable accuracy. This would give for a day's work, from 5 A.M. to 7 P.M., with two hours stoppage, 12 hours, at 3600 strokes per hour, or 43,200 strokes a day.

We soon began to find that no feeling of sympathy existed between our Iroquois and Ojibway voyageurs; nor was any effort made by individuals of either nation to assist or enliven those of the other. As an instance of their utter indifference and selfishness, the following trait may be mentioned. One of the Iroquois from Caughnawaga was very ill at Fort William, and not only incapable of working, but unable to walk without assistance. He suffered much from chills and cramps, and was thought to be in a very dangerous condition; with care, medicine, and a good constitution, he grew a little better, and was able to eat, but the salt pork and salt beef we had with us were not very well adapted for a man so much reduced in flesh and slowly recovering from a severe illness. I shot a pigeon, and in the hearing of the Ojibways stated that it was to be cooked for the sick man. The following day I found that the Ojibway who picked up the pigeon had not only eaten it in the presence of the invalid Iroquois, but he and his companions had devoured two partridges which had also been reserved for the Iroquois. Another day, passing near the fire of the Ojibways, I found them roasting a pike which was intended for the dinner of the invalid. In future we committed the pigeons into the sick man's charge, and they were cooked by his friends, but the Ojibways laughed loud and long at the excellent practical joke they had enjoyed, and for many days after they reminded Lambert, the interpreter, about the sick man and his pigeons and partridges. "Tell him," said one waggish fellow, pointing to me, "shoot pigeon for Iroquois, Ojibway eat it, do Iroquois much good." A joke lasts an Indian a long time, and is continually repeated, both in canoe and in camp; it never appears to lose interest or grow stale.

The former extension of Dog Lake in a westerly direction for fourteen or fifteen miles, up the valley of the river of the same name, is shown by numerous sand ridges which intersect it nearly at right angles to its course, as well as by the probable former extension of a prolongation of the sand ridge which has been described as occurring at the Great Dog Portage, across the valley of the Little Dog River.

Great Dog Lake appears to be a centre of communication to which some degree of speculative interest may be attached; from one of the deep westerly bays, our guides pointed out the direction through which a communication with Thousand Lakes, on the other side of the water shed, has long been known to exist. No doubt the country through which this communication passes embraces extensive marshes, yet, if it avoids the objectionable ascent of Prairie River and Portage, it may be worthy of attention. Thousand Lakes, or Milles Lacs, as it is more commonly called, is eight hundred and thirty-two feet above Lake Superior, consequently one hundred and twenty-two feet above Dog Lake.

This route has long been known to the voyageurs and

Indians about Fort William, and the same may be remarked of many other routes of which the Indian guides speak and attempt to describe. Thirty-three years ago it was an old "path," and may have been one for centuries to the Indians of this region. Communications superior to those now travelled may yet be found, but it seems clear that until the water-shed of Rainy Lake is reached, no connection possessing sufficient water to form a boat route exists, or can be made without numerous dams.

Mr. Keating, so far back as 1823, relates that his party were shown an arm of the lake which extends to the south-west, and which they were informed connects Great Dog Lake by an uninterrupted water communication with the Thousand Lakes. The route is shorter than that by Prairie Portage, but broken by rapids. The same authority says that there is a communication between the Kaministiquia and Thousand Lakes passing more to the south than that from Dog Lake.* This is doubtless the Matawan River which joins the Kaministiquia at Couteau Portage, and rises within five miles of Milles Lacs.

So sluggish is the flow of Dog River, that a rise of ten feet in the level of the lake would push back its waters to a distance of thirty-five miles up the tortuous course of that stream, and the voyageurs relate that in the spring of the year they are accustomed to paddle their canoes over the tops of the willows which fringe its banks below the first rapids, fourteen miles in an air line from the mouth of the river; the greater portion of the intervening valley being then under water.

The banks of Dog River are altogether alluvial for some distance up the valley, with the occasional excep-

* Narrative of an Expedition to the source of the St. Peter's River, &c., &c., by Wm. H. Keating, A.M.S., 1824.

tion of abrupt sand-cliffs already noticed, which come upon the river and seem to form the termination of ridges, traversing the valley at nearly right angles to the course of the stream.

These ridges of sand are probably of very ancient date, and point to the period when the waters of Dog Lake were many feet higher than at present. The wearing away of the barrier at the mouth of Little Dog River would be sufficient to account for the former higher elevation. Recent water marks showed a rise of 5 feet within three miles of the mouth of the river. Further up the stream a rise of 6 feet was indicated. Its average breadth is about 80 feet in ordinary seasons; its general depth at this period of the year cannot be above 2 or 3 feet, indeed we were informed by our steersman, that he has often known canoes to be constantly impeded by shallows and sand bars when the level was probably much lower than during the present extraordinary season. The banks showed alder bushes, willows, dogwood, and tamarack.

The average height of the banks rises from 4 feet, a short distance from the mouth of the river, to 10 feet, 14 miles further up. At nearly every turn, newly formed oval and elongated ridges of sand protrude and show a general elevation of 5 feet above the present level. Low hills of granite begin to narrow the valley after passing a small stream coming from the north, and said to lead to a communication with the Nipigon.

From the summit of a low granite hill, perhaps 200 feet above the river bed, the surrounding country was distinctly mapped at our feet. The valley of the river appeared to have a breadth of a mile at our point of view, widening out in the direction of Dog Lake, and contracting towards the Height of Land between low ranges of hills, which did not seem anywhere to exceed 200—280 feet in altitude.

Some of the hills consisted of bare rock, others were covered with a young forest growth, which appeared to consist chiefly of Banksian pine and aspen. In the distance the tops of a few hills showed clumps of red pine standing erect and tall above the surrounding forest. They may be the remnants of an ancient growth which probably once covered a large portion of this region, having been destroyed by fire at different epochs; wide areas were still strewed with the blackened trunks of trees, and in the young forest, which seems fresh and green at a distance, the ground was found to sustain the charred remains of what had once been a far more vigorous vegetation.

The low ranges of hills bear a great outward resemblance to those which surround Dog Lake. No precipitous escarpments are visible, but most of them have a rounded dome-like aspect, and close inspection of some of them gave indications of the abrading action of ice. Large quantities of Labrador tea (*Ledum palustre*), were seen wherever we landed. The flow of the river for a distance of twenty-five miles from Dog Lake, varies from half a mile to one mile an hour.

The general character of this valley is very uniform, and the idea presented to the mind, in endeavouring to picture its aspect when covered with water in the spring, was that a general rise of 20 or 25 feet would give it an appearance very similar to Great Dog Lake, with analogous deep bays formed by the valleys of its tributaries, and having on its shores hills of the same altitude and similar formation as are found bordering the lake below; in fact, a high (25 feet) dam, as has already been hinted, at the source of Little Dog River, might perhaps convert Dog Lake into a magnificent sheet of water, having in a westerly direction a farther extension of at least 15 miles. It remains, however, to be ascertained whether Dog Lake has not other outlets than the one which leads through Little Dog River. It is not at all improbable that this may be the case.

At our camp on the 9th of August, near the head of a fall of 3½ feet named Barriere Portage, about 3 miles below the mouth of Prairie River, blue berries, not yet ripe, were very abundant, showing a marked difference in the climate of this spot and the Grand Falls, where some days before we had found them perfectly ripe and in the greatest profusion. The difference in elevation is about 542 feet. A quarter of a mile from the camp, in our course up the river, we came upon a bare granite hill, about 250 feet high, rising from the water's edge at an angle of nearly 45°. Its surface consisted of smooth rounded ridges, and 15 feet above the river a collection of water-worn boulders, from 6 inches to 2 feet in diameter, were deposited upon a ledge, leading to the inference that they had been left there by ice during spring freshets, and so far showing some confirmation of the statements of the Indians respecting the remarkable rise of water in the valley during the spring months.

The last portage on Dog River, following the canoe route to Fort Francis, is the Jourdain Portage, four miles in an air line from the height of land, and thirty-seven miles from Dog Lake by the windings of the river, according to Mr. Gaudet, who measured the distance in 1858. It involves an ascent of 8·60 feet by a portage 6½ chains long. A very short distance above it, the mouth and windings of Prairie River are seen with difficulty through the tall rushes which seek to conceal its course for a distance of 200 or 300 yards. Up this little streamlet, scarcely ten feet broad, the canoe route lies, while Dog River, still measuring a breadth of forty feet, can be traced far to the north by a succession of small lakes and ponds which mark its course.

Mr. Murray, of the Canadian Geological Survey, ascended Dog River up to its feeding marsh in 1847, and describes its course after receiving Prairie River, through which our route lay, "as turning off nearly due north, and widening out into a long narrow lake for about two or three miles, after which there follows in the same line a chain of twelve small lakes, or ponds, connected by short rapid streams, comprised within the distance of ten to twelve miles. The uppermost pond appeared at its northern extremity to terminate in a great marsh, which was supposed to be the ultimate source of the river, and to extend far and wide along the height of land, probably joining the Great Marsh of the Savannah Portage on the Red River route."*

Prairie River for a few hundred yards is so thickly fringed with rushes that two canoes cannot proceed side by side, or even pass one another with facility. The distance to Cold Water Lake is about 1¾ mile in an air line, and perhaps nearly double that distance by the windings of Prairie River, whose general course is a few degrees to the south of west. Much of the route towards the high barrier of land at Cold Water Lake, which now comes into view, lies through small marshy lakes or ponds, three in number, very shallow, and much encumbered with aquatic plants. The third or last lake, called Muddy Lake, is about 200 yards long and 100 yards wide. The voyageurs all complained of the great difficulty they experienced in paddling through this small

* Report of Progress, 1846-7.

and shallow sheet of water. It has long been celebrated as the Viscous Lake. Mr. Keating notices this supposed property of the Viscous Lake, but remarks at the same time that no such character was observed when he passed through it. A lake with a similar reputation occurs at the height of land on the Pigeon River route. During our voyage through it in 1858, the voyageurs then persisted in the statement that they experienced great difficulty in urging the canoes forward; one of the gentlemen attached to the expedition, after practically testing the resistance, expressed a strong opinion in unison with that of the voyageurs. The lake was only about three feet deep, and a paddle could be thrust into the soft slime as far as it would reach.

The barrier behind Cold Water Lake, stretching far to the north and south, may rise 220 feet, the western extremity of the portage path, according to measurement, being 157 feet above the lake. It constitutes the great and formidable prairie, or height of land Portage, two miles and five-eighths of a mile long. Cold Water Lake is well named on account of its temperature. Careful observation made it 41·5°, and the large spring or source which feeds it, and gives rise to the Prairie River, one of the sources of the great St. Lawrence, gushes out of the rocky side of the barrier, about 50 feet above the lake, with a temperature of 39·5°. Prairie Portage passes over the height of land, but not the highest land on the route, and its course lies first south-west up a steep wooded hill, without rock exposure, but composed of drift clays, sand, and numerous boulders; it then enters a narrow valley, which terminates in a small lake, about five acres in area and 20 feet deep, occupying a hollow among the hills on the height of land. The portage path continues on in the same direction until the Height of

Land Lake is reached, a small sheet of water, about a square mile in area, and 157 feet above Cold Water Lake. The utmost elevation attained on the Prairie Portage is probably 190 feet above Cold Water Lake, or nearly 900 feet above Lake Superior. No hill within view appeared to possess an elevation exceeding 20 or 30 feet above this limit.

Prairie Portage sustains some spruce and pine of fair dimensions; one *Pinus Banksiana* measured 5 feet 9 inches in circumference four feet from the ground, and many of equal dimensions were seen in the neighbourhood. A considerable portion of the timber is burnt, and the underbrush everywhere shows a profusion of hazel nut, with small shrubs and plants, such as raspberries, blue berries, gooseberries, and strawberries, all of which were here gathered ripe. The Labrador tea (*Ledum palustre*) grew in great profusion in particular spots, and at the termination of the portage, near the Height of Land Lake, the fragrant Indian tea plant (*Ledum latifolium*) abounded in the moss bordering this elevated sheet of water, which is 885 feet above Lake Superior, or 1,485 above the sea.

The following estimates of the heights of Prairie Portage above the sea, are taken from Sir John Richardson's "Arctic Searching Expedition."*

	Feet.
Dog Lake, above Lake Superior	657
Ascent of Dog River	14
Portage to Cold Water Lake	2
West end of Prairie Portage, and Middle Portage	161
Lake Superior above the sea	641
Height of Prairie or Middle Portage above the sea	1475

"In 1849, the height of the upper end of Dog Portage

* Arctic Searching Expedition; a Journal of a Boat Voyage through Rupert's Land and the Arctic Sea, in search of the Discovery Ships under Sir J. Franklin, by Sir John Richardson, C. B.

was ascertained by me with Delcro's barometer. In the previous season the aneroid barometer gave 328 feet as the height, which was a greater degree of accordance between the instruments than I generally found. Major Long estimates the watershed between Lakes Winnipeg and Superior, at 1200 feet above the tide. Major Delafield calculates the height of Cold Water Lake at 505, to which if 161 be added for the Prairie Portage, and 641 for Lake Superior, we have 1307 feet for the height of Prairie Portage over the sea. Captain Lefroy, by barometrical measurements, made in connection with the Observatory at Toronto, makes the west end of Prairie Portage 1361 feet above the sea; but the distance between the two places of observation renders the result liable to some error."

At our camp on the Height of Land (Aug. 12th) an atmospheric phenomenon of singular beauty occurred. The night was very beautiful and calm. The moon shone with great clearness and brilliancy, and numerous meteors darted through the sky in the south and west. Early in the morning, before daylight, I noticed a distinct arch of what at first sight I mistook for an aurora, but, observing its position to be nearly due west, referred it to very elevated clouds illumined by the sun's light. Its appearance was like that of a dim auroral arch, well defined, and forming the complete segment of a circle to the height of 45°, its form being persistent as long as observed. The remaining portion of the sky was clear, the moon and planets shining at the time with a very brilliant lustre. It occurred to me that it might be the forerunner of a storm, an idea which the rising sun, lighting up the tops of the trees beneath a perfectly cloudless sky about an hour afterwards, banished for a few hours. Towards noon the sky became overcast from the south-west. About half-past 3 thunder was heard in the distance, and at 4, scud from the south-east began to traverse the sky. At 5 P. M. the clouds in the south-west presented a very magnificent spectacle; they seemed like gigantic waves setting towards the north east. This wave-like appearance occurred in different parts of the heavens, and almost every variety of cloud passed in review. A few minutes before 5 P. M. a very long and vivid flash of lightning shot across the sky in a direction from south to north, succeeded by a distinct snap like that produced by an electrifying machine. About ten seconds afterwards the loud rolling thunder recorded the flash, and at 5 P. M. the rain commenced; the lightning was intensely vivid, and the thunder unusually loud.*

The scarcity of animal life at this season of the year on the canoe route has several times been remarked by travelers. It is probable that the noise inseparable from the passage of several canoes through the lakes and rivers would drive away the game into the interior, but their tracks would be seen if they existed in large numbers. On the Kaministiquia the following animals or their fresh tracks were seen: Of quadrupeds: Cariboos, bears, foxes, hares, minks, otters, squirrels, muskrats, and fieldmice. Of birds: eagles, hawks, ducks, pigeons, plover (two varieties), sandpipers, cherry birds, loons, partridges (two varieties), jays, magpies, blackcaps, nighthawks, Canadian nightingales, swallows, humming birds, kingfishers, and owls. There were shot by different members of the expedition, the ruffled grouse or hardwood partridge, spruce or cedar partridge, pigeons, plover (two varieties), squirrels, and one jay. Of fish we

* A thunder storm occurred at Toronto this day from 8.30 to 11.30 P.M. See remarks in the Toronto Meteorological Register for August. Canadian Journal, November 1857.

caught one pike. There is no doubt that we might have procured an abundance of pike, but a fear of retarding progress by the drag of the trolling line and its interference with the steersman in the tortuous course of the rivers, as well as in the lakes, led to the discontinuance of its use, until necessity should advise us to adopt it as a means of procuring food.

Mr. Keating states, in 1823, that from Rainy Lake to Lake Superior they did not meet with a single quadruped. The only animals they saw were about thirty or forty birds, chiefly ducks Among the birds observed were the Canadian jay (*Garrulus Canadensis*); blue jay (*Garrulus cristatus*); hairy woodpecker, Indian hen, golden plover, and woodcock. Partridges (*Tetrao umbellus*) were killed; a whip-poor-will was heard, and a rail seen.

The Bishop of Montreal in 1844 saw one wolf, some of the smaller quadrupeds, innumerable ducks, many loons, some other aquatic birds and a few of the heron tribe.

CHAP. III.

THE HEIGHT OF LAND TO RAINY LAKE.

The Height of Land Lake.—Savanne Lake.—Savanne Portage.—Savanne River.—Mr. Bell from the Mackenzie.—A youthful Traveller.—Milles Lacs.—The Seine River.— Character of the Seine.—Baril Lake.—Ancient Forest.— Cannibal Lake.— Cannibalism.— Brulé Lake.— Pickerel Lake.—Portage des Morts.—Death of a Voyageur.—Doré Lake.—French Portage.—Sturgeon Lake and River.—Picturesque Character of Sturgeon Lake.—Lac de la Croix.—Rattlesnake Portage.—Morning in the Wilderness.—Nameaukan Rapids.—Narrow Escape.—Nameaukan River.—Nameaukan Lake.—Rainy Lake.

THE marshy lake which stretches along the narrow level plateau forming the Height of Land in this region, is about one-third of a mile broad, but its length from the north-west to the south-east could not be determined on account of the vast expanse of rushes, with islands of tamarack, which seemed to blend it with an extensive marsh stretching far in both directions. Its elevation above the sea is 1485 feet. A portage about half a mile in length, letting us down 16⅓ feet, brings Savanne Lake into view. The shores of this small but reedy expanse of water are fringed with Labrador and Indian tea, and here, for the first time, the beautiful Indian Cup or Pitcher Plant (*Sarracenia purpurea*), once so common at the Grenadier's Pond near Humber Bay, Lake Ontario, was seen in great profusion. From the topmost branches of a pine tree which I ascended, a slight depression to the north and north-east of the dividing ridge was observed in the generally level outline of the horizon; by this depression it is not improbable that the waters of the Height of Land Lake and its connecting swamps drain into Dog River. With this exception the horizon appeared to be perfectly uniform, the slight difference in the height of the tamaracks and spruces, which seemed most to abound, furnishing the only deviation from a level expanse in all other directions.

The Savanne Lake with its feeding swamps may therefore be considered to be the source of the waters which, in this latitude, send tributaries to Hudson's Bay; although the Indians say that there exists a connection between the Height of Land Lake and Savanne Lake; the portage between them is named Portage de Milieu; it passes over a low sandy ridge supporting small pine, with tamarack and spruce at its foot. The connections, indeed, which exist between different water-sheds by means of swamps at the Height of Land, impassable even to a small canoe, are by no means of rare occurrence. In the present case we have the Height of Land Lake sending water both to the St. Lawrence and to Hudson's Bay; but if we go a little farther south, we find that in the territory of the United States these interlockages are numerous and complex.* The St. Croix Lake, connecting the Mississippi with Lake Superior; the west fork of Bad River and the Nemakagon at Long Lake, establishing the same connection; and the Big Fork, which flows into Rainy River, thence into Hudson's Bay, is connected with the Ondodawanoan River, a tributary of Lake Winibigoshish, through which the Mississippi flows.

Savanne Lake is about one mile broad; at its southwesterly termination begins the Great Savanne Portage, near the mouth of a small stream, flowing into Savanne

* See Dr. Norwood on this subject, in the Geological Survey of Iowa, Wisconsin, &c. &c.

River, and much encumbered with fallen trees; by this small stream canoes pass when the water is high, and thus avoid the troubles of the Great Savanne Portage.

This common dread of the voyageurs is one mile and forty-one chains in length; it descends $31\frac{1}{2}$ feet to Savanne River, and consists of a wet tamarack swamp, in which moss grows everywhere to the depth of one foot or eighteen inches; the moss is supported by a retentive buff clay, which is exposed at the western extremity of the portage. The remains of an old road formed of the split trunks of trees, probably constructed in the time of the North-West Company, passes through it; it is now in a thorough condition of decay. The same may be said of all the swampy portages along this line of route. In the time of the North-West Company, this portage was doubtless one of the best, considering its length and general character, but now a false step from a rotten or half floating log, precipitates the voyageur into eighteen inches of moss, mud, and water. No physical impediment appears to exist which would prevent this portage from being drained at a very small cost, and converted into one of the best on the whole line of route.

Savanne River, to which it leads, is very rapid a little above the landing place; but on wading up the stream for about a quarter of a mile, the occurrence of dead water without froth or bubbles, showed that the feeding swamp or lake was near at hand. Savanne River is about twenty-five feet broad here, and it continues a very meandering and crooked westerly course of about eighteen miles to Milles Lacs, or Lake of the Thousand Islands, as it is sometimes termed.

Mr. Gaudet, one of Mr. S. Dawson's assistants, ran a line between Jourdain's Rapids on Dog River and the Savanne River in 1858. The first two miles of the line were swampy, but not considered impassable even in their present condition, the depth of black mould over clay being from six to eighteen inches. From the second mile post to the seventh the country is well adapted for a road, consisting of a sandy ridge clothed with Banksian pine (*Cyprès*). The remaining distance, about one mile and three-quarters, is a gradual descent to the Savanne River.

The banks of this river are altogether alluvial, and diminish gradually from ten feet in altitude, near its source, to the level of Milles Lacs, at its entrance into that extensive and beautiful sheet of water. The immediate banks of Savanne River are clothed with alder, willow, and dogwood; behind these are seen tamarack, pine, spruce, and aspen. Near its mouth much marshy land prevails, and at its confluence with Milles Lacs it is characterised by a large expanse of rushes and other water plants common in such situations.

While descending this tortuous stream we were surprised and delighted at hearing the exclamation, "canoes! canoes!" from the lips of our keen-eyed voyageurs, and soon, sweeping round a distant bend, we observed a north canoe rapidly approaching. It contained Mr. Bell, an officer in the service of the Hudson's Bay Company, who had started from his post on the Mackenzie River in May of this year, and was on his journey to Montreal, where he proposed to spend two years of furlough. Mr. Bell was accompanied by his daughter, a child of about twelve years, who had journeyed for three months with her father through the trackless wilderness separating Mackenzie's River from Savanne River. Such early experience of life in the wilderness it is the lot of few to suffer or enjoy.

Milles Lacs was described by the Indians as extending in a direction due west much farther than was visible from the canoe route, on account of the numerous islands with which it is everywhere studded. In the lower portion of the Savanne River many large ponds and reedy lakes, connected together by small water courses, join with the main river, and indicate the great extension which Milles Lacs assumes in an easterly direction during spring freshets. It appears very probable that a length of twenty-five miles, with an average breadth of four miles may be taken as a fair representation of this remote sheet of water; the canoe route through it is twenty-one miles in length, from the mouth of the Savanne to Keg or Baril Portage. Granitic dome-shaped islands are very numerous, and occasional exposures of clay and sand banks come into view on the points and islands along the line of route.

Milles Lacs is drained by the River Seine, which empties into Rainy Lake. The Seine was examined in 1858 by Mr. S. Dawson, and the following notice of its general features is abbreviated from his report published in Toronto in May, 1859. Where the Seine issues from Milles Lacs it is more than one hundred feet wide. It pursues a winding course in a westerly direction through a narrow valley ("flat") thickly wooded with Banksian pine and poplar of large size. On either side low hills rise gradually; they are covered for the most part with a dense growth of poplar, interspersed here and there with tall pines, which rise singly or in groves above the surrounding forest.

"At times the valley contracts, and where it does so the river presents cascades past which a portage has to be made, or little rapids which can be run with a canoe; but between these there is generally a considerable extent of navigable water. This description will apply to the country for about forty miles below Lac des Milles Lacs. The lower part of the valley presents a succession of lakes, varying from a mile to fifteen miles in length, until near

Rainy Lake, into which the river, much increased in volume, discharges itself in a series of cascades, making a plunge of over 112 feet in the distance of five miles and a half. The lakes just referred to are bounded, for the most part, by low hills, generally wooded, but in some cases rocky, with an occasional valley between them presenting a less barren appearance."

The Seine River enters Rainy Lake at Seine Bay, one of the deep north-easterly expansions of that irregular body of water. Seine Bay and a part of the river are shown on Thompson's map of the Boundary Survey executed in 1826. A beautiful reduced copy of the geographical outlines of part of this map is published in the Quarterly Journal of the Geological Society for May, 1854, accompanying a paper on the Geology of Rainy Lake, by Dr. J. J. Bigsby. Seine Bay and the mouth of Seine River are both shown on this reduced copy.

The Seine River receives an affluent called Fire Steel River, which rises in the wide spreading marshes at the height of land, from which also Dog River, flowing into Lake Superior, issues.

Soon after leaving Milles Lacs, the Seine falls by a series of rapids, seven in number, a depth of thirty-six feet in a distance of nine miles; its waters are then precipitated twenty-four feet in two steps at Little Falls, and before reaching Rainy Lake, a distance of sixty-seven miles in a direct line, it falls 350 feet by twenty-nine steps varying in altitude from three to thirty-six feet.

The hills surrounding Milles Lacs here and there bear pine of fair dimensions, while in the narrow and shallow valleys between them there is every indication of hardwood over large areas. Exposures of white quartz are repeatedly seen on the islands and main land at the western extremity of the lake; and not unfrequently are they taken by travellers during their first voyage for the sails of distant boats. The name "sail rock," given to them by the voyageurs, is derived from this erroneous impression. Where the lake narrows on approaching Baril Portage, gneissoid hills and islands about 100 feet high show a well defined stratification dipping north, at an angle of about 15°, and on that side smooth, and sometimes roughly polished; on the south side they are precipitous and abrupt. The same character was noticed at the Baril Portage, which has a length of sixteen chains eighty-five links, with an altitude of 72½ feet, and an *ascent* of 1·86 feet. The north-eastern exposure of the rocks here was smooth, the southern rugged and often precipitous.

Baril Lake is seven and a half miles long, and is the counterpart of the western extremity of Milles Lacs, but it belongs to a different water system, being 1 ft. 10 in. higher than Milles Lacs. It is terminated by the Brulé or Side Hill Path Portage twenty-one chains long, leading to Brulé Lake forty-seven feet below Baril Lake. At Brulé Portage I ascended a steep hill bordering a small rapid stream called Brulé River, and from an altitude of fully 200 feet, had a fine view of the surrounding country. The vegetation upon the hill side and summit was truly astonishing, and the term Brulé Portage received an unexpected interpretation on finding hidden by a rich profusion of brushwood, the dead trunks of many noble pines. Throughout the day the tall trunks of white pine, branchless and dead, rising in clumps or in single loneliness far above the forest, had attracted attention, and on the side of the Brulé Hill we observed many prostrate half-burnt trees of the largest size. One dead trunk was measured, and found to be twelve feet in circumference, five feet from the ground. A living tree, tall, clean, and apparently quite sound, measured nearly ten feet in circumference, and many of the prostrate pines were of equal dimensions.

There can be little doubt that these were the remains of a magnificent white pine forest, which formerly extended over a vast area in this region, since from the summit of the hill the forms of scattered living trees, or tall, branchless and scathed trunks, met the eye in every direction. The young second growth indicated a soil not incapable of sustaining pine trees of the largest proportions; black cherry, birch (both the white and black), alder, small clumps of sugar maple, and a thick undergrowth of hazel nut now occupies the domain of the ancient forest. The south west side of this hill formed a precipitous escarpment 150 feet above the waters of a long clear lake. All around the eye rested upon low dome-shaped hills dipping towards the north-east, and covered with a rich profusion of second growth. The vast wilderness of green was studded with black islands of burnt pine, and a few isolated living trees, serving by their surprising dimensions to tell of the splendid forest which must have once covered the country.

The soil wherever examined consisted of a red sandy loam, covered with a thin coating of vegetable mould. Occasionally bare rock exposures protruded, and granitic boulders were numerous. The uniform size of the second growth timber on the Brule Hill, seemed to prove that the great fire which devastated this region may have occurred about thirty years since. The hill round which the portage path winds is considerably higher than any observed range on the height of land, and its summit, from which a view of the surrounding country was obtained, is probably about 100 feet above the Height of Land Lake, or 1,585 feet above the ocean; McKay's mountain having an elevation of 1,600 feet above the same level. Brulé Lake bears another name of terrible import. It is called Win-de-go or Cannibal Lake, a term applied to it in commemoration of an unnatural deed committed here by a band of Ojibways in 1811. Although not less, it is stated, than forty in number, yet they were unable to procure sufficient food to preserve them from famine. Many perished with hunger and the survivors sustained existence by feeding upon the dead bodies of their companions. The whole of the band with one exception perished, the survivor, a woman, preserved existence by murder; she, however, was not long permitted to live. Meeting with another party of Indians, who drew the dreadful secret from her, she was put to death, under the impression that those who have once fed on human flesh always retain a desire for it, which they are not unscrupulous in gratifying when opportunity offers. Several instances of cannibalism were mentioned to us by the voyageurs as having occurred on this route; and in the following summer noted spots in the basin of Lake Winnipeg were pointed out, which preserve a similar dreadful reputation. Both voyageurs and Indians always spoke of these horrible deeds in subdued tones and with an expression of anxiety and alarm.

The impression produced by a survey of the solitudes about the western extremity of Milles Lacs and of Baril Lake was rather of a favourable character. If in the course of time mineral wealth should be found to exist in profitable distribution about Milles Lacs, there would be no scarcity of arable soil between the low hill ranges of that beautiful but desolate lake to supply the wants of a mining population; or, in the event of a line of communication between Thunder Bay and Rainy Lake being established, its western shores and those of Baril and Brulé Lakes offer suitable localities for village depôts.

From Brulé Lake to French Portage, a distance of four miles, the canoe route lies through a series of lovely lakelets and short rapid streams fringed with cedar and spruce; behind these are fair-sized red pine, birch, aspen, and large spruce. French Portage bearing due west, is $1\frac{3}{4}$ miles long, and lets us down $99\frac{3}{4}$ feet into French Portage or Pickerel Lake. The timber on this portage consists of aspen, red pine, and spruce. On the shores of the lake low hills appear, wooded with an extensive forest of red pine, varied with patches of spruce, aspen, and birch. Ice formed on the upturned canoes during the night of August 16th.

Pickerel Lake, through which in a direction nearly due south-west the canoe route continues, is a fine sheet of water, thirteen miles long by two to four broad; its shores consist of low hills covered with a thick forest of pine, with spruce, aspen, and birch in the valleys. On the east side of the lake, the remains of an ancient pine forest are often visible in the forms of noble, isolated trees. These occur about six miles from its head, and further on there may be observed small groups of the same trees rising far above the comparatively young growth which now surrounds them. The half-burned standing trunks of huge dimensions, show the extent and character of the earlier forest, and the cause which destroyed their companions. White pines in large numbers still remain at the foot of the lake, and were seen at the portage, which is called Portage du Pin, also Portage des Morts. The first name is evidently derived from the prevalence of large red and white pine; the second has a melancholy reference. It commemorates the death of a voyageur, who being over anxious to cross the portage while supporting the bow of a north canoe, lost his footing, and was so much injured by the heavy burden crushing him as he sank to the ground, that he died after the lapse of a few hours. A north canoe often weighs between three and four hundred pounds when soaked by long immersion in water; this unwieldy burden is borne by two voyageurs, one at the bow, the other at the stern, when crossing the portages; and bruises, sprains, or ruptures are the frequent consequences of over-exertion, rendered necessary, however, by the present condition of the portage paths.

Portage des Morts is twenty-six chains long, and it overcomes a descent of seven feet in the small stream connecting Pickerel Lake with Doré Lake, a sheet of water about a mile across, but extending much further in a north-westerly direction.

Among the trees observed here remarkable for their size, cedar, ash, white and red pine, with birch of two kinds, may be enumerated. The cedar is far superior to any before seen. A clay sub-soil is found in the valley of a small river running near the portage path, and the upturned roots of trees on the hill-side showed fine washed white sand upon which a sandy loam rested. The foot of Doré Lake brings us to the Portage des Deux Rivières, which lets us down 117·21 feet into Sturgeon Lake, in a distance of 32 chains.

The whole country seems to sink with the French and the Deux Rivières Portage. The hills about Sturgeon Lake at its upper end are not above 100 feet high, and if the valleys and lakes were filled up between the tract of country south-west of French Portage, it would be nearly a level plain, with a slight south-westerly descent. In Sturgeon River, leading to the lake of that name, we met with the first marshy place since leaving the mouth of the

Savanne River. The canoes here were forced through a profusion of aquatic plants, among which the beautiful white water lily, with its golden-hued companion, frequently occurred. Willows, small aspen and alder, grew on the banks, but no hill or elevated tableland was visible from the shallow but tortuous river. Once on the open lake, hills about 200 feet high rose into view at some distance on the eastern side. The bushy tops of what appeared to be a grove of elms, were seen near the head of this large and beautiful sheet of water; again wide tracts of burnt land attracted attention, with a few white pines, remains of a forest long since destroyed. The north-eastern termini of hill ranges slope to the water's edge, and, when bare, are found to be evenly smoothed and ground down. Everywhere on the shores of the first large expansion of the lake, remains of an ancient forest lay black and branchless, or still flourished green and erect amidst a vigorous undergrowth of spruce and aspen.

Sturgeon Lake and River, or rather a succession of lakes and rivers bearing the above names, extend for thirty-six miles from the Portage des Deux Rivières to Island Portage, which leads into Pine Lake, a small sheet of water connected by means of a broad river about three and a half miles long, with the great Nequauquon Lake or Lac la Croix.

Nine miles from its head, Sturgeon Lake was found to have forty-five feet depth of water, with a muddy bottom. The temperature of the lake was 68° at six P.M.; the pines and balsams growing near the shore were seen to be scraped or barked for about a foot near the ground by Indians, for the purpose of procuring gum or resin.

No lake yet seen on the route can bear comparison for picturesque scenery with Sturgeon Lake. The numerous deep bays, backed by high-wooded hills or rocks, rugged or smooth, according to their aspects, its sudden contraction into a river breadth for a few yards between large islands and the equally abrupt breaking out into open stretches of water, offered a constant and most pleasing variety of scene. The high jutting points of granite rock, which here and there confine the channel, offer rare opportunities for beholding on one side an intricate maze of island scenery, and on the other an open expanse of lake, with deep and gloomy bays, stretching seemingly into the dark forest as far as the eye can reach.

Here we met several Ojibways in their elegant birch bark canoes. They were very friendly, and apparently delighted with a small present of tobacco and tea. One young hunter with his squaw hurried to the shore as we approached, but soon returned gaudily painted with patches of vermilion on his cheeks and in bars across his forehead.

The fourth large expanse of Sturgeon Lake is bounded by low, densely-wooded shores, with high hill ranges in the far distance. The first cascades, with a fall of four and a half feet, occur at the foot of this last expansion; these are quickly followed by the second falls of six and a quarter feet descent, then occurs a narrow reach of river for three miles, which is terminated by the third rapids of two and a half feet fall, leading to another expanse with a general direction nearly due west; the fourth and fifth rapids then occur within four miles of one another, and are followed by Island Portage two miles further on.

Island Portage lets us down ten feet, and involves a portage of fifty yards. Crossing the small Pine Lake, the river now assumes a course nearly due west, and within a distance of four miles, brings us to a north-eastern arm of Lac la Croix. The canoe route passes near the north shore of this extensive and beautiful lake. High precipitous rock exposures begin to show themselves, often clothed with dense groves of pine rising above the mass of light green aspen foliage which prevails. Although Lac la Croix is fourteen or fifteen miles long, yet our traverse did not exceed eight, for we entered the Nameaukan River, which issues from the north-western coast, and takes a circuitous north-westerly direction, bringing us to Rattlesnake Portage, where the river descends by a beautiful cascade 12·14 feet, involving a portage of 110 yards.

We camped at the edge of the cascade, the portage path offering the only even spot where our blankets could be spread. The guide pointed significantly to the surging waters at the foot of the falls, and with a quiet smile said, "better not walk much in night." Three steps from my resting-place would have precipitated me into the rapid, and as a somnambulist happened to be one of the party, he was carefully warned not to indulge in midnight explorations. The noise of the cascade effectually drove sleep from my eyes, and although the night was really short, it seemed an interminable age. Generally my sleep was excellent, however hard the bed or stormy the night, yet if rain did not penetrate the canvas tent, I slept soundly and well, invariably awakening with the first streak of day.

The dawn of morning and the early start in this rocky wilderness possess some characteristics peculiar to the country and the strange companions with whom necessity compels you to associate. Rising from a bed on the hard rock, which you have softened by a couple of rugs or a north blanket, and if time and opportunity permitted by fresh spruce or pine boughs, the aspect of the sky first claims and almost invariably receives attention. The morning is probably calm, the stars are slightly paling, cold yellow light begins to show itself in the east; on the river or lake rests a screen of dense fog, landwards a wall of forest impenetrable to the eye. Walking a step or two from the camp a sudden rush through the underbrush tells of a fox, mink, or marten prowling close by, probably attracted by the remains of last night's meal. From the dying camp fires a thin column of smoke rises high above the trees, or spreads lakewards to join the damp misty veil which hides the quiet waters from view. Around the fires are silent forms like shrouded corpses stretched at full length on the bare rock or on spruce branches carefully arranged. These are the Indians, they have completely enveloped themselves in their blankets, and lie motionless on their backs. Beneath upturned canoes, or lying like the Indians, with their feet to the fire, the French voyageurs are found scattered about the camp; generally the servant attached to each tent stretches himself before the canvas door. No sound at this season of the year disturbs the silence of the early dawn if the night has been cold and calm. The dull music of a distant waterfall is sometimes heard, or its unceasing roar when camped close to it as on the Rattlesnake Portage, but these are exceptional cases, in general all nature seems sunk in perfect repose, and the silence is almost oppressive. As the dawn advances an Indian awakes, uncovers his face, sits on his haunches and looks around from beneath the folds of his blanket which he has drawn over his head. After a few minutes have thus passed, not observing his companions show any sign of waking or disposition to rise, he utters a low "waugh"; slowly other forms unroll themselves, sit on their haunches and look around in silence. Three or four minutes are allowed to pass away when one of them rises and arranges the fire, adding fresh wood and blowing the embers into a flame. He calls a French voyageur by name, who leaps from his couch, and in a low voice utters "lève, lève." Two or three of his companions quickly rise, remain for a few minutes on their knees in prayer, and then shout lustily "lève, messieurs, lève." In another minute all is life, the motionless forms under the canoes, by the camp fires, under trees, or stretched before the tent doors, spring to their feet. The canvas is shaken and ten minutes given to dress, the tent pins are then unloosened and the half dressed laggard rushes into the open air to escape the damp folds of the tent now threatening to envelope him. Meanwhile the canoes are launched and the baggage stowed away. The voyageurs and travellers take their seats, a hasty look is thrown around to see that no stray frying pan or hatchet is left behind, and the start is made. An effort to be cheerful and sprightly is soon damped by the mist into which we plunge, and no sound but the measured stroke of the paddle greets the ear. The sun begins to glimmer above the horizon, the fog clears slowly away, a loon or a flock of ducks fly wildly across the bow of the first canoe, the Indians and voyageurs shout at the frightened birds or imitate their cry with admirable accuracy, the guide stops, pipes are lit, and a cheerful day is begun.

After leaving Rattlesnake Portage, rapids and falls follow one another in quick succession. The most important are Crow Portage, with 9·88 feet fall; the Grand Falls Portage 16 feet, and the great and dangerous Nameaukan Rapids, letting the river down in steps between fifteen and sixteen feet. In descending the Grand Rapids, my canoe had a narrow escape. Lambert acted as steersman, and

GRAND FALLS OF THE NAMEAUKAN RIVER

Charley, an Ojibway Indian, as bow-man Lambert was not strong enough to give the proper direction to the canoe in order to avoid a rock jutting out at the head of the rapid. Just as we made the leap, the stern, borne swiftly round by the current, grazed the rock and tore the bark, without, however, doing serious damage. The moment Charley felt the graze, he turned round, brandished his paddle, and shook it at the unfortunate Lambert; we shot down the rapid with great velocity, and embraced the opportunity afforded by the first safe eddy to examine the bark of the canoe. We were deeply laden, and the bottom of the canoe was so covered with our baggage, that no part was visible. "Put your fingers to the bottom of the canoe, monsieur," said Lambert to me; "how much water?" "Two inches," I replied. "That will do, we shall not make more water now we are out of the rapid, it is only a crack, and the bark is tough." We made, however, three inches of water in a short time, and as the baggage was in danger of being wetted, it was deemed advisable to gum the leak without unnecessary delay.

The shores of Nameaukan River are fringed with the Banksian pine, and where an alluvial soil has accumulated, the aspen grows to a large size. Where the river debouches into Nameaukan Lake, there is a fine grove of ancient elms, and underneath their wide-spreading branches we found a large encampment of Indians.

The traverse across Nameaukan Lake is six and a half miles in length, the lake itself extending for more than double that distance in a direction due west. At the extremity of the traverse is the Nu Portage, where the descent is eight and a half feet, leading us into a narrow circuitous river, without perceptible current, which meanders through a reedy expanse, fringed with low willows for about three miles. The canoe route then takes a winding course, whose general direction is nearly due north for a distance of two and a half miles, when turning westward we suddenly arrive at the open and beautiful, but indescribably barren and desolate region of Rainy Lake.

The canoe route followed by the North-West Company, commonly called the Pigeon River route, joins the chain of communication which has just been described in Nequauquon Lake; from this point they both pursue the same course by the Winnipeg to Red River. The difficulties of the Grand Portage induced the North-West Company to establish their chief depôt at Fort Charlotte, nine miles west of Grand Portage Bay. Fort Charlotte was connected with Point des Meurons by a traveled road in the time of the North-West Company, but owing to neglect and the diversion of traffic, it has long been choked up with young trees, and now only serves as an Indian path. There is also an old cart road between Point des Meurons and White Fish Lake, which is close to Arrow Lake.

In 1858 I took the expedition under my charge by the Pigeon River route, with a view to compare its facilities with those of the more northern communication by the Kaministiquia. The following brief notes of this route will complete the description of the available water communications between Lake Superior and Rainy Lake within British territory.

The Grand Portage made to overcome the falls of Pigeon River, 120 feet high, has been often cited as the chief obstruction to the Pigeon River route. Its length is eight miles fifteen chains. The road is dry, and in comparison with some of the portages on the Kaministiquia route, in good condition. It is passable for an ox team, which is employed by the people in charge of the American trading post in forwarding their supplies.

I endeavoured to procure a waggon and team from the American traders at Grand Portage Bay to transfer the heavy baggage from the east to the west end of the portage, but although the vehicle was available the team was not; one ox having died during the winter, and the other was in such a miserable condition that he could scarcely draw the empty waggon.

The passage of the Grand Portage consequently occupied five days instead of two, and in making a comparison between the two canoe routes to Lake Winnipeg, these facts must be borne in mind. In 1857 the Red River expedition landed at Fort William on the 31st of July, and reached the Settlements on the 4th of September, having been thirty-four days on the road, or forty from Toronto. The expedition of 1858 reached Grand Portage on the 5th of May, and arrived at the Stone Fort on the 2nd of June, a period of twenty-eight days, or thirty-four from Toronto. The Grand Portage lying within the territory of the United States loses all interest as the terminus of a Canadian route; but that part of the water communication which forms the boundary line, and the country between Arrow Lake, White Fish Lake, and Fort William, seems to acquire importance in proportion to the extension of our knowledge respecting its capabilities and resources.

The waters on the rivers and lakes on the east side of the Height of Land, the Lake Superior water-shed, were high in 1858, while those on the west side, or the tributaries to Lake Winnipeg, were unprecedently low. In many of the lakes recent water-marks, four and five feet above the present level, were frequently observed. This remarkable lowness of the water was attributed by the half-breeds and

Indians to the very small quantity of snow which fell on the western slope during the winter of 1857.

It is important to bear in mind that the voyage of the expedition of 1858 was made under the great disadvantages inseparable from unusually low water, and whatever superiority the route appears to possess over that of the Kaministiquia by Fort William, will be much more apparent in ordinary seasons, when the lake and river levels are from two to five feet above their present altitude.

On our arrival at Moose Lake, May 12th, a glistening sheet of solid ice overspread its surface, and seemed to threaten a long delay; but by noon on the following day, under the influence of a hot sun and a gentle breeze, lanes of water opened, through which we succeeded in passing the canoes, and on the evening of the same day a high wind, accompanied by rain, completely broke up the ice in the higher lakes, and opened the communication.

The part of the Pigeon River route to which this notice refers, commences at Arrow Lake, a fine expanse of water close to White Fish Lake, lying in a north-easterly direction, and within thirty miles of the Kaministiquia.

From Arrow Lake, a short portage brings us into Rose Lake on the course of the old North-West Company's route, following the boundary line.

The portages between Rose Lake and the Height of Land are short and low, while the Height of Land Portage is not 500 yards long, and does not rise above fifty feet. The passage from the St. Lawrence water-shed to that of Lake Winnipeg is short, easy, and dry, incomparably superior to the Prairie Portage, and the Great Savanne on the Kaministiquia route. In consequence of the very low stage of the water that year, numerous small rapids were formed in the rivers connecting Gun Flint

Lake with Lake Seiganagah; during ordinary seasons, these rapids are passed without difficulty, but in 1858 they involved the portage of a portion of the baggage, and the letting of the canoes down them by rope.

From Lake Seiganagah * an Indian route passes into Little Seiganagah Lake, which is connected with Sturgeon Lake on the northern route. The Little Seiganagah is a favourite wintering place of numerous families of Indians; it abounds in fish, and near its shores the winter road to Fort William runs.

Between Knife Lake and Birch Lake there are two routes, one coinciding with the boundary line, the other, which we followed, passing in a north-westerly direction, causing us to make two portages instead of one, but escaping some rapids.

From Nequauquon Lake one route passes into the Nameaukan River, and another, turning south, follows the boundary line through Loon's Narrows, and then north into Nameaukan Lake. Our guide preferred going by Loon's Narrows, fearing that the always dangerous Nameaukan Rapids would be almost impassable for heavily laden canoes, on account of the low stage of the water.

In Loon's Narrows we found a shallow river with a strong current and many boulders, and in making the north-westerly turn, instead of the broad channel shown on the map of the Boundary Commissioners, a very tortuous, sluggish and shallow stream, led us into the south arm of Sand Point Lake.

Sand Point Lake is connected with the Nameaukan Lake by a broad channel, and it is at this point that the route through Loon's Narrows coincides with the more

* Seiganagah or "Full of Islands."

northern route, and follows the boundary line through Rainy Lake to Fort Frances.

At the close of the 2nd volume there is a table of the portages, décharges, rapids, lakes, lake straits, and navigable channels on the Pigeon River route, from Lake Superior to Rainy Lake, showing their lengths and distances from Lake Superior.

CHAP. IV.

RAINY LAKE TO THE SOURCE OF THE WINNIPEG RIVER.

Rainy Lake.—Description of.—Rainy River.—Affluents of Rainy River.—Fort Frances.—Lac la Pluie Indians.—Valley of Rainy River.—Character of the Valley.—The Winter Road to the Lake of the Woods.—Arrangement for crossing the Swamps to Red River direct from the Lake of the Woods.—Fertility of Rainy River.—The Manitou Rapids.—Obstructions to Navigation.—The Long Rapids.—Indian Encampments.—Tumuli.—Graves.—Banks of Rainy River.—Caterpillars.—The Lake of the Woods.—Beauty of the Lake of the Woods.—Confervæ.—Garden Island.—Refraction.—Indians.—A Council.—Its Results.—Grasshoppers.—Shoal Lake.—North West Corner of the Lake.—Monument Bay.—Route to Rat Portage.—Indians.—Sturgeon.—Polished Rocks.

In 1826 a map of Rainy Lake, as part of the survey under the seventh article of the treaty of Ghent between Great Britain and the United States, was constructed by David Thompson, Astronomer and Surveyor. No labour was spared in producing a correct delineation of the geographical features of this part of the country, and the portion of the map accompanying this narrative, which includes Rainy Lake, Rainy River, and the Lake of the Woods, besides the Pigeon River referred to in the previous chapter, is reduced from an authorised copy of those parts of the survey. Dr. Bigsby, who accompanied the Commissioners as Geologist, communicates the chief facts in the following enumeration of the geographical position, &c., of Rainy Lake, in the Quarterly Journal of the Geological Society for May, 1851.*

* On the Geology of Rainy Lake, South Hudson's Bay. By Dr. J. J. Bigsby, F.G.S., &c.

Rainy Lake, or Lac la Pluie, as it is more frequently called by the voyageurs, is 225 miles west of Lake Superior, and from the head of Rainy River eighty-five southeast of the Lake of the Woods. It is fifty miles long by thirty-eight and a half broad, and is 294 miles round by canoe route. Its form is that of three equal troughs, with deep lateral indents, the main one running in an east and west direction, the other two northerly from it. It is through the main trough that the canoe route lies from the mouth of Nameaukan River, in latitude 48° 30′ N., longitude 92° 40′ W. to the source of Rainy River, thirty-eight miles distant, in a direction a few degrees to the north of west.

The shores of Rainy Lake are generally low, and often consist of naked shapeless masses of rock, with marshy intervals, or they rise in ridges which become hills 300 to 500 feet high, half a mile to four miles from the lake. The timber seems to be very small and thin in the marshes, and on the islands, which exceed 500 in number, the largest growth was observed. Taken as a whole, the general aspect of the shores is forbidding, and furnishes on the ridges and hill flanks a picture of hopeless sterility and desolate waste. Dr. Bigsby says that there is but little loose débris about Rainy Lake, the earth or gravel banks being scarce, and seldom exceeding a few feet in thickness. Whenever the land rises, for the most part bleached and naked rocks occur for many square miles together.

Colonel Lefroy made Rainy Lake 1,160 feet above the sea, by barometrical measurement. Its height deduced from the levels taken at the portages, and the estimated rise and fall in the current of the rivers along the line of route, was 1,035 feet. In this calculation the level of Lake Superior is taken at 600 feet above the ocean.

FALLS ON RAINY RIVER OPPOSITE FORT FRANCES

Major Long estimated it to be 1,200 feet above the same level.

Rainy Lake freezes about the 1st December, and is open about the 1st of May; as is usually the case where large rivers issue from spacious lakes, the discharging stream is not frozen for a number of miles from its source. The warm waters coming from beneath a shelter of ice in their capacious feeding lake, retain their heat so as to enable them to resist the cold of these regions for many miles below the Chaudière Falls.

At the entrance of Rainy River on the evening of August 19th, the delightful odour of the balsam poplar (*Populus balsamifera*) loaded the air, and seemed to welcome our arrival in a region differing altogether from those through which we had lately passed. Where Rainy River issues from Rainy Lake, it is a broad and rapid stream, with low alluvial banks, clothed with a rich second growth. The fine forests with which they were once covered had long since been stripped of their ornaments by the occupants of the old North-West and the present Hudson's Bay Company's Fort.

The general course of Rainy River is a few degrees to the north of west for a distance of eighty miles by the windings of the river, and in an air line sixty miles. The rapids at its source offer no impediment to skilful navigation, nor do the whirlpools which usually accompany the passage of such a large body of water, in consequence of their being distributed over a wide area. Two miles below the head of the river, Fort Frances is situated on a high bank, just below the Chaudière Falls. These magnificent cascades let the river down 22·88 feet, and at their foot is a famous fishing ground, from which the Lac La Pluie Indians obtain an abundant supply of their staple food. Three miles from Fort Frances, the river takes a sudden southerly bend, which it maintains for a distance of four miles ; it then again assumes a course due west for about sixteen miles, and receives the Pekan, or Little Fork, the Missatchabe, or Big Fork, and the Kakmaskatawagan Rivers, on the south or United States side ; the course then turns abruptly due north, and continues for a distance of six and a half miles, when it again resumes a westerly direction for eighteen miles, its otherwise gentle and uniform current is here broken by the Manitou Rapids and Long Rapids, which let the river down about two and a half feet, and three feet respectively. Six miles from the Long Rapids a short northerly bend again occurs, after which the river, with slight meanderings, pursues a north-west by west direction, until it debouches into the Lake of the Woods. In this part of its course it receives on the British side, small sluggish streams known by the names of the Kiskarko or Pine River, the Kahlawakalk, and Kawawakissinweek, and from the territory of the United States, the Muttontine, or River of the Rapids, the Wishahkepekas, and Kapowenekenow, or Winter Road River. Its affluents, on the British side, are insignificant outlets to the swamps which occupy the region north of Rainy River valley ; but some of those on the United States side are of important dimensions.

Fort Frances, two miles from the source of Rainy River, is situated on the right bank, in lat. 48° 36′, and longitude 93° 33′ W. Mr. Pether, the gentleman then in charge, stated that the river never freezes between the falls and the Little Fork, a distance of twelve miles, nor between the falls and its source in Rainy Lake. Wheat is sown at this establishment of the Honourable Hudson's Bay Company, from the 20th to the 23rd of May ; it ripens about 1st September. Potatoes, turnips, carrots, and indeed all common culinary vegetables succeed well.

Potatoes are dug in the first week of October, and barley is ripe by the middle of August. Snow falls here to the depth of four feet.

The great enemies to extended cultivation are the Lac la Pluie Indians. They are not only numerous, but very independent; and although diminishing in numbers, they frequently hold near Fort Frances their grand medicine ceremonies, at which 500 and 600 individuals sometimes assemble. The number of Indians visiting this

Ojibways at Fort Frances, Rainy River.

fort for the purpose of trade, reaches 1,500. They do not scruple to jump over the fences, and run through the growing crops, if the ball in their games is driven in that direction.

In the immediate neighbourhood of Fort Frances, the swamp or morass bounding the valley of Rainy River on the right bank, is about half a mile in its rear. This swamp, which extends from Rainy Lake to the Lake of the Woods, is described by Mr. Pether, and the Indians who were questioned about it, as consisting of a springy, movable surface, overlying a vast deposit of peat, through which a pole might frequently be pushed to the depth of thirty feet without reaching the bottom. The surface sustains low bushes, with here and there islands of small pine. Its borders approach and recede from Rainy River with the windings of that stream ; the breadth of the dry wooded and fertile valley varying from half a mile in the rear of Fort Frances, to six or eight miles in the direction of the Lake of the Woods. The average breadth of superior land for a distance of seventy miles might perhaps, with propriety, be assumed to be not less than four miles, giving an area of available soil of high fertility, exceeding 170,000 acres; and there can be little doubt, that with the progress of clearing, much that is now included in the area occupied by swamps would be reclaimed without difficulty or great expense.

In 1858 Mr. S. Dawson was instructed to examine the country on the right bank of Rainy River, with a view to ascertain the extent of surface available for the purposes of an agricultural settlement. He reports as follows :—

"The land immediately bordering on Rainy River, on the British side, is of an alluvial description, and almost as uniformly level as the prairies at Red River. For a mile or so inland from the main stream the ground is dry, and a dense growth of large timber, consisting of poplar, elm, oak, basswood, and occasional white pines, indicates a productive soil. For a mile or two beyond this, however, swampy ground predominates, while beyond that again the land gradually rises to a range of hills of no great eminence, which, as far as we could observe, seemed to run parallel to the river, at a distance of from four to eight miles back. The distance from Rainy Lake to the Lake of the Woods, following the windings of the stream, is about eighty miles, and throughout the whole of this extent the land fronting on the river is fit for settlement."

At Fort Frances the party was separated into three divisions, with a view to explore different routes. Mr. Napier was furnished with a guide to conduct him by the Rivière du Bois, from one of the north-westerly bays of Rainy Lake to the Lake of the Woods. The result of this exploration established the fact, that however advantageous this route may be for Indians, in their small canoes, it is far inferior to that by Rainy River, and the Lake of the Woods, as a boat communication. In describing it Mr. Napier says:—

"This is the winter road, and is preferred to the route by the Rainy River, as being more sheltered and free from the long open traverses necessary in crossing to the Rat Portage from the mouth of Rainy River. From Rainy Lake this road follows a chain of small lakes and connecting creeks, with occasional portages, until the north-east corner of the Lake of the Woods is reached, where the route continues through the numerous islands to the Rat Portage. The land throughout is rugged, rocky, and timbered with spruce and birch."

Mr. Gladman, accompanied by the remainder of the party, with the exception of Mr. S. Dawson, myself, and two French Canadian voyageurs, took the old route by Rainy River, the Lake of the Woods, and the Winnipeg, to Red River.

On Saturday, August 22nd, I started from Fort Frances at noon, in company with Mr. S. Dawson, for Muskeg River, a small stream flowing easterly from the swamps which occupy the summit of the water-shed between the valley of Red River and the Lake of the Woods; our object being to ascend the Muskeg River, cross the swamps, and descend the Roseau River to its junction with Red River. We were provided with two small canoes fit for transportation through the swamps, and supplied ourselves with provisions to last for ten days, one change of clothing, a small tent, and a pair of blankets each.

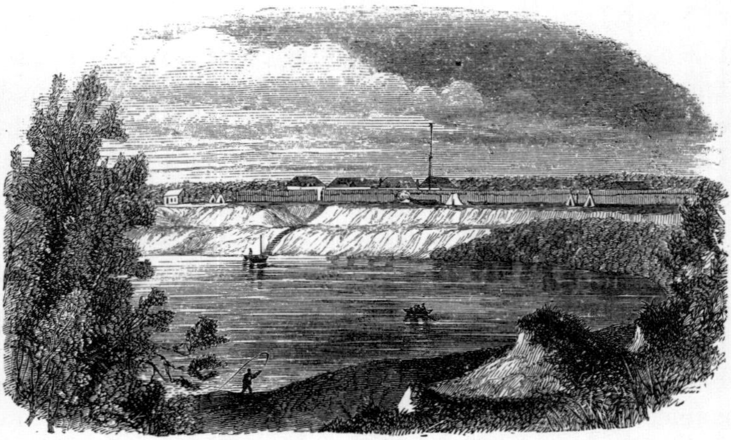

Fort Frances, Rainy River.

In Mr. Dawson's canoe were a French Canadian (François) and an Iroquois (Pierre). In my canoe an Indian guide from Garden Island, Lake of the Woods, and Lambert, who acted as interpreter.

In describing the general aspect of the banks and valley of Rainy River, it will be advantageous to sketch with considerable minuteness the features of the soil and vegetation at the different stopping places, where very excellent opportunities were offered for acquiring information on these particulars.

The ground at our camp, twelve miles below Fort Frances, was covered with the richest profusion of rose bushes, woodbine, convolvulus in bloom, helianthii just beginning to flower, and vetches of the largest dimensions. Fringing this open interval of perhaps 280 acres in extent, were elms, balsam-poplars, ashes, and oaks. One elm tree measured three feet in diameter, or nine feet eight inches in circumference; and there is no exaggeration in saying that our temporary camping place was like a rich, overgrown, and long neglected garden.

Similar intervals to the one just described were noticed occasionally as we descended the river; the banks preserving an average altitude of about forty feet, and sustaining a fine growth of the trees before enumerated. No part of the country through which we have passed west of Lake Superior can bear comparison with the rich banks of Rainy River which everywhere preserves a very uniform breadth, varying only from 200 to 300 yards. The soil is a sandy loam at the surface, much mixed with vegetable matter, but where the bank has recently fallen away, clay may be seen stratified in layers of about two inches in thickness, following in all respects the contour of what appears to be unstratified drift clay below. Basswood is not uncommon, and sturdy oaks, whose trunks are from eighteen inches to two feet in diameter, were found in open groves with luxuriant grasses and climbing plants growing beneath them. The lodge poles of an Indian camp of former seasons were covered with convolvulus in bloom, and the honeysuckle twined its long and tenacious stems around the nearest support, living or dead.

The banks of the river maintain for twenty miles an altitude varying from fifteen to sixty feet. Occasionally, they show the abrupt boundaries of two terraces, the lower boundary having the form of a sloping bank or an abrupt cliff from fifteen to thirty feet in altitude; in its rear the upper terrace rises gradually or abruptly from fifteen to twenty feet higher, according to its position with reference to the river. There is every appearance in places of fire having destroyed a former larger growth of trees than those which now occupy these areas.

The extraordinary height of the water in August 1857, was seen by the lodge poles of former Indian encampments at the foot of the bank. They were under water to the depth of one and even two feet. The river does not appear to rise high in the spring, as the trees fringing the banks to the water's edge show no action of ice. The difference between the highest and the lowest water levels may be seven feet, and no record of recent higher levels meets the eye.

The Manitou Rapids let us down about two and a half feet, and appear to be caused by a belt of rock crossing the river at nearly right angles to its course. On the American side the hill range has an altitude of about eighty feet, on the British side it is much lower, and appears to subside rapidly in gentle undulations. The Manitou Rapids are capable of being ascended by a small steamer of high power without difficulty, and cannot be considered as presenting an obstacle to the navigation of this important stream as long as the water maintains its present altitude, which is about three feet higher than is usual at this season of the year, but often exceeded in the spring and fall. Two locks of ten feet lift, with one guard lock, would overcome the falls at the mouth of the river, and thus form a splendid water communication between the head of Rainy Lake and Rat Portage, Lake of the Woods, by the north-west coast, a distance of 190 miles, or between the head of Rainy Lake and the north-west point of the Lake of the Woods, a distance of 170 miles.

High clay banks are exposed above and below the Rapids, and many hundred acres are very scantily timbered with second growth.

Ascending the bank two miles below the Rapids, I was much surprised at the number of birds of different kinds chirruping and singing in the light and warmth of a bright morning sun. I heard more birds in ten minutes here than during the whole journey from the Kakabeka Falls on the Kaministiquia to the mouth of Rainy River.

At the second or Long Rapids an extensive area denuded of trees presents a very beautiful prairie appearance. Here we landed to examine two immense mounds which appeared to be tumuli. We forced our way to them through a dense growth of grasses, nettles, and helianthii, twisted together by the wild convolvulus. On our way to the mounds we passed through a neglected Indian garden, and near it observed the lodge poles of an extensive encampment. The garden was partially fenced, and contained a patch of helianthii, six and seven feet high in the stalk, and just beginning to show their flowers. The wild oat attained an astonishing size, and all the vegetation exhibited the utmost luxuriance. The mound ascended was about forty feet high and one hundred broad at the base. It was composed of a rich black sandy loam, containing a large quantity of vegetable matter. On digging a foot deep no change in the character of the soil was observable. The Indian guide called them underground houses; he informed Lambert that a tradition existed regarding the origin of these mounds which he had often heard his father repeat in the spring of the year when his tribe assembled at the foot of the Rapids to catch sturgeon. About two hundred and fifty years ago, so runs the tradition, a large party of Sioux had penetrated from the south-west into the hunting grounds of the Ojibways, to make war upon them in the heart of their country and at their best fishing station. The Sioux were driven back, but with terrible loss to the Ojibways. A grand council was held after the defeat of the enemy, and it was resolved to erect a number of mounds or underground houses, wherein a store of provisions might be laid up and the women and children retreat in case of a sudden invasion. These mounds were the result of that determination, and are regarded by the Lac la Pluie Indians, according to my guide, as the fortifications which their ancestors erected to protect their families from the invading Sioux, to enable the warriors, freed from the embarrassment occasioned by the presence of their wives and children, to harass the enemy in all directions, and possibly cut off his retreat.

The Rainy River mounds are far larger than the burying-places or ossuaries which are scattered over Canada, and especially on the south-eastern shores of Georgian Bay, Lake Huron. It might be supposed by many familiar with those curious mounds containing the bones and relics of the barbarous people who occupied Canada some centuries since, that, like the smaller ossuaries distributed in the forests of the Lake Region, the gigantic mounds of Rainy River were places of sepulture. The custom of burying the remains of many individuals in one spot and heaping over them a mound of earth was common in remote times among the wandering tribes who hunted over the rocky and barren plateau north of Lakes Huron and Superior. The dead were laid upon the bare rock and covered with stones to protect the body from wild animals. After a certain number of years the tribe made a gathering of their dead, and bore the bones to a suitable resting-place where earth existed in sufficient abundance to admit of a mound being made without difficulty. Over very extensive areas on the north shore of Lake Huron this could not be easily accomplished, in consequence of the rocky character of the country and the general absence of loose earth, except in the valleys of rivers liable to annual overflow.

The modern graves of Indians are numerous on Rainy River; they have each a little birch bark roof placed over them, and facing the south a small opening is left, through which the relatives introduce tobacco, rice, or other offerings at their periodical visits.

About three hundred yards below the second rapids, which have a fall of three feet at the present high stage of water, twenty-three skeletons of Indian lodges were seen, all clothed with the wild convolvulus, and now serving as records of the love of change which seems to form a leading characteristic in the habits of the barbarous race who possess, without appreciating or enjoying them, the riches of this beautiful and most fertile valley. Limestone fragments and boulders, more or less water worn, with pebbles of the same rock, are found everywhere on the beach, at the foot of the clay or loamy banks.

As we approached the Lake of the Woods the river increased in breadth, and at each bend a third low terrace was in process of formation, often from 200 to 300 acres in area, and elevated above the present high water level from one to three feet. Coarse grasses grow in great abundance upon many of these rich outlying alluvial deposits, and it appeared very probable that in ordinary seasons they would furnish some thousand acres of rich pasture land, as the grasses they sustain are like those which on the Kaministiquia the settlers cut for their winter supply of fodder for cattle. Near the mouth of the river the tall tops of a few red and white pine may be seen, which rise far above the aspens occupying the lower plateau, while a vast reedy expanse, probably in ordinary seasons available for grazing purposes, marks the junction of Rainy River with the Lake of the Woods.

Rainy River flows upon an alluvial bed partly of its own formation, the materials being probably derived in a great measure from the cutting away of the clay and sand which constitute the higher of the two terraces by which its boundary is so well defined. The first or lowest terrace being generally from twelve to fifteen feet above the present water level, frequently terminates on the river in abrupt, low, clay bluffs, capped with loam and sand, or rich alluvial deposits; sometimes both terraces come upon the river together in one bold bluff, often forty feet in altitude, and again the lower terrace is found to occupy the bank without the higher one in the rear being visible from a canoe.

The separation of these terraces is an important item in the description of the topography and general characteristics of Rainy River. Where the lower terrace is alone visible, the vegetation it sustains is often characteristic of a poor and sandy soil. Red pines, some of them of fair dimensions, red cedar and small poplars occupy it, and if any passer by were to draw an inference from the prevailing timber which, in such situations, meets the eye, he would at once form the opinion that the land was comparatively worthless. But let him cross the lower terrace until he reaches, at a distance of 200 yards or perhaps a quarter or half a mile, the higher one, and the magnificent growth of poplar, elm, and basswood would quickly reverse such judgments. As far as I penetrated in different places back from the river, the soil of the higher terrace was of admirable quality, and supported a heavy growth of timber. The clay upon which it rested was often exposed by the steep banks of numerous sluggish streams which cut the terrace to nearly the level of

Rainy River, and evidently form channels by which the swamps in the rear are drained.

On the 6th June, 1846, Mr. Kane* saw the trees on each side of Rainy River, and part of the Lake of the Woods, for full 150 miles of the route, literally stripped of foliage by myriads of green caterpillars, which had left nothing but the bare branches. In such extraordinary numbers were these destructive insects that Mr. Kane's party found it impossible to breakfast on land unless they submitted to the unpleasant addition of numbers of these caterpillars to their meal, dropping as they did from the trees without intermission and covering the ground.

Sir John Richardson relates that in 1847 multitudes of caterpillars spread like locusts over the neighbourhood of Rainy River. They travelled in a straight line, crawling over houses, across rivers, and into large fires kindled to arrest them. Throughout the whole length of Rainy River, on the Lake of the Woods, and on the River Winnipeg, they stripped the leaves from the trees and ate up the herbage. They destroyed the *Folle avoine* (wild rice) on Rainy Lake, but left untouched some wheat that was just coming into ear. This was the first time that Fort Frances had experienced such a visitation. The following year Sir John Richardson found the still leafless trees covered with the cocoons of the previous year, in each of which there remained the hairy skin of a caterpillar. In 1858 we noticed the trees in the Bad Woods, on the north bank of the Assinniboine, covered with an incredible number of small green caterpillars, resembling the palmer worm, so destructive in the United States in some seasons.

The Lake of the Woods is about seventy-five miles in

* Wanderings of an Artist among the Indians of North America, by Paul Kane.

length, and the same in breadth. It is 400 miles round by canoe route*, and is broken up into three distinct lakes by a long promontory, which in periods of high water becomes an island. The southern part is termed the Lake of the Sand Hills, the eastern portion White Fish Lake, and the northern division the Lake of the Woods. White Fish Lake and Lake of the Woods are separated from Sand Hill Lake by the broad promontory before referred to, respecting which little is known. The name of the latter division is derived from vast numbers of low sand hills, which occupy its south-western coast. The distance of the Lake of the Woods from Lake Superior, is north-west 325 miles by the Pigeon River route, and 381 by the route from Fort William, followed by the expedition. The north-west corner of the lake is only about ninety miles from Red River, in an air line. Its elevation above Lake Superior is 377 feet, or 977 feet above the sea. Major Long made it 1,040 feet above the ocean level.

The scenery among the islands towards the north-west corner of the lake is of the most lovely description, and presents in constantly recurring succession every variety of bare, precipitous rock, abrupt timbered hills, gentle wooded slopes, and open grassy areas. Some of the islands are large and well timbered, others show much devastation by fire, and often a vigorous second growth of a different kind of tree under the blackened trunks of branchless pines.

The ordinary course of the canoe route to Red River lies in a north-easterly direction, following the trend of the coast towards Turtle Portage, which leads from the Lake of the Sand Hills to White Fish Lake. In pursuance of our intention to endeavour to pass from the west

* See vol. viii of the Quarterly Journal of the Geological Society, for an account of the Lake of the Woods, by Dr. Bigsby.

side of the Lake of the Sand Hills across the country, in as direct a line as possible to Red River, we made a traverse in a north-westerly direction towards the south point of Keating Island, a distance of sixteen miles. The surface of the lake was perfectly smooth, reflecting the sun's rays with extraordinary power and brilliancy. As we receded from the shores the low sand dunes to the south-west were refracted into the similitude of distant mountain ranges, and the rocky coast of the eastern side, as seen through a glass, into high, precipitous, half wooded cliffs.

The origin of the sand dunes is interesting; they probably point to the existence in this neighbourhood of the Chazy formation (Lower Silurian) so characteristic of the west coast of Lake Winnipeg.

About four miles from land the water became tinged with green, deriving its colour from a minute vegetable growth (confervæ), which increased as we progressed, until it gave an appearance to the lake like that of a vast expanse of dirty green mud. On lifting up a quantity of water in a tin cup, or on looking closely over the side of the canoe, the water was seen to be clear, yet sustaining an infinite quantity of minute tubular needle-shaped organisms, sometimes detached, and sometimes clustered together in the form of small spherical stars, varying from a quarter to half an inch in diameter. Five miles from the shore the lead showed thirty-five feet of water, and four miles further on thirty-six feet; the green confervæ increased in quantity, and the little aggregations assumed larger dimensions, some of them exceeding one inch in diameter.

The temperature of the lake near the mouth of Rainy River was 67° at half-past eleven, A.M.; yet five miles from land it was found to be 76°, six inches below the surface; an hour afterwards, repeated and careful observations showed the temperature to be $77\frac{1}{2}$°. At one, P.M. the temperature two feet below the surface was 71°, and at the surface 78°. The depth of water was here thirty-six feet, and the green confervæ uniformly abundant, so that it was impossible to obtain a tablespoonful of liquid free from their minute forms. The presence of this "weed," as the voyageurs termed it, was the probable cause of the unusual temperature of the lake. Occasionally grasshoppers were seen resting on its calm glistening surface, and as we approached Keating Island they increased in number, all of them preserving, with singular uniformity, a direction towards the south-east. The Indians think the "weed" proves destructive to fish; they had seen it on Lake Winnipeg, where also we recognised it in September of the following year.

After passing the south point of Keating Island we steered for Garden Island, distant from us about nine miles. On the west side of Keating Island the Indian guide pointed out one of their fishing grounds, where he stated the water was thirty fathoms deep, and illustrated the manner in which he arrived at that estimate of the depth by explaining, through the interpreter, the mode of fishing during the winter months, showing the length of a fathom and the number of these in the lines his people employed to reach with their nets the feeding grounds of the white fish at that period of the year. He also described the thickness of the ice through which they had to break before they arrived at the water as sometimes exceeding five feet.

Wild rice (*Zizania aquatica*) grows very abundantly in the marshes bordering the Lake of the Woods. It is an important article of food to the Indians. They gather it about the end of August and beginning of September, and lay up a store for winter consumption. A soup made of wild rice and blue berries is a very palatable dish, and eagerly sought after by those who have been living on salt food for several weeks.

On approaching and receding from Keating's Island, the effects of refraction were most astonishing, elevating low detached island rocks into huge precipitous promontories, and giving to a shore, a few feet above the level of the water, the appearance of a high rock-bound coast. On nearing a small island about four miles east of Garden or Cornfield Island, the grasshoppers on the surface of the lake became more numerous, the green confervæ was visibly less in quantity, and before we landed to dine it had disappeared altogether, but the grasshoppers were found in great numbers on the shore. The island on which we rested for an hour was about three acres in extent, and sustained some fine old oaks and elms, with a profusion of long grass, not much destroyed by the grasshoppers, which had evidently only just arrived there, as was afterwards inferred, while those which had been observed scattered over the surface of the lake were probably stragglers from a vast flight of these insects, whose main body we saw subsequently on Garden Island.

During the morning the sky had been cloudless, the air still, and the sun oppressively hot; but in the afternoon a long gentle swell began to rise upon the lake, and when we put off for our destination, a wind arose which gradually increased to a gale before we landed during the evening of the 24th August, on a low gravelly beach, at the north-west corner of Garden Island.

We camped near a well-cultivated field of Indian corn, and a rapid exploration of the island revealed to us a large potato patch, and a small area devoted to squashes and pumpkins of different kinds. We ascertained that the island had been cultivated by the Lake of the Woods Ojibway Indians for generations.

My companion and the Iroquois, Pierre, both complained that evening for the first time of being unwell.

Our camp fire evidently soon attracted the attention of a number of Indians, who were then living on a neighbouring island about four miles from us, for at midnight we were aroused by the sudden appearance at the door of the tent of two of these people, and in half an hour twenty or more had arrived. In the morning we answered their inquiries, and were requested to visit their chief, who remained with his tribe on the island already referred to. Declining their invitation, as we were anxious to hasten to the mouth of the Muskeg River, they told us they would send for their chief, who would arrive as soon as the wind fell. We made the necessary preparations for a long council, and about noon the chief's son, who was one of the first arrivals on the evening previous, announced that the canoes were coming.

We counted thirteen canoes, and found that they contained in all fifty-three men and boys, there being seven of the latter; the others were the chief and warriors of the tribe. A portion of them had just returned from an expedition against the Sioux, and were decorated or disfigured, according to taste, with whatever advantages paint, feathers, and ornaments could confer. As the object of their visit was to ascertain the reasons why we wished to pass through this part of their country, a long council or "talk" was the result of the visit.

The council terminated by a distinct refusal, on the part of the chief, to allow any of the tribe to guide us through the swamps which separate the Lake of the Woods from the prairie country to the west. The replies and objections of the chief were often couched in very poetical language, with a few satirical touches, which were warmly applauded by the audience. The following is a specimen of the colloquy: —

"What reason can we offer to those who have sent us, for your having refused to allow us to travel through your country?"

Chief.—"The reason why we stop you is because we think you do not tell us why you want to go that way, and what you want to do with those paths. You say that all the white men we have seen belong to one party, and yet they go by three different roads, why is that? Do they want to see the Indian's land? Remember, if the white man comes to the Indian's house, he must walk through the door, and not steal in by the window. That way, the old road, is the door, and by that way you must go. You gathered corn in our gardens and put it away; did you never see corn before? Why did you not note it down in your book? Did your people want to see our corn? Would they not be satisfied with your noting it down? You cannot pass through those paths."

"We ask you now to send us one of your young men to show us the road; we shall pay him well, and send back presents to you: what do you wish for?"

Chief.—"It is hard to deny your request; but we see how the Indians are treated far away. The white man comes, looks at their flowers, their trees, and their rivers; others soon follow; the lands of the Indians pass from their hands, and they have nowhere a home. You must go by the way the white man has hitherto gone. I have told you all."

At the close of the council, the chief said to the interpreter, "Let not these men think bad of us for taking away their guides. Let them send us no presents; we do not want them. They have no right to pass that way.

We have hearts, and love our lives and our country. If twenty men came we would not let them pass to-day. We do not want the white man; when the white man comes, he brings disease and sickness, and our people perish; we do not wish to die. Many white men would bring death to us, and our people would pass away; we wish to love and to hold the land our fathers won, and the Great Spirit has given to us. Tell these men this, and the talk is finished."

A hasty consultation with my companion as to what we should do in this dilemma, was abruptly closed by being informed that the Iroquois Pierre was very ill at the back of the tent. Without his paddle, without guide, and Mr. Dawson feeling much worse than on the previous evening, we determined at once not to attempt to cross the swamps between the Lake of the Woods and Red River alone, but decided to go to Red River by the Rat Portage and Winnipeg.

We told this determination to the chief, and asked for assistance to take the canoes to the settlements.

He pointed out two young men, who received orders to take us down the Winnipeg. One was to return from Islington Mission, the other to go on to Red River. We then told the chief that we would send him presents from Red River, at which he expressed satisfaction, and suggested tea and tobacco.

Garden Island is about a mile and a half long and a mile broad at its widest part. Its western half is thickly wooded, but the greater portion of the eastern half is cleared, cultivated, and planted with Indian corn. Near the centre of the field were several graves before referred to, with neatly constructed birch bark coverings. Only one lodge was seen on the island, and that was placed about 100 yards from the graves. Near the space devoted to Indian corn, were several small patches of potatoes, pumpkins, and squashes. An air of great neatness prevailed over the whole of the cultivated portion of the island, and in the part still remaining in its natural state, thickets of raspberry, black currant, and gooseberry bushes grew in the intervals between groves of elm, basswood, and oak; and on the sandy beach an abundance of the sand cherry (*Cerasus pumila*), the favourite Nekaumina of the Indians. Large flocks of passenger pigeons (*Columba migratoria*) flew backwards and forwards over the island, occasionally alighting in dense masses in the small groves. The shores were covered to the depth of two or three inches with countless millions of grasshoppers, which had been washed there during the gale of the preceding night. The greater number of the grasshoppers were alive, and as the rising sun warmed and invigorated them they spread with much regularity over the field of Indian corn and the potato patches; their progress across the potato patches was like that of an invading army of insects, eating and destroying every living green thing in their way. Before we left the island they had advanced, here and there, some thirty or forty yards from the beach, in a well defined and undulating line, leaving behind them nothing but the bare and blackened stalks of the plants over which they had spread themselves and destroyed. By inclining the head, and seeking shelter from the wind under the lee of a bush, the noise of their jaws could be distinctly perceived; and had it been calm, I have no doubt it would have been heard with the greatest ease for a distance of several hundred yards. The Indians had seen the grasshoppers before, but never in such alarming numbers; they appeared, however, quite indifferent to their progress, and quietly amused themselves as they squatted or lay on the ground, by jerking the intruders off their arms and legs with a thin piece of wood, bent by the fingers so as to act as a spring. In another chapter it will be shown that this grasshopper is a true locust (*Acrydium femur-rubrum*). The term grasshopper is retained in the narrative because it is universally applied by the half-breeds to this destructive insect.

From Garden Island to the north-west corner of the lake, in longitude 95° 15′, is about twenty miles, but the westerly limits of navigation are not yet found there. It is possible to proceed without difficulty some miles in a due west direction, through a narrow, shallow channel, into a small lake called Shoal Lake. Although no facts derived from personal observations can be here adduced respecting the general features of Shoal Lake, yet the importance which it derives from its position requires special mention to be made of it. From our Indian guide, permitted to take us to Rat Portage by the chief on Garden Island, I learned that Shoal Lake is a reedy expanse of water, eight or ten miles long, connected with the Lake of the Woods. The north and west sides of Shoal Lake were represented to be blended with a vast marsh or muskeg, which stretches from near Rat Portage to far south of the Lake of the Woods, and is the source of numerous rivers which flow from it both eastward and westward. It is this great muskeg or marsh which forms the barrier between the Lake of the Woods and Red River valley.

On part of the south shore of Shoal Lake, and all along the west coast of the Lake of the Woods, there is a considerable area of dry land timbered with spruce and small pine. Shoal Lake is about eighty-seven miles in a direct line from Fort Garry, but by the very dangerous and circuitous Winnipeg route, it is at least 320 miles. The South shore of Shoal Lake is in latitude 49° 23′, and the same parallel cuts Red River at a spot twenty-five miles north of the boundary line. The importance of the north-west corner of the Lake of the Woods, and possibly also of Shoal Lake at the terminus of a communication by land with Red River cannot fail to be appreciated.

The north-west corner of the lake is styled Monument Bay; it marks the point where the boundary line between British America and the United States, after passing down the middle of Rainy River and striking across the Lake of the Sand Hills reaches the north-west corner of Monument Bay; from this point the boundary line pursues a due south direction, intersecting the 49th parallel in longitude 95° 15′, according to Dr. Bigsby's map published in the Quarterly Journal of the Geological Society, but on the map of the Boundary Survey in the Crown Lands Office at Quebec, the longitude of the Monument at the extremity of the north-west corner of the Lake of the Woods is represented to be 95° 24′.

Shoal Lake, or Lac Plat, as it is also termed, was examined during the winter of 1857-8 by some of the assistants attached to the Surveying Department of the Expedition. The exploration having been made during the winter months, when the deep bogs and wide spreading marshes were frozen and covered with snow, loses much of the interest it would doubtless have possessed had it been executed during the summer season. An effort to penetrate from Fort Garry to the Lake of the Woods in the direction indicated by the exploring party of the previous year, was wholly unsuccessful in the autumn of 1858. A number of half-breeds from the settlements at Red River also made the attempt to reach the Lake of the Woods in September 1858, but were defeated in the attempt by unpassable swamps. In 1859 some half-breeds belonging to Mr. Dawson's party succeeded in passing through with horses.

The route we pursued, from a point near the north-west corner of the lake, lay through a labyrinth of islands in a north-east by north direction for a distance of twenty-eight miles. Six miles more, nearly due north, through scenery of the same description but of a bolder character, brought us to Rat Portage, on one of the numerous mouths of the rocky Winnipeg. Much good pine timber was seen on the larger islands, near the northern part of the Lake of the Woods, and if conclusions may be drawn from the accounts which the Indians gave us of their gardens, it is very probable that extensive areas of excellent land exist on the great promontory, and on some of the large islands. They spoke of growing Indian corn to a far greater extent than seen by us on Garden Island.

During our voyage from Garden Island to Rat Portage we met with six small canoes one after the other; four of them contained only women and children, who had been gathering berries on the islands. We exchanged tea and tobacco for a birch bark basket of blue-berries (*Vaccinium Pennsylvanicum*), of which they had collected a large supply.

Sturgeon are very numerous in the Lake of the Woods; they were repeatedly seen leaping out of the water in their gambols at the approach of evening. The Indian in my canoe told me that in calm weather they could be seen lying on the bottom in deep water, among the beautiful islands with which the narrows between the Lake of the Sand Hills and the true Lake of the Woods are studded. Whenever we landed to dine or camp before reaching Rat Portage we found hosts of grasshoppers; they covered the bare rocks, and rose in small clouds from the grass at my approach. We succeeded in securing a large pike which lay basking in the sun at the surface of the water, by cautiously approaching and striking him on the head with a paddle; he proved to be a splendid fish, probably not less than ten pounds in weight. A thunder storm of great violence detained us for a few hours on a small island, where the rock was so smoothly worn and polished by the action of ice, that we found some difficulty in keep-

Falls at Rat Portage.

ing our feet. The recent action of ice is often beautifully shown on the islands in this lake, and instances of ancient glacial action are by no means uncommon here as elsewhere on the canoe route. We arrived at Rat Portage, where the Great Winnipeg issues from the Lake of the Woods, on the morning of the 27th of August.

CHAP. V.

THE WINNIPEG RIVER.

Character of the Winnipeg.— Rat Portage.— Thunder Storm.— Thunder Storms in the North-West.—A View on the Winnipeg.—Islington Mission.—Cultivable Areas on the River.—Rev. Robert Macdonald.—Church Service.—State of Islington Mission.—Indian Superstitions.—Farm at the Mission.—The School-House the Hope of the Mission.—En route for Red River.—James' Falls.—Animal Life.—Rice Grounds.—Mr. Clouston.—Otter Falls.—The Pennawa River.—Scarcity of Food on the Winnipeg.—Bonnet Lake.—Indian Cache.— The Silver Falls.— Fort Alexander.—Lake Winnipeg.— Character of the Coast.—Camp in the Marshes.—Mouth of Red River.—Indian Village.—Christian Indians.

Issuing from the Lake of the Woods through several gaps in the northern rim of the lake, the River Winnipeg flows through numerous tortuous channels for many miles of its course in a north-easterly direction. Some of the channels unite with the main stream ten to fifteen miles below Rat Portage, and one pursues nearly a straight course for a distance of sixty-five miles, and joins the Winnipeg below the Barrière Falls. The windings of this immense river are very abrupt and opposite, suddenly changing from north-west to south-west, and from south-west to north-west, for distances exceeding twenty miles.

In its course of 163 miles, it descends 349 feet by a succession of magnificent cataracts. Some of the falls and rapids present the wildest and most picturesque scenery, displaying every variety of tumultuous cascade, with foaming rapids, treacherous eddies, and huge swelling waves, rising massive and green over hidden rocks. The pencil of a skilful artist may succeed in conveying an impression of the beauty and grandeur which belong to the cascades and rapids of the Winnipeg, but neither sketch nor language can portray the astonishing variety they present under different aspects; in the grey dawn of morning, or rose-coloured by the setting sun, or flashing in the brightness of noon day, or silvered by the soft light of the moon.

Hudson's Bay Company's Post at Rat Portage.

The river frequently expands into large deep lakes full of islands, bounded by precipitous cliffs or rounded hills of granite. The Fort in the occupation of the Honourable Hudson's Bay Company at Rat Portage is beautifully situated on an island at one outlet of the Lake of the Woods. It is surrounded with hills about 200 feet high, and near it some tall white and red pine, the remains of an ancient forest, are standing amidst a vigorous second growth. The rock about Rat Portage is chloritic slate, which soon gives place to granite, without any covering of drift, so that no area capable of cultivation was seen until we arrived at Islington Mission. We did not pursue the usual canoe route, but in the hope of overtaking the other members of the expedition, followed an Indian route for some miles, which was said by our guide to be half a day's journey shorter than that by the main river.

In descending this branch of the Winnipeg, a terrific thunder storm, accompanied by a hurricane of wind and an extraordinary fall of hailstones, approached us from the south as we reached the high portage which connects this route with the great river. Turning round in my canoe I saw, about half a mile in our rear, a white line of foam advancing rapidly towards us. Directing the attention of Lambert who occupied the stern, to the approaching squall, he instantly changed the course of the canoe to the opposite bank. The river here was about 300 yards broad, and we were swiftly paddling close to the west side, which was bounded by high and precipitous granite rocks. I had barely time to stretch a gutta-percha cloth over the canoe, before the hurricane came down upon us. Lying, as directed, at full length on the floor of our small craft, I left her to the dexterity of Lambert and the Indian. They met the shock skilfully, and paddled before the storm with great rapidity, as did also our fellow-travellers in the other canoe. We continued on our way for some minutes, gradually drawing near to the right bank, where we intended to land. Suddenly, however, large hailstones began to descend with such force as to bruise my hands severely in endeavouring to retain the covering in its place. The Indian in the bow laughed heartily at first, but having no covering on his head beyond his thick and matted hair, he soon crouched, and drew a part of the gutta-percha cloth over him. Lambert being provided with a thick fur cap held bravely on, although he loudly exclaimed that the hailstones were bruising his hands, and he would not be able to paddle much longer. Fortunately we were now close to the bank, and Lambert called out to the Indian to keep the canoe from striking against the rocks. A few strokes of the paddles brought us within a yard of the shore, when the Indian, lightly springing out of the canoe, caught her bow as she was about to strike the rock. I succeeded in disentangling myself from the covering, which was pressed down by an accumulation of hailstones, enough to have filled at least three buckets, and looking over the side of the canoe I saw the Indian in Mr. Dawson's tiny craft leap out on the bank, and catch the fragile vessel as it was about to strike with wonderful agility. Lambert's hands were much bruised, and soon became swollen and very painful; he wrapped them in wet cloths, and on the following morning, with his usual lightheartedness, declared they were as sound as ever. We made a large fire and a comfortable tent of spruce bows, soon forgetting, as the storm cleared away and allowed us to enjoy a most gorgeous sunset, the imminent danger from which we had escaped. Violent thunder storms are very common during the summer months on the Winnipeg; hailstones have been known to descend with such force as to pass through the thin birch-bark of the small canoes constructed by the Indians who hunt on this river. On the grand Coteau de Missouri, the tough buffalo-skin tents of the Crees and Sioux are sometimes penetrated by the small angular masses of ice which fall on that elevated plateau. In the summer of 1858 the canoes we carried across the prairies to the elbow of the south branch of the Saskatchewan were indented, and in some instances penetrated by hailstones during a severe storm; in several places, many miles apart, we observed the long grass laid smooth and uniform over wide areas as if pressed by a heavy roller.

Just before sunset I ascended a hill about 250 feet high, and obtained from its summit a very extensive view of the surrounding country. The broad river, with its numerous deep and spreading bays, was seen stretching far to the north, and all around dome-shaped hills, similar to the one on which I stood, showed their bare and scantily-wooded summits; generally they seemed to be thinly covered with stunted pine, but in the hollows or valleys between them, pine and spruce of large dimensions, interspersed with fair sized aspen and birch, flourished abundantly. The pine on the granite hills grew in little hollows, or in crevices of the rock. The general surface was either bare, and so smooth and polished as to make walking dangerous, or else thickly covered with cariboo moss and tripe de roche. The aspect of the country was similar in its outline to the region about Milles Lacs, but the vegetation was far inferior. Until we arrived at Islington Mission, the general features of the country maintained an appearance of hopeless sterility, and inhospitable seclusion.

Islington Mission, or the White Dog, or Chien Blanc, for by these names it is known to the voyageurs, occupies an area of what seems to be drift clay extending over 250 acres, surrounded by granite hills. The soil of this small oasis is very fertile, and all kinds of farm and garden crops succeed well. Wheat sown on the 20th of May was reaped on the 26th of August; in general it requires but ninety-three days to mature. Potatoes have not been attacked by spring or fall frosts during a period of five years; Indian corn ripens well, and may become a valuable crop on the Lower Winnipeg. Spring opens and vegetation commences at Islington about the 10th of May, and winter sets in generally about the 1st of November. These facts are noticed in connection with the small cultivable tract at the Mission, on account of the occurrence of other available areas, varying from fifty to three hundred acres in extent, between the Mission and Silver Falls, about eighteen miles from the mouth of the river. From Silver Falls to where the river flows into Lake Winnipeg, poor and rocky land is the exception, alluvial and fertile tracts, bearing groves of heavy aspens and other trees, prevailing.

The cultivable areas on the river banks may yet acquire importance, for they may be regarded in the light of productive islands in a sterile waste of rock and marsh.

The Rev. Robert Macdonald, the missionary in charge at Islington, informed me that the Mission was formerly held by the Roman Catholics for a period of several years, but was abandoned on account of the opposition of the Indians, who drove them away in consequence of the death of a young girl in the nunnery at Red River settlement. The heathen Indians persuaded the converts that all who embraced Christianity would soon die. The Mission was left vacant for a period of six years, after which Mr. Philip Kennedy was appointed catechist in 1850, a post which he held until the 20th October 1853, when the Rev. Robert Macdonald was enabled to revive the Mission by the generous and Christian liberality of an English lady.

This Mission is at present sustained by a munificent gift from Mrs. Landon of Bath of 1,000*l.* sterling for its establishment, and 100*l.* a year for its maintenance. Its present prospects are favourable, and it will eventually become an important station in the wilderness by which it is surrounded. I attended divine service in the school-house, where it was celebrated in the Ojibway language; at my request the heathen Indians, who had assisted in conducting us from Garden Island, were present. They maintained a respectful silence during the service, and a favourable impression may have been produced. Mr. Macdonald has given away ten hundred weight of flour and forty bushels of wheat since May last, to the wandering Indians who occasionally touch at the White Dog. Their attention and sympathy must first be enlisted by continually renewed presents; this shows the advantage of agricultural operations being associated with spiritual labour at remote stations. In seasons of scarcity the Indians assemble at the Mission and expect to be assisted by material contributions; these failing, they lose confidence and heart, and the influence which the missionary may have succeeded in acquiring is soon lost when they disperse in search of the means of sustaining life. The present congregation numbers about forty-five when the entire body of Christians attached to the Mission are assembled. They belong to the Swampy Crees, and hunt on the lower Winnipeg.

Mr. Macdonald divides the Indians who hunt north-east and north of the Lake of Woods from those who inhabit the shores, islands, and the country east and south-east of that beautiful lake. The former belong to the Muskeg nation (Muscaigoes) or Swampy Crees, the latter to the great Ojibway nation. The Swampys have a tradition that at a remote period they drove the Ojibways from the lower Winnipeg to the country bordering on the Lake of the Woods, and since that time they have maintained their footing in the conquered territory.

ISLINGTON MISSION—WINNIPEG RIVER

The heathen Muskegs or Swampys address their invocations to the Evil Spirit, but they acknowledge the existence of a Supreme and Good Being. The Bishop of Rupert's Land had an opportunity of witnessing the idolatrous worship of the conjurors of the Swampys on the Lower Winnipeg in 1852. He describes the scene in the following words: — "There were two or three tents. I entered the largest, and there found the son of Wassacheese sitting in solitary state. I was about to sit down where I saw some articles expanded, and where at first I thought he had prepared a seat for me; but I found, on a second look, that these were the idols of the chamber of imagery, the instruments of his art as conjurer, and the feast spread out for spirits. I asked him to explain his magic art, which he said he would if I would give him some flour. I gave him instead a little tobacco, and then heard his tale. He showed me, as a special favour, that which gave him his power — a bag with some reddish powder in it. He allowed me to handle and smell this mysterious stuff, and pointed out two little dolls or images which, he said, gave him authority over the souls of others; it was for their support that flour and water were placed in small birch-rind saucers in front. . . . The altar was raised a little on some Indian matting, and on it, ranged in order, the bags and images and all the instruments of his craft. O what is man without the grace of God!" *

At Islington there are now (1857) five houses, besides the mission-house and a school-house, which is used as a chapel. With the exception of the hill of drift clay, on which the Mission is situated, the country is wholly granitic in the immediate neighbourhood, and resembles the

* The Net in the Bay, p. 19.

bold but dreary region through which the Winnipeg flows from Rat Portage to this, the first small area on or near the banks of the river where farming operations on a small scale can be conducted with any prospect of success. The river is about 250 yards broad, and flows past this little oasis with a rapid current. Mr. Macdonald has about six acres of wheat and four acres of potatoes in a very promising condition. In a garden surrounding the dwelling of one of the Christian Indians, I observed an acre and a half of potatoes, some Indian corn, and a patch of wheat. Missionary labour here, as elsewhere, is not encouraging; the adults are generally very indolent, unsettled, and improvident; they make but little improvement, and their conduct tends to confirm the impression which I find to prevail among missionaries, that many embrace and adhere to Christianity for the sake of the temporal benefits it secures, often relapsing when material advantages diminish or cease, into their original condition of barbarous idolatry. Mr. Macdonald looks to the schoolhouse as the main hope for the future. When the Mission was established the Indians who visited this favourite camping ground made many promises; but even in 1852, when the Bishop of Rupert's Land visited Islington, he had occasion to remark, "that the Indian promised more at that time than he has since performed will not appear surprising to those who know the Indian character, and therefore all would stand prepared for some little disappointment in the carrying out of the plan."*

I started from Islington Mission on the 31st August, with Lambert, a young Indian lad, and the Ojibway from the Lake of the Woods whom I had persuaded to accompany me to the settlements at Red River, leaving Mr.

* The Net in the Bay, by the Bishop of Rupert's Land.

Dawson in the charge of the Rev. Robert Macdonald and his sister, in a fair way of recovery, although far too much indisposed and feeble to attempt the rough and dangerous voyage down the Winnipeg in a small canoe, weakly manned. I arranged with Mr. Macdonald to send supplies, nourishing food, and a north canoe, as soon as I arrived at the settlements, which I hoped to reach in five days.

The poles of wigwams are numerous in the flat country bordering the shores of Grand Turn Lake; the timber is aspen and poplar, with a few Banksian pine. Eagles and fish-hawks were the only birds visible in this part of the Winnipeg, the other species of the feathered tribe being in the rice-fields. Near James' Falls there is a small area of clay sustaining aspen and poplar; it may embrace 200 or 300 acres. Here I observed some gigantic ant-hills close to the river. The largest were from five to seven feet in diameter at the base, and from four to five feet high. I counted thirteen of these curious habitations of the "wise" ant. James' Falls are about thirteen feet high, and are always approached with great caution in descending the stream; they are considered to be among the most dangerous obstacles on this broken and tumultuous river. The foaming torrent passes over three steps, each about four feet in height and ten apart, caused by a ledge of rock stretching across the river in a direction W. 10° S.

Below James' Falls the poles of wigwams are numerous, and many Indians were seen at the foot of the different rapids engaged in fishing. The scarcity of animal life of all kinds was very remarkable, eagles and fish-hawks, ducks and rabbits, being the only representatives seen. This scarcity is, however, confined to the autumnal months, as to the time, and to the Great Winnipeg River in respect of area. Some distance from the river the extensive rice grounds cover many thousand acres, and continue for miles on either bank. Here the game congregates, and reveling in the midst of such an abundant supply of nutritious food, vast flocks of ducks, geese, and all kinds of aquatic birds common in these regions, are to be found. It is here too that the Indians assemble at stated periods amidst the rice-grounds, procuring without any difficulty, in favourable seasons, a large supply of food for winter consumption.

The falls of the Portage du Bois, where we arrived on the following day, are exceedingly beautiful; they let the river down ten and a half feet, and are divided into four parts by three prettily wooded islands; the whole breadth of the river here is about one-third of a mile. The portage path passes over a fertile oasis sustaining oak, ash, and cherry, with rosebushes, honeysuckle, and a great variety of sweet scented flowering plants, which fill the air early in the morning with delicious fragrance. Succeeding this romantic break in the river, the Point aux Chiens falls and rapids let the river down nineteen feet. It is succeeded by the Roche Brulé Portage and then by the Slave Falls, the scene of one of those terrible incidents in Indian life formerly of frequent occurrence in these inhospitable wilds. Tradition tells that a slave of a ferocious master, maddened by long continued cruelty, calmly stepped into a canoe above these falls in the presence of the tribe, and suddenly pushing off from shore, wrapped her deer-skin robe round her face and glided over the crest of the cataract, to find rest in the surging waters below.

Above the Barrière Falls, which are next approached, a clay cliff comes on the river, and exposes a section about fourteen feet high; its breadth is not great, as in the rear bare rocks can be seen from the river; it occupies, however, a considerable area, being also observed below the falls, where it attains an altitude of eighteen feet. At this spot we met Mr. Clouston, of the Hudson Bay Company's service, in charge of supplies for Rat Portage and Fort Frances. He kindly consented to open his cases and allow Mr. Macdonald to select some medicine and whatever necessaries he might require for Mr. Dawson's use. The motley crew of Indians and half-breeds were engaged at the time of our meeting in hauling a bateau

Slave Falls, Winnipeg River.

over the falls, having deposited the cargo on the rocks. Twenty to twenty-five men were pulling at a rope attached to the bateau, and with the utmost exertion slowly lifting the unwieldy craft up the wall of water, five feet high, which constitutes the Barrière Falls. Late in the evening we arrived at the mouth of the Pennawa, a small branch of the Winnipeg, which joins it again at Bonnet Lake. At and below the Otter Falls were a number of Indians engaged in fishing; as we hurried past them they shouted to us to stop and camp with them, but not being provided with presents, and having only a very small stock of provisions, I thought it advisable to decline the invitation; leaving a few pieces of tobacco on the portage path we hastened to the mouth of the Pennawa, and camped about half a mile down the river. Being anxious to reach Fort Alexander, I awoke the men an hour before daylight, and whilst breakfast was preparing, strolled over the rugged and fissured rocks through which the little river finds its way. The almost oppressive silence was broken only by the occasional splash of a pike, or the distant howling of a wolf, or the subdued roar of the Winnipeg wafted by a very gentle breath of wind which now and then stole from the west. The musk rats were busily engaged swimming across the little ponds which separate the rapids into which the Pennawa is broken, near where it branches off from the Winnipeg.

Instead of following the course of the Great Winnipeg, after arriving at the Otter Falls, I passed down the Pennawa River into Bonnet Lake, in order to avoid the dangerous "Seven Portages," and save several miles of route. Near the entrance of the Pennawa into Bonnet Lake, the little river winds through an immense marshy area covered with wild rice, and I succeeded in collecting a considerable quantity as the voyageurs paddled through its light and yielding stalks with undiminished speed. There, too, were seen vast numbers of different species of duck, and many other kinds of birds, such as herons, pigeons, woodpeckers, cedar birds, jays, &c.

The Indians we met lamented the failure of the rice this year; they described the appearance in favourable seasons of the ground through which we were hurrying, as a vast expanse of waving grain, from which they could soon fill their small canoes, by beating the heads with a stick. The waters of the river and marshes were unusually high, so as to check the growth of the rice to an extent which, when coupled with other deficiencies, threatened them with famine during the coming winter.

The same cause which originated the partial failure of the wild rice led also to a great scarcity in fish. In general, the Winnipeg teems with fish, among which are sturgeon, pike, two kinds of white fish, perch, suckers, &c., affording a bountiful supply to the Indians who hunt and live on or near the lower portion of this majestic river. The extraordinary height of its waters during the summer of 1857 had so extended the feeding grounds of the fish, that they were with difficulty caught in sufficient numbers to provide the Indians with their staple food.

The unlooked-for short supply of wild rice and fish were more severely felt in consequence of the unaccountable disappearance and death of the rabbits, which are generally found in vast multitudes in the region of the Lake of the Woods and Winnipeg River. During the spring and summer, large numbers of rabbits were found dead in the woods, owing probably to the exhaustion which followed a severe winter, prolonged this year to an unprecedented length in these regions. With a partial failure in the rice, great scarcity of fish, and the prospect of a very limited supply of rabbits, the anticipations of the coming winter on the part of those who cared to think of the sufferings of the wretched Indians of the Winnipeg, were gloomy indeed.

The Pennawa enters Bonnet Lake between high and rugged rocks, the termination of a range of dome-shaped hills, which decline from an altitude of 250 feet to small island bosses in the lake. The strike of the range is nearly north and south, curving slightly to the south-east; the summit of the range is bare, rounded, and apparently polished. Camping on an island in the lake to escape an approaching squall, I found an Indian cache elevated on a stage in the centre of the island. The stage was about seven feet above the ground, and nine feet long by four broad. It was covered with birch bark, and the treasures it held consisted of rabbit-skin robes, rolls of birch bark, a ragged blanket, leather leggings, and other articles of winter apparel, probably the greater part of the worldly wealth of an Indian family. At the great Bonnet Portage there are many acres of good farming land, and, indeed, from this point to Lake Winnipeg, a strip on the river, widening as we descend, possesses all the requirements as far as soil and timber is concerned for a large settlement. The Silver Falls, the last but one, are perhaps the most imposing and beautiful of all the cascades on the Winnipeg. The volume of water precipitated here is immense; all the inosculating branches of the Winnipeg uniting some distance above the magnificent Silver Falls. The vast torrent descends a slope about 200 yards long with an inclination of nearly sixteen feet, in the form of five or six gigantic swells. The observer may stand close to the huge heaving waves and watch them rush past him with astonishing velocity and ever-changing form. Sometimes they send a thin sheet of water over the smooth rock on which he is standing at the edge of the torrent; in another minute there may be a gulf ten or fifteen feet deep, with a terrible whirlpool raging below, between him and the crested swell fifty feet from the shore; suddenly the gulf is filled, and the turbulent waters, dashing against the rocks, send a shower of spray far and wide over the polished gneiss which confines them. We reached Fort Alexander at four in the afternoon, stayed half an hour to procure some flour and potatoes, and then hurried on to Lake Winnipeg, camping at the mouth of the river.

Fort Alexander is situated within one mile and a half from Lake Winnipeg. In fields near the fort I saw wheat in process of being harvested on the 3rd of September, and obtained some new potatoes of great size and excellent quality. I was informed by the officer in charge of the fort that Indian corn succeeded well in many parts of the south-eastern rim of the lake, and that it was very rarely touched by late spring frosts; it is cultivated by the Indians. On the following morning I aroused the men at three, and made the first lake traverse by moonlight. The

Fort Alexander, Mouth of the Winnipeg River.

west shore of Traverse Bay is high, and shows an excellent soil, thickly covered with balsam-poplar, aspen and birch. The lodges of Indians are very numerous on the west point, which forms one of their most important fishing stations. When we landed to breakfast or dine, opportunities were afforded of examining the precipitous but unstable cliffs which were occasionally exposed. At a point on the east coast of the Grand Traverse, a section showed one feature of interest, which is common to all the great lakes of the St. Lawrence basin. The summit of the cliff, clothed with an inch or two of sandy loam, exhibits an ancient lake beach, composed of water-worn boulders, pebbles, and stratified sand, two feet thick. This is underlaid by sixteen feet of stratified sand, containing limestone fragments, and boulders of the unfossiliferous rocks; it is flanked by a talus of shingle, slabs and boulders, among which, bright yellow, cream-coloured, and beautifully variegated limestones are numerous. This talus is the present shore of the lake, and the shingle, slabs, and boulders have probably been washed out of the unstable cliff. Its breadth may reach sixty feet, with an inclination of three to five feet from the level of the lake, giving to the ancient beach at the summit an elevation of twenty-one feet above the present level of Lake Winnipeg.

About five miles further south, I ascended a cliff fifty feet high, consisting of stratified sand and marl, in which were embedded primitive boulders of most gigantic dimensions: some of them measured twelve to fifteen feet in diameter; they were all water-worn, and distributed throughout the cliff. On the surface, walking was exceedingly difficult on account of their numbers and size. Many of them were covered with the Virginian creeper (*Ampelopsis quinquefolia*). The base of the cliff was well protected by an immense accumulation of these erratics, which had fallen from the loose sand of the cliff. The temperature of the lake, six miles beyond this point, was 64° 5′. A heavy squall from the north-west compelled us to approach the shore when within three miles of the mouth of the Red River; the waves rose with great rapidity, as is usual in large, open, and shallow sheets of water, compelling a hasty retreat among the willows and rushes, where, notwithstanding exposure to the discomfort of the waves washing over our camp during the night, we were compelled to remain in a damp maze of reeds until the wind and waves subsided. Here I had an opportunity of observing the vast number of duck, geese, and plover, which congregated amongst the rushes at the approach of night. Early in the morning, flights swept backwards and forwards close to our camp in constant succession.

Red River enters Lake Winnipeg by six distinct channels, which will be described in the second volume. Its junction with the lake by the branch through which we entered is marked by a low spit of sand, which was the only piece of land visible amidst the tall reeds extending far to the south.

Land which is dry during the summer months at the stage of water in the river on the 5th September, 1857, begins five miles from the mouth of the main channel. Half a mile above this point, Netley Creek comes in from the west, and by means of this small affluent, much of the water from the upper country during floods reaches Lake Winnipeg. Large numbers of haystacks were seen here in September 1857. An immense area is flooded during the spring, and produces a very rank profusion of those grasses which delight in a rich and marshy soil.

A little below the Indian village, fourteen miles from the mouths of the river, the whole country rises, the banks are about thirty feet high, the timber is imposing, and all the aspects of a level, fertile region, gradually invest the scene; but the sameness in the general appearance of the banks at this season of the year soon becomes monotonous, after the wild and varying beauties of the Winnipeg. The sight of clearings, however, with the neat white houses of settlers at the Indian missionary village, speedily creates other impressions, aroused by such fair comparisons between the humanizing influence of civilization, and the degraded, brutal condition of a barbarous heathen race.

These familiar and suggestive signs of improvement in moral and social position, rapidly create a healthy tone of feeling in passing from the cascades and rapids of the Winnipeg, where half clad savages fish and hunt for daily food, to the even flow of Red River, where Christian men and women, once heathen and wild, now live in hopeful security on its banks.

I arrived at the Stone or Lower Fort about six P.M., and after a little delay succeeded in procuring an Indian horse, which carried me to Fort Garry, a distance of twenty miles, in a little less than three hours, where I found the other members of the expedition camped under the walls of Fort Garry.*

* The Blue Book, published by order of the Legislative Assembly, dated 10th May 1858, contains an account of the steps which were taken to send supplies to Islington Mission.

CHAP. VI.

RED RIVER SETTLEMENTS.

The Red River of the North.— Its Tributaries.—The Red Fork.—The Red River within British Territory.— Its physical Features.— Objects seen on ascending the River.— Section of the River and Prairie.— Objects on the Banks. — The Settlement.— The King's Road.— Character of the Country north of Fort Garry.—Aspect of the Prairies.—Beauty of the Prairies.—The Assinniboine River.—Effect of Evaporation on the Volume of Water in the Assinniboine.—Description of the Assinniboine.—Prairie Portage.—Mud and Sand Flats in the River.—Timber.—John Spence. —Lignite reported to exist on the Assinniboine.—Sioux.—Indian Corn. —The Big Ridge.—An Overturn.—The Prairies of the Assinniboine.— Mr. Lane.—Mr. George Flett.—Mr. Gowler.—Mr. Gowler's Farm.—His Opinions respecting the Prairies on the Assinniboine. — Melons. — Old Associations.—Independence. — Mr. Gowler's Success. — The Nor'wester. — A Newspaper published at Red River Settlements.

THE Red River of the North rises in Ottertail Lake, State of Minnesota.* The north-east end of Ottertail Lake is in latitude 46° 24′ 1″, and the general course of the river is westerly, through an attractive undulating country, until it makes its great bend to the north, in latitude 46° 9″. It then meanders through a boundless prairie, which gradually declines in elevation until it forms a vast level plain, elevated above the water of the river only about one and a half to two feet, at its ordinary stage in June. The distance of this great bend is 110 miles from Ottertail Lake, and the vast low prairie through which the river subsequently flows, in an exceedingly tortuous

* The description of that part of Red River which lies within the territory of the United States is abbreviated from Dr. D. D. Owen's account in his geological survey of Wisconsin, Iowa, and Minnesota.

channel, is level as a floor. In latitude 46° 23′ 30″ a belt of timber sets in, and continues with some interruption along its banks, on one side or the other, to Pembina, near the boundary line. In latitude 46° 41′ 12″ the level of the prairie above the river is thirty feet; this depression is probably due to the gradual erosion of the river channel in soft clay. Red River receives some important tributaries south of the 49th parallel: on the left bank the Shayenne sweeps round the north-east flank of the Grand Coteau de Missouri, and joins Red River in latitude 47° 3′. Numerous streams, draining a lower terrace of tableland than the Grand Coteau, cut the fertile prairies through which Red River meanders at right angles to its course, but no affluent of importance is received between the Shayenne and the Pembina, near the international boundary line, or 525 miles from Ottertail Lake, by the windings of the main stream.

On the right bank, the Red Fork, issuing from Red Lake, and joining the main river in latitude 47° 50′, is the most important; in consequence of its being on the line of water communication between Rainy River and Red River, and also separated by a low-water parting from the Mississippi and Lake Superior, it may hereafter acquire some degree of prominence.

Dr. Owen remarks of the country through which Red River flows in the United States territory, that it possesses features both geologically and physically of great sameness and flatness, without the least indication of minerals of any value, except salt, which may be crystallised out of saline springs.

The length of Red River within British territory is about 140 miles by the windings of the stream. It debouches into Lake Winnipeg, in latitude 50° 28′, longitude 96° 50′. Its most important affluents on the east side are Roseau River and German Creek. On the west side it receives, in latitude 49° 53′ 24″, and longitude 96° 52′, the Assinniboine* River; at the confluence of these streams Fort Garry, the capital of Assinniboia, and the head quarters of the fur trade in British America, is situated.

The following description in detail of Red River, within British territory, supposes the observer to ascend the stream from Lake Winnipeg in a bateau or canoe, and is confined to those objects which come under observation during the voyage.

Fourteen miles from the mouths of the river, the Indian missionary village occupies a terrace thirty feet above the summer level of the stream. Above the village the banks are fringed with oak, elm, and maple, which soon give way to aspen, and then to open prairie land, the trees of larger growth appearing at intervals on the points and on the insides of the bends.

About four miles above the Indian missionary village, a remarkable bend in the course of the stream gives rise to a sharp projection of the level plateau of the prairie, called Sugar Point, from the groves of maple which cover it. It has been preserved from the abrading action of the stream by numerous fragments of limestone which lie at the bottom of the river bank and continually increase in number and size in its ascending course, as far as the exposed strata of rock in position, at and above the Stone Fort, where their place is supplied in part by the parent rock.

The maple, which at one time grew in considerable quantities near Sugar Point, is not the true sugar maple (*Acer saccharinum*) so common in Western Canada, but

* Assinniboine, from "assinni" a stone — Cree. Howse, in his grammar of the Cree language spells the name of the tribe, "Assinneboigne."

another species, generally known as the ash-leaved maple (*Negundo fraxinifolium*), also furnishing an abundance of juice from which sugar is made as far north as the Saskatchewan. Near to Sugar Point is an Indian school, in connection with the Indian Mission below, situated north of the line which divides the parish of St. Peter from that of St. Andrew, and marking the northern limits of the Red River Settlement. The banks on both sides are very heavily timbered close to the river; but between this place and the Stone Fort there are very few farmhouses. The general direction of the river from Sugar Point to Fort Garry, is a few degrees to the west of south. In an air line the distance is twenty miles; by the road on the left or west bank twenty-one, and by the river itself twenty-three and a half miles. The scenery and objects which meet the eye in ascending the river between the Lower Fort and 49th parallel are uniform, but singular and interesting.

In order to arrive at a true conception of its physical features, it is merely necessary to imagine a river from 200 to 350 feet broad, with a moderately rapid current, having in the course of ages excavated a winding trench or cut to the depth of from thirty to forty feet in tenacious clay, through a nearly level country for a distance exceeding 100 miles, and the general physical aspect of Red River, within British territory, is reproduced. Here and there local diversities occur which give some appearance of variety. Such are noticed at the Grand Rapids, where the even flow is broken and disturbed by a ledge of limestone, which may occasion a fall of six feet within a mile. A lower terrace has here and there been excavated perhaps ten feet below the general level of the prairie banks. Occasionally sand, mud, and gravel bars are formed at the numerous sharp turns in the general

RED RIVER FROM ST. ANDREW'S CHURCH

course of the stream, before the volume of its waters is augmented by the Assinniboine which comes in at Fort Garry. These projecting bars or points are often covered with fragments of limestone, primitive boulders, and vast numbers of large fresh-water shells. The current round them is rapid, and during low stages they present an obstacle to the navigation of the river by means of steamers exceeding 120 feet in length. Often, too, on one side or the other, and sometimes on both sides, a narrow belt of heavy forest timber closes upon the river, and seems suddenly to narrow and darken its abrupt windings. The most uniform character, however, and one which is more frequently found on the west side, is a clean and steep line of bank about thirty feet in altitude, perfectly level to the eye, and forming the boundary of a vast ocean of prairie, whose horizon or intermediate surface is rarely broken by small islands of poplar or willow, and whose long, rank, and luxuriant grasses show everywhere a uniform distribution, and indicate the character of the soil they cover so profusely. A subsequent closer inspection of the soil never failed to establish its fertility and abundance, as well as its distribution over areas as far as the eye can reach, both eastward and westward, from the banks of this remarkable river.

The objects which arrest attention in ascending the river between Sugar Point and the Stone Fort, are limited to precipitous clay banks, fringed with elm, poplar, maple, oak, and ash, all of large growth, but not fair representatives of the forest which once skirted the stream, so long subjected to a destructive culling process, in order to supply the necessities of the settlement above. Among the underbrush, the Virginian creeper, and occasionally a wild grape, with a profusion of convolvulus twining round hazel, alder, and rose bushes are most conspicuous. At the Stone Fort, massive layers of limestone crop out, which have been extensively quarried, and their application is seen in the walls and bastions of the fort built upon the bank, here about forty feet in altitude, and forming the abrupt termination of the prairie stretching westward, which for some distance sustains a small but dense growth of aspens. At each turn of the river above this point the houses of the inhabitants of Red River Settlement come in sight, and occupy the banks, at short intervals, all the way up to Fort Garry, a distance of twenty-three and a half miles by the windings of the river. Two miles above the Stone Fort, is the so-called Whirlpool Point, and immediately above it the Big Eddy; these are obstacles to further progress, formidable only in name, and like most other local descriptive titles on this river, must be accepted with the mildest interpretation, and only understood to designate marked differences from the general even flow of the river. A small brook on which a water-mill is situated enters the river at the Big Eddy. A short distance above the same locality limestone crops out in heavy layers on the west bank, and detached fragments in great abundance protect the base of the cliff, which in no instance, from the mouth to the 49th parallel, as far as my observation permitted me to judge, rises above forty feet from the water level. Some very substantial illustrations of the adaptation of the limestone for building purposes occur here, and particularly at the Grand Rapids, two and a quarter miles farther up the stream. The east side of the river is wooded to a depth varying from a few yards to a mile, and generally this feature prevails along the eastern bank as far as Fort Garry; the timber is similar to that already described. At the Grand Rapids, which even during the low stage of water in

September, offer no formidable obstacle to the Company's and freighters' boats carrying four and five tons, an assemblage of well-built stone buildings are grouped, which create a very favourable impression of Red River resources and comfort, not unfrequently repeated in ascending the stream. There is erected a very substantial stone church, capable of seating 500 people, and surrounded with a stone wall enclosing an extensive burying-ground. About 300 yards south of the church, the parsonage house is seen from the river, and a visit to its interior, to be more fully noticed subsequently, proved that every desirable comfort was enjoyed by the kind and hospitable incumbent, Archdeacon Hunter. Adjoining the parsonage is the residence of the curate, Mr. Kirkby (now on Mackenzie's River), and next to it a capacious and well-built school-house of wood. Four miles above the Grand Rapids, Water-mill Creek enters the river, having cut its way through the yielding clay substratum of the prairie to a depth of twenty-five feet, half a mile from its mouth.

Above Mill Creek the river banks break off abruptly from the prairie level, and are well wooded on the east side, the houses of the inhabitants occurring at regular intervals upon the west bank. At a short distance above the very commodious and comfortable residence of Mrs. Bird, a lower terrace, caused by denudation, commences, and its prairie boundary passes in the rear of the house occupied by the Expedition, and comes upon the river again before reaching the Presbyterian church. The following section shows the relation of the lower terrace to the general level of the Great Prairie, the relation of the Big Swamp, noticed hereafter, to the river, and also of the ancient beach or ridge of Lake Winnipeg to the general level of the country.

Distance from water mark, West.	Height above level of river, 18th September.	
Water mark 0 feet.	0 feet.	
66 „	18·48 „	
109 „	11·36 „	
152 „	20·74 „	Bench-mark.
233 „	20·06 „	
830 „	16·52 „	
1230 „	19·07 „	
1330 „	25·76 „	
1853 „	27·52 „	King's Road.
2482 „	23·80 „	Small shallow bed of creek.
2667 „	27·38 „	Grand Prairie Level.
2988 „	27·30 „	
4212 „	26·31 „	Commencement of marsh.
East.		
4 miles nearly N. E.	8 feet.	Ancient beach of Lake Winnipeg.

Above St. Paul's church, in the Middle Settlement, eight miles north of Fort Garry, the river winds between high prairie banks, which generally maintain an altitude of about thirty feet; houses and windmills occur at regular intervals, until the steeple of St. John's church, the peaked roof of St. John's College, the school-house, the Bishop's residence, &c., offer the appearance of a large village, which is again re-produced after a sharp turn at Point Douglas, by the imposing Roman Catholic church, dedicated to St. Boniface, the spacious nunnery and the parish school, with other buildings on the left, and a group of several commodious private dwelling houses just below Fort Garry, on the right. About half way between these small centres of population, German Creek, a small meandering stream, comes in from the south-east. A quarter of a mile above the Roman Catholic church, the Assinniboine enters Red River, and a short distance up this stream the bastions of Fort Garry come into view. Above the mouth of the Assinniboine the course of the river is exceedingly tortuous. An idea of its meanderings may be obtained from the comparison between distances by the river from Fort Garry to the mouth of La Rivière Sale, and the relative position of the same places by the road, the former being sixteen, and the latter nine miles. The houses of settlers appear at intervals on the banks for several miles above La Rivière Sale, the last house being situated thirteen miles from Fort Garry, or fifty-seven from the 49th parallel. Above this the river windings are fringed with forest, varying in depth from a few yards to half a mile. Here and there naked bends are exposed to the prairie, the peninsula portion on the opposite side being generally clothed with trees of large dimensions; this character is preserved far south of the 49th parallel.

Returning to the Indian village, and following the road to Fort Garry, thence to the 49th parallel, the following description refers to the west or most thickly inhabited parts of Selkirk Settlement.

From that part of the Indian village which lies on the west bank of the river to the Stone Fort, little can be seen of the surrounding country, as the road traverses a forest of small aspens, and farms are few in number and small in extent.

The Stone Fort covers an area of about four acres, and encloses within its walls numerous buildings of which several excellent photographs were taken in 1858. The main or King's Road does not follow the windings of the river, but stretches from point to point, sometimes approaching it at these places within a quarter or half a mile. Where the river windings throw it back to a distance exceeding a mile, inner roads, as they are termed, branch off to the river for the convenience of settlers, and there is a bridle path all the way from the Lower to the Upper Fort, on the immediate bank of the river. Aspen woods continue to shut out the view until we arrive within a mile or two of Water-mill Creek, when a scene opens upon the right which discloses on the one hand the white houses and cottages of the inhabitants, with their barns, haystacks, and cattle yards, grouped at short distances from one another, and stretching away in a thin vanishing line to the south ; while on the other hand, a boundless, treeless ocean of grass, seemingly a perfect level, meets the horizon on the west. The same kind of scenery, varied only on the left hand as the road approaches or recedes from the farmhouses, on the river banks, or passes near neat and substantial churches, which at almost regular distances intervene, prevails without interruption until within four or five miles of Fort Garry. Here, stretching away, until lost in the western horizon, the belts of wood on the banks of the Assinniboine rise above the general level, while from the Assinniboine toward the north again is an uninterrupted expanse of long waving prairie grass, sprinkled with herds of cattle, and in the fall of the year with clusters of stacks of hay. This is the ordinary aspect of the country comprising that portion of Red River Settlement which lies between Water-mill Creek and Fort Garry. Remove the farmhouses and churches, replacing them on the river banks by forest trees of the largest growth, and the country between Fort Garry and the 49th parallel, as seen along the road to Pembina, a distance of seventy miles, is continually reproduced in its ordinary aspect of sameness and immensity.

The vast ocean of level prairie which lies to the west of Red River must be seen in its extraordinary aspects, before it can be rightly valued and understood in reference to its future occupation by an energetic and civilised race, able to improve its vast capabilities and appreciate its marvellous beauties. It must be seen at sunrise, when

THE PRAIRIE LOOKING WEST

the boundless plain suddenly flashes with rose-coloured light, as the first rays of the sun sparkle in the dew on the long rich grass, gently stirred by the unfailing morning breeze. It must be seen at noon-day, when refraction swells into the forms of distant hill ranges the ancient beaches and ridges of Lake Winnipeg, which mark its former extension; when each willow bush is magnified into a grove, each distant clump of aspens, not seen before, into wide forests, and the outline of wooded river banks, far beyond unassisted vision, rise into view. It must be seen at sunset, when, just as the huge ball of fire is dipping below the horizon, he throws a flood of red light, indescribably magnificent, upon the illimitable waving green, the colours blending and separating with the gentle roll of the long grass in the evening breeze, and seemingly magnified towards the horizon into the distant heaving swell of a parti-coloured sea. It must be seen, too, by moonlight, when the summits of the low green grass waves are tipped with silver, and the stars in the west disappear suddenly as they touch the earth. Finally, it must be seen at night, when the distant prairies are in a blaze, thirty, fifty, or seventy miles away; when the fire reaches clumps of aspen, and the forked tips of the flames, magnified by refraction, flash and quiver in the horizon, and the reflected lights from rolling clouds of smoke above tell of the havoc which is raging below.

These are some of the scenes which must be witnessed and felt before the mind forms a true conception of the Red River prairies in that unrelieved immensity which belongs to them in common with the ocean, but which, unlike the ever-changing and unstable sea, seem to promise a bountiful recompence to millions of our fellow-men.

On the 10th September I started for Prairie Portage,

on the Assinniboine, about sixty-five miles west of Fort Garry, with Mr. Napier and a Cree half-breed named John Hallet. Our conveyance, which resembled a very shaky old fashioned light cart, was furnished by Mr. M'Dermott, the most enterprising and wealthy merchant and freighter in the settlements. Mr. M'Dermott became a kind of purveyor-general to the Expedition, and supplied us with whatever was to be procured for money in this remote region. Hallet provided two horses, one of which he declared to be an excellent buffalo runner, but not to be trusted in shafts, as we found to our cost. In the following general description of the Assinniboine, a few facts are incorporated which were acquired during the exploration of the following year.

The Assinniboine rises in latitude 51° 40′, and pursues a south-easterly course for a distance of about 260 miles parallel to the basins of the Great Lakes on the east of the Riding and Duck Mountains. Within eighteen miles south of the 50th parallel it takes a sudden bend to the east, which general direction is preserved until it falls into Red River, a distance of about 240 miles from the great bend. At Lane's Post, twenty-two miles from Fort Garry, the Assinniboine is 120 feet broad (June 28th, 1858), with a mean sectional depth of six feet. Its greatest depth here is seven and a half feet, and the rate of its current is one and a half mile an hour. Near Prairie Portage, sixty-seven miles from Fort Garry, the speed of the current is two miles an hour, and its fall, as ascertained by levelling, is 1·18 feet in a mile. At its junction with the Little Souris, an affluent which it receives 140 miles from its mouth, the breadth of the river is 230 feet, its greatest depth twelve feet, and its mean sectional depth 8·6, the speed of its current being one and a quarter mile an hour; this river is apparently considerably larger 140 miles from its outlet than twenty-two miles from the same place. Even at Fort Ellice, 280 miles from its junction with Red River, the Assinniboine is 135 feet wide, 11·9 feet deep in the channel, with a mean sectional depth of eight feet, and a current flowing at the rate of one and three quarters of a mile an hour; in other words, this river, 280 miles from its mouth, carries a larger body of water than at a point twenty-two miles from it.

The following table shows the quantity of water which the Assinniboine carries at three different points, distant respectively in round numbers, twenty-two miles, 140 miles, and 280 miles from its outlet by the windings of the river valley, but not by the windings of the river itself, which will be at least double the length of the river valley.

	Cubic feet per hour.	Distance from outlet at Fort Garry.
Lane's Post	5,702,400	22 miles.
Mouth of Little Souris . . .	12,899,040	140 „
Opposite Fort Ellice . . .	9,979,200	280 „

It thus appears that the volume of water in the Assinniboine is nearly twice as large at Fort Ellice as 258 miles lower down the river, if the foregoing table affords sufficient data on which to rest an opinion. It is very probable that the character of the season would modify these results in different years. The measurements were not made simultaneously, and the rainfall in the neighbourhood of the Touchwood Hills and in the region about Fort Pelly in 1858 was represented to be more in the extreme, than is usual during the summer months. But judging from the appearance of the river bank, and the statements of Indians and half-breeds, familiar with the summer level at the localities where the sections were made, there is no reason to suppose that its waters were in excess of their ordinary summer level. It is, therefore, very probable that evaporation during a long and tortuous course through an open valley, is adequate to diminish the volume of water in the Assinniboine very much in excess of the supply which it receives from tributaries or springs during its course to Red River.

East of Prairie Portage the Assinniboine flows through a flat, open, prairie country, not more than sixteen or twenty feet above the general level of the stream, until within a few miles of Fort Garry. The whole country rising in steps west of Prairie Portage, the Assinniboine has excavated a deep and broad valley through it, in which it meanders with a rapid current.

At the mouth of the Little Souris, or Mouse River, this valley is 880 yards across and eighty-three feet below the general level of the prairie. At Fort Ellice the valley is one mile and thirty chains broad, and 240 feet below the prairie.

The Assinniboine receives numerous and important affluents. On its eastern water-shed are the Two Creeks, Pine Creek, Shell River, Birdstail River, and Rapid River or the Little Saskatchewan. The distances of the rivers from Fort Pelly, which may be considered as lying at the head of bateau navigation, will be noticed hereafter when the country they drain is described. From its western water-shed it receives the White Sand River from the Touchwood Hills; the Qu'Appelle or Calling River, inosculating with the south branch of the Saskatchewan; Beaver Creek, a small rivulet on which Fort Ellice is situated; and the Little Souris or Mouse River, sweeping round the flanks of the Grand Coteau de Missouri. The Crees of the Sandy Hills on the south branch, state that Elbow Bone Creek, an affluent of the Qu'Appelle River, inosculates by a deep valley with the Mouse River, or an arm of it, and is connected continuously with the Assinniboine.

For a distance of several miles above Fort Garry the Assinniboine flows in a trench excavated through a level prairie to the same depth as the river it feeds, or from twenty-five to forty feet. Differences due to local variations in the height of the bank are often referable to very slight undulations in the level of the prairie, and to the occurrence of ancient lake beaches or ridges, the first of which is cut by the river, near St. James' church. This ridge continues in a direction nearly due north, until it dies away in the general rise of the prairie. It is near this spot that the rapids occur which, in the summer months, when the water is low, offer the only impediment to the continuous boat navigation of the Assinniboine for many miles.

Some short distance above the rapids the river widens; at its mouth it may be 130 feet in breadth, and four miles from its mouth, 150 feet, a breadth which it preserves with considerable uniformity for a distance of 130 miles. About six and a half miles from Fort Garry the Assinniboine receives a small affluent called Sturgeon Creek, coming from the north-west. The general direction of the river up to this point is nearly due west, and its course comparatively straight. The south bank is heavily timbered to a small depth, but the north bank is chiefly taken up by farms, and devoid of timber.

From Sturgeon Creek the course against the stream continues still westerly, but with more decided meanderings, and the wooded points on both sides of the river rarely penetrate a quarter of a mile into the vast prairie on either side. The distance from Fort Garry to where it makes its north-westerly bend is twenty-three miles by the river's windings, but by the road through the prairies and settlements only sixteen miles. The river banks are here about eighteen feet high, and their height imperceptibly diminishes until, at Prairie Portage, they were found by measurement not to exceed sixteen feet, during the stage of water, on the 7th of September, 1857.

After making its north-west turn, the Assinniboine is so remarkably crooked that a straight line drawn through the tract of country in which it meanders for a distance of twelve miles, would be cut eighteen times by the river, and these windings are confined within such a limited breadth that in a strip of the same length, and 1,000 yards broad, the curves of the river would just overlap this boundary four times.

The physical features of the Assinniboine as far as Prairie Portage, resemble in every important particular those of Red River. The tortuous sinuosities of the larger stream are reproduced with curious fidelity in the magnificent prairies through which its western rival flows.

There is little or no variety in the character of the banks either of Red River or the Assinniboine; they consist of Post Tertiary stratified clays and marls, overlaid by vegetable mould. At Lane's Post, twenty-two miles west of Fort Garry, a fresh exposure of the bank, which, by the way, is continually breaking down in small patches and changing, during the lapse of many years, the channel of the river, exhibited stratified whitish marly clay, and dark drab coloured clay from the water's edge to within five feet of the prairie level, which here, as is frequently the case, comes abruptly upon the river. Dark alluvial clay succeeds, having an average thickness of about four feet; this is followed by from six to eighteen inches of black prairie mould.

Beyond Lane's Post the river course is westerly for a few miles, it then makes a bend towards the north-west, until Long Lake, an old bed of the river, is reached, after which it turns towards the south-west for about sixteen miles, thence westerly, ten miles further to Prairie Portage. Nine miles beyond Lane's Post the settlements cease; they recommence thirty miles further up the stream by the road, and although the distance from Lane's to Prairie Portage is not more than forty-three miles, the course of the winding Assinniboine would probably exceed ninety miles. The river banks are heavily timbered, and sustain trees of very large dimensions. The distance between the top of the banks on either side of the river is variable, but it appears to be generally between 600 and 800 feet, at sharp turns it was often not more than 400, and whenever it exceeded that distance one side was steep and washed by the water, the other occupied by a sand spit or mud-flat at the foot of the opposite bank.

During my stay at Prairie Portage, in September, 1857, I had an excellent opportunity of examining the relation of the sand and mud flats to the river banks, as well as the forest which fringed it to the depth of half a mile. The river is here about 180 feet broad, and with a rapid current sweeps under the south bank, which forms the outer arc of a very beautiful curve extending over 120 degrees. The cord of this arc is well defined by the old north bank of the river, under which, probably, it once swept, but now only touches when the channel is full during spring freshets; the length of this cord is perhaps 700 yards, and at each extremity the river is seen sweeping between steep banks, sixteen feet high, until, a little lower down or a little higher up, similar curves, with their accompanying sand and mud-flats recur. These sand and mud-flats are arranged in the order of the specific gravities of the materials which compose them, but with such singular regularity, and with such curious and interesting admixtures, that I have considered it worth while to describe them with some degree of particularity.

At the western extremity of the curve, a few rounded boulders were seen, not exceeding eight inches in diameter;

these were followed by gravel spits as the area opened; beyond the gravel spits, which extended perhaps over a quarter of the segment, flats of coarse sand showed themselves; these were strangely filled and strewed with the decayed and broken horns of the elk, the bones and horns of the elk, buffalo, deer, and just beyond these a human skull, with two or three scattered and water-worn skulls of buffalo; the sandy areas ceased in curved lines, with a small steep descent of about two feet, and were succeeded by mud partly covered here and there with fine sand, probably drifted by wind. The sandy mud was followed by fine compact mud, with numerous deep cracks partially filled with fine sand. Another fall of about three feet occurred in the form of a bank, and recent mud, smooth and treacherous, occupied the remaining portion of the segment a few inches above the present water level. This arrangement of mud, sand, and gravelly spits was noticed elsewhere, and probably frequently occurs in the Low Prairie region through which the Assinniboine is continually changing its course by excavating new channels in the soft and yielding clays.

The timber on the banks of this river is perhaps not so heavy as on Red River, nevertheless some very fine oak and elm, with white wood and poplar of extraordinary dimensions, were seen near the Prairie Portage. A fair quantity of sugar is made by the Assinniboine half-breeds, but not in comparison with what might be easily obtained if systematic habits and a proper appreciation of the fruits of industry existed here. A species of grape grows in profusion on the banks of this river. I suppose it to be the frost grape (*Vites Andifolia*). The fruit when first gathered is not very palatable, but after hanging in the open air for forty-eight hours it acquires a sweet taste and a very delicious flavour.

The name Prairie Portage is derived from the existence of a carrying place nine miles long, between this part of the Assinniboine and Lake Manitobah. It is stated by half-breeds at the settlement, that at seasons of extraordinary high water, canoes can approach each other from the Assinniboine and Lake Manitobah, so as to leave but a very short distance for the portage; and instances have occurred of water, during periods of high floods, flowing from the Assinniboine into Lake Manitobah by the valley of Rat River.

I had an opportunity of meeting, at this isolated settlement, with one John Spence, a Cree half-breed of great experience in Rupert's Land. He drew a small chart for me, showing the position of what he called "coal" on the Assinniboine. I saw and conversed with a half-breed who had brought "a few bushels" of this coal to the settlement, for the purpose of ascertaining its fitness for the forge; he stated that he was a blacksmith, had used the coal and found it answer, but it required a strong draft. I procured from another half-breed several specimens, and ascertained that it was lignite, and not the true coal of the coal measures. On the Little Souris, a tributary of the Assinniboine, the lignite was described as cropping out in bands exceeding a foot in thickness, and occupying a large area on its banks, a statement which the exploration of the succeeding year did not verify, at least in the locality pointed out to me. Dr. Hector, however, found lignite in the valley of the Souris, three hundred miles west of Prairie Portage. I endeavoured to induce John Spence to accompany me and point out the locality where the lignite cropped out on the Assinniboine; he expressed perfect willingness to do so, if I could procure for the trip ten men in all, so that watches might be established by night, in consequence of the presence of several bands of Sioux Indians on the trail of the buffalo hunters, who were then coming in from the Great Prairies, after their summer hunt. The Sioux had succeeded in driving off ten horses from the tail of the caravan, about half a day's journey from Prairie Portage the night preceding my arrival there; and this incident led John Spence and others to decline going with me, unless the number of the party amounted to ten in all. So large an addition I found it impossible to procure at Prairie Portage, and after my return to the settlements, the time at my disposal was too short to admit of the exploration. In carefully searching the recent mud-flats of the Assinniboine at, and a little above Prairie Portage, I found numerous small fragments of lignite, from which it might be inferred that an exposure of the parent rock was situated some distance up the river; but beyond this and the reiterated statements of many who had descended the stream in a batteau, I found no proof of the existence of lignite in available quantities.

In the settlements on Red River and the Assinniboine, small specimens of lignite were frequently shown to me by different people, who stated that they procured them from the crossing-place on the Little Souris; and an Indian had a bag containing about half a bushel of the same material, together with specimens of silver mica, carefully treasured up in many folds of dressed buffalo skin. Many intelligent people in the settlements appeared to be much impressed with the importance of ascertaining the true nature and extent of the lignite beds on the Little Souris. The great scarcity of wood in the prairie country, and all through the valleys of Red River and the Assinniboine, making the question of a permanently increasing settlement in a measure dependent upon the supply of fuel which may be obtained from other sources than those offered by the aspen-covered ridges, or the thin strips of timber on the immediate banks of the rivers.

In order to reach John Spence's house, I passed through a field of Indian corn, and from the proprietor I obtained the following statement respecting the cultivation of this valuable grain. The kind of Indian corn most common in the settlements is called the Horse-teeth corn, and it does not always ripen. The variety sown by Spence he termed the Mandril corn, the seed was procured from the Indians near the head waters of the Missouri; probably the "Mandan corn" would be the correct name. He had cultivated it for two years; it ripened

Prairie Portage, Assinniboine River.

well both years. One of his neighbours, a Cree Indian, had cultivated it for four years and had not met with any failure. Spence sowed his corn on the 1st June, and gathered it 10th September, or after a period of 102 days. In dry seasons it ripens earlier, and is planted about the 20th of May; the wet spring of the present year retarded all agricultural operations. A small house adjoining the one in which Spence resided I found filled with a portion of his corn crop.

The road from the village of Prairie Portage follows

a general north-easterly direction for a distance of twenty-nine miles, before it turns south-westerly towards Fort Garry. This deviation is necessary in order to avoid Long Lake, an ancient bed of the river, now converted into a narrow, winding lake of great length.

On each side of the road is a very magnificent prairie, bounded on the right by the wooded banks of the Assinniboine, and on the left by the horizon; a few scattered clumps of poplar are seen here and there, but no trees, until the "Big Ridge" comes in sight. The ridge is probably an ancient beach of Lake Winnipeg; its elevation does not appear to be more than sixty feet above the prairie level. Where the road touches Long Lake, a spur of the Big Ridge is distant about three miles. I made a diversion from the main track for the purpose of examining the character of the ridge. It rises almost imperceptibly from the prairie, and at its base are numerous small fragments of limestone. Ascending the ridge, the limestone débris was found to increase in quantity, and near its summit the slabs were of large dimensions.

Our cart-horse beginning to show signs of breaking down soon after we left the Big Ridge, Hallet proposed to put his buffalo runner into the shafts, thinking that the journey had so far subdued his spirit, that he would submit to the indignity of drawing a light cart. He travelled for a few hundred yards quietly enough, but when we came into the open prairie, he started off at a gallop, and swerving suddenly, overthrew the cart, projecting its contents into the grass. Mr. Napier received a serious sprain in the wrist, but I fortunately escaped without injury, having fallen on my feet. The buffalo runner soon broke loose from the shafts, and after a short gallop stopped to survey the ruin he had made. Hallet caught him without much trouble, but did not venture to harness him again. A few strips of buffalo hide placed our cart in travelling condition, and permitted us slowly to resume our journey.

Regaining the main road, well marked by the deep ruts formed by the buffalo hunters' carts, we soon arrived at the White Horse Plain, a vast, slightly undulating prairie, bounded by the horizon in every direction but the south, where the distant wooded banks of the Assinniboine afford some relief to the eye. The grass is long and rank, and the soil a black mould of great depth, often exceeding eighteen inches. In many places it is thrown up into conical heaps by moles, and uniformly displays the same rich appearance, truly represented by the bountiful profusion of verdure it sustains. In 1857 the edges of the White Horse Plain unfortunately teemed with another kind of life. The grasshoppers (locusts) appeared in countless millions just before my arrival; every bare patch of ground in the road was filled with their eggs, the living insects were leaping through the tall grass in infinite multitudes, yet, notwithstanding, failing to change the appearance of the country in the midst of so great a profusion of food. What the next year's brood may do remains to be seen, their progenitors had come in swarming clouds from the south side of the Assinniboine, but no one could tell of their origin, or of the devastations they must have created before they took their flight, and alighted on the White Horse Plain.

The last house of the settlement, westward of White Horse Plain, is about thirty-three miles from Fort Garry, and between it and the Company's Post, in charge of Mr. Lane, there are nine houses and farms. The Prairie Portage road, however, does not pass near them, it touches the river only at those bends which do not necessarily compel much deviation from a straight course. The farmhouses are similar to those on Red River, but the soil appears to be, if possible, of a better description.

We were very hospitably entertained by Mr. Lane, the gentleman in charge of the Hudson's Bay Company's Post on the Assinniboine, twenty-two miles west of Fort Garry.

Mr. Lane informed me that Indian corn did not always ripen on that part of the river. Spring frosts rarely affect it, but autumn frosts sometimes cut it off. He thought that careless cultivation was the reason why it did not progress fast enough to escape the early autumnal frosts. Indian corn sown on dry points of the river arrived at maturity much sooner than that which was sown on the rich and moist prairie mould.

Leaving Lane's Post, the river is touched again at the Roman Catholic Mission of St. François Xavier. The road now follows the general course of the river, in the rear of the farms which, from this point to Fort Garry, are not far apart.

On the night of the 15th September, we stayed at the house of Mr. Geo. Flett, fifteen miles west of Fort Garry. Mr. Flett's turnips had been altogether consumed by the grasshoppers, but his wheat was safe and good; he says that Indian corn succeeds well, and almost always ripens; it is his opinion that it may always be relied upon when care is taken; it does not progress quick enough on the open prairie to escape every season the early autumnal frosts; on the points of the river where the soil is lighter and dryer than in the open prairie, and where some shelter may be obtained from the neighbouring timber, he has never known it to fail. Mr. Flett finds the cut worm the great enemy to his turnips; his potatoes for the summer crop are planted 1st June, and ready for eating from the 10th to the 15th August; the winter supply he does not lift until October. Over the whole of the White Horse Plain district, thirty bushels to the acre is an average crop of wheat, but on new land forty bushels is not only common but generally expected.

On the morning of the 16th we paid a visit to Mr. Gowler, whose farm is situated on the immediate banks of the Assinniboine, about nine miles from Fort Garry. Nearly all farming operations were over, but an inspection of his farmyard and garden enabled me to form an opinion of his success and prospects as an agriculturist on the Assinniboine.

A small stack-yard was filled with stacks of wheat and hay; his barn, which was very roomy, was crammed with wheat, barley, potatoes, pumpkins, turnips and carrots. The root crops were shortly to be transferred to root houses, which he had constructed by excavating chambers near the high bank of the Assinniboine, and draining them into the river. The drain was supplied with a close and tightly fitting trap, which was closed when the water rose during the spring above its mouth, at that time eight feet above the level of the river. The chambers were about nine feet high, and their ceilings three feet below the prairie level. Access was obtained through a hole in the ceiling, which was covered with a neat little movable roof. There were three of these cellars or root-houses before the dwelling-house, and between it and the river. Frost never entered them, and he found no difficulty in preserving a large stock of potatoes and turnips through the severe winters of this region.

Mr. Gowler farmed fifty acres in white and green crops, hay and pasture being furnished by the prairie. He owned much more land, but found it useless to crop it, as no market for surplus produce existed. In 1856 he had sold many bushels of potatoes at sixpence per bushel, and had carted them nine miles. I had been previously informed of the extraordinary success of Mr. Gowler in growing wheat, but I found upon inquiry that the practice he employed was simply not to grow wheat after wheat; he had grown fifty-six measured bushels to the acre. The price of wheat at the time of my departure was 4*s.* 5*d.* sterling a bushel, but last year at the same time it had been 3*s.* 6*d.* sterling. His turnips (Swedes) were magnificent; four of them weighed 70 lbs., two weighing 39 lbs., and two others 31 lbs. Whatever manure his yard and stables supplied he gave to green crops and the garden. A portion of the potato crop was still in the ground; they far surpassed in quantity, quality and size, any I had ever seen before. Mr. Gowler very kindly turned them up out of the soil wherever I pointed out. I counted thirteen, fourteen, and sixteen potatoes, averaging three and a-half inches in diameter, at each root respectively. They were a round white-skinned variety, like those known in Canada as the "English White." The potatoes were planted on the 1st June, and were ready for eating on the 16th or 18th August. The winter supply was rarely taken out of the ground before the beginning of October. The greatest enemy to the turnip crop is the cut-worm (the grub of an elater).

Indian corn succeeds well on Mr. Gowler's farm, and onions of rare dimensions were growing in his garden. He had had this year a splendid crop of melons, the seed being sown in the open air at the end of May, and the fruit gathered about the 1st September. At the time of my visit the melons had all been consumed, but I had several opportunities of tasting and enjoying this fruit, at Fort Garry and elsewhere, on the Assinniboine and Red River. In every instance they were grown in the open air, without any artificial aid beyond weeding, from the time the seed was planted to the maturation of the fruit.

Mr. Gowler insisted on my tasting his wife's cheese, and smoking his tobacco, before I departed. The cheese was tolerable; the tobacco, which was grown in the neighbourhood and highly prized by Mr. Gowler, was dreadfully strong, and would involve long training in order to acquire a taste for its qualities. Nevertheless, Mr. Gowler preferred it to some excellent fig-leaf which I offered him; he remarked that he had grown and prepared it himself, and knew what it was.

I may here relate, with a view to show how long old associations linger in the recollections of the European portion of the population in this remote region, that when I sat down to table Mr. Gowler turned inquiringly to his wife, saying, "And where is my plate?" "Oh, John! you would not think of sitting at table with gentlemen?" Mr. John seemed puzzled for a moment; his son-in-law and children were looking in silence from different corners of the room. He cast a hasty glance around, and the true feelings of independence and manly right showed themselves, as he exclaimed, "Give me a chair and a plate; am I not a gentleman, too? Is not this my house, my farm, and these my victuals? Give me a plate."

As Mr. Gowler accompanied me to the gate, where my horse was tied, he expressed, with much warmth of feeling and manner, the following opinion of husbandry and its prospects in Assinniboia: —

"Look at that prairie; 10,000 head of cattle might feed and fatten there for nothing. If I found it worth my while, I could enclose 50, 100, or 500 acres, and from every acre get 30 to 40 bushels of wheat, year after year. I could grow Indian corn, barley, oats, flax, hemp, hops, turnips, tobacco, anything you wish, and to any amount, but what would be the use? There are no markets, it's a chance if my wheat is taken, and my potatoes I may have to give to the pigs. If we had only a market, you'd have to travel long before you would see the like of these prairies about the Assinniboine."

The substantial character of the barn, stables, and piggeries, constructed of wood, their neatness and cleanliness, the admirable arrangements of the hammels for cattle, and the sheds for sheep, all showed how far a little energy and determination, instructed by the experience of earlier years, would go in reproducing amidst the boundless prairies of Assinniboia, the comforts and enjoyments which are by no means the rule among the small farmers of Great Britain. I regretted to find that a few days before my visit the grasshoppers had arrived from the south-west, and consumed in a single day every green leaf in the garden which remained exposed to their attacks.

The "Nor'-Wester," a newspaper published for the first time at the Red River settlement on the 28th December, 1859, mentions Mr. Gowler's success as an agriculturist in the following terms: —

"At seed-time of the present year (1859), all traces of the pestilence (the grasshoppers) had disappeared, and Mr. Gowler having before his eyes the pretty sure prospect of a good market, brought under cultivation a greater breadth of land than any year previously. He sowed 63 bushels of wheat, 36 of barley, 24 of oats, and 101 of potatoes, and from these he realised 700 bushels of wheat, 350 of barley, 480 of oats, and 2,100 of potatoes. The cost of the seed was 50*l.*; in preparing and tilling the soil, about 25*l.* more were expended; and the cost of gathering in and thrashing the crops is set down at 100*l.* – making a total expenditure of 175*l.* Place against that the sums representing the sale of the wheat at 6*s.*, the barley at 3*s.* 9*d.*, the oats at 2*s.* 6*d.*, and the potatoes at 1*s.* 3*d.* per bushel (average prices, which the produce will easily command), and an argument more strong and convincing than could be wrought out by any other process of reasoning, stands stubbornly forth in favour of the claims of the settlement as being one of

Confluence of the Assinniboine and Red River.

the best agricultural countries on the face of the globe. It should be added that Mr. Gowler's profits have already enabled him to enlarge the bounds of his estate to 600 acres; to stock it with a noble herd of cattle and horses, and to make the necessary preparations for erecting thereon, next summer, a snug and comfortable mansion."

CHAP. VII.

THE WEST AND EAST BANKS OF RED RIVER, BETWEEN FORT GARRY AND THE BOUNDARY LINE.

La Rivière Sale.—Pembina Mountain.—Scratching River.—Pembina Fort. —Pembina.—The Roseau River.—Ancient Lake Ridge.—Roseau Lake. —Route to the Lake of the Woods.—Meet an Indian.—Indian idea of Money.—Crossing place at the Roseau.—Indian Wigwams.—Ancient Lake Ridge.—Prairie Hens.—Indian Snares.—Still Water Creek.—Rat River.—The Nine Mile Swamp.—Mr. Pierre Gladieux.—Half-breed Politeness and Hospitality.—La Rivière Seine.—Character of the Country. —Indian Scruples.

As soon as I returned from Prairie Portage preparations were made for an exploration of the Roseau or Reed-grass River, with a view to ascertain whether a communication might be effected between the Lake of the Woods and Red River, as well as to ascertain the limit of the palæozoic rocks, and their junction with the metamorphic series on which they rest. On the 21st September, accompanied by my assistant, Mr. Fleming, I started from our quarters in the Middle Settlement. Our equipment consisted of three men, five horses, and a Red River cart. We crossed the Assinniboine by the ferry at Fort Garry, and took the road on the west side of the river to Pembina.

The country lying to the west of Red River was examined by Mr. Dickinson in 1858, and, for the sake of uniformity, his observations are incorporated in the description which follows, of La Rivière Sale, and an extensive range of table-land called Pembina Mountain.

Nine miles above Fort Garry, La Rivière Sale joins with the main stream. The buffalo hunters' trail to the great south-western prairies on the Grand Coteau de Missouri passes up the south side of this river for a distance of thirty miles, cutting across the large and winding bends of the valley.

The country lying between it and the Assinniboine is very marshy, and is covered with willows and clumps of small aspen. In the valley of La Rivière Sale, and along both sides, grow oak and elm and some fine ash, many of the trees being two feet in diameter; this narrow forest extends the whole way up the river on the north bank. On the south side there is a prairie apparently as level and boundless as the ocean; the grass on it is most beautiful and luxuriant, indicating the richness of the soil.

The valley is about a quarter of a mile wide and forty feet deep, abounding in salt springs, which make the water in the river quite brackish, from which it derives its name. The river higher up opens out into small lakes, and rises from an extensive marsh. The track here joins the hunters' track from the White Horse Plain; it then turns to the south, in which direction it goes for about twenty-five miles through open prairies, until it crosses "La Rivière des Iles des Bois," a river fifteen feet wide and two deep, flowing into the Scratching River. This portion of the country is all a level prairie, the greater part of it being wet and marshy, except near the last-named river, where it is quite dry for five miles; the land is a rich sandy loam, yielding most luxuriant grass. On both sides of the river there is a skirting of trees, chiefly of oak, averaging one foot six inches in diameter.

The buffalo hunters, when they have crossed this little river, begin to keep a sharp look-out for the Sioux, and to take their usual precautions, such as setting watches by night and placing their carts in a ring.

The track, continuing in the same direction, crosses a

prairie twenty miles wide, of light sandy soil, with clumps of aspen and willows growing here and there; it is intersected by many small valleys, in all of which, with one exception, the brooks that formed them are now dried up. The valley of "La Rivière Tabac" is seven chains wide and twenty feet deep, with very little water in the fall of the year, where in spring time there is a rapid flow.

The prairie on the south and west is bounded by what is generally called the "Pembina Mountain," which is rather a series of steps rising up from the prairie below to one above. There are three steps, from twenty to twenty-five feet high, together with a gradual ascent for two miles; the whole of it is thickly strewn with granitic boulders. This "mountain," which consists of clay, gravel and sand, runs in a south-easterly direction, from a little above Prairie Portage to Pembina. Where we crossed it there is no timber, but on both sides it is well covered, particularly on the south, where the trees seemed large and good. Here the forest is said to begin which reaches to the Assinniboine, but with the exception of some oak on the mountain there is no good timber, nothing but young aspen from twenty to thirty feet high, growing very close together, and forming a dense thicket.

Scratching River joins the main stream thirty-seven miles from Fort Garry. The postman who carries the mail between Pembina and the settlements lives here, and has established an apology for a tavern and a ferry. Scratching River winds for many miles through a boundless prairie, without a tree or shrub on its banks.

We arrived at Pembina Fort on the 24th, just in time to partake of an excellent dinner with Mr. Murray, the gentleman in charge. In the afternoon we crossed Red River, passing through the miserable cluster of houses which bears a prominent position on maps of the north-western states, figuring under the name of the town of Pembina. Most of the inhabitants of Pembina have moved to St. Joseph, so that the population of this frontier village does not now number more than 100 souls. On the evening of the 25th we camped on the banks of the Roseau, after a hard march of thirty-one miles through a very fine and promising country.

The general course of this stream from its confluence with Red River to Roseau Lake is a few degrees to the south of east. It enters Red River about ten miles north of the 49th parallel, and it is probable that Roseau Lake is on the boundary line between Rupert's Land and the State of Minnesota. The course of the Roseau is very tortuous, and for the first twenty miles it meanders through a beautiful prairie, with a belt of heavy forest trees on its banks. Near the mouth of the river, on the south side, there is a large area of low land, but above that point the banks vary from fifteen to twenty feet in height until, at the crossing place, the ancient Lake Ridge is reached. Here the banks are from fifty to fifty-five feet above the level of the river. Near the crossing place, the ridge has probably an elevation exceeding sixty feet above Red River ; it, with its offsets, form a very singular and most interesting feature in the topography of the valley of this river.

The ridge once past, the whole face of the country changes. The soil becomes poor and sandy, although still preserving a prairie or plain character. The timber on the banks of the river fast dwindles to small sized oak, elm, birch, and poplar, until it gives place, about forty-six miles from the mouth, and perhaps seventy or eighty by the winding of the stream, to extensive marshes, in which there are islands of small pine. At the commencement of these marshes the Roseau River moves sluggishly, and its stream soon becomes dead water, with a vast expanse of flooded land on either side, extending, according to our guide, fifty miles to the right hand and to the left.

Having found it impossible to proceed further on horseback than the beginning of the great marshy tract of the Roseau, and not being provided with a canoe, the following description of the country rests upon the authority of the guide who accompanied us, and who had resided at Roseau Lake for a year and a half when in the service of the Hudson's Bay Company. The river channel can be traced through a marsh ten miles long, nearly on a level with the water in the river. The depth of the marsh does not exceed three feet, and it is quite possible to wade on horseback through it. The Hudson's Bay Company's route to their post on Roseau Lake (in 1851) retired from the banks of the river when the waters began to flow sluggishly, and pursued a direction some miles to the south of the channel, probably within the United States territory. In 1847, a very dry season, it was possible to proceed with carts in a direct line near the banks of the river from the beginning of the marsh to the post, one mile and a half from Roseau Lake.

An idea of the character of the country about this post may be inferred from the guide's description of his attempts to destroy the monotony of his life, when stationed at Roseau Lake. He informed me that when he wished "to see anything" beyond the four walls of his log shanty, and the rushes by which it was surrounded, he was in the habit of mounting to the roof, and from the top of the mud chimney enjoying the view; which consisted of reeds to the north, reeds to the south, and reeds to the west, as far as the eye could reach, and to the east, Roseau Lake, fifteen miles long by ten broad, with a deep fringe of reeds. On the bosom of this retired sheet of water, in the spring and the fall, he was enabled to watch countless millions of ducks and geese; and the noise of their shrill cries, with the flapping of wings as they would rise to take their morning flight to the north or south, according to the season of the year, were almost the only sounds he heard, save the sighing of the wind through the reeds, during his dreary abode in the waste of Roseau Lake. The altitude of Roseau Lake above Lake Winnipeg probably does not exceed 170 feet; and as the elevation of the Lake of the Woods is at least 370 feet above the same level, there must still be a rise of 200 feet to be overcome before reaching the height of land. Our guide described the Roseau River, before it enters Roseau Lake, as stretching to the south in the territories of the United States. He also said, that issuing from the Great Muskeg, or swamp, occupying so much of the height of land between Red River and the Lake of the Woods, was a narrow rapid stream fifty miles long, emptying into Roseau Lake, thus forming a route by means of which none but the smallest sized canoes can pass from Roseau Lake, through the Great Muskeg, to the Lake of the Woods.

The ancient Lake Ridge is a continuation of the one referred to on page 132; it extends in an unbroken line, except where the river from the higher level in the rear has cut channels through it, from near Lake Winnipeg, far beyond the international boundary. At the crossing-place on the Roseau, about forty-six miles from Red River, its height was estimated to be the same as at the Middle Settlement; it forms a beautiful dry gravel road wherever traversed, and suffers only from the drawback of being the favourite haunt of numerous badgers, whose holes on the flank, and sometimes also on the summit, are dangerous to horses; it is, apparently, perfectly level for a hundred miles, and everywhere, as far as my observation enabled me to judge, shows the same even rounded summit; it may yet form an admirable means of communication through the country, and it marks the limit of the good land on the east of Red River. This ridge is a favourite resort of the prairie hen (*Tetrao cupido*), when they perform their curious circular dances in the early spring months. We frequently met with a ring of sticks, placed in a circle about ten feet in diameter, to each of which a noose of sinew was attached. Our half-breeds informed us that they were snares which the Indians set to catch the prairie hens In the spring the males congregate on dry gravelly ridges, frequenting the same spot year after year, and march round and round, with feathers erect and wings rubbing the ground as a preliminary to a general combat. The Indians observe the spot where the birds congregate, and after night-fall set their snares on the edge of the ring, which the male birds have selected to try their strength, and to attest their claims to the favour of the females who are perched on the neighbouring bushes. In the battle which ensues, or during their solemn march, some of them are caught and strangled. The following spirited description from Audubon's delightful "Birds of America" will, doubtless, be read with interest:—"Their love season commences, and a spot is pitched upon to which they daily resort until incubation is established. Inspired by love, the male birds, before the first glimpse of day lightens the horizon, fly swiftly and singly from their grassy beds, to meet, to challenge, and to fight the various rivals led by the same impulse to the arena. The male is at this season attired in his full dress, and enacts his part in a manner not surpassed in pomposity by any other bird. Imagine them assembled to the number of twenty by daybreak; see them all strutting in the presence of each other; mark their consequential gestures, their looks of disdain, and their angry pride as they pass each other. Their tails are spread out and inclined forwards, to meet the expanded feathers of their neck, which now, like stiffened frills, lie supported by the globular orange-coloured receptacles of air, from which their singular booming sounds proceed. Their wings, like those of the turkey cock, are stiffened and declined so as to rub and rustle on the ground, as the bird passes rapidly along. Their bodies are depressed towards the ground; the fire of their eyes evinces the pugnacious workings of the mind; their notes fill the air around, and at the very first answer from some coy female, the heated blood of the feathered warriors swells every vein, and presently the battle rages. Like game cocks, they strike and rise in the air to meet their assailants to better advantage. Now many close in the encounter; feathers are seen whirling in the agitated air, or falling around them tinged with blood. The weaker begin to give way, and one after another seek refuge in the neighbouring bushes. The remaining few, greatly exhausted, maintain their ground, and withdraw slowly and proudly, as if each claimed the honours of victory The vanquished and the victors then search for the females, who, believing each to have returned from the field in triumph, receive them with joy."

At noon on the 26th September, when discussing with the guide the possibility of proceeding further up the banks of the Roseau River on horseback, we heard the sound of a gun, proceeding apparently from the river. Having fired one in return, we were not surprised some time afterwards to see an Indian approach. He had just arrived with his family from the Lake of the Woods, by the route proposed to be taken by Mr. S. Dawson and myself some weeks before. He described the route in the same way as the guide, and in no material respect differed from the accounts we had before received from the Lac la Pluie Indians, who had been engaged to convey us through it, before the intervention of the tribe at Garden Island, narrated in Chapter IV. He had been ten days on the road, but might have accomplished the journey thus far in shorter time, had he not found it necessary to hunt for his family, who accompanied him. At my request he drew a chart of the route, which was, in almost all particulars, similar to that furnished by the Indian at Fort Frances. He ascended a small river, marked on the map Reed River, from the Lake of the Woods, for a distance of thirty miles to the Great Muskeg at the height of land. He was two days dragging his canoe through the Muskeg, which is here nine miles broad. He then descended the rapid stream, forty to fifty miles long, before noticed, which is called by the Indians Muskeg River, and found himself among the rushes or reeds of Roseau Lake. In his canoe we found his wife and two children. The half-naked little savages were busily engaged in plucking a goose for their noon-day meal. I offered him some tea in exchange for the bird, and when the transfer was made, asked him what they intended to eat for their own dinner; he replied by pointing to the bow of his canoe, addressing at the same time a word or two to his wife, who raised a piece of birch bark and disclosed two more geese, which he had shot a few minutes before we saw him. Having bartered for them also, with a small plug of tobacco, I asked the guide what he would take for a new stone pipe which one of the children was playing with; to my astonishment the Indian replied, three beaver skins (about five shillings), but at the same time casting his eyes upon our cups and saucers which lay on the grass, he said he would prefer a cup, worth about four-pence. He really knew nothing of the value of money or of cups, although he was quite aware of the worth of a beaver-skin in ordinary articles of trade, such as powder, shot, tobacco, or tin-ware, but a painted earthenware cup was something new to him; and his wife expressed great delight as she examined with much minuteness the addition to her household goods.

Returning nearly in our steps to the crossing-place, we went over to the right bank of the Roseau, and after threading through a forest of fine oaks, about one quarter of a mile deep, found ourselves emerging upon an open, dry prairie, bounded on the east by the low wooded ridge before noticed as occurring on the south side of the river. The distant belt of woods, fringing Red River, might just be seen in the far western horizon, the whole intervening space being a rich and level prairie, without shrubs or willows. On the bank at the crossing place the skeletons of Indian wigwams and sweating-houses were grouped in a prominent position, just above a fishing weir where the Ojibways of this region take large quantities of fish in the spring. The framework of a large medicine wigwam measured twenty-five feet in length by fifteen in breadth; the sweating-houses were large enough to hold one man in a sitting position, and differed in no respect from those frequently seen on the canoe route between Lakes Superior and Winnipeg, and which have been often described by travellers.

Six miles from the Roseau, Still Water Creek occurs; its waters are deep, and, as its name implies, sluggish, or almost stagnating. Between Still Water Creek and Rat River some marshy spots occur, while on the right the ridge, wooded with aspen, continues in the direction of the Rapids of Red River, near which spot it is found within four miles of the banks of the main stream. Rat River is an insignificant brook coming from the Great Muskeg, which occupies the height of land to the east of the Valley of Red River. At the crossing place it is fifteen feet broad.

Four miles from Big Rat River, Little Rat River was crossed, and the trail then led to the point of junction of the two streams, until it came upon a ridge, which it followed for a distance of ten miles, after which the great Nine Mile Swamp occurs, where water lodges in marshy intervals, for the distance which has given its name to this wet prairie. A strong Scotch plough, drawn by a stout team of oxen, would soon effect the drainage of the Nine Mile Swamp. It partly originates from the excessive luxuriousness of the grasses growing upon the level expanse, which, in a humid season, hold up sufficient water to give permanency to the wetness of this portion of the prairie. Hay, in considerable abundance, as exemplified by the stacks which were seen in all directions, is made on the dry intervals of the Nine Mile Swamp. A French settlement commences immediately on the northern extremity of this characteristic illustration of Red River indifference and unconcern in regard to the improvement of the country. A very little well-directed labour would convert these extensive marshy areas into the richest pasture and hay privileges, and drive to more congenial haunts the myriads of snipe and plover we disturbed in our passage through them.

We arrived at Mr. Pierre Gladieux's house an hour after sunset on the evening of September the 29th. We were soon provided with an excellent supper, and our horses, seven in number, well supplied with hay in the yard. Before starting next morning an almost sumptuous breakfast was given to us, and while the horses were being saddled, I begged permission to see the farmyard, &c. Under a small shed there was a neat, light, four-wheeled carriage, which as we passed Mr. Gladieux very politely and kindly placed at my disposal during the remaining period of my stay at Red River. He remarked that on the morrow he was going to the plains to hunt buffalo, and should not require the carriage for several weeks after my proposed departure. I requested the guide to ask what I had to pay for the entertainment of the party. The polite answer returned was as follows: "Nothing; it is not the custom of the people of this country to charge strangers who may honour them with a visit."

Mr. Gladieux is a French "native," he resides on the right bank of the Red River, five miles south of Fort Garry. He showed me his farmyard, barns, garden and cattle.

The Red River at Pierre Gladieux's.

Four pea stacks, several wheat stacks, and five or six hay stacks, all of fair dimensions, were neatly arranged in the stack yard, while the cattle yard was tenanted by a number of cows, pigs, horses, and poultry. His peas were sown on the 7th May, and reaped on the 25th September. Before Mr. Gladieux's house, the trunk of an immense liard lay ready for splitting into firewood; the size appeared to be so unusual that I measured it carefully, and found it to be four feet ten inches in diameter six feet from the base, and four feet eight inches in diameter ten feet from the base; at the base it measured 16·5 feet in circumference, and showed 150 well defined rings.

German Creek, or La Rivière Seine, flows into Red River about two miles below Fort Garry. In 1858 this river was explored by Mr. Dickinson, whom I had requested to try and penetrate to the Lake of the Woods. Mr. Dickinson set out on the 16th September: his description of the country is published in my Report on the Assinniboine and Saskatchewan Expedition, and the introduction of an abstract here, while completing the view of the Valley of Red River within British territory, will not materially affect the narrative of our explorations in the succeeding year. I give the following description nearly in Mr. Dickinson's own words:—

"As the country east of Red River, extending to the Lake of the Woods, is quite unknown, except for a few miles back from the river, to any but Indians, I was anxious to procure one of them as a guide. Having succeeded in doing so after some little delay, I was obliged to examine this part of the country first, as the Indian guide was about to leave the settlement in a few days for his winter quarters.

"Considering that one of the objects of this exploration should be that of ascertaining where a *summer road* could be most easily made from Red River to the Lake of the Woods, that being now a subject of great interest among the settlers, who were about sending a party out for that special purpose, I thought it advisable first to go along the straight picket line made by Mr. Dawson last winter, in which direction, I understand, he reports that a road can be made for some miles, in order that I might be able to institute a comparison between this and any other portion of the adjacent country through which the Indian might guide me.

"The first day I was able only to go about fourteen miles — two-thirds of this distance at least being through marsh and wet prairie. The general course was along the picket-line, from which I was obliged to diverge frequently — sometimes a mile or more, but always keeping it in view — in order to avoid, when possible, the wide marshes through which it passes. The next day I continued in the same direction, and having reached a point opposite the 22nd mile-post, on the picket-line, I could go no further, being stopped by a swamp or quagmire, impassable for horses, or even men, extending in front for many miles, and on both sides as far as the eye could reach. Though taking advantage of all the dry places within reach, ten miles of the course I took lay through marsh and wet land, and five miles at least through swamp. There are a few small clumps of young aspens along the line, and low willows in some of the marshes; but far away towards the north may be seen some clumps of larger trees.

"The land is, for the most part, a rich loam, with a subsoil of sandy clay; but the difficulty, or rather the impossibility of draining the numerous swamps and marshes, and the want of timber, render this tract of country unfit for settlement; and for the same reasons, the difficulty of constructing a suitable road through it would be very considerable, and the expenses enormous.

"Judging, then, that I had seen enough of this part of the country for my purposes, I retraced my steps to the settlement, from which I set out again, under the guidance of the Indian, who promised to conduct me by the only dry path towards the Lake of the Woods, as far as the boundary of his hunting grounds.

"On the morning of the 23rd, I proceeded along the south side of 'La Rivière Seine,' or German Creek, which flows into the Red River a little below its junction with the Assinniboine. There are farmhouses and a good road along it for a distance of five miles, when an Indian track begins, which keeps close to the valley of the creek for eight miles, between it and the marsh.

"This dry space varies from half a mile to a quarter mile wide, and is crossed by two small sluggish creeks, which, if widened and deepened, would effectually drain the marsh. There is plenty of good timber along the valley, consisting of poplar, elm, and black ash, with small oaks. Leaving German Creek here on our left, we went along a low ridge about one foot above the level of the marsh, and varying in width from fifty to a hundred yards; it runs in a south-easterly direction for about three miles, and then widens out on the left as far as I could see, and on the right to half a mile. At this point we were about three miles from German Creek, which we lose sight of now for some time. Continuing in the same direction for three miles more through beautiful rich grass, with clumps of aspens on the left, and high willows on the right, we came to Oak Creek, which is about two chains wide, but so still and sluggish that it rather resembles a long lake. Our course lay along it nearly due east for two and a half miles, when the creek turns to the south. This would be an admirable place for a settlement, the land being as rich as any in the whole country, and there being a large supply of oak, averaging one foot six inches in diameter, and poplars suitable for fencing.

"On the south side of Oak Creek the open prairie stretches away to the horizon, the greater part of that which was within view being dry, there being only a few patches of wet land. Leaving Oak Creek we went through a country of this character for about nine miles in a south-easterly direction, our track winding, however, a little to avoid the wet places, a few of which we had to cross ; but none of them were more than seven or eight chains wide, and easy of crossing. There are numerous clumps of small aspens and willows in every direction. We then proceeded nearly due east for about seven miles, German Creek being from one and a half to two miles on the north, a beautiful and rich prairie lying between us and it, and on the south, one mile distant, runs a well wooded ridge, parallel with our course; then turning to the south-east we wound round numerous large clumps of aspen from five to thirty feet high, and willows for seven miles, when we came to a rising ground so densely covered with young aspen and fallen timber, that it was impossible for carts to go further; we therefore left them here, and made packs of a few things for the horses to carry. Here the land becomes of a lighter description, being of a light sandy and clay loam. The timber has been all burnt, and the ground was so thickly strewed with the fallen logs, that it was with much difficulty the horses could travel. Two miles further on we came to the banks of German Creek ; its valley here is from fifteen to twenty chains wide, and about forty feet deep ; it is full of excellent timber, elm, oak, poplar, and black ash, all large enough for building purposes. The creek, which is here very rapid, is thirty feet wide, and about one foot six inches deep. We followed its course for twenty-seven miles, never being more than half a mile away from it. The country through which we passed is for the most part covered with trees of various kinds growing in large clumps, such as balsam poplar, aspen, tamarack, balsam spruce, cedar, and oak. The whole country has been burnt some years ago; the remains of the timber everywhere to be found indicate that there was once a vast forest of large trees.

"The Indian guide now said he had come to the boundary of his own country, and could not bring me further; and though I tried to induce him by every available means, he remained firm to his resolution. He was unwilling for some time even to give me a description of the country beyond; but finally I procured from him the following account:—

"At half a day's journey on snow shoes, or a distance of fifteen miles from where we were, there is a mountain or ridge thickly covered with trees stretching towards the Lake of the Woods. A part of this intervening space is a swamp in which grow tamarack, cedar, and spruce; the remainder is dry ground covered with small aspens and willows. Passing along the 'mountain' you come to a marsh which extends to the 'Lake of the Woods;' but through it there flows a river up which large canoes could come within the hearing of a gun-shot, or about two miles from the mountain. The entire length of the way I had come was seventy miles; fifty miles, at least, of this distance, being fit for settlement, and throughout the whole of it a road could be made without the slightest difficulty, and at little cost. If time and means had permitted, I would have pushed through to the lake, but under the circumstances I considered it better not to attempt it.

"From the description given by the Indians of the country, and which I think may be relied on as correct, I am of the opinion that a road can be easily made through it."

On the 29th September I arrived at the settlement again, and during the week following, while Mr. Fleming was making preparations for our journey to Canada, *viâ* St. Paul, and Mr. Napier and his assistants were engaged in preparing maps and a report on the canoe route, I traversed the settlement from the Indian Mission to the last French house on the road to Pembina, gathering whatever information was accessible respecting the capabilities, condition, and resources of the Valley of Red River, and its singular inhabitants.

CHAP. VIII.

BRIEF HISTORY OF THE COLONY.—STATISTICS OF POPULATION.—ADMINISTRATION OF JUSTICE.—TRADE AND OCCUPATIONS.

Lord Selkirk.—First Emigrants.—Difficulties of the Emigrants.—The De Meurons.—Mr. West.—First Missionary.—The Census.—European and Native Population.—Statistical Table.—Population by Families.—"Natives."—Character of the Half-breeds.—Occupations.—Improvidence of the Half-breeds.—Aids to Improvement.—Administration of Justice.—Governor and Council.—Quarterly Courts.—Council of Assinniboia.—Trade and Occupations.—Absence of Trades.—Mills.—Merchants.—Freighters.—Land.—Leases.—Unoccupied Area fit for Settlement.

The first attempt to found a colony in the country now occupied by the Red River settlements was made in the year 1812, under the patronage of Lord Selkirk. A large tract of country, extending from the sources of the Winnipeg to Lake Winnipego-sis*, and stretching from Lake Winnipeg far beyond the United States boundary, was purchased from the Hudson Bay Company by Lord Selkirk in 1811, for the establishment of his contemplated colony. The colonists consisted of several Scotch families, who, after they had reached the spot which was to be their future home, were met by a large party of half-breeds and Indians in the service of the North-West Company, and warned not to attempt to establish a permanent settlement. They were conducted by a number

* "Winnipego-sis," Little Winnipeg. The affix "sis" signifies in Cree "little."

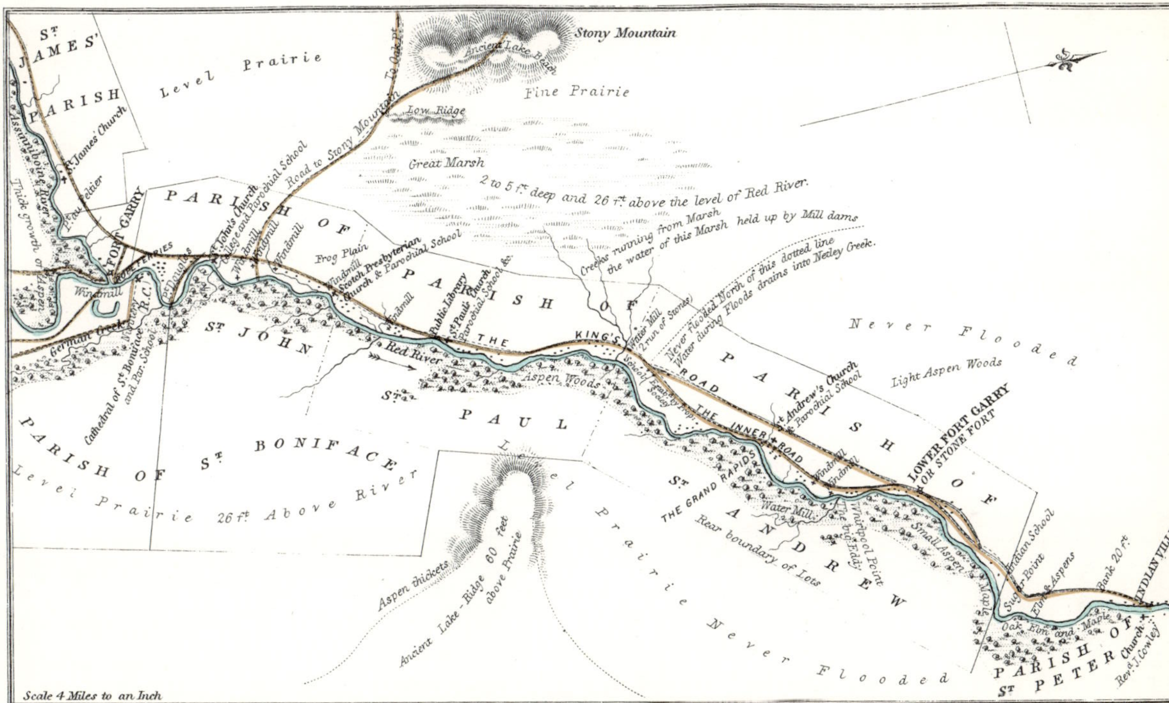

Plan of Selkirk Settlement

of these wild and reckless children of the prairies to Fort Pembina, a post of the Hudson Bay Company, where they passed the winter in buffalo-skin tents, and soon adopted the habits of life belonging to the savage and half-savage natives by whom they were surrounded. In May, 1813, the emigrants returned to the neighbourhood of Fort Douglas, about two miles below the present site of Fort Garry, and here commenced their agricultural labours. In the fall of the year they again sought refuge at Fort Pembina, and after a winter of much suffering, revisited, in the spring of 1814, the scene of the previous year's attempt to plant themselves on the banks of Red River, with a determination to make it a permanent resting-place. During the summer, however, their houses were destroyed by the wandering half-breeds, who were opposed to the establishment of a colony; and when, in October, 1815, the main body of emigrants arrived from Scotland, they found poverty, ruin, and despondency prevailing where they had hoped to meet with a warm reception and comfortable homes. The provident care of Lord Selkirk prevented the colonists from suffering all the horrors of starvation during the inclement winters of this region. His lordship had established a general store of goods, implements, arms, ammunition, clothing, and food at Fort Douglas, from which the impoverished emigrants were supplied on credit. This store was erected in the first year of the colony, and regularly replenished from time to time by shipments from England.*

In 1816 a serious conflict took place between the colonists and the native employées of the North-West Company. Many were killed on both sides, and the settlement

* The Red River Settlement; its Rise, Progress, and Present State, by Alexander Ross. London, 1846.

was again destroyed, the settlers dispersed, and some of them banished, by the half-breeds, to Norway House.

At the time when these disastrous occurrences were taking place, Lord Selkirk was on his way to the Red River with about 100 disbanded soldiers of the De Meuron regiment, composed chiefly of Germans, French, and Swiss. After Lord Selkirk's arrival order was restored, the Scottish emigrants recalled, the De Meuron soldiers rewarded with grants of land on German Creek, a town was laid out on Point Douglas, and such arrangements completed for the government of the colony as the position of the Hudson's Bay Company and the interests of the fur trade would admit of. The social condition of Rupert's Land at this period may be gathered from the following brief description by Governor Semple, who was killed in the unfortunate conflict just referred to:—"I have trodden the burnt ruins of houses, barns, a mill, a fort, and sharpened stockades, but none of a place of worship, even on the smallest scale. I blush to say that, throughout the whole extent of the Hudson's Bay territories, no such building exists."*

On the 16th July, 1818, several French-Canadian families, under the guidance of two priests, Messrs. Provencher and Dumoulin, arrived in the colony, and in the same year, and almost at the same period, innumerable hosts of grasshoppers came from the south-western prairies, and in a few hours destroyed every green thing, threatening the young colony with famine. In 1820 the foundation of a Roman Catholic church was laid near the site of the present Cathedral of St. Boniface; and in the fall of that year Mr. West, a minister of the Church of England, visited the colony as chaplain to the Hudson's Bay Com-

* Governor Semple. Quoted by Tucker, in the "Rainbow of the North."

pany, aided and encouraged by the Church Missionary Society. Mr. West's instructions were to reside at Red River Settlement, and endeavour to meliorate the condition of the native Indians.* The following extract from Mr. West's journal shows the state of the settlement at this period : —

"On the 14th of October we reached the settlement, consisting of a number of huts widely scattered along the margin of the river; in vain did I look for a cluster of cottages, where the hum of a small population at least might be heard, as in a village. I saw but few marks of human industry in the cultivation of the soil. Almost every inhabitant we passed bore a gun upon his shoulder, and all appeared in a wild and hunter-like state. The colonists were a compound of individuals of various countries. They were principally Canadians and Germans of the Meuron regiment, who were discharged in Canada after the conclusion of the American war, and were mostly Catholics. There was a large population of Scotch emigrants also, who with some retired servants of the Hudson's Bay Company were chiefly Protestants, and by far the most industrious in agricultural pursuits. There was an unfinished building as a Catholic church, and a small house adjoining, the residence of the priest ; but no Protestant manse, church, or school-house, which obliged me to take up my abode at the Colony Fort (Fort Douglas), where the charge-d'affaires of the settlement resided, and who kindly afforded the accommodation of a room for Divine worship on the Sabbath. My ministry was generally well attended by the settlers, and soon after my arrival I got a log-house repaired about three miles below the fort, among the Scotch population, where the school-

* The Substance of a Journal during a Residence at the Red River Colony, &c. &c., by John West, A.M. London, 1827.

master took up his abode, and began teaching from twenty to twenty-five of the children."

In 1821 the North-West and Hudson's Bay Companies united, and from that time the condition and prospects of the Red River settlements became more encouraging, and their progress slow but sure.

In 1823 the population of the colony was about 600, twenty years afterwards it had increased to 5143, and thus assumed an important, though not a prominent, position among Christian communities, in the midst of barbarous and savage races.

POPULATION OF THE SETTLEMENTS.

The census upon which the statements contained in the following pages are founded was taken in the years 1843, 1849, and 1856, and the copies were kindly furnished me by Mr. W. R. Smith, the clerk to the Council of Assinniboia.

The total population at the settlements on Red River and the Assinniboine amounted to 5143 in 1843 ; 5291 in 1849 ; and 6523 in 1856, showing an increase in the first six years of only 148, and in the last seven years of 1232 souls. This great difference in the apparent rates of increase is one which may be easily explained, by enumerating the offsets from Red River Settlement, which have occurred since the period when the census was taken. These consist of a number of families, embracing 120 persons, forming a settlement at Prairie Portage. St. Joseph's at Turtle Mountain has absorbed a very considerable number, exceeding 500 persons, and many families have left the settlement to seek a home in other localities. At the same time the population of Red River has received very few accessions from distant countries ; indeed, the foreign element, as it may be termed, shows a very decided diminution in one important source of supply.

Between the periods of the census taken in 1843 and 1849, there was an increase in the European and Canadian element to the extent of 74 families, and of the half-breed of 113 families. During the seven years which elapsed between 1849 and 1856, a decrease in the numbers of Europeans or Canadians—that is, of people not born in Rupert's Land, although British subjects, and originally coming from England, Scotland, Ireland, or Canada, has taken place to the extent of 102 families. The increase in native or half-breed families during the same period was 132. The diminution in the number of European settlers has already worked a change for the worse in the habits and customs of the half-breeds or natives. The tendency of the native population is gradually to lose many of the humanities of civilisation, and approach nearer to the savage wildness of Indian life. An influx of European or Canadian blood had a very good effect in arresting this tendency, which circumstances, far more than disposition, have induced and fostered.

According to origin, the population of Red River now stands as follows :—

	Families.	Families.	Families.	Period of comparison, 13 years.	
	1856.	1849.	1843.		
Rupert's Land— Half breeds . } Natives . . }	816	684	571	Increase in half-breed families	245
Scotland . . .	116	129	110	" Scotch "	6
Canada. . . .	92	161	152	Decrease of Canadian "	60
England . . .	40	46	22	Increase of English "	18
Ireland. . . .	13	27	5	" Irish "	8
Switzerland . .	2	2	2	" Swiss "	0
Norway . . .	1	3	0	" Norwegian "	1

In 1843, or thirteen years before the census of 1856.

there were twenty-seven more European or Canadian families than there were at Red River in May, 1856. These numbers show that in place of an introduction of emigrants of a character likely to refine and elevate the rough natures of the natives, endowed as they are with many peculiar and valuable qualities, those who have been from their youth familiar with the advantages and blessings of civilisation have gradually left the settlement and sought a home elsewhere. The increase of poverty, or incapability of supporting families, is seen by the average number of individuals belonging to each family.

In 1849 the average of each family was $5\frac{31}{1052}$.
In 1856 „ „ „ $6\frac{43}{1080}$.

The difference in the whole population of 1856 and 1849 being 1232 souls, while the difference in the number of families was 28 only. This very extraordinary discrepancy was stated by Mr. Smith, under whose direction the census was taken, to arise from the generally depressed circumstances in which many families found themselves. Numbers were unable to live in separate houses, and it now happens that two, and sometimes three families, formerly occupying distinct houses, and cultivating distinct farms, are crowded together in one house for the sake of economy. In 1849 there were 137 more males than females in the settlements; in 1856 there were 73 more females than males. The reason of this remarkable change in the relative numbers of males and females in so small a community, and in such a short period of time, was stated to arise from the circumstance that during the past five or six years, many young men have gone to seek recompence for industry in the United States, which the district of Assinniboia has not yet offered to them.

The term "native," distinguishing the half-breeds from

the European and Canadian element on the one hand, and the Indian on the other, appears to be desired by many of the better class, who naturally look upon the epithet "half-breed" as applied to a race of Christian men, scarcely appropriate. There is a strong and growing feeling among the few who have turned their attention to such matters, that in the event of an organic change occurring in the Government of the country, the "native" or half-breed population should not be neglected, or thrust on one side.

The half-breeds of the north-west are a race endowed with some remarkable qualities, which they derive in great part from their Indian descent, but softened and improved by the admixture of the European element. It is, however, much to be regretted that, from the singular necessities of their position, many of them are fast subsiding into the primitive Indian state; naturally improvident, and perhaps indolent, they prefer the wild life of the prairies to the tamer duties of a settled home; this is the character of many, but it belongs more to those of French descent than of Scotch or English origin.

About the 15th of June they start for their summer hunt of the buffalo. There are now two distinct bands of buffalo hunters, one being those of Red River, the other of the White Horse Plain, on the Assinniboine. Formerly these bands were united, but, owing to a difference which sprung up between them, they now maintain a separate organisation, and proceed to different hunting-grounds. The Red River hunters go to the Coteau de Missouri, and even as far as the Yellow Stone River; the White Horse Plain settlers generally hunt west of the Souris River, and between the branches of the Saskatchewan, but also over the same grounds as their Red River brethren.

The improvidence of many of the half-breeds is remarkable. During the winter before the last, those of the White Horse Plain camped out on the distant prairies, and killed many thousand buffalo in wanton revelry, taking only their skins and tongues, little caring that the reckless destruction of these animals must exercise a very important change for the worse in their own condition. As the buffalo diminish and go farther away towards the Rocky Mountains, the half-breeds are compelled to travel much greater distances in search of them, and consume more time in the hunt; it necessarily follows that they have less time to devote to farming, and many of them can be regarded in no other light than men slowly subjecting themselves to a process of degradation, by which they approach nearer and nearer to Indian habits and character, refusing to adopt or relinquishing the tame pursuit of agriculture, for the wild excitement and precarious independence of a hunter's life. The fascination of a camp in the high prairies, compared with the hitherto almost hopeless monotony of the farms of Red River, can easily be understood by those who have tasted the careless freedom of prairie life. I was often told that the half-breeds generally sigh for the hunting season when in the settlements, and form but a feeble attachment to a permanent home, which cannot offer to the majority a comfortable maintenance under present circumstances, or secure the consciousness of possessing a free and manly spirit, with rational aspirations and hopes.

But few simple aids are required at Red River to ameliorate and vastly improve the condition of the more improvident and careless half-breeds. They frequently bring in a large quantity of buffalo meat or robes to the trading posts, and receive a considerable sum of money in exchange, or if they insist upon it, a certain quantity of rum. The money is spent at once in simple necessaries, dress, and ornaments. The establishment of a savings bank would have an excellent effect, and doubtless become the source of much permanent good.

There are several hundred half-breeds who, like their ancestors, pass their lives on the prairies, visiting the settlements occasionally, according as they may be in want of ammunition or clothing. It is impossible to arrive at an accurate estimate of their numbers, but there is no doubt that collectively they form a numerous and influential body.

The half-breed hunters, with their splendid organisation when on the prairies, their matchless power of providing themselves with all necessary wants for many months together, and now, since a trade with the Americans has sprung up, if they should choose, for years; their perfect knowledge of the country, and their full appreciation and enjoyment of a home in the prairie wilds during winter or summer, would render them a very formidable enemy in case of disturbance or open rebellion against constituted authorities. The half-breed hunters of Red River could pass into the open prairies at a day's notice, and find themselves perfectly at home and secure, where men not accustomed to such a life would soon become powerless against them, and exposed to continual peril.

The causes which have led to the present condition and prospects of this people are truly a painful subject. It is one which cannot escape the attention and care of philanthropists. Men will inquire how it is, that a race giving evidence of admirable discipline, self-government, and courage, when in the open prairies, should subside into indifferent and indolent husbandmen when in the settlements. Considered as the native population of Red River, how is it, it will be asked, that so few among the many have succeeded in the course of years in acquiring comfortable homesteads, and well stocked granaries and farm yards? and why has the European and Canadian element disappeared? The chances of nearly all have been equal,—land of admirable fertility everywhere surrounds them; with unsurpassed advantages for rearing horses, cattle, and sheep, yet little or no progress has been made; and in respect of sheep, which might soon in a measure supply the place of the buffalo, a serious diminution in numbers has taken place. It is true that within the last few years many hundred head of cattle have been driven across the prairies of Minnesota to St. Paul and sold well there. But this new export trade should have given encouragement to raising stock; yet stock, with unlimited pasture, is diminishing. The distant hunt consumes the time which might be given to far more profitable home industry; and those who really enjoy a settled life, and know the advantages which industry confers, from experience gained in Canada or Europe, leave the country and seek their fortunes elsewhere.

Every stranger is struck with surprise that the houses of half-breed hunters generally show no signs of recent improvement, show no signs of care and attention devoted to gardens or the cultivation of fruit. Plums grow wild in the forest, but none are seen in the settlements. Apple trees are only now beginning to be tried at the Stone Fort. No effort of manufacturing industry is visible beyond the windmills for grinding wheat.

It must not be supposed that this stationary, or rather retrograde, condition, is unnoticed by the mass of the people. They see the comfort by which the retired factors, the clergy, and the traders of the settlement are surrounded, and the comparative luxury which exists at the forts; but they do not rightly understand how their own condition might be remedied, for the majority cannot discover in what way the reward of industry may be won, or where a market for labour is to be found, except that kind of wild labour in the distant prairie, or in the woods, which they love instinctively, and which they have always been taught to consider most profitable, and alone capable of securing their comfort and happiness. Under such circumstances it cannot cause surprise that discontent prevails in the settlements. Much disappointment and dissatisfaction is everywhere seen, and wrongs, real or imaginary, for which they have no redress, form the constant subject of complaint in daily conversation. In these repinings, all who were not in the service of the Hudson's Bay Company, or in some way connected with them, as far as my experience enabled me to judge, uniformly agreed.

Let the condition of the half-breed hunters, generally, be contrasted with the present prosperity of the Gowlers, Gladieux, Fletts, the McKays, and several others that might be named, who farm with industry and economy, and the capabilities of Red River and the Assinniboine will not be overlooked in surveying the paralysed efforts of those who are taught to rely chiefly upon the hunter's precarious gains.

ADMINISTRATION OF JUSTICE.

The mode in which justice has been, and is administered in the settlements, is of rather an undetermined character. In 1839 the first Recorder was appointed, and in some instances the office of Governor of Assinniboia, a district comprised within a circle of fifty miles radius round Fort Garry, has been associated with that of Recorder. The Governor has a council of twelve of the principal inhabitants of the settlement to assist and advise him in performing the duties of his office. All the chief factors of the Company are magistrates, *ex officio*, and although instances have occurred of a court of magistrates trying cases without a jury, yet in general, and probably at the present day, a jury is always empannelled.

The reports in the recently established newspaper at Red River exhibit the mode in which justice is now administered; and as this is the first instance in which a reporter has enjoyed the opportunity of supplying the Red River public with a full account of the proceedings of these courts of justice, it possesses an interest apart from and superior to the subjects to which it refers.

RED RIVER SETTLEMENT, Dec. 28, 1859.

GENERAL QUARTERLY COURT.

A sitting of the above Court took place within the Court House, at the Upper Fort, on Thursday, the 15th instant, before William Mactavish, Esq., Governor; Dr. Bunn, Sheriff; Thos. Sinclair, Esq., Robert McBeath, Esq., and François Bruneau, Esq., Chairmen of the several District Courts. The building was crowded throughout the day, and the liveliest interest appeared to be taken in the proceedings, which, on the whole, as will be seen from the report, were of an animated character. Mr. James Ross acted as French and Mr. James McKay as Indian interpreter.

THE ROBBERY AT THE STONE FORT.

Catherine Daniel and Mary Daniel, aged respectively 13 and 16, were charged, the former with having stolen two several sums of money from the Stone Fort, the property of the Hudson's Bay Company, and the latter with having received part of the cash at the time, well knowing it to have been stolen. They both pleaded not guilty.

Margaret Daniel, a still younger girl than either of the prisoners, stated that on the 10th of October Catherine Daniel stole money from the shop at the Lower Fort, which she entered by the window. Witness saw her get through the window, go to the drawer, and take the money out. When she returned, Catherine told witness she had taken £6. Part of the money was spent by Catherine in the shop, and of the balance she gave witness one pound. Subsequently, Catherine a second time entered the store by the same window, witness being with her. On this occasion Catherine went to the counter drawer and abstracted five £1 notes. The drawer had been left unlocked. They left the shop by the same way that they had entered it. Catherine again spent a portion of the money in the shop, Afterwards, Catherine told witness that she had given some of the money to Mary Daniel; and Mary acknowledged to witness that she had received it. The first time Catherine spoke to Mary about the money she told her she had found it; but when the money had been spent, Catherine confessed to her that she had stolen it. Mary likewise spent her money at the Stone Fort. Witness only once heard Catherine tell Mary that she had stolen the money, and that was after the money had been spent.

There being no further testimony adduced, the prisoners were asked if they had anything to say in defence. They made no reply. Dr. Bunn then summed up, and the jury, after a short deliberation, returned into Court, with a verdict of guilty against Catherine Daniel. Mary was acquitted.

Dr. Bunn, addressing the prisoners, said: Catherine Daniel, after a fair and impartial trial, the jury have found you guilty of felony. The offence you have committed is one of a very serious nature, and in any other country you would in all probability have been condemned to seven or perhaps fourteen years' confinement. The Court, however, taking into consideration the fact that you are young, and hoping that you will never commit such an act again, are inclined to be merciful to you. Already you have been nearly three weeks in prison; we were disposed to have committed you other three months; but the Governor, who has in his power to remit your term of imprisonment to any period he pleases, thinks it will be sufficient punishment to you and warning to others if he treats you with great leniency. Therefore, the sentence of the Court is that you be imprisoned for two weeks from this day. You, Mary Daniel, are discharged. But take care. You have had a narrow escape. There is a strong impression on the minds of every one present that you have acted dishonestly. Avoid being brought up again; for if you come hither a second time, the evidence which has been given to-day will tell heavily against you.

LOWER DISTRICT COURT.

The usual bi-mensal meeting of this Court was held at the Rapids, on Monday, November 28th, before Thomas Sinclair, Esq. (chairman), Donald Gunn, Esq., John Inkster, Esq., and Donald Murray, Esq.

COUNCIL OF ASSINNIBOIA.

The Governor and Council of Assinniboia held a general meeting at Fort Garry, on the 7th instant, at which were present—William Mactavish, Esq., Governor of Assinniboia, President; and the following Councillors of Assinniboia—Right Rev. the Lord Bishop of Rupert's Land, Right Rev. the Lord Bishop of St. Boniface, John Bunn, Esq., John Inkster, Esq., Pascal Berland, Esq., Solomon Emlyn, Esq., H. Fisher, Esq., Maximilian Genton, Esq., Robert McBeath, Esq., Thomas Sinclair, Esq., and John E. Harriott, Esq.

TRADE AND OCCUPATIONS.

Upon making inquiry of Mr. Smith, under whose superintendence the census was taken, why no enumeration of trades and occupations was introduced into the census roll, I was informed that no kind of industry or a distinct trade or occupation existed in the settlements. Almost every man was his own wheelwright, carpenter, or mason; carpenters, blacksmiths, masons, &c., could be found, but they were also engaged in other occupations, either as small farmers or hunters. Mr. Smith did not think that one man could be found in Assinniboia who pursued any particular trade, or limited his industry to one special branch. The present condition of the settlements would not, it was thought, afford a living to any distinct class of artificers. A horse-shoe imported from England could be purchased as cheap as the unmanufactured iron required to make one; every article, no matter of what description, was imported in its manufactured condition. Even the ponderous and unwieldy grindstone was conveyed across the portages from Hudson's Bay, although material well adapted for grindstones existed on the shores of Lake Winnipeg, not one hundred miles from Red River. Grindstones had, I was informed upon authority I could not doubt, been made from the rock in question, and brought to the settlement, but they could not compete commercially with those imported by the Hudson's Bay Company, which, for a time, were sold little above cost, even after their long and expensive journey.

In 1858 I had occasion to send to a blacksmith near Fort Garry for some hasps which I wished to have made, to replace those which had been broken from the provision boxes in crossing the portages on the Winnipeg.

The hasps were made according to my directions as regards form, but the material was copper instead of iron, the blacksmith alleging that he had no iron from which he could make the hasps. It is not improbable that this was an excuse arising from indolence rather than inability to execute the work, or the want of the material, for I saw a few days afterwards plenty of iron suitable for the purpose in the storehouse at Fort Garry.

The Bishop of Rupert's land says: "After all, our grand want is division of labour. We have no separate trade; all are engaged in everything, farmers and carpenters at the same time, and so on. At a meeting held two years ago, for the promotion of social improvement, I endeavoured to press this upon them, but they are slow in understanding the 'philosophy of improvement.' We want one skilful in tanning, for the hides of the domestic animals are wasted at present. We want one to instruct them in making soap, to save the importation of this bulky and necessary article from Britain. We want, too, improvement in the fulling of cloth, to bring the wool into use, and provide clothing cheaper than what is imported. We have country cloth now, but the fabric is imperfectly fulled, and therefore not sufficiently warm. Young men coming among us, who could guide and instruct the people in any of these branches, would be a great gain."*

The mechanical force employed in preparing food is represented by sixteen windmills, nine water-mills, and one steam-mill, which is also used as a saw-mill. Articles of pottery, notwithstanding their fragile nature, are imported, although, as if reflecting upon the industry and enterprise of the modern inhabitants of Red River, ancient

* Report on the Red River Expedition of 1857. Letter to the Author. Appendix, No. 9.

articles of pottery, in the form of broken fragments, are sometimes found in making excavations on the banks of the river. Speaking in general terms, it may be said that trades and occupations, as representing special branches of industry, do not exist in Assinniboia.

Under the head of merchant-shops, we find no less than fifty-six enumerated in the last census, a heading which, it will be observed, is not represented in the census of 1849. In fact, the class of merchants, including petty traders, has almost sprung into existence during the last ten years. They obtain their goods from St. Paul on the Mississippi, and purchase them in exchange for gold or peltries. This trade with the United States is fast growing into importance, and from the immense extent of frontier, it is not easily checked by fiscal regulations ; its continuance must effect to a most serious extent the position of the Hudson's Bay Company in the valley of Lake Winnipeg.

Some of the merchants at Red River import largely from England by the Company's vessels, and almost any article of common necessity or ornament can be procured at the stores ; which, by the way, are of the rudest description, without the least effort being made by their owners to display the wares, but rather showing an endeavour to conceal from outward view whatever goods they may contain.

Besides being merchant or trader, in the ordinary acceptation of the term, some of the inhabitants are freighters, conveying goods between Hudson's Bay and the valley of Lake Winnipeg. They employ Indians and half-breeds to row their boats of 3 to 5 tons burden, and haul them and their freights over the portages. Fifty-five of these boats are enumerated in the census as belonging to Red River, but whether the Hudson's Bay Company's fleet is included in the number is not stated. The employment of Indians by the freighters has, at times, given rise to some little difficulty between them and the Hudson's Bay Company, as introducing a species of industry not compatible with a hunter's pursuit, and likely to divert attention from the great objects of the fur trade.

Among numerous documents which are in the possession of many of the most respectable people of Red River, treasured up, perhaps, as memorials of bygone but not forgotten difficulties in gaining a livelihood by pursuits not connected with the fur trade or its interests,—the following brief note may or may not possess some little historic interest, and, if rightly understood and interpreted, offer a clue to the present condition of the Red River settlements, and of the Indian missionary stations.

FORT GARRY, June 5th, 1844.

SIR,—I am informed that private freighters from Red River frequently employ and afford passages to Indians along the line of communication to York Factory in their boats, which is highly objectionable in many points of view. I have therefore to desire you will not in future receive as passengers, or employ Indians in your craft, on the line of communication between York and Red River.—I am, &c.

(Signed) G. SIMPSON.

Mr. Edward Mowat, &c. &c.

Copied, 30th July, 1844. (Signed) A.

TENURE OF LAND.

Land in Assinniboia is sometimes sold to purchasers at the rate of 7*s.* 6*d.* sterling per acre. The title is conveyed under the form of a lease for 999 years. The conditions in the lease are: 1st. That one-tenth of the land is to be brought into cultivation within five years. 2nd. That trading or dealing with Indians or others, so as to violate the chartered privileges of the Company, be forsworn.

3rd. Obedience to all laws of the Company. 4th. Contributions to expenses of public establishments in due proportion. 5th. All trade or traffic in any kind of skins, furs, peltry, or dressed leather, except under licence of the Company, forbidden. 6th. Land not to be disposed of, or let, or assigned, without the consent of the Company. These are the main features of the lease ; the document is long, otherwise it would have been inserted in full; it is contained in the minutes of evidence taken before the Select Committee on the Hudson's Bay Company.

It is necessary here to remark, that I did not see this lease in the hands of any one of the settlers of whom I made inquiries respecting their tenure. I heard of its existence, and saw a copy, through one of the resident clergy, but in no single instance could I find any half-breed, in possession of a farm, acquainted with its existence. In very many instances the settlers did not know the number of their lots, and had no paper or document of any kind to show that they held possession of their land from the Company, or any other authority. These inquiries were necessary for the purpose of ascertaining the exact position of a line of section across the valley of Red River, which I caused to be made for the purpose of ascertaining the level of the swamps, &c. The required information was obtained through Mr. Smith, the Clerk of the Council, but from the people themselves no information of the kind could be obtained. They knew that they had paid a certain sum for their land, or it had been given them in return for services, or that they had squatted upon it, and that they were now in possession, but of title-deeds or receipts they knew nothing. These remarks refer only to those from whom the information was sought for the purposes mentioned above.

When passing from Fort Garry towards the 49th parallel, with a view to explore the Roseau River, our guide pointed out a number of hay stacks occupying a delightful bend on the west side of Red River, about 25 miles from the settlements; he informed us that the hay stacks were made by himself and some friends, a few weeks ago, and that they intended to "move there" during the winter and form a new settlement. I inquired how much he had paid for his land; the reply was, "Nothing; we are not required to pay anything for land lying beyond the present limits of settlement on the river." I may add, that many million acres of land, which cannot be surpassed for fertility, being composed of rich prairie mould, from 18 inches to 2 feet deep, lie free and unoccupied on the banks of Red River, the Assinniboine, and their tributaries, inviting settlement.

REVENUE AND EXPENDITURE.

The following abstract of the public accounts of the Red River Settlement, ending May 31, 1859, will show the condition of the revenue and expenditure of the colony:—

DISBURSEMENTS.

PUBLIC SERVANTS' SALARIES.

	£	s.	d.
Governor of Gaol and Sheriff	30	0	0
Executive Officer	100	0	0
Presidents of Petty Courts	26	0	0
Collector of American Duty	15	0	0
Petty Magistrates	50	0	0
Constables	108	0	0
Total amount of Salaries	329	0	0

GAOLER AND GAOL.

	£	s.	d.
Gaoler's Salary, Rations, and Advances	32	12	6
Gaoler's Wood	4	7	2
Gaol Expenses	3	18	3
Prisoners' Expenses	4	1	6
Total, Gaoler and Gaol	44	19	5

PUBLIC WORKS.

	£	s.	d
Labour performed	34	11	6
Timber	29	1	3
Roads	282	4	2
Bridges	64	3	6
Total amount expended on Public Works	410	0	5
Sundries	5	0	6
Rent of Court Houses	6	0	0
FERRY AND SCOWS.			
Ferry	1	8	0
Scows	10	0	0
Total	11	8	0
Premiums paid for Wolves' Heads	35	10	6
Post Office	154	11	11½
Grand total	996	15	4½

REVENUE.

	£	s.	d
By Outstanding Creditor Balance, June 1, 1859	869	9	6½
IMPORT DUTY.			
Hudson's Bay Company's European	788	3	10
Hudson's Bay Company's American	129	6	7
Settlers, European	351	11	0½
Settlers, American	*35	15	4
Total amount of Import Duty	1,304	17	4½
Interest on £186 8*s.* at 4 per cent.	7	9	1
Ferry	6	0	0
Advanced Cash returned	5	0	0
Debtors' Maintenance	0	2	4
Old Materials sold by the Board of Works	2	5	0
Fines	1	0	0
Post Office	149	16	8
Marriage Licence	1	0	0
Grand Total	2,347	0	0

* There is £30 of American duty in the hands of the Collector, which came in too late to be entered in this year's accounts.

		£	s.	d.
May 31, 1859.	To Disbursements	996	15	4½
"	To Balance	1,350	4	7½
		2,347	0	0
May 31, 1859.	By Revenue	2,347	0	0
"	By Balance carried to New Account . .	1,350	4	7½

W. R. Smith, E. O.
John Inkster, Auditor.

CHAP. IX.

THE MISSIONS AT RED RIVER.

Religious Denominations.—Missionaries, Stations, and Congregations.—The Protestant Congregations.— St. John's Church.— St. Andrew's Church and Parsonage.—The Parish of St. Andrew.—Its History.— St. Paul's Church. — St. James's Church.— Church at the Indian Settlement. — Service.—A Novel Night Bell. —A Contrast.—Peguis.—Prairie Portage.— A Congregation.—Wild Indians.—The Presbyterian Church.—The Presbyterian Congregation.—The Roman Catholic Churches.—The Cathedral of St. Boniface.—St. Norbert.—St. François Xavier.—The Congregations at Red River.—Their Demeanour and Appearance.—Protestant and Roman Catholic Parishes.—Extent of the Charities of the Home Missionary Societies. —Apathy of the wealthy at Red River.— Difficulties of Missionary Enterprise at present.—Privations and Difficulties at remote Stations.

There are three religious denominations in Assinniboia —Church of England, Presbyterian, and Roman Catholic. In the census of 1843 and 1849, two divisions only were recognised—Protestant and Roman Catholic—and the numbers of members were stated to be 2798 Roman Catholics and 2345 Protestants. In 1849 the Episcopalian families were stated to number 539, and the Roman Catholic families 513. In 1856, a division in the enumeration of the Protestant element was made, probably on account of the advent of a Presbyterian minister, who responded to the call of a numerous body belonging to that denomination, yet in the absence of a minister, formerly enumerated with the Episcopalians. In 1856, the census, according to religions, stood thus :

FAMILIES AND CHURCHES.

Roman Catholics,	534 families,	with	3 churches.
Episcopalian,	488 „	„	4 „
Presbyterian,	60 „	„	2 „

The settlement at Prairie Portage, and the Indian missionary village, are not included in this enumeration, and, in addition to the churches enumerated, services are performed in two or three school-houses.

Subjoined is a table of the missionaries, stations, congregations, income and sources of income, belonging to the Protestant and Roman Catholic Missions in Assinniboia:—(See Appendix, Vol. II. for the latest statistics.)

PROTESTANT MISSIONS.

	Missionaries	Stations.	Congregations.	Income.	Sources of Income.	Remarks.
				Sterl. £		
		Red River.				
1	The Right Rev. the Lord Bishop of Rupert's Land	St. John's .	500	700	£300 Hon. Hudson's Bay Company. £400 funded property.	
2	Rev. T. Cochrane	„	. .	100	Society for Propagation of the Gospel.	
3	Rev. J. Chapman .	St. Paul's .	300	200	£150 Hon. Hudson's Bay Company. £50 the Bishop.	The Hon. Co.'s Chaplain.
4	Rev. Arch. Hunter . . .	St. Andrew's	1200	250	Church Missionary Society.	
5	Rev. W. W. Kirkby . . .	„	. .	200	. . . . .	Curate.
6	Rev. A. Cowley .	Indian Settlement.	600	200	. . . . .	Indian Missionary.
		Assinniboine River.				
7	Rev. W. H. Taylor	St. James .	250	200	£100 Society for Propagation of the Gospel. £100 Bishop.	
8	Rev. Arch. Cochrane	Portage la Prairie.	200	200	Church Missionary Society.	

PRESBYTERIAN CHURCH.

	Missionaries.	Stations.	Congregations.	Income.	Sources of Income.	Remarks.
	Rev. John Black .	*Red River.* Middle Settlement.	400	150	£50 Hon. Hudson's Bay Company. Remainder by the Congregations.	

ROMAN CATHOLIC MISSIONS.

	Missionaries.	Stations.	Congregations.	Income.	Sources of Income.	Remarks.
1	The Right Rev. the Lord Bishop of the North-West, and 5 to 7 Clergy.	*Red River.* St. Boniface . St. Norbert. De la Rivière Salé.	1500 Included in the above.	. .	£100 from the Hon. Hudson's Bay Co.	A spacious Nunnery and Schools attached.
2	Rev. M. Thibeault, Grand Vicar	*Assinniboine River.* St. François Xavier	1000	. .	. . . . .	A Nunnery attached.

THE PROTESTANT CONGREGATIONS.

St. John's Church is in a very unstable condition, the walls being supported with wooden props. A large quantity of stone is now lying near it for the construction of a cathedral, which is estimated to cost £5000 sterling.

The Bishop of Rupert's Land returned to Red River in the autumn of 1857 from a visit to England, where he succeeded in obtaining a large sum towards the erection of his cathedral, but not sufficient to defray the entire cost. St. John's Church, and the proposed site of the new cathedral, are within two miles of Fort Garry, near where it is most desirable that a substantial and commodious Church of England cathedral should be erected, in the centre of what will soon become the capital of

Assinniboia, and nucleus of Christianity and civilisation in the great North-West.

St. Andrew's Church, called also the Rapids' Church, is a new and very substantial structure of stone, well buttressed, and very conveniently and neatly furnished: all its interior arrangements are attractive and substantial. It is surrounded by a thick stone wall enclosing a capacious churchyard. The parsonage house, also recently completed, is in every respect fitted for the severities of the winter climate of the country. The size is 50 feet by 30, and two stories high; the walls, of limestone, are 2 feet 8 inches thick, the rooms lofty and capacious, and in its internal arrangements it leaves nothing to be desired.

The Rev. W. W. Kirkby's (now of Mackenzie's River) house is roomy and comfortable, but its architectural points are far from being attractive. A school-house, constructed of wood, is admirably arranged, and in it I saw sixty children pursuing their studies, under the instruction of Mr. Mayhew, lately from Dublin, with a decorum and attention not often to be surpassed in the primary schools of this or the European continent.

The parish of St. Andrew is the most populous on Red River, and the Church of England element is very largely in excess of the Roman Catholic, there being two hundred and six Protestant and eight Roman Catholic families.

The commencement of the mission of St. Andrew may be dated as far back as the year 1824, when four or five retired servants of the Hudson's Bay Company, with their native wives, first sought a permanent home at the Grand Rapids. To these, year by year, a few families were added until 1829, when the Rev. W. Cochrane (now

Archdeacon), was placed there, and from that time the population of the Grand Rapids' district increased by immigration with great rapidity. After the erection of his own house, the first act of Mr. Cochrane was to build a room about 30 feet by 18, which served the double purpose of school-house and church. As soon as it became known in the far-distant outposts of the Company that a school for the young, and a church, with a resident minister, were established at Red River, many of the retired servants of the Company, half-breeds and Europeans, came from the Saskatchewan, Albany, and even Moose, to avail themselves of the advantages which were now for the first time offered to them, with a guarantee for their continuance.

In two years the congregation had outgrown the school-house, and found themselves obliged to build a larger church, to accommodate their growing numbers.

This they completed in the autumn of 1831. It was a neat wooden building, 60 feet by 24, and cost about £200, which was paid by the people themselves in labour, materials, &c. No sooner was the church built, and things began to wear an aspect of stability and permanency, than others came in from Norway House, and the southern department. The population of the parish went on steadily increasing until every lot of land was taken up for about five miles on either side of the church, and a ten-mile line of white-washed cottages and pleasant homesteads enlivened the banks of the river.

In the year 1844, the church was again too small for the congregation, not more than three-fourths of whom could find admittance. Mr. Cochrane now determined to make an effort to build one of stone, instead of wood, which should be large enough for the increasing wants of the settlement. Fortunately there was abundance of limestone about four miles down the river, so that when the people were consulted, all were found to be unanimous on that point. It was commenced in the autumn of 1845, and finished in the autumn of 1849, when it was consecrated by the Bishop of Rupert's Land, who had just arrived in the country. Its dimensions are 88 feet by 44, with a tower 20 feet square, and 100 feet high, that is, to the vane; the stone work is only 50 feet. The tower contains three bells, and it is proposed by the inhabitants to increase their number to five.

The entire cost was about 1600*l.* which, with the exception of 100*l.* kindly given by the Hon. Hudson's Bay Co., 50*l.* by Duncan Finlayson, Esq., of Lachine, 30*l.* from a clergyman in England, were collected on the spot in the shape of money, labour, materials, &c. It is a plain, unpretending, at the same time a solid and substantial building; and one that reflects credit upon the piety and liberality of the inhabitants.

In 1852-3, the substantial parsonage before referred to was built by the Church Missionary Society at a cost of 700*l.*, and during the same year the people built the new school-house, at a cost of 120*l.*

The average attendance of divine worship is about 500, and the number of communicants 207.

St. Paul's church, parsonage, and school-house are substantial and serviceable buildings, with no pretensions to architectural display, but well fitted for the object of their construction. They are built a few hundred yards from Red River, and at the edge of a boundless ocean of prairie, which, when illuminated by the setting sun, seems in its bright and gorgeous vastness to be emblematic of eternity according to the hope and faith of a Christian, but when dull and cold and grey in the dawn, it symbolizes the gloomy ignorance of its heathen wanderers.

St. James's Church, on the Assinniboine, is a pleasing object at a distance. I had no opportunity of ascertaining how far its internal arrangements comported with its external aspects. The congregation is the smallest in the Red River settlements.

Birch-bark Tents, west bank of Red River, Middle Settlement.

The church at the Indian settlement is also a new and spacious building of stone, with a wall of the same material enclosing the churchyard, in which is a wooden school-house, where I saw about fifty Ojibway Indian young men, young women, and children, receiving instructions from the Rev. A. Cowley, Mrs. Cowley, and a native schoolmaster. The young Indian women read the Testament in soft, low voices, but with ease and intelligence. During service (Sunday, October 4th, 1857), the church was about three-fourths full. The congregation appeared to be exclusively Indian; in their behaviour they were most decorous and attentive. The singing was very sweet, and all the forms of the service appeared to be understood, and practised quietly and in order by the dusky worshippers. A seraphino was played by Mrs. Cowley to accompany the singers; the responses were well and exactly made, and the utmost attention was given to the sermon. The prayers were read in English, the lessons in Ojibway, and the sermon was delivered in Cree. After service an Indian child, neatly dressed in white, was baptized. A few of the women and girls wore bonnets, but the greater number drew their shawls over the head.

The minister and part of the congregation suffer under the mutual disadvantage of being separated by the river. The settlement is chiefly on the left, the church, school, and parsonage on the right bank of the river. A good ferry, which will probably soon be procured, would enable the congregation to cross with ease. The Rev. Mr. Cowley enjoys no sinecure; he is not only missionary, but the doctor, magistrate, and arbitrator of the settlement. During my short visit of a day and a half, he was sent for three times to visit sick children, and he says that when the Indians require his services during the night, they come into the parsonage, the door of which is never locked, and tap gently at the stove-pipe, which passes from the sitting-room into his bed-room above, to arouse him. They agreed among themselves that they would adopt this novel kind of night bell, and he has never known them endeavour to call him, after retiring to rest, in any other way; they open the outer door and steal without the slightest noise, in the darkest night, to the well known stove-pipe, give two or three low Indian taps, and quietly await the result.

A wonderful contrast do the subdued Indian worshipers in this missionary village present on Sunday, to the heathen revelers of the prairies, who perform their disgusting ceremonies within a mile and a half from some of the christian altars of Red River. On two Sundays during my visit, at the time when Divine service was being celebrated in all the churches of the settlement, the heathen Indians held their dog feasts and medicine dances on the open plain. In one instance five dogs were

Ojibway Tents on the banks of Red River, near the Middle Settlement.

slaughtered, cooked and devoured; in another instance three; the evil spirit was invoked, the conjuror's arts used to inspire his savage spectators with awe, and all the revolting ceremonies belonging to the most degraded heathen superstition practised within a mile and a half of the spot where the stones are now gathered for the Bishop of Rupert's Land cathedral, and nearly the same distance from two capacious churches, Protestant and Roman Catholic, where Divine service was at the same time being solemnized to orderly resident congregations.

The farm attached to the Indian mission is cultivated with more than ordinary care, not only being intended to serve as a model for the Christian Indians settled in the vicinity, but also to provide them with seed and supplies in the event of their own stock failing, a contingency by no means improbable, since habits of forethought or economy are rarely acquired by these people until the second generation. In part of the garden allotted to vegetables, a small area was devoted to wheat in 1857, for the purpose of raising seed from an early variety, which Mr. Cowley had procured from Scotland the year before. The "Scotch wheat" was sown on the 16th and 18th of May. It was ready for the sickle and reaped on the 24th of August, having been 97 days in arriving at maturity. The common wheat of the country was sown May 5th, and harvested August 18th, having required 105 days to grow and ripen. Barley was sown May 28th, and reaped August 18th. Indian corn is planted about the 23rd May, and ripens every year. Potatoes are planted from the 22nd to the 26th of May. The potato crop is here truly magnificent. I was favoured with an inspection of the produce of a small field, afterwards visited, and certainly no finer or more plentiful returns could be desired. All were perfectly clean and sound, and of very unusual size and weight. With the permission of Mr. Cowley I took four potatoes which lay close at hand, on the top of a large heap, containing very many equalling in size those I had taken without special selection; when carefully weighed, they were found to average ten ounces each (10·1 ounces); a practical experiment proved them to be an excellent table variety. I may here mention that in the garden I noticed asparagus growing luxuriantly, beet, cabbages, brocoli, shallots, and indeed most culinary vegetables. In the farmyard were ducks, fowls, turkeys, pigs, sheep, with some excellent milking cows, and, through the politeness of Mrs. Cowley, I was enabled to form a very favourable opinion of several varieties of preserve from the wild strawberry, cranberries and plums, which grew in profusion not far from the village. Among many kinds of wild fruits common here and much sought after by the Indians, are red and black currants, high and low bush cranberries, two kinds of raspberries, two kinds of gooseberries, moss-berries, blueberries, summer berry, choke cherry, stone cherry, &c.; these are the common names by which they are known in the settlements.

In the garden around the house some flowering shrubs and annuals were still in bloom on the 3rd October. The air was fragrant with the perfume of mignonette, and the bright orange-yellow eschscholtzia shone pre-eminent among asters and sweet peas, which had escaped the autumn frosts.

I was introduced to Peguis, the great Ojibway chief, who at one time commanded three hundred warriors, and about whom so much has been written by the missionaries. He is now a quiet old man, a good Christian, and happy, as he states, in his belief.

Up to the day of my visit, October 4th, there had been fifty-one baptisms, exclusively Indian, in Mr. Cowley's mission, during 1857; and in the same period, twenty-six deaths, six of the number being adults. The population of the mission in 1855 was 473 baptized Indians, and 203 heathens; only four adult baptisms were celebrated in 1855.

Next to the Indian settlement, Prairie Portage is the most interesting illustration of a Christian settlement, in a wilderness still inhabited by roving bands of Indians, who, as of old, occupy themselves in barbarous warfare, hunt for daily food, or submit with abject humility to the conjuror's malignant influence. Prairie Portage owes its existence to the untiring energy and undaunted zeal of Archdeacon Cochrane. The church at this most westerly limit of civilization in the Far West is constructed of wood, and contains twenty-five or thirty very substantial family seats, but is capable of holding three times that number; each seat is manufactured by the owner, according to a pattern supplied by the archdeacon. The congregation (on Sunday 13th) was composed of Plain and Swampy Cree Indians and half-breeds: one Plain Cree woman's home was three hundred miles to the west; she was a fine specimen of the race, and neatly habited in the dress of the half-breeds. Near the door of the church, inside the building, a number of heathen Indians from the prairies stationed themselves to indulge their curiosity; they squatted on the floor, remaining quiet and grave, and conducted themselves with the utmost propriety during the service. They were Plain Crees, followers of the Buffalo hunters, with whom they had lately arrived from the high prairies; some were clothed in dressed skins, others robed in blankets, with head and neck decorations; and one young heathen girl, wild, and almost beautiful, triumphed in a robe of scarlet military cloth. Who can say what benign influence the sight of Christian worshipers may have upon many of these savage children of the prairies, who saunter in during the services of the Church, and with characteristic decorum always maintain a respectful demeanour, and a grave and earnest look.

The Upper Presbyterian Church is a neat building of stone, situated in the middle of the settlement. The cost of its erection exceeded 1,000*l.* sterling, and it has sittings for 500. The manse is delightfully placed on the river bank, which here slopes uniformly to the water's edge from the great prairie, some thirty feet above the stream during its summer level.

In the letter referred to on page 219, the Rev. Mr. Black says, "As to church matters, we have here two congregations, or rather a congregation and a mission station belonging to this congregation. In the one where I live there are about sixty families; in the other (situated at Mr. Gunn's, Stone Fort) there are ten or eleven in all. There are somewhat upwards of 120 members in full communion. The people are mostly Scotch, or of Scotch parentage. There are a few Orkney men, whom our Highlanders scarcely recognise as Scotch, a few half-breeds, one Englishman, and one Swiss. We have sabbath schools at both places: here the attendance may just now average eighty-five; below about thirty. Here we have divine service every Sabbath forenoon, and in the afternoon alternately here and below. We have also week lectures on Thursdays, and prayer meetings on Tuesday evenings. In regard to temporalities, the congregation below have no property but their small meeting-house; the one here has about 300 acres of good land, a stone church which cost about 1,000*l.*, and the cottage in which I live. My stipend is 150*l.* sterling a year, 100*l.* of which is raised by voluntary contributions, and 50*l.* is allowed me by the Hudson's Bay Company. My people are mostly all farmers in comfortable circumstances, but none rich. They are, however, allowed to be the most steady and industrious portion of our population."

THE ROMAN CATHOLIC CHURCHES.

By far the most imposing ecclesiastical building in the settlement is the Roman Catholic Cathedral of St. Boniface, near Fort Garry. The external appearance is neither pleasing nor tasteful, although at a distance the two tinned spires, 100 feet high glittering in the sunlight, give an imposing aspect to the building. They can be seen from a great distance, and with the spire of St. James's Church on the Assinniboine, are well-known landmarks. The internal decorations of St. Boniface, for so remote a region, are very striking, and must necessarily exercise a potent influence upon the large and singular congregation who worship every Sunday within its walls. Two or three very sweet-toned bells ring at matins and vespers, and to a stranger just arrived from a long journey through unpeopled wastes, no sight or sound in Red River creates such surprise and melancholy pleasure as the sweet tones of the bells of St. Boniface, breaking the stillness of the morning or evening air.

The body of the cathedral is 100 feet in length, 45 in breadth, and 40 in height. The three bells weigh upwards of sixteen hundred pounds, and their chimes are deservedly listened to with pride and emotion by the Roman Catholic population of Red River.

The parish of St. Norbert is thinly peopled, but the chapel, which is constructed of wood, is 90 feet long, and 33 broad. The chapel of St. François Xavier is in a very dilapidated condition, but there is every prospect that a new and commodious structure will soon be erected on the White Horse Plain. (See Appendix, Vol. II.)

The appearance of the congregations of the different churches in the settlements, produces a very favourable impression upon a stranger. From whatever direction Red River may be approached, the journey has to be made through a wilderness in which no signs of civilization are to be seen for several hundred miles, much less a church in which divine service is celebrated to attentive and intelligent worshipers.

The congregations at Red River consist of resident and retired officers of the Hudson's Bay Company, some merchants, farmers, and the natives or half-breeds of the respective parishes. The services are conducted in strict accordance with customary forms, and the demeanour of the congregations is very attentive and decorous. A fair proportion of the congregations come to and go from church in neat carriages, or on horseback, and the external appearance of the assemblages, taken as a whole, in relation to dress, is superior to what we are accustomed to see in Canada, or in the country parishes of Great Britain. The young men wear handsome blue cloth frock coats, with brass buttons, and round their waist a long scarlet woollen sash; the young women are neatly dressed like the country girls at home, but in place of a bonnet they wear the far more becoming shawl or coloured handkerchief thrown over or tied round the head; sometimes they allow their long black hair to serve the purpose of a covering and ornament for which, from its profusion, it is admirably fitted. In this particular many of the half-breed girls follow the custom of their Indian ancestry, who, as a general rule, never cover the head.

There is a distinct and well-preserved difference in faith between the populations of the different parishes into which the settlement is divided. Some are almost exclusively Protestant, others equally Roman Catholic. In the parish of St. Norbert there is not one Protestant family, but 101 Roman Catholic families. In the parish of St. Boniface there are 178 Roman Catholic families and five Protestant; so also in the parish of St. François Xavier, on the Assinniboine, there are 175 Roman Catholics to three Protestant families. On the other hand, in the parish of

St. Peter there are 116 Protestants and but two Roman Catholic families, and in the parishes of Upper and Lower St. Andrew's there are 206 Protestant to eight Roman Catholic families.

A very short stay in Red River is sufficient to create both admiration and surprise at what may not be inaptly termed, the condition of religion in Assinniboia. Admiration is aroused by the extent and design of the charities of the different societies in England, who sustain such a large ecclesiastical corps in connection with the Church of England as resident missionaries in the settlement, and who have contributed very munificently to the erection of the excellent churches which are now constructed. In addition to these demands upon their liberality, the home societies give large sums towards the maintenance of missions in different parts of Rupert's Land, so that at the present time there are scattered over this immense country nineteen clergymen of the Church of England, costing between 6000*l.* and 7000*l.* sterling, annually. The Church Missionary Society have expended, up to the date of their last report, very nearly the sum of 60,000*l.* sterling upon missionary operations in Rupert's Land. While, however, so much is done by those in England for charity's sake, it is much to be wondered that so little is contributed by the wealthy residents of Red River, such as the retired factors of the Hudson's Bay Company, the merchants, traders, and better class of farmers, towards the maintenance of the clergy, the support and extension of schools, and to the christianising of the heathen Indians, whose medicine-drum, accompanying the monotonous song of the conjuror, can almost always be heard in summer during the hours of service.

The outward appearance of many among the congregations of the episcopal churches as they come and go in neat little carriages, or on horseback from comfortable well-furnished homes, enforces the expectation that in proportion to their means, they would at least endeavour to prepare the way for the spread of Christianity among the thousands of heathens who frequent the settlement; that such is not the case there is too strong ground for belief.

In the present condition of the country, with the interests of the fur trade to be upheld, the whole subject of missionary enterprise in Rupert's Land is full of difficulty. This much appears certain: the Indians must be induced to settle in one place for a few months of the year at least; schools must be founded, and their children taught the truths of Christianity; missionaries must learn the Indian language; and then the spread of Christianity among the heathens may be in some degree commensurate with the charity which animates the different supporting societies in Great Britain and Ireland. In the settlements at Red River, and on the Assinniboine, all the services are conducted in the English tongue; and among the clergy of the Church of England at Red River, but one only speaks one Indian language with the fluency and ease necessary to make himself understood by the natives. Of course the Indian Mission below the settlements is not included in this enumeration.

The Hudson's Bay Company continue to be very liberal in their support of missionaries, as far as money is concerned; their contributions will be seen in the foregoing table; but the impression was irresistibly forced upon me, and I found it strongly felt by some residents in Red River, that the progress of Christianity among Indians would be rather aided than otherwise, if missionaries were not to receive any assistance in the form of an annual stipend from the Hudson's Bay Company. Perfect freedom of action in inducing Indians to settle; in the education of Indian orphan children, and in teaching them and adults the blessings of a settled, Christian home, as opposed to a heathen hunter's life, are essentially necessary before much satisfactory progress can be made. Can the ministrations of the Church in the English tongue to orderly resident congregations of European, Canadian, or half-breed origin, be missionary labour in the sense in which that highest of all duties is understood by those who contribute to the spread of the truths of Christianity among a wandering, degraded, and barbarous heathen race?

Missionary work at the outposts is altogether different to missionary work in the settlements. In no respect do the material means, necessary to make life comfortable, fall short of rational requirements at Red River, if proper forethought be exercised; but at the distant stations, isolated and often alone, the missionary's work becomes a labour which must spring from the heart in order to secure even an outward show even of promise or success. Many country parishes in England are far less attractive, remunerative, or desirable than the missions at Red River Settlement.

Those who have not experienced the privations resulting to missionaries in remote outposts from the non-arrival of their supplies by the customary route and at the expected season, can form but a feeble conception of the troubles and anxieties which chequer the life of a zealous missionary in the wilds of Rupert's Land. It is not mere personal inconvenience which causes him care and embarrassment; it is the impossibility of taking advantage of many opportunities for inducing wandering Indians to settle around the mission, of clothing and feeding the children entrusted to his charge, and of securing, by aid judiciously applied, the respect and affection of those he is endeavouring to christianise and educate, or seeking to draw from their faith in strange and imaginary gods.

The Indian generally, from his habits and precarious mode of subsistence, requires something tangible in the first instance to arrest his attention, and practical encouragement, often repeated, to secure his good-will, before an impression can be made on his heart. If the missionary be cut off from his supplies in the infancy of a Mission, much of his work has to be done over again. Indian wants are few and simple, but they must be supplied without fail at new stations; hence the importance, if success is to be secured, of effecting and sustaining a tolerably regular communication once or twice a year with the settlements at Red River.

It has sometimes happened that this is not convenient, or perhaps quite impossible; it is natural to suppose that when, from missing a season or from other causes, the supplies for the service of the different posts of the Company are in arrears, and the brigade of boats can take only a certain quantity of goods, those for the purposes of the trade will first receive attention. It has happened two or three times that one year's supplies for the whole population, of tea, sugar, articles of clothing and other important necessaries, have been unavoidably left at York Factory, causing no little inconvenience and trouble to the settlers as well as the missionaries. At Red River their wants can be in part supplied from the stores at Fort Garry, but at the missionary outposts such relief cannot be looked for.

Now that rapid communication may be said to be established between Fort Garry and St. Paul by steamboat and stage coach, there will always be an abundant supply of necessaries at the settlements, which was not the case when the chief means of communication with the outer world lay through York Factory. Opportunities may now be embraced for furnishing distant outposts, which did not exist before Georgetown in the State of Minnesota was connected by steam with Fort Garry, and by stage-coach with St. Paul on the Mississippi.

CHAP. X.

EDUCATION IN THE SETTLEMENT.—AGRICULTURAL INDUSTRY.

Schools.—Protestant Schools in the Settlement.—Subjects taught.—Collegiate School.—Distinguished Scholars.—School Attendance.—Sources of Income.—School Wants.—The Presbyterian School.—The Roman Catholic Schools.—Agricultural Industry.—The Farms.—Want of Improvement manifest.—Cause of the Absence of Progress.—Cultivated Crops.—Indian Corn.—Wheat.—Hay.—Barley and Oats.—Root Crops.—Sugar.—Hemp and Flax.—Live Stock.—Agricultural Implements.—Facilities for raising Stock.—Timber.—Country west of Red River.

THE CONDITION OF EDUCATION AT RED RIVER.

EDUCATION is in a far more advanced state in the colony than its isolation and brief career might claim for it under the peculiar circumstances in which the country has been so long placed.

There are seventeen schools in the settlements, generally under the supervision of the ministers of the denomination to which they belong. The following enumeration is nearly accurate :—

STATISTICS AND ENUMERATION OF SCHOOLS.

1. St. John's College, including a boarding school for boys and girls, under the immediate supervision of the Bishop of Rupert's Land.

2. Archdeacon Hunter's School, "Model Training School," conducted by Mr. Mayhew, recently from the Normal School, Dublin.

3. Mr. Gunn's Commercial Boarding School, more particularly in connection with the Presbyterians.

4. The Rev. Mr. Taylor's Parochial School on the Assinniboine.

5. The Rev. Mr. Chapman's Parochial School, near the middle settlement.

6. The Presbyterian School, under the superintendence of the Rev. Mr. Black.

7. Three minor schools, under the supervision of the Episcopal ministers in different parishes.

8. The Roman Catholic schools are three in number, one of them occupying a very spacious and imposing building, near the church of St. Boniface, and providing ample accommodation for female boarders.

9. At the Indian missionary village, an excellent school is under the control of the Rev. Mr. Cowley. All of the foregoing establishments are independent of the Sunday schools, properly so called, in connection with the different churches.

10. A private ladies' school is now established in the parish of St. Andrew, near the rapids. The house is commodious, and the boarders are under excellent supervision.

PROTESTANT SCHOOLS IN THE SETTLEMENT.

The present condition of Protestant education has been so ably described by the Bishop of Rupert's Land, in a letter with which he favoured me, in reply to inquiries on the subject, that I avail myself of some extracts, which embody as full and exact information as could be desired.*

" We may perhaps take the limits of the settlement as extending from Portage La Prairie to the Indian settle-

* This letter was published at length in the Canadian Blue Book, " On the Exploration of the Country between Lake Superior and the Red River Settlements," 1858. Also in the Imperial Blue Book on the same subject, June 1859.

ment. Within these boundaries the schools connected with the Church of England are thirteen. They are necessarily more numerous than would under any other circumstances be required by the population; from the houses of the settlers lying along the banks of the two rivers, and not being in the form of a town or village, the children cannot go to school above a certain distance, and the schools have been in consequence multiplied to suit the convenience of the inhabitants. The thirteen are exclusive of the two higher academies for young ladies and for boys.

"The subjects taught must vary considerably, from the great difference of capacity in the pupils. The two leading schools would be "St. John's Parochial School," in the upper part of the settlement, and the "Model Training School," connected with St. Andrew's Church. In the former, in addition to the usual branches, the upper pupils have the opportunity of studying Latin, French, and mathematics. In the model school, which is taught by a certified master from Highbury, the senior pupils have also the advantage of instruction in Latin, Euclid, and algebra. They are thus an approach to the grammar schools in Canada. In the other schools, of which St. Paul's is the best example, there is an excellent education afforded in British history, grammar, geography, arithmetic, with the elements of general history. Of course we must be content with much less where the pupils are the children of Indian parents. With them it is difficult to go beyond reading, writing, and arithmetic.

"In the Collegiate School many of the pupils make very great progress both in classics and mathematics. Soon after my arrival in the country I was induced to found some scholarships as an incentive to study, and an approximation to what takes place in other countries. To the scholars elected from year to year, was assigned a free board, and the sum of 10*l.* a year, or in all about 30*l.* per annum. Of these so elected, some have done well elsewhere, and reflected credit on their early training. I would only specify among these Mr. Colin C. McKenzie, B.A., of St. Peter's College, Cambridge; Mr. Jas. Ross, B.A., who has distinguished himself very highly at the University of Toronto; the Rev. Peter Jacobs, ordained by the Bishop of Toronto to labour among the Indians on Lake Huron; and the Rev. Robert McDonald, ordained by myself to the missionary station of Islington, on the Winnipeg River. With more advanced pupils the higher classics have been read, such as Æschylus, Herodotus, and Thucydides. The turn of the native mind is, however, more towards mathematics. All attain to excellence in algebra, and acquire it with great ease. All, too, have naturally imitative power, and write and draw well. While I have had great pleasure in carrying on these branches of education, my one feeling of disappointment has been, that there is comparatively little opening for those who distinguish themselves in this country in after life. Yet I have felt that the duty is ours; the event was with God. In the young ladies' school the want of adequate motive to excite to study is felt more than in the collegiate school. They have the opportunity of learning every branch usually taught in such establishments elsewhere, such as French and music, and there is a very great change perceptible in the seven years. Their education is all-important with a view to the training of the next generation; and although the progress may not be visible in their case, the effects will, I trust, be fully acknowledged when they are settled in life.

"In the thirteen schools there may be about six hundred, from that to seven hundred children. In one or two there may be above fifty in attendance in winter, but the average will not exceed forty. The students at the Collegiate School have been as many as twenty-four, but as the standard of education rises in the parochial schools, the Collegiate School, as such, will be comparatively unnecessary, and it will ultimately be limited to those who may be under preparation for holy orders. For such, and for the clergy generally, there is a library, possessing now one thousand books of standard divinity, as well as other useful subjects.

"The sources of income vary much; ten out of the thirteen schools are connected with the Church Missionary Society. The masters of such schools have all a salary from the Society. The model training master is entirely paid by them, and also the masters of the pure Indian schools. In the other schools about one half may be paid by the Society, sometimes less, and the rest made up by the parents of the children. In the three parochial schools unconnected with the Church Missionary Society: in St. John's parochial school, a portion of the salary is paid by my own college, Exeter College, Oxford; in St. James's, by some christian friends in Edinburgh; and at Headingly, by the congregation of the Rev. T. M. McDonald, Trinity Church, Nottingham.

"The sum paid by parents is 15*s.* a year; where Latin is taught, 1*l.* In some parishes they prefer to pay the pound, or thirty shillings a family, and to send as many as they choose for that sum.

"We want much, school apparatus, books, and maps. A very large quantity of books have been imported, and the Society for the Propagation of Christian Knowledge has given many valuable sets of maps to several parishes; but scattered over thirteen schools, they are still insufficient. * * *

"On the ground of education, let none fear to make trial of the country. The parochial school connected with my own church is equal to most parochial schools which I have known in England; in range of subjects superior to most, though in method and in the apparatus of the school necessarily a little inferior."

The Rev. John Black, the Presbyterian minister at Red River, also favoured me with an account of his church and school. The following extract from that gentleman's letter conveys the necessary information: *—

"First, then, as to the school: This is entirely supported by the people of the district, or rather by those of them who send their children to it. There is no endowment, no public money, nor any allowance by any missionary or other society. The salaries of the different teachers have varied from 22*l.* to 40*l.* sterling a year. The branches taught are English reading, writing and grammar, geography, arithmetic, and the elements of algebra and geometry. In the last two branches I think there are no pupils at present. The average attendance will be from thirty-five to forty. The school is kept open for the whole year, excepting a month in harvest, and the usual holidays. The school is not exclusively composed of the children of Presbyterian families, neither do all the children of such families attend it; some of these, residing at the extremities of the parish, attend the Church of England schools at the upper and middle churches, whilst some of the Church of England people who reside amongst us send their children here. You are aware that we have no public school system in the colony, and this, like the rest, is therefore

* This letter was published at length in the Canadian Blue Book, "On the Exploration of the Country between Lake Superior and the Red River Settlements," 1858. Also in the Imperial Blue Book on the same subject, June 1859.

essentially a denominational school. We would like to raise its character, but owing to other burdens lying upon them, and to their being left without assistance, the people are not able to hold out sufficient inducement in the way of salary to secure the services of an able teacher, at least permanently."

ROMAN CATHOLIC SCHOOLS IN THE SETTLEMENT.

The Bishop of St. Boniface furnished Mr. Dawson, as late as February 7th, 1859, with a statistical account of the schools and missions of the Roman Catholic Church.* The parishes on the banks of Red River and the Assinniboine are four in number, St. Boniface, St. Norbert, St. Francois Xavier, and St. Charles. Fifty-eight children receive education in the school of the brothers of the Christian doctrine, in the parish of St. Boniface. In the convent belonging to the Sisters of Charity, commonly known in Canada as the Grey Nuns, twenty young ladies are boarded, and receive an excellent education, suitable to their station in life. Besides the boarders, the sisters maintain and educate fifteen poor orphan girls, and keep a day school for the benefit of the poorer portion of the parishioners. In the parish of St. Norbert, thirty-one boys and twenty-nine girls attended the schools kept by a priest and the Sisters of Charity. In the parish of St. François Xavier thirteen boys and twenty-six girls receive instruction from the Sisters of Charity. In the parish of St. Charles there is no school or chapel. With reference to the zeal shown by the Roman Catholic population, in matters relating to education, the Bishop says :—

"Considering the sparse character of the settlements,

* Letter from the Bishop of St. Boniface. Red River Settlement; published in Mr. Dawson's Report, 1859.

the schools would need to be increased in number in certain districts, but the absence of any law relative to education, and of zeal in the people themselves, renders it utterly impossible to do more. I venture to assert what all reasonable and impartial persons must, in view of what is done, acknowledge, that the result far transcends the means which we can command. The truth is that, but for the unselfish zeal of some who devote themselves without fee or earthly reward to the arduous and meritorious task, it would be absolutely impossible to keep up the schools. So far, scarcely one child in ten has paid for his schooling, although the charge does not exceed ten shillings per annum, and I am certain that if we insisted on the payment of even that trifling sum, many of the scholars would leave the schools; such is the carelessness and indifference of the parents in that respect, notwithstanding our oft-repeated entreaties and the sacrifices which are made in that behalf. This indifference concerning the education of their children and neglect of the many advantages afforded them is a standing reproach which may be justly cast on our population."

AGRICULTURAL INDUSTRY.

Immediately on the banks of Red River and the Assinniboine, and extending rearwards to the public road, thence into the prairies beyond, the farms of the settlers are laid out in narrow strips, so as to give to each a small frontage on the river.

The houses are generally built close to the edge of the prairie, where it is abruptly cut by the winding channel of the river, and is thought to be high enough to protect them from occasional floods; but where the boundaries of the prairie retire from the present river channel, they are sometimes placed near the public road, and rarely in the depression formed by the ancient course of the stream. Above Mill Creek, on Red River, there does not appear to be any rise of land sufficient to afford security against extraordinary floods, such as those of 1826 and 1852, when the waters rose above the road, or more than thirty feet above the present river level. On the west of the road, as already remarked, is a boundless expanse, in which are enclosures, offering to the eye perfectly level fields of waving grain, or luxuriant pasture. Where no enclosures west of the road have been made, the prairie often passes into what are locally termed swamps or marshes; but which are so susceptible of drainage, and conversion into the richest pasture lands, that they do not deserve the title which has been assigned to them.

Familiarity with the settlements dispels the favourable impression with which a stranger at first regards them. At a distance, the neat whitewashed houses, with their gardens and farmyards, continuing without interruption for twenty miles between the forts, the herds of cattle, horses, and sheep, feeding on the plains, the vast expanse of what seems to be meadow of the richest description, lead one to suppose that universal prosperity and contentment would here be won without anxiety or trouble. Nevertheless, no one can fail to be struck with the indifference to the future, which seems habitually to characterise the people, especially the French portion of the population, and to show itself in their unfinished dwellings, neglected farms, and extravagant indulgence in dress, or in articles they covet. Many of the apparent efforts of industry which, seen from a distance, excite admiration, shrink upon a nearer approach into sluggish and irregular attempts at improvement, often abandoned before completion. The farms and farm buildings in the occupation of the majority afford no sign of recent improvement; and in general it may be said, that the buildings which in Canada would be considered good, roomy country houses are exclusively possessed and occupied by the retired officers of the Hudson's Bay Company, the traders or merchants of the settlement, and the clergy.

The farmers' homesteads, and the hunters' and trappers' cottages, if these classes here can with propriety be separated, bear rather the appearance of slow decay, and a decline in fortune, than a healthy, hopeful, progressive condition.

With few exceptions, and these are chiefly among the Scotch, farming operations are conducted in a very slovenly manner. Weeds abound in most of the fields appropriated to grain; some fields are seen here and there to be altogether abandoned, and the out-houses wear a neglected aspect, or one of ruinous decay. As might be supposed in this primitive part of the world, manure is allowed to accumulate in the front of the stables and cattle sheds, or is sometimes thrown into the river, or heaped in such a position that it may be swept away by spring freshets. All these drawbacks and indications of negligence and imprudence are not uncommon, within certain limits, in every new country, indeed in any locality remote from markets, and wherever ignorance of the first principles of rural economy prevails; but where such marked neglect and seeming dulness abound in the midst of very general intelligence and acuteness, and are limited to the so-called agricultural class, in possession of a soil of unsurpassed excellence, the enjoyment of an admirable summer climate for agricultural purposes, and no greater share of periodical contingencies than those to which every other country is liable, the causes which induce these evils must be sought for in other directions than those which may be said to spring from a dislike for agricultural operations, or a characteristic inability to take advantage of the boundless appliances promoting happiness and comfort which lie within their reach.

The Bishop of St. Boniface, in his letter before referred to, page 220, points to the results of this apathy in relation to agriculture, in the following remarks on the means and resources of a population of about 7000; occupying a country possessing a soil of extraordinary fertility, to supply an unexpected demand for the ordinary necessaries of life.

"In the present condition of the Red River settlement, those who have large families are not the persons who should come; we are more in want of arms than of mouths. A company of soldiers and the exploring parties who visit the settlement suffice, as it is, to create a famine. The price of many of the necessaries of life has doubled since last year, and although the harvest was pretty good, and hunting and fishing abundantly successful, nevertheless, there is an extreme scarcity of everything."

The description which has been given of the general aspect of the farms and farm-houses in the settlements is not such as to create a favourable impression of the condition of husbandry in this remote region; but it would be very unfair to form an opinion of the agricultural capabilities of the country from the results obtained by the majority, under its present state of isolation, and the direction of the best efforts of the inhabitants to objects the reverse of those which belong to a pastoral life.

The farm, as an exclusive object of industry and attention, is recognised by very few of the people of Red River.

Hitherto remuneration for agricultural industry has been impossible as a general rule, on account of the want of a market. Where, however, due attention has been given to husbandry, it has secured comfort and solid independence. The fruits yet remain to be reaped; for now that immigration is taking place to a limited extent, prices of all kinds of farm productions have doubled; and those who have looked to the soil as their means of support in Red River, will be in a position to benefit by the industry and care of former years, and probably find a remunerative market for all they can produce during the next two or three years, until the market becomes overstocked, which is extremely probable in respect of grain, on account of the remarkable facilities which the country offers for bringing large areas of land into immediate cultivation.

CULTIVATED CROPS.

1. *Indian Corn.*—Varieties of corn exist, which may always be expected to ripen in Assinniboia. In order to secure this result, the rich and moist prairie soil requires draining, which may be accomplished without difficulty or expense, by running deep furrows with a common plough, at certain distances apart through the field devoted to Indian corn. This grain is a sure crop on the dry points of the Assinniboine and Red River, where the absence of superabundant moisture permits it to ripen within a certain period, so as to be secure against the early autumnal frosts. No doubt varieties of Indian corn are to be found in New England and in Lower Canada, which would ripen several days earlier in Assinniboia than

the horse teeth or even the Mandan corn, which are cultivated there.

2. *Wheat.*—This is the staple crop of Red River; its cultivation is so general, and the good quality of the grain is so well and widely known, that very little need be said respecting it. In favourable years, that is, in years which have not been distinguished by so wet and backward a spring for farming operations as that of 1857, wheat ripens and is ready for the sickle three months from the day of sowing. I think it very probable that new varieties from Canada, or the New England States, would ripen in less than three months, and this is the opinion of several of the best farmers in Red River. No fact, however, is more satisfactorily determined than the admirable adaptation of the climate and soil of Assinniboia to the culture of wheat. Forty bushels to the acre is a common return on new land; and I have elsewhere stated that Mr. Gowler has obtained fifty-six bushels to the acre, without the introduction of any artifice beyond deep land furrows to keep the rich vegetable mould of the prairie dry.

The great drawback to the cultivation of wheat is the want of a market. Asking a native to show me his wheat field, he said that he had grown enough the year before to last for two years, and the chances of his being able to dispose of any surplus were so small, that he determined not to trouble himself this year with growing wheat. As it happened, he would have been well repaid for any surplus, the expected arrival of the troops, and other circumstances, created a temporary market for wheat, which, however, could not have been foreseen by the easy going half-breed.

None of those diseases, with the exception of smut or rust, or insect enemies, to which the wheat crops in Canada and the United States are subject, occur, it is said, at Red River. Of this fact I cannot speak from personal experience; all I can say is that I heard no complaints of rust, nor did I see a single instance of its presence; yet it would be very unwise to infer from so short an experience that rust is not an enemy to the wheat crops there; the character of rust leads to the supposition that it will be found wherever wheat is grown, if the climate be favourable to its production. The absence of rust is probably more a question of summer climate than of peculiarities in the soil which prevent its attacks. Although I made numerous inquiries respecting destructive insects, yet I could hear of none similar to the Hessian fly or wheat fly, as having been observed there. The grasshoppers, from 1817 to 1820, were the most destructive enemies known; in 1857 and 1858 they destroyed the wheat crop at Prairie Portage, and to a small extent in the settlements.

3. *Hay.*—Quantity unlimited, and quality excellent. The prairies for hundreds of miles, through which Red River, the Assinniboine, Rat, and Roseau Rivers flow, offer everywhere a bountiful supply of grass and hay. Hay ground privileges have been established on the banks of the larger rivers, and the right of making hay within particular limits is recognised by the inhabitants.

Barley and oats are not much cultivated; hops grow wild and in the greatest luxuriance. In 1858 they were seen on the Little Souris River, at the Qu'appelle Lakes and on the Little Saskatchewan. Beet yields very abundantly. Tobacco is cultivated to a small extent, but from trial of its qualities, I infer that it is susceptible of great improvement in the manufacturing process to which it is subjected. The season is, perhaps, too short for it to acquire maturity, and produce a good article.

All kinds of root crops grow well, and attain large dimensions. All common garden vegetables, which are cultivated in Canada, are equalled, if not surpassed by the productions of the rich prairie soil of Assinniboia. Considerable quantities of sugar are made from the ash-leaved maple on the Assinniboine. As no care is taken of the trees furnishing this useful article, it is probable that the supply from this source will soon cease. In cutting wood for fuel, the "natives" do not seem to have any special regard for the valuable trees.

Some years since, at the instance, it is stated, of Sir Geo. Simpson, flax and hemp were cultivated to a considerable extent by the settlers at Red River. The product was of excellent quality, and gave every promise of furnishing a very valuable commodity for home manufacture, and for exportation. The cultivation of these important crops was stimulated for a few years by premiums given by the Hon. Hudson's Bay Company, but when the premiums were withdrawn the cultivation soon ceased. Many settlers with whom I conversed had grown both of these vegetables, but that universal complaint, the want of a market, or of machinery to work up the raw product, led them to discontinue this very important and profitable branch of husbandry.

Live Stock.—The live stock of the settlements are represented by 2,799 horses, 2,726 oxen, 3,883 cattle, 2,644 calves, 4,674 pigs, and 2,429 sheep. Since the census of 1849 an increase has taken place in all of the foregoing items, with the exception of sheep: this useful animal appears to be fast diminishing at Red River, and little wonder, when only one carding mill, and that not in operation, as I was informed, exists in the settlement. In 1856 there were 667 fewer sheep in Assinniboia than in 1849, and 1130 less than in 1843. This decrease is very much to be lamented; it is said to arise from the want of a market for the wool, or means to manufacture it in the settlement. During the winter of 1855—6, the number of animals lost amounted to 184.

The Rev. Mr. Black expresses a strong opinion in favour of the rearing of sheep.* "You saw what a splendid country it is for sheep pasture, and were there means of making wool into cloths, blankets, &c., greater attention would be given to the rearing of sheep; great quantities of such goods are also required for the fur trade, and it would be an advantage to have them manufactured here. Among the emigrants coming up to take possession of the land, it would be a great advantage were there somebody to establish machinery for carding, fulling and dyeing, perhaps spinning and weaving also."

AGRICULTURAL IMPLEMENTS.

The agricultural implements are English and American ploughs, of which 590 are now to be found in the settlement; together with 672 harrows, eight threshing machines, two reaping machines, and six winnowing machines. Produce is hauled in the celebrated Red River carts, of which there were 2,108 in the settlement in 1856. They are admirably constructed of wood; no iron is employed, but sometimes buffalo-hide is made to serve as a tire; these carts will last for several years; and one which conveyed some heavy boxes of geological specimens from Red River to Crow Wing last autumn, had previously been twice near to the foot of the Rocky Mountains, and was still in good condition.

The vast prairies of Red River and the Assinniboine, clothed with a rich profusion of most nutritious grasses,

* Vide letter referred to on page 219.

offer unrivalled advantages for rearing stock. The introduction of mowing machines would enable the settlers to lay in any required quantity of hay for winter consumption. Few of the better class of farmers keep more than thirty or forty head of cattle, in consequence of the want of a market for beef, tallow, hides, &c. The answer I received on all hands to the question, "Why do you not raise more cattle?" was always the same in substance: "Find us a market for beef, tallow, and hides, and we will soon furnish any quantity of cattle you may require." There does not appear to be any good reason why sheep and cattle should not supply the place of the buffalo; the experience of many years shows that no physical impediments arising from climate or soil exist to prevent the prairies of Red River from becoming one of the best grazing countries in the world. Two reasons for the neglect of this important branch of industry are soon apparent, even to a stranger at Red River. Buffalo meat, pemmican made from buffalo meat and fat, together with the robes and sinews, are always a cash article at the Hon. Company's stores; whereas beef, mutton, hides, tallow, and wool, are a mere drug in the market; again, the habits of the half-breeds, who have long been trained to the hunt, are opposed to the quiet monotony of a pastoral life. Introduce European or Canadian emigrants into the settlement, with the simple machinery they have been accustomed to employ in the manufacture of homespun, and in a very few years the beautiful prairies of Red River and the Assinniboine will be white with flocks and herds, and the cattle trade, already springing into importance between the settlements and St. Paul, will rapidly increase, or without much difficulty be diverted into an easterly channel. Such are the ideas of many with whom I discussed the subject when in the settlements, and my own observations lead me to the opinion that no real difficulty exists in the least degree likely to hinder Red River from becoming a grazing country of the first class, when other interests shall be permitted to exist in the presence of that all absorbing, all-controlling service—the Fur Trade.

Timber fit for lumbering purposes is only found in narrow strips on the Red and Assinniboine rivers, and in still less quantities on the Roseau and Rat rivers; the timber consists of elm, oak, maple, and poplar of very large growth, as is stated elsewhere; but if the settlements progress,—and why should they not?—these supplies will soon be consumed. The ridges afford small aspen and pine; and in the rear of the great ridge, on the east side of the Red River, good pine is to be found towards the Lake of the Woods; the Winnipeg would doubtless furnish some good pine, but the difficulty would lie in bringing it up Red River, in its unmanufactured state. Saw-mills are unknown in the settlement, but the rapids of the Winnipeg could afford any required power there. The question of a supply of timber for building purposes is not so important as the requirements of the same material for fuel; hence it is that those who interest themselves in the future of Red River are anxiously turning their inquiries in the direction of the upper Assinniboine and the little Souris, to those supposed beds of lignite or tertiary coal which are so often spoken of by the Buffalo hunters who have occasion to cross these rivers in their progress to the high prairies. The value of the expectations of the settlers in this respect will be discussed in the second volume.

Whatever may be the future of Red River, it is quite evident that it will depend to a large extent upon the character of the country lying to the west of the present area occupied by settlements. With a view to supply some data on which an opinion may be based regarding the probable political and commercial status this country may acquire in process of time, the succeeding chapter is devoted to a very general description of the physical character and resources of the region included between the Saskatchewan, from the elbow of the South Branch to the Grand Rapids, the west coast of Lake Winnipeg, and the 49th parallel or international boundary. It is intended to serve as an introduction to the narrative of the Assinniboine and Saskatchewan Expedition of 1858, which follows the description of the Journey to Canada in the autumn of 1857 by Crow Wing and St. Paul.

CHAP. XI.

SKETCH OF THE COUNTRY WEST OF RED RIVER.

General Surface. — Elevation of the Prairies of Red River. — Pembina Mountain.— Terraces.— Mountains.— Lakes and Rivers.— East of the South Branch of the Saskatchewan.—North-East of the Assinniboine.—Riding and Duck Mountains.—The Great Lakes.—Geological Formations.—The Touchwood Hills.—Turtle Mountain.—Lake Winnipeg. — Lakes Manitobah and Winnipego-sis. — The Qu'appelle Lakes.— The South Branch.— The Main Saskatchewan.— The Grand Rapid.— The Little Saskatchewan.— The Qu'appelle, or Calling River.—The Little Souris.—Wooded and Prairie Land.—Areas fit for Settlement.—Valley of the Assinniboine.—Valley of the Saskatchewan.—East of the Riding and Duck Mountains.

GENERAL SURFACE.

THE prairies of Red River at Fort Garry are about eighty feet above the level of Lake Winnipeg. They form the southern portion of a vast region of lake, swamp, and marsh, which is bounded in a very well defined manner by the Pembina Mountain, and its continuation to the Saskatchewan, which river it crosses a few miles below the Nepowewin Mission, opposite Fort à la Corne. Pembina Mountain forms the western limit of an ancient sea or lake coast; its direction is partly shown on the map as far as the Assinniboine. On the precipitous eastern flanks of the Riding and Duck Mountains, it occurs in the form of a ridge, of which a description is given in the second volume. From the Pasquia Mountain the course of the ancient sea or lake coast is westerly towards Fort à la Corne. The whole of the country east of Pembina Mountain, and its continuation as described above, with the exception of the Assinniboine and Red River prairies, is low, swampy, and in great part occupied by Lakes Winnipeg, Winnipego-sis, Manitobah, and other bodies of water of less magnitude, having an area exceeding in the aggregate thirteen thousand square miles.

Rising above Pembina Mountain in the form of steps, are two other terraces, best seen on the east and west flanks of the Riding and Duck Mountains, but obliterated in the valley of the Saskatchewan and Assinniboine by the denuding forces which have swept over the whole of this region. The south-western limit of these table-lands is marked by the boundary of the Grand Coteau de Missouri, which forms the highest terrace or plateau of the series.

MOUNTAINS, LAKES, AND RIVERS.

Surveying the country in the direction in which the great rivers flow, these vast plains slope gently from a low height of land near the south branch of the Saskatchewan with an easterly trend to the Assinniboine. This slope is continued throughout the valley of the Assinniboine to Red River, after an abrupt declension, where the Assinniboine descends the flank of Pembina Mountain.

North-east of the Assinniboine the country rises almost imperceptibly for a distance of fifteen to thirty-five miles, as far as the base of a series of hill-ranges lying parallel to the general direction of the river valley, before it makes its easterly bend; it then rises by successive steps and sloping plateaux to a summit altitude of about one thousand feet above Lake Winnipeg, or sixteen hundred feet above the sea.

These hill-ranges are known by the names of the Riding Mountain and the Duck Mountain. On their eastern and south-eastern flanks they show an abrupt and broken escarpment, and within the space of five miles the country sinks from sixteen hundred to six hundred and eighty feet above the sea, or within eighty feet of the level of Lake Winnipeg.

At the foot of these hill-ranges, and east of them, lie the great Lakes Winnipego-sis and Manitobah, which are separated from Lake Winnipeg by a low, marshy, and nearly level tract, having an elevation rarely exceeding eighty feet above it.

The uniformity which obtains in the geographical distribution of the great lakes of the Winnipeg Basin is a beautiful illustration of the dependance of geographical features upon geological structure. It is equalled only by the relations of the great Canadian lakes, whose form and general features have been shown to be determined by the formations in which they are excavated.*

The outcrop of the different formations in the valley of Lake Winnipeg, as far as they are known, follows the general direction of the rim of the basin in which they are deposited with remarkable uniformity. Conforming to the direction of the Laurentian system exposed on the east side of Lake Winnipeg, and constituting the Laurentide Mountains, the Silurian series stretches from Pembina on the 49th parallel, to the Saskatchewan on the 54th, and thence towards the Arctic Sea. Following its outcrop, the Devonian series is symmetrically developed between the same distant boundaries; but the most singular feature of this region is, that the soft Cretaceous shales should also conform, with tolerable exactness, to the exposed edges of the unfossiliferous rim of the great

* On the Physical Structure of the Western District of Upper Canada, by Sir W. E. Logan, F.R.S.

basin in which they lie. The present nucleus of the fossiliferous basin is occupied by the great lignite formation of the Tertiaries on the Grand Coteau de Missouri; and so symmetrical is the arrangement in this part of the north-west, that a line drawn through any part of the country from the Grand Coteau de Missouri, where it strikes British territory, to any point between Pembina and Cumberland on the Saskatchewan, would pass over proportionally extensive areas of the Tertiary, Cretaceous, Devonian, Silurian, and Laurentian series.

Besides the imposing Riding and Duck Mountains, the Touchwood Hills may be enumerated as very important and striking in a region whose marked characteristic is that of a gently sloping plain. These hills lie between the head waters of the Assinniboine and the South Branch; the elevation of the highest peak, the Heart Hill, probably does not exceed 700 feet above the general level of the great plain. The course of this range is from north-east to south-west, and it forms the most prominent of several ranges which lie parallel to one another.

South of the Assinniboine the Turtle Mountain is a prominent and important feature. It is cut by the 49th parallel. The Blue Hills of the Souris serve to destroy the general sameness of the prairie level on the river after which they are named; while the Blue Hills south of the Assinniboine, and east of the little Souris River, offer perhaps the wildest and most picturesque scenery in the area here referred to. The Porcupine Hill, Thunder Mountain, and Pasquia Hill, were not included within the area explored. They are eminences which lie between the Grand Rapids of the Saskatchewan and the head waters of the Assinniboine, all of them probably forming, in connection with the Riding and Duck Mountains, at a former epoch, a continuation of a vast tableland, now broken into detached mountain ranges by denudation, with bold abrupt escarpments on their eastern exposures, and gently sloping terraces, separated by steps, on their western aspects.

Prominent among the physical features of this region are the vast expanses of water which occupy the larger portion of its low eastern area. Lake Winnipeg is two hundred and eighty miles long, and in several parts more than fifty miles broad. Lakes Manitobah and Winnipegosis together are nearly of the same length, and the broadest part of the first named is not less than twenty-six miles across. Nearly the whole country between Lake Winnipeg and its western rivals is occupied by smaller lakes, so that between the valley of the Assinniboine and the eastern shore of Lake Winnipeg fully one third is permanently under water. These lakes, both large and small, are shallow, and in the same water area show much uniformity in depth and coast line. Some of the smaller lakes are of dimensions which entitle them to notice. Such are St. Martin's Lake, with an area exceeding three hundred square miles: Water-hen Lake, Ebb and Flow Lake, and Dauphin Lake, each covering an area exceeding one hundred and fifty square miles.

West of the Assinniboine are the Qu'appelle Lakes, situated in the Qu'appelle valley, eight in number, and with an aggregate length of fifty-three miles. Besides these, the last Mountain Lake is probably forty miles long, and varies from three quarters of a mile to two miles in width. The Qu'appelle Lakes are very deep, eleven fathoms or sixty-six feet having been recorded.

North-east of the Touchwood Hills there are numerous large lakes, having areas varying from one hundred and twenty to one hundred and thirty square miles. Some of these are strongly impregnated with saline ingredients, and are the haunts of innumerable hosts of geese and other aquatic birds. On the south-east flank of the same range, and throughout the plain stretching towards the Assinniboine, lakes and ponds are everywhere distributed; the same remark applies to the western flanks of the Riding Mountain and Duck Mountains, as well as to a large area south of the Assinniboine and east of the Little Souris.

Lake Winnipeg receives the waters of numerous rivers which, in the aggregate, drain an area of about 400,000 square miles. The Saskatchewan (the river that runs swift) is its most important tributary. The South Branch of this magnificent river flows for fully two hundred miles below the Elbow, at the foot of a continuation of the Eyebrow Hill Range, a low offset of the Grand Coteau, in a north-easterly direction, and its deep excavated valley appears to lie at an average distance of twelve miles from it. This range is cut by several narrow but deep valleys, and from the small lakes or ponds which occupy their summits, water flows during spring freshets to the Saskatchewan and Assinniboine.

The valley of the Qu'appelle River is a remarkable and important instance of this interlockage, but not the only one which connects two different drainage slopes in this region. Within fifty miles south-west of the Grand Forks, and a short distance south of the Lumpy Hill, there is another deep valley in the dividing ridge, from whose summit lakelets water flows in the spring to the South Branch, a distance of ten or twelve miles, and also to the main Saskatchewan, which it reaches below Pine Island Lake, a distance exceeding 160 miles. One other interlockage between the South Branch and the valley of the Assinniboine will be noticed in the proper place.

The South Branch, eighteen miles below the Elbow,

and 584 miles from its mouth, is 600 yards broad. The rate of the current is here $2\frac{3}{4}$ miles per hour, the greatest depth 10 feet in the main channel, the mean depth across being 4·6 feet. There are channels on both sides of the river, one being 6 and the other 10 feet deep. After passing the Moose Woods, about ninety miles from the Elbow, the river channel is much contracted, its current is uniform and swift, varying from $2\frac{3}{4}$ to $3\frac{1}{4}$ miles per hour : mud and sandbars disappear, and it flows between high banks of drift clay, with a treeless, arid prairie or plain on either hand. At the Moose Woods, where the river is very broad and sandbars numerous, the paddles of canoes have touched the bottom from one side to the other with the ordinary stroke of the voyageurs ; this occurred during a season of low water. In August 1858, Indians were crossing on horseback from the right to the left bank above the Elbow, the depth not exceeding four feet. Before joining the North Branch the current becomes very strong, often from $3\frac{1}{2}$ to 4 miles an hour. The river winds between high precipitous banks, with forests of oak, elm, ash, aspen and birch covering the low points, the opposite hill banks being clothed chiefly with birch and aspen. Groves of spruce show themselves on approaching the North Branch, but the soil on the prairie plateau maintains the most luxuriant growth of vetches, roses, and berry-bearing bushes of different kinds wherever the aspen forests have been burnt and open areas formed. From the Elbow to the Grand Forks the distance is 250 miles, and generally throughout the last fifty miles of its course the South Branch flows through a thinly wooded country, but possessing a soil of great depth and fertility.

The main Saskatchewan, opposite Fort à la Corne, is 320 yards broad, 20 feet deep in the channel, and flows at the rate of 3 miles an hour. The mean depth across the river here is 14 feet, but it is in the memory of those living at the fort, when the river was crossed on horseback during a very dry season.

About 158 miles below Fort à la Corne, near Tearing River, the main Saskatchewan is 330 yards broad, 22 feet deep in the channel, has a mean sectional depth of 20 feet, and flows at the rate of two miles an hour. Two hundred and ninety one miles below the Grand Forks the main Saskatchewan enters Cedar Lake, a dilatation of the river thirty miles long. Issuing from this large body of water, it expands into a small lake, but soon again contracting its channel, the Cross Lake Rapids come into view; these rapids have a fall of 5¾ feet. Hudson's Bay Company's boats of four or five tons are tracked up them with half cargo, but loaded boats, descending, run the rapids. The Saskatchewan then enters Cross Lake, and after issuing from this elongated expanse of water begins a rapid course to Lake Winnipeg, with a current often 3 and sometimes 3½ miles an hour. The head of the Grand Rapid is about four miles from the mouth of the river. The length of the portage is one mile seven chains, and the rapids below the portage are about 1½ mile long, so that the total length of the Grand Rapid exceeds 2½ miles. The fall from the west to the east end of the portage, as ascertained by leveling, is 28½ feet; the fall below the portage is estimated to be 15 feet; consequently the total fall is about 43 feet. The Grand Rapid is run by Hudson's Bay Company's loaded boats; in ascending from the foot of the Rapids to the east end of the portage, boats are tracked or towed up with half cargo; they are then run back again, and again tracked up with the other half of their freight. From east to west end of the portage boats are tracked up on the south side of the river, with a load of fifteen pieces (1350 lbs.), the remainder of the freight is carried over the portage. The distance from the Grand Forks to the mouth of the Saskatchewan is 342 miles; the distance from the Elbow of the South Branch to the mouth is 603 miles.

The Saskatchewan receives several affluents on its south side which are important only on account of the fertile tracts of country they drain.

About 110 miles in an air line south from the Grand Rapid, and 136 miles by the canoe route along the coast, Lake Winnipeg receives the Little Saskatchewan or Dauphin River, through which Lakes Manitobah and Winnipego-sis discharge themselves. During ordinary summer levels, the Dauphin River offers no impediment to small steamers of light draft; it thus forms a valuable and direct communication between the vast water areas which it links together. It flows through a flat and swampy country offering very few inducements, or indeed opportunities, for settlement. The mission of Fairford is situated on that part of this river which lies between St. Martin Lake and Lake Manitobah, having been removed to its present position from the lower part of Dauphin River in consequence of the occurrence of destructive floods, the surface of the country not being above eight feet over the summer level of the river. Dauphin Lake is connected with Lake Winnipego-sis by Moss River, navigable in high water by Red River freighters' boats. The tributaries received by Dauphin Lake scarcely require notice here, although they may become useful as affording means for transporting the valuable spruce of the Riding and Duck Mountains to Lake Manitobah; the most important of these tributaries is the Valley River, which separates the Duck from the Riding Mountain.

Lake Winnipego-sis receives the Red Deer River and Swan River, which open communication to an important tract of country east and north-east of the head waters of the Assinniboine. The south-western extremity of Lake Manitobah is distinguished by the extent and richness of the prairies which at a higher lake level it has assisted in forming. The White Mud River which meanders through them may be classed among the most valuable of the lesser tributaries of the great lakes of the Winnipeg basin.

The Red River of the north and the Assinniboine having been already described, require no further notice. Some of the affluents of the last-named river are sufficiently important to deserve a separate notice.

The Qu'appelle or Calling River falls into the Assinniboine about five miles below Fort Ellice. At its mouth this stream is 88 feet broad, 12 feet deep in the main channel, and shows a mean sectional depth of 8 feet; its current is at the rate of 1½ miles an hour. The valley in which it flows inosculates with the South Branch of the Saskatchewan at the Elbow. It is 269 miles long, and seventy miles from the Assinniboine about one mile broad, and 310 feet below the prairie, which stretches north and south from its abrupt edges as far as the eye can reach. At the Qu'appelle Mission, 119 miles from the Assinniboine, the valley is one mile and a quarter broad, and 250 feet deep. The river here is 48 feet wide, 6 feet deep in the channel, with a mean sectional depth of 3 feet 6 inches, and a current of one mile an hour. The lakes at this point have a depth of 57 feet, so that the total excavation below the prairie on either hand is 307 feet.

Near the first or Qu'appelle Forks the valley is one mile and one third broad, and 220 feet deep. At the east end of Sand Hill Lake, 239 miles from the Assiniboine, and thirty-one miles from the South Branch, the valley is one mile and five chains broad, with a depth of 140 feet below the prairie. Eight miles from the west end of Sand Hill Lake, or fifteen miles from the Saskatchewan, the excavation is one mile and seventy chains broad and 150 feet deep. At the height of land where it has been invaded by sand dunes from the west and south-west, it is still nearly one mile broad and 110 feet deep, estimated from the well defined edge of the valley, where a low escarpment of rock still uncovered by the advancing sand of the dunes, serves to mark its limit and the power of the forces which excavated it. The level of the prairie studded with sand hills and dunes is some 30 feet above the edge of the rock noticed above.

The Little Souris, or Mouse River, joins the Assinniboine 140 miles from Fort Garry, by the windings of the river valley, and 116 by the buffalo hunters' trail. At its mouth the Little Souris is 121 feet broad, 3 feet 6 inches deep in the channel, with a mean sectional depth of 2 feet 4 inches, and a current of half a mile an hour. Its valley, at the Back-fat Creek, twenty-five miles from the Assinniboine, is one mile and a half broad (8,016 feet), and 225 feet deep, with a level prairie on either hand. Near Snake Hill, sixty-one miles from the outlet, the valley is only 110 yards broad, and 66 feet deep, with open prairie on both sides. The river here is 100 feet broad, and 4 feet deep in the channel. In its passages through the Blue Hills of the Souris, the river has excavated a profound valley between 400 and 500 feet deep, making a sudden turn from a due easterly course to one almost northerly, and avoiding what appears to be an ancient channel but slightly elevated above its present level. This old channel pursues a straight course to Pembina River, with which, on the authority of half-breeds familiar with the country, it is said to be connected. Entering the territory of the United States against the course of the stream, the Little Souris may be traced as far south as the 48th parallel of latitude, when suddenly turning to the northwest it pursues a course parallel to the flanks of the Grand Coteau de Missouri, re-entering British territory about the 102nd degree of longitude.

WOODED AND PRAIRIE LAND.

The western and south-western slopes of the Riding and Duck Mountains support heavy forests of white spruce, birch, aspen, and poplar. The trees are of a large size, and often exceed $1\frac{1}{2}$ and 2 feet in diameter, with an available length of 30 to 50 feet. On the summit plateau of the Riding Mountain the white spruce is the largest tree; here it attains dimensions, and is found in quantity sufficient to give to this region a great economic value. The wooded area over which timber consisting of the four kinds of trees enumerated, is found on the Riding and Duck Mountains, has a length of one hundred and twenty miles, with a breadth exceeding thirty miles. The affluents of the Assinniboine will serve during spring freshets, to bear these valuable forest productions to areas which will probably first attract settlement, and where they will be most required.

In the valley of the Assinniboine is an extensive and valuable forest of oak, elm, ash, maple, poplar, and aspen, with an average breadth of four miles; its length is about thirty miles. The flats and hill sides of the deep eroded valley, through which this river flows above Prairie Portage, sustain a fine forest, in which aspen, oak, birch, elm, and maple appear to prevail in numbers corresponding with the order in which they are enumerated; but this forest does not extend beyond the excavated valley of the river or its tributaries. All the affluents of the Assinniboine flow through deep ravines, which they have cut in the great plain they drain; these narrow valleys are well clothed with timber, consisting chiefly of aspen and balsam-poplar, but often varied with bottoms of oak, elm, ash, and the ash-leaved maple. On the west side of the main river, the valleys of the tributaries, such as the Little Souris and the Qu'appelle River, are timbered continuously for a distance of thirty to seventy miles from their outlets, and at intervals only, further up stream. On the Qu'appelle River good timber is found as far as the mission; but in progressing westward it is seen gradually to diminish in size, and finally to disappear altogether.

The Touchwood Hill Range, together with small parallel ranges, such as the Pheasant Mountain and the File Hill, averaging twenty miles in length by ten in breadth, are in great part covered with aspen forests, but the trees are generally small. At the Moose Woods, on the South Branch of the Saskatchewan, forests of aspen begin to appear; they continue, with occasional admixtures of birch and oak, more rarely of oak and elm, as far as the Grand Forks; here the spruce becomes common, and, with aspens, occupies the excavated valley of the main Saskatchewan for many miles. The hill-banks and the plateau on the south side of the river, for a distance of three or four miles south, sustain the Banksian pine, which disappears as the soil changes from a light sand to a rich and deep vegetable mould, supporting detached groves of aspen and clumps of willows.

On the Little Souris, especially in the neighbourhood

of the Blue Hills, the country is fertile and beautiful, but the areas adapted for settlement lose much of the value which would otherwise belong to them, from the absence of wood. West of the Souris is a boundless, treeless prairie; so that in crossing from Red Deer's Head River to Fort Ellice it was found necessary to carry wood for fuel for a distance of sixty miles. At Sand Hill Lake, on the Qu'appelle, timber is so scarce in the river valley and gullies leading to it, that bois de vache is the only available fuel. The South Branch, from the Elbow to the Moose Woods, flows through a treeless region, as far as relates to the prairie on either side; but in the ravines leading to the river detached groves of small timber occur. The boundary of the prairie country, properly so called, may be roughly shown by a line drawn from the great bend of the Little Souris, or Mouse River, to the Qu'appelle mission, and from the Mission to the Moose Woods, on the South Branch.

AREAS FIT FOR SETTLEMENT.

Valley of the Assinniboine.—Issuing from the Duck Mountain are numerous streams which meander through a beautiful and fertile country. This area may be said to commence at the Two Creeks, ten miles from Fort Pelly, thence on to Pine Creek, fifteen miles further. The vegetation is everywhere luxuriant and beautiful, from the great abundance of rose-bushes, vetches, and gaudy wild flowers of many species. After passing Pine Creek the trail to Shell River pursues a circuitous route through a country of equal richness and fertility. Shell River is forty-two miles from Pine Creek, and in its valley small oak appear, with balsam-poplar and aspen, covering a thick undergrowth of raspberry, currant, roses, and dogwood. Between Shell River and Birdstail River, a distance of thirty-nine miles, the country is level and often marshy, with numerous ponds and small lakes, but where the soil is dry the herbage is very luxuriant, and groves of aspen thirty feet high vary the monotony of the plain.

Between the trail and the Assinniboine the soil is light, and almost invariably as the river is approached it partakes of a sandy and gravelly nature, with boulders strewn over its surface.

The flanks of the Riding Mountain are covered with a dense growth of aspen and poplar, and cut by numerous small rivulets. From Birdstail River to the Little Saskatchewan, or Rapid River, a distance of thirty-three miles, the same kinds of soil, timber, and vegetation prevail. About one hundred miles from its mouth the Rapid River issues from the densely wooded flanks of the Riding Mountain through a narrow excavated valley filled with balsam-poplar, and an undergrowth of cherry and dogwood, with roses, convolvuli, vetches, and various creepers. The slopes are covered with poplar eighteen inches in diameter. Descending the river, groves of poplar and spruce show themselves, with thick forests of aspen and balsam-poplar covering the terrace on either hand. The river is here forty feet wide, with a very rapid current. Before it makes its easterly bend the ash-leaved maple shows itself in groves, and on both sides is an open undulating country, attractive and fertile, with detached clumps of young trees springing up in all directions. The region drained by the Rapid River continues beautiful and rich until within twenty-five miles of the Assinniboine, so that it may with propriety be stated, that for a distance of seventy-five miles this river meanders through a country admirably adapted for settlement. Ponds and lakes are numerous, wild fowl in great numbers breed on their borders, and the waters of the Rapid River abound in fish. It will probably become important as a means of conveying to the settlements on the Assinniboine and Red River supplies of lumber from its valley and the Riding Mountain.

From the Rapid River to White Mud River the distance is thirty-three miles, and the country continues to preserve the same general character with respect to fertility and fitness for settlement which has now been traced out for a space of 164 miles. White Mud River flows into Lake Manitobah at its south-western extremity. This river drains an extensive area of the richest prairie land, similar in all respects to the White Horse Plain on the Assinniboine, or the rich wastes on Red River. White Mud River is connected with Prairie Portage by an excellent dry road, the crossing place being about eighteen miles from the Portage. The river banks are well timbered with oak, elm, ash, maple, aspen, and balsam-poplar. It possesses valuable fisheries, and communicates by an uninterrupted canoe navigation with Lake Manitobah for a length of thirty miles. The soil on its banks and far on either side is of the finest quality. At the mouth of the river a fishing establishment has been maintained by the people of the Portage for several years.

The valley of La Rivière Sale has a general direction parallel to that of the Assinniboine, and about sixteen miles south of it. The country between the two rivers is wet and marshy, with large areas covered with willow thickets and clumps of small aspen. South of the valley of the first-named river the prairie is magnificent, and not surpassed by any area of equal extent on Red River.

The area of the region well adapted for settlement on

the east and north of the Assinniboine, and in the valley of La Rivière Sale may be assumed to occupy 3,500,000 acres. In the valleys of Mouse River, the Qu'appelle River and White Sand River, the area of land likely to invite settlement does not exceed one million acres. The lakes in the valley of the Qu'appelle River are important, they abound in fish, among which white fish are numerous, large in size, and of excellent quality; the gray and red suckers, pike and pickerel are also abundant.

Valley of the Saskatchewan.—1. The country between the Lumpy Hill of the Woods and Fort à la Corne, or the Nepowewin Mission, including the valley of Long Creek and the region west of it, bounded by the South Branch and the Main Saskatchewan. This area may contain about 600,000 acres of land of the first quality.

2. The valley of Carrot River and the country included between it and the Main Saskatchewan, bounded on the south by the Birch Hill range. There is a narrow strip on the great river, about five miles broad, where the soil is light and of an indifferent quality. The area of available land probably does not exceed 3,000,000 acres.

3. The country about the Moose Woods on the South Branch.

4. The Touchwood Hill range.

5. The Pheasant Hill and the File Hill.

The aggregate area of these fertile districts may be stated to extend over 500,000 acres.

Assuming that the prairies of Red River and the Assinniboine east of Prairie Portage contain an available area of 1,500,000 acres of fertile soil, the total quantity of arable land included between Red River and the

Moose Woods on the South Branch of the Saskatchewan will be as follows :—

	ACRES.
Red River and the Assinniboine Prairies east of Prairie Portage	1,500,000
Eastern watershed of the Assinniboine and La Rivière Sale .	3,500,000
Long Creek and the Forks of the Saskatchewan . . .	600,000
Between Carrot River and the Main Saskatchewan . . .	3,000,000
The Touchwood Hill range, the Moose Woods, &c., &c. . .	500,000
Mouse River, Qu'appelle River, White Sand River . . .	1,000,000
The region about the head waters of the Assinniboine, including the valley of Swan River	1,000,000
Total area of arable land of first quality	11,100,000

Of land fit for grazing purposes, the area is much more considerable, and may be assumed equal in extent to the above estimate of arable land.

East of the Riding and Duck Mountains.—The region east of the Riding and Duck Mountains, when viewed as a whole, will furnish a very insignificant field for settlement. Where permanently dry land exists, the limestone rock generally approaches so near to the surface, as to be exposed whenever small trees are blown down, or the soil is penetrated to the depth of six or eight inches. The greater portion of the area on the shores of Lake Winnipeg, Lake Manitobah, the Little Saskatchewan, Moss River, Dauphin Lake, and St. Martin Lake, together with the region between Lakes Winnipeg and Manitobah, always excepting the southern shore of the latter lake, is not generally fitted in its present condition for the habitation of civilized man.

CHAP. XII.

THE JOURNEY TO CANADA VIÂ ST. PAUL.

Isolation of the Valley of Lake Winnipeg.—The Country drained by the Saskatchewan.—Routes to the Valley of Lake Winnipeg.—The Northern Route.—The Southern Route.—Pembina.—St. Joseph.—Deux Rivières. —Pine River.— The Mail.— The Red River Post Office.— Red Lake River.—War Path of the Sioux and Ojibways.—Turtle Creek.—Burning Prairies.— Height of Land Hills.—Caravans.—The Southern Slope.— Leaf River.—Crow Wing River.—Table of estimated Distances between Camps.—Crow Wing.—St. Paul.—Toronto.

THE valley of Lake Winnipeg is separated from the valleys of the Mississippi and St. Lawrence by extensive barriers, which have hitherto been instrumental in preserving it from the approach and intrusion of civilized races. The time has now arrived when this secluded region is likely to attract a wide-spread attention, and inquiry will naturally be turned not only to its resources, but also to its relations in point of geographical position and means of communication with the commercial world, as well as the opportunities it may supply for establishing a direct line of communication across the continent of America between the Pacific and Atlantic Oceans.*

* The Missouri is navigable as far as Fort Benton in long. 110° 30′, lat. 47°40′; 3,120 miles from its junction with the Mississipi. The flat-bottommed steamer "Chippewa" left St. Louis on the 1st June, 1857, and arrived at Fort Benton on the 17th July. Returning she reached St. Louis on the 19th August, performing a distance of 6,240 miles in 80 days, or 78 miles a day. At Dauphin's Rapids, above the Yellow Stone, the steamer was hauled by line, with this exception no difficulty in navigating the Missouri was experienced. This pioneer steam vessel carried 130 packages for the American Fur Company.

The Saskatchewan together with the Red River of the North drain an area exceeding 350,000 square miles, and the part of the valley they drain included within British territory, lies between the 49th and 55th parallels of latitude, and the 93rd and the 118th degrees of longitude west of Greenwich.* A European area, similarly situated east of the 10th degree of longitude, would comprehend very nearly the whole of England and Ireland, part of the German Ocean, the English Channel, the north-eastern corner of France, the whole of Belgium and Holland, and the greater part of the valley of the Rhine, together with the kingdom of Hanover.

The routes by which access is obtained to this great valley lie in the courses of three different watersheds. First, the present Hudson's Bay Company's route from York Factory, Hudson's Bay, *viâ* Hayes River, &c. to Lake Winnipeg; second, the Lake Superior route, *viâ* Dog Lake and Rainy Lake; third, the Mississippi Valley route from St. Paul to Red River.

THE NORTHERN ROUTE VIÂ YORK FACTORY.

As it is not in the least degree probable that the Northern or Hudson's Bay route will ever be selected as a permanent line of communication between Canada and the country drained by the Saskatchewan, a very brief notice of its most prominent features will be sufficient.

This route begins at York Factory, Hudson's Bay, and goods are forwarded in boats of the same size and build as those employed by the Red River freighters, of which a description is given in the second volume. York Factory is situated on the left bank of Hayes River, up which the boats are tracked, poled, or sail as far as Rock

* A description of the boundaries of the valley of Lake Winnipeg is given in the Second Volume.

Portage, a distance of 124 miles.* From Rock Portage to Lake Winnipeg the route lies through a low region, involving numerous portages, and occupied by the primary unfossiliferous rocks. The distance between Hudson's Bay and Lake Winnipeg is about 400 miles, the difference in level being 628 feet. Vessels cannot remain with safety longer than six weeks at York Factory, in consequence of the probability of ice arresting their passage to the Atlantic. Nelson River, through which the drainage waters of Lake Winnipeg find their way to the sea, is characterized by heavy falls and rapids which effectually oppose communication even by canoes or boats.

THE SOUTHERN OR ST. PAUL ROUTE.

On the 8th October, 1857, I despatched those members of the expedition who were to return to Canada by the road on the west side of Red River to Pembina, remaining behind myself in order to see Mr. Dawson, who, having partially recovered his health, was coming up the river with the Rev. J. Macdonald from Islington Mission, where, it will be recollected, he had remained in consequence of illness. Mr. Gladman had returned to Lake Superior in canoe early in September, Mr. Napier and his assistants were comfortably lodged in the Middle Settlement, Mr. Dawson had secured excellent accommodation within half a mile from Fort Garry, and nearly opposite to the Roman Catholic Cathedral of St. Boniface, where he could with ease enjoy the opportunity of attending the religious services of that communion, of which he is a warm and zealous member. Their winter's work before them was to explore the country between Red

* See Report by Lieut. T. Blakiston, R.A.

River and the Lake of the Woods, write reports on the route from Lake Superior, and collect information respecting the resources of the country and the condition of its inhabitants.

On the 9th I overtook my party about six miles from Fort Garry. All told, we were five gentlemen, five half-breeds, six saddle-horses, and five carts, to which were respectively attached four poor horses and one refractory mule. During the afternoon we met the Bishop of Rupert's Land and Miss Anderson, who were returning to Red River from England.

On the 11th we camped at Pembina, near the mouth of the river of the same name. Mr. Murray, the gentleman in charge of the Hudson's Bay Company's post, two miles north of the boundary line, gave us an excellent dinner, thus maintaining to the last the reputation for hospitality which the officers in charge of the posts of the company have justly earned. Whatever may have been the former condition of the village of Pembina, it is now only a small and scattered collection of log-houses, situated on the right bank of the Red River, in the new territory of Dakotah. The ruins of several good houses, formerly occupied by the Roman Catholic mission, are still to be seen, but in all other respects the town and port of Pembina exist only on paper. The few log-houses which have given it a name and a certain reputation, derived probably from its being formerly a frontier post of far more pretensions than at the present time, still serve for an excuse to attract public attention to the fancied progress of the Americans in this part of the Red River valley. In the late returns for the election of officers in the new state of Minnesota (October, 1857), the names of many resident voters are recorded, but it would be a matter of great difficulty to discover their abode now. The presence of some United States dragoons, forming part of an exploring party which camped near Pembina two years ago, gave rise to a report which has often appeared in print and on maps, that Pembina is a post garrisoned by United States troops, instead of being a small village containing about a dozen scattered log-houses. About a day's journey west of Pembina the village of St. Joseph is situated, in the territory of Dakotah, seven miles south of the boundary line. It was founded by the Red River half-breeds, who were induced to settle there in order to escape the floods of Red River, from which they had suffered or anticipated severe losses. The village has already acquired considerable importance as a depôt for articles of trade which are brought by the citizens of the United States from St. Paul.

The village of Pembina, Dakotah Territory.

The country about Pembina is remarkably fertile, and so free from undulations, that towards the west the flanks of Pembina Mountain, already briefly described, may be seen in clear weather thirty miles distant.

Our course lay on the east side of the river, through a

beautiful level prairie studded with willow bushes, for about nineteen miles in a south-east direction, when we struck the first of the "Deux Rivières" at sunset, crossed the river and camped, having traveled twenty-two miles. A very perceptible change was observed in the prairie the next day on approaching Pine or Tamarac River; the soil consisted of a light vegetable mould, and wherever rain had fallen and collected in little hollows, sand showed itself. Hummocks of aspen and willow relieved the sameness of the scenery, and a distinct rise by ridges, at the base of which the river flowed, was easily recognized. Pine River at the crossing-place is about 25 feet broad with a rapid current. Between Pine River and Middle River the soil preserves its light character, the trail running for many miles on ancient lake ridges or beaches which are similar in every respect to those observed between the Roseau and Fort Garry. The night of the 13th October was cold and fine; a few grasshoppers still lingered on the prairies, and their eggs in many places lay in vast numbers on the surface of the ground. The following day was clear, beautiful and warm, but as night approached, the sky in the north-west began to assume a ruddy tinge, and finally a lurid red, produced by the fires in the rich prairies beyond the Assinniboine, at least ninety miles in an air line from Pine River.

During the morning of this day we traveled along an ancient lake ridge, doubtless a continuation of one of those which appear some miles east of the settlements on Red River. Prairie hens were seen here in great abundance, and numerous flocks of wild geese passed overhead. Near Pine River we met "*The Mail*," borne on the back of a half-breed, who was accompanied by a boy fifteen or sixteen years old carrying the blankets and cooking utensils. The mail bearer was ill, and had not eaten food for two days, having been longer on his journey than he expected. He had no means of killing the prairie hens which were so abundant on the trail, and which might have provided him with food. He carried the mail in a large leather bag, by means of a strap passing round his head; he was poorly clothed, wet, and miserable, and had been fifteen days coming from Crow Wing. We gave him some buffalo meat and pemmican, on the strength of which he hoped to reach Pembina in two days.* Serpent or Snake River, where we arrived during the afternoon, flows between steep sand-banks and hills; the soil is very light, and after passing Serpent River is scarcely fitted for arable farms, but might fur-

* The subjoined notice of the "Red River Post Office" is from the Nor'wester of the 28th January, 1860:—

THE POST OFFICE.

"The year 1858 witnessed a new feature in the postal arrangements of the country. In that year, the Canadian Government authorised the conveyance of mails to and from the settlement, *viâ* Fort William. Since that time, therefore, and up to the beginning of the present winter, we have had two lines of mail communication—one through American and one through British territory. The former has hitherto on the whole given great satisfaction, and we doubt not the latter will also give satisfaction, after some more experience of the route and its requirements. In June, 1859, the two lines together brought in 713 papers and 400 letters, besides a number of magazines and reviews. The last mail, which arrived on the evening of the 19th instant, brought in 880 newspapers and 210 letters. This is the largest number of papers ever brought in by a single mail. The number of letters is smaller than by previous mails; but there will of course be a fluctuation. Our mails are only monthly. We may, it is true, send to Pembina twice a month; but as there is but a monthly mail from there, the arrangement is but unsatisfactory. Let us have a genuine fortnightly mail. There was one from July, 1858, to July, 1859: why was it given up?

"The outgoing mail on the 28th ultimo conveyed 350 letters and a large numbers of newspapers. As that was the first instance of newspapers being sent abroad from this settlement, it will mark an important era in the history of the Red River Post Office."

nish very extensive and excellent sheep pasturage. The prairies here are altogether denuded of timber, so that this day we were compelled to carry our fuel for cooking purposes from Serpent River to the middle of the plain where we encamped for the night.

In the morning ice was found in the kettles, but the coolness of the night was not unpleasant. The trail continued for many miles on a level and rounded Lake Ridge, and then descended into a low, rich, wet prairie, towards Red Lake River, 186 miles from Fort Garry by our estimate. Across this fine stream the baggage was passed in two small canoes, the horses swam across, and the carts were hauled with ropes. The valley of Red Lake River is heavily timbered, and will probably become an important stream as the settlements from the south begin to descend the valley of Red River. Already a trading-house has been built at the crossing-place on Red Lake River, and on some of the American Government maps it figures as Douglas, the name of the town which it is proposed to build there. The valley of Red Lake River is the war path of the Sioux and Ojibways, and our half-breeds asked us not unnecessarily to fire off any guns or pistols as long as we were within ten or twelve miles of Red Lake River, that we might not attract the attention of any stray parties of Sioux who might possibly be within hearing.

On the 16th we passed over a high prairie, rising in steps at long intervals apart and level at its marshy summit. The breadth of this prairie is about twenty-three miles, Turtle Brook forming its southern termination. No trees are visible, the soil is generally light, and the higher portions gravelly, but in depressions the soil is of the first quality. Boulders of the primary unfossiliferous rocks were observed in great numbers on the north flanks of the ancient lake ridges. On this desolate prairie we met a caravan of nine Red River carts containing merchandise, which the owners had purchased at St. Paul; they had been twenty-one days coming a distance of 320 miles. In the afternoon we arrived at a part of the prairie where the fire had run; as far as the eye could see westward the country looked brown, or black, and desolate. The strong north-westerly wind which had been blowing during the day drove the smoke from the burning prairies beyond Red River, in the form of a massive wall towards us; a sight more marvellously grand, and at the same time gloomy and imposing, could scarcely be conceived than that approaching wall of smoke over the burnt expanse of prairie stretching far away to the west. The upper edge was fringed with rose-colour by the rays of the sun it had just obscured; and as it swept slowly on, the rich rose-tints faded into a burnt sienna hue, which gradually died away as the obscuration became more complete, until, though early in the afternoon, with a bright cloudless sky towards the east, a twilight gloom began to settle around us, and the rolling folds of smoke sweeping over the prairie, rapidly enveloped all things in a thin but impenetrable haze. When the sun was still some degrees above the horizon the light was that of dim twilight; the prairie hens flew wildly across the trail, and without, as is usual with them, any determined and uniform direction; our horses appeared to be uneasy or alarmed, and the whole scene began to wear an aspect of singular solemnity and gloom. Night came on suddenly, and with a darkness which might be "felt," as we reached the valley of Sand Hill River; here, trusting to the sagacity of our horses, we let them find their way to the stream, on the banks of which we encamped. During the night the horses were very restless, often galloping suddenly among the carts and tents, and at no time appearing to venture far from the camp.

The wind changed during the night, and the morning of the 17th brought a bright and brilliant sky, with a glistening hoar frost on the prairie. Ice was observed in the ponds, and at our camp it was found about a quarter of an inch thick in the kettles which were exposed. Numerous pelicans and geese were seen flying south, and all the customary indications of approaching winter were observed from time to time. The trail this day lay through a fertile rolling prairie intersected by sandy ridges, the slopes were very rich, but the valleys wet. Here we saw the Height of Land Hills, apparently about twenty-five miles off, and having arrived at Marsh River we encamped on a hill near it, on the west side of the dividing ridge.

During the day we met a caravan of six carts, nineteen days from St. Paul; they belonged to private Red River speculators, and were laden with ploughs, whisky, stoves, scythes, &c.

Rising half an hour before daybreak we found ice in the kettles, a strong wind from the north, and a snow storm approaching, which just whitened the ground at nine A.M. but soon passed away. From Wild Rice River, we crossed an undulating prairie about twenty miles broad, to the foot of a low range of hills constituting a spur of the Height of Land, and camped on the north side of the undulating plateau which forms the dividing ridge. A heavy snow storm occurred during the night, and on the morning of the 19th we found the wind strong and very cold; ice half an inch thick having formed in the kettles two yards from the fire. The trail continued through a very beautiful rolling plateau, with clumps of wood here and there, and lakelets between the hills. Camped at noon near the edge of the southern slope, which however does not send water into the Gulf of Mexico except during floods, when a communication is established between Otter Tail Lake and Leaf Lake. Our course lay in the valley of Otter Tail River, a tributary of Otter Tail Lake from which Red River takes its rise. The direction of this stream is due south, and although our route descended rapidly towards the Mississippi, yet we were still in the valley of a river flowing into Hudson's Bay. The distribution of the drift on this dividing ridge frequently determines the course of a stream in a direction diametrically opposite to the general trend of the surface. Hauteur de Terre River, or the upper part of Otter Tail River and Red Eye River, for instance, flow for a distance of forty miles nearly parallel to one another, in a general direction to the S.S.E., but one flows into Otter Tail Lake and thence into Hudson's Bay, the other is an affluent of Crow Wing River, which joins the Mississippi near Fort Ripley.

Even on the summit of the southern slope the aspect of the country begins to change, and prettily wooded lakes become numerous, affording in summer a most delightful variety of scenery. The soil, however, is light, and not favourable for cultivation. Camped at Forty-fourth Lake, about 110 miles from Crow Wing.

The country passed through on the 20th was extremely beautiful, the soil good, timber and prairie land being distributed in about equal quantities. The grackle in countless numbers were seen passing south ; the lakes were alive with ducks, geese, and several other kinds of water fowl, recalling to mind the appearance of the ponds in Red River and the Assinniboine valleys. In the woods we met sixteen carts from St. Paul, bound to St. Joseph's, and laden with tea, sugar, powder, and dry goods. We descended the successive steps of the southern slope rapidly, and soon reached a warmer climate; passed little Red River at noon, and camped in the middle of the prairie, hearing, not without pleasurable feelings, the barking of dogs during the night, which, besides indicating our approach to settlements, was suggestive of pleasant thoughts and delightful anticipations. The prevailing character of the soil, hitherto, is light, but the country is truly beautiful.

On Wednesday, 21st, we arrived at a house near Leaf River, called by its occupants Leaf City, and so represented on the official map of Minnesota; it is within a few miles of Otter Tail city, on Otter Tail Lake. Otter Tail city contains half a dozen log-houses, and is intended by its present proprietors to become a town of importance. Leaf River connects the waters which flow into Red River with those which seek the Mississippi basin, and during floods a canoe can pass from one watershed to the other without difficulty. South of Leaf River the country becomes rolling with deep valleys and extensive swamps between the hills. Leaf River is fringed with a magnificent forest, in which we recognised many trees not seen since leaving Canada.

Camped on the 22nd seven miles from Crow Wing River. During the day we met the Roman Catholic bishop and two priests, together with some French Canadian emigrants bound to Red River. Early the next morning, after passing through a poor country, we arrived at Crow Wing River, where we found a new store well-stocked with goods, which the enterprising owner said he had built and furnished for the benefit of the Red River people; he thinks he will be able to drive a very profitable trade with them. Our road now lay through pine woods and bad swamps, which continue for eight miles, until within twenty-five miles of Crow Wing. The communication through these swamps is wretched, but there is every prospect of the State constructing a new road next year.* Reached Crow Wing at sunset, Saturday, the 24th October, having been sixteen days out from Fort Garry.

The subjoined table of distances affords a close approximation to each day's journey:—

TABLE OF ESTIMATED DISTANCES BETWEEN CAMPS.

Place	Miles
Fort Garry	0
Stinking River	9½
Scratching River	37½
Plum River	51
Pembina	70
First of the Two Rivers to the upper crossing	95
Little Bridge Creek	104
Middle River	110
Second of the Two Rivers	114
Pine or Tamarac River	136
Rock River	142
Serpent or Snake River	147
Middle of Prairie	160
Red Lake River	186
Turtle Brook	212
Sand Hill River	216
Marsh River	242
Wild Rice River	247
Spur of dividing ridge	270
Buffalo Creek	279
Forty-fourth Lake	310
Little Red River	320
109th milestone from Fort Ripley	329
Rush Lake	338
Seventy-seventh-mile Lake	361
Seven miles east of Leaf River, 62½ miles from Crow Wing	376
Twenty-four miles from Crow Wing	403
Crow Wing	428
St. Paul	558

Crow Wing is a small, new town, depending chiefly upon the pineries in its neighbourhood for support, as

* A new road has been completed, and all the rivers bridged as far as Graham's Point on Red River, opposite Fort Abercrombie. (1859.)

well as upon the prospect of a road between it and Superior City. Its position in relation to Lake Superior and the valley of Red River, is thought to be very favourable, and the people say that a plank road from Superior City to Crow Wing, which need not exceed one hundred miles in length, would secure the trade of the valley of Lake Winnipeg.

From Crow Wing we travelled by stage to St. Paul, thence by steamer down the Mississippi to Prairie du Chien. From this far western town there is direct communication, by rail, with Toronto, where we arrived on the 4th of November, three months and a half from the day of our departure on this exploration.

THE

ASSINNIBOINE & SASKATCHEWAN EXPLORING EXPEDITION

OF 1858

INTRODUCTION.

THE following instructions will explain the object of the Assinniboine and Saskatchewan exploring expedition.

No. I.

Secretary's Office, Toronto, 14th April, 1858.

SIR,—During the last week I communicated to you, verbally, instructions in reference to the proposed expedition to the neighbourhood of the Red River during the present year.

2. It has been decided, as you are aware, with a view to keep down as much as possible the expenditure this year, to dispense with the services of Mr. Gladman as its general manager.

3. The exploration party this year will consist of two divisions, one to be placed under your direction and control, and the other under the direction of Mr. Dawson.

4. His Excellency in Council has been pleased to place under your charge the topographical and geological portion of the exploration, respecting which full instructions will be given in another letter, while Mr. Dawson will continue to perform the same duties as last year, viz., those of surveyor, &c.

5. The estimate of the probable expenditure of the expedition, submitted by you on the 6th instant, was laid before his Excellency in Council, and has been approved of by them, and I have accordingly now to direct you to be guided as much as possible by that estimate in engaging your assistants, hiring your

men, as well as in the other necessary expenditures of the expedition.

6. It is hardly necessary to say that his Excellency relies upon your exercising a due economy in all matters connected with the expedition.

7. As soon as you have completed your contemplated party, you will furnish me with a schedule, giving the names of all the persons composing it, and stating their rates of pay, and the dates from which their pay is to commence. Such a schedule will be necessary to supply the auditor with the means of auditing your accounts.

8. Having organised your party, you will lose no time in repairing with them to Red River, taking with you the supplies (referred to in the estimate) required for Mr. Dawson.

9. On your way to the Red River, you will take possession of the canoes, provisions, and other articles belonging to the Government, either at Collingwood or Sault Ste. Marie. These, with the men intended for Mr. Dawson, you will deliver over to that gentleman when you meet him, either at Red River or on his way back.

10. You are to consider all the articles and materials of any description belonging to the Canadian Government, connected with the late expedition, as available for the purposes of the present expedition, and you and Mr. Dawson may therefore divide them between you in whatever way you may think most advantageous. Such articles, if any, as may not be required by either of you, should be left in the custody of some trustworthy person to await the orders of the Government.

11. As soon as you shall have put Mr. Dawson in possession of the men and canoes intended for him, each of you will be held separately responsible for the expenses of his own party. You will therefore be careful to keep an accurate account of your expenditure.

12. The auditor-general of public accounts will give you any information you may require as to the most convenient mode of making out and furnishing your accounts, &c.

13. On your return from Montreal I shall be prepared to give you your instructions with reference to the localities in which

your explorations are to be conducted, and as to the objects to which your attention is to be more especially directed.

I have the honour to be, Sir,

Your obedient servant,

T. J. J. Loranger,

Secretary.

Henry Y. Hind, Esq., Toronto.

No. II.

Secretary's Office, Toronto, 27th April, 1858.

Sir,—I have the honour to communicate to you the instructions promised in the last paragraph of my letter to you of the 14th instant, for your guidance in connection with the branch of the expedition to the west of Red River, which has been committed to your charge.

2. The instructions contained in that letter will suffice for your guidance up to the time of your arrival at the Red River settlement, and the present instructions therefore have reference merely to your operations after having left that settlement.

3. The region of country to which your explorations are to be then directed is that lying to the west of Lake Winnipeg and Red River, and embraced (or nearly so) between the river Saskatchewan and Assinniboine, as far west as "South Branch House," on the former river, which latter place will be the most westerly point of your exploration.

4. It will be your endeavour to procure all the information in your power respecting the Geology, Natural History, Topography, and Meteorology of the region above indicated.

5. As to the general character of the geological portion of your labours, it is unnecessary to add anything to the instructions

communicated to you last year, and which, so far as this point is concerned, will serve for your guidance for the present season.

6. There are, however, two matters to which I am to request you to direct your particular attention, namely, the salt region in the neighbourhood of Lake Manitobah, adverted to in your report for last year, and the deposit of Tertiary coal or lignite, reported to exist in the valley of Mouse River.

7. It is most important that you should ascertain, by actual examination, as far as possible, the existence, extent, and character of these deposits.

8. In ascending or descending the different rivers you may have occasion to explore, it is advisable that you should note with care, their breadth, depth, rate of current, and the probable quantity of water discharged by them at different points, and at different seasons of the year; their facilities for navigation by boats or steamers, and whether they overflow their banks to any extent at any season of the year.

9. The general aspect of the whole region should be carefully described. The character of the timber and soil observed, and the general fitness of the latter for agricultural purposes ascertained as far as may be from observation and inquiry.

10. It is desirable that your meteorological observations should be made with the maximum and minimum thermometer, and with the wet and dry bulb. The temperature of the rivers, lakes, and springs should also be recorded, and the rainfall observed.

Any reliable information you can obtain as to the quantity of snow precipitated during the winter, would also be of interest.

11. Your topographical explorations should be made with reference to the construction of a map (as complete as possible) of the region explored, on a scale of two miles to one inch—and your operations should be conducted in view of a possible extension, at some future time, of the exploration, so as to embrace the entire valley of Lake Winnipeg and its feeders.

12. With a view to illustrate the natural history of the country, you will avail yourself of such opportunities as may present themselves to collect any objects that may be useful for that purpose.

13. Any geological or natural history specimens which you

may have collected during your explorations, may be left by you at Red River, on your return, with the other property of the Government belonging to the expedition, to await the orders of the Government, with the other articles referred to in the tenth paragraph of my letter of the 14th inst.

14. I am to add that his Excellency, having every confidence in your judgment and discretion, does not wish to trammel you with more detailed instructions, and that you are left at liberty to make any other exploration, in addition to those particularly named therein, should you, upon information obtained in the locality, deem it desirable for the general purposes of the expedition.

15. It is hardly necessary to state that you will be held responsible for the conduct, diligence, and fidelity of the party under your charge.

16. With a view to distinguish your branch of the expedition for the present year, it will be convenient to designate it as the "Assinniboine and Saskatchewan Exploring Expedition;" by this title therefore you will describe it in your reports.

I have the honour to be, Sir,

Your obedient servant,

T. J. J. Loranger,
Secretary.

Henry Y. Hind, Esq., Toronto.

I am indebted to Mr. F. B. Meek, of Washington, U.S., for the description of the Cretaceous fossils collected during the exploration. Mr. Meek's remarks and descriptions are contained in Chapter XIX. of my Report on the expedition to the Canadian Government.

The Silurian and Devonian fossils were examined and described by Mr. Billings, the palæontologist to the Canadian Geological Survey. The descriptions and remarks of Mr. Billings are contained in Chapter XX. of the Report referred to in the preceding paragraph. I am glad to have another opportunity of expressing my obligations to these gentlemen.

I have also much pleasure in tendering my warmest thanks to Sir George Simpson, not only for the letters of introduction with which he favoured me to the officers of the Hon. Hudson's Bay Company's service in Rupert's Land, but also for his personal efforts when at Fort Garry, to facilitate the progress of the expedition by every means in his power. The assistance rendered by Sir George Simpson was of the greatest use to me, and the courteous manner in which it was granted increases my indebtedness to him.

From the officers of the Hon. Hudson's Bay Company's service in charge of the different posts, I received without any exception kind attention and valuable assistance; in the following pages many friendly actions are faithfully recorded.

CHAP. XIII.

FORT GARRY TO THE BOUNDARY LINE, VIÂ THE ASSINNIBOINE AND LITTLE SOURIS.

Members of the Expedition.—Iroquois Indians from Caughnawaga.—Detroit.—Sault Ste. Marie.—Grand Portage.—Fort Frances.—Red River.—Expedition into the Interior.—The Start.—Supplies.—Prairie Ridges.—Pigeon Traps.— Stony Mountain.—Birds.— Saline Efflorescence.—Character of the Big Ridge.—The Assinniboine.—Grasshoppers.— Ojibway Encampment.—Archdeacon Cochrane.—Prairie Portage.—Cliff Swallow.—Thunder Storms.—Ojibways.—The Bad Woods.—Assinniboine Forest.—River.—Rabbits.—Sandy Hills of the Assinniboine.—Latitude.—Dimensions of Valley. — Variation of Compass. — Sand Dunes. — Aspect of Country.—Hail Storm.—"Smokes."— Balsam Spruce.— Pine Creek.—The Little Souris.—Grasshoppers.—Fish.—Sioux.—Cretaceous Rocks.—Blue Hills.— Pembina River.— Backfat Lakes.— Vast Prairie.— Prairie Fires.—Horizontal Rocks.—Inoceramus.— Guelder Rose.—Lignite.—Ancient Lake Beaches.—Sand Dunes.—Oak Lake.—Souris Sand Hills.—Night Hawk.—Bog Iron Ore.—Floods in 1852.—Bois de Vache.—Grasshoppers, infinite Multitude of.—Appearance of the Sky, of Prairie.—Little Souris Valley.—Tracks.—Turtle Mountain.—Sioux.—Character of Prairie.—Souris Lakes.— Boulders.—Mandan Village.—Probable Source of the Little Souris.—Character of the Souris south of the 49th Parallel.—Absence of Timber on the 47th Parallel.

THE gentlemen whose names are subjoined composed the expedition which started from Toronto for Red River on the 29th April, 1858.

HENRY YOULE HIND, M.A., in charge of the Expedition.
JAMES AUSTEN DICKINSON, B.A., Surveyor and Engineer.
JOHN FLEMING, Assistant Surveyor and Draughtsman.
HUMPHREY LLOYD HIME, Photographer.

The following Iroquois Indians were engaged at Caughnawaga near Montreal to man the canoes:—

Charles Skanasati, guide.
Martin Takatsitsienseré.
Louis Pekageiaien.
Ignace Tior-ateken.
Lazare Aneratentka.
Mathias Shaitikarenes.
Thomas Orite.
Louis Atioksisaks.
Thomas Shakashetstha.
Mathias Asinrathon.
Ignace Taseraren.
Thomas Tekarenhonte.
Pierre Aronhiakenra.

Fortunately the steamer "Illinois" from Detroit to Lake Superior Ports had some freight for Grand Portage, the Lake Superior termination of the Pigeon River route to Red River. By this long-neglected communication I determined to go. Having secured a passage to Grand Portage direct, and embarked our baggage and stores, which weighed nearly six thousand pounds, we left Detroit on the 30th April, and reached the Sault Ste. Marie on the 2nd of May.

A week before leaving Toronto I wrote to Mr. Simpson, the officer in charge of the Hudson's Bay Company's post at Sault Ste. Marie, enclosing a copy of a letter from Sir George Simpson, and requesting him to send two new North canoes, fully appointed, to the United States side of the St. Mary's river, in order that their embarkation might occasion as little delay as possible.

The canoes were lying in readiness on the wharf at the Sault together with a supply of gum, watap *, cod-lines, and other indispensables to canoe navigation.

* The root of the tamarac, used for sewing together the pieces of birch bark of which the canoes are constructed.

We arrived at the Grand Portage on the 5th of May and immediately made preparations for crossing the Portage. I found it necessary to engage three more voyageurs to man a third canoe which the large quantity of baggage rendered necessary; these were Wigwam, an Ojibway half-breed; Charles Louis, a French Canadian, and François Chabot, also of French Canadian origin.

We started from the west end of the Grand Portage on the 9th May, reached Fort Frances on the 24th, and the mouth of the Winnipeg on the 29th.*

We camped off the mouth of Red River seven days after leaving Fort Frances, and might easily have reached the settlements on the first day of June, but in view of our rapid voyage from Rainy Lake, and being in advance of Sir George Simpson, I did not think it necessary to press the guide, we therefore waited for a few hours at Fort Alexander, and enjoyed the hospitality of Mr. Sinclair, the gentleman in charge.

The exact time the expedition spent in canoes between Lake Superior and Red River, after deducting the delays at the forts before mentioned, was twenty-one days and six hours, as opposed to twenty-seven days and six hours by the Kaministiquia route the year before. The average daily progress was twenty-eight and a half miles against twenty-five miles in 1857.

On my arrival at the Middle Settlement, where Mr. Dawson and his party had their quarters, I found Mr. Russell in charge of the house and effects, Mr. Dawson with other members of his party having started some days previously for the Saskatchewan, whence they were not expected to return until the end of June. I therefore placed Mr. Russell in possession of the canoes and men intended for

* For an estimate of the cost of opening the Pigeon River Route, see Appendix, Vol. II.

Mr. Dawson, and immediately commenced to organize a party to explore the country drained by the Assinniboine and Little Souris.

On the morning of the 14th June, the half-breeds engaged for the expedition into the prairie country west of Red River, assembled at our temporary quarters and began to load five Red River carts and a waggon of American manufacture, with two canoes, camp equipage, instruments, and provisions for a three months' journey. At noon the start was made, and the train proceeded to Fort Garry*, a distance of eight miles, to take in a supply of flour and pemmican. We camped about half a mile from the fort and took an inventory of our baggage, and made such regulations and arrangements as are considered necessary at the commencement of a long journey through a country partly inhabited by hostile tribes of Indians, and not always affording a supply of food even to skilled hunters.

The party was composed of the gentlemen already named, six Cree half-breeds, a native of Red River of Scotch descent, one Blackfoot half-breed, one Ojibway half-breed, and one French Canadian. Our provisions consisted of one thousand pounds of flour, four hundred pounds of pemmican, one thousand rations of Crimean vegetables, a sheep, three hams, and a supply of tea for three months, with a few luxuries, such as pickles, chocolate, a gallon of port wine, and a gallon of brandy. Each cart was loaded with about 450 lbs. and the waggon with double that amount. The birch bark canoes were 18 feet

* "The mean of five observations at Upper Fort Garry, at the mouth of the Assinniboine, for latitude, three meridian by altitude of the sun and two by Polaris, gave for the latitude 49° 53′ 24″. Mr. Calhoun, who was attached to Major Long's expedition in 1823, made it 49° 53′ 35″, but according to a record in the possession of one of the officers of the fort, Lefroy placed it in latitude 49° 58′." Owen's "Geological Survey of Wisconsin, Iowa, and Minnesota," p. 180. Capt. Palliser places Fort Garry in latitude 49° 52′ 6″. Longitude, 96° 52′ 27″.

long, and weighed 150 lbs. each. At the White Horse Plain, twenty-two miles from Fort Garry, I procured an additional cart, and purchased an ox to serve as a *dernier ressort* in case we should not meet with buffalo; and at Prairie Portage, the last settlement on the Assinniboine, I engaged the services of an old hunter of Cree origin, who had been from his youth familiar with Indian habits and stratagems. This addition increased the party and equipment, before we left the last settlement, to fifteen men, fifteen horses, six Red River carts, one waggon, and one ox.

Leaving our camp early on the morning of the fifteenth, we ascertained by leveling the altitude of an ancient lake ridge, near to St. James's Church*, to be eleven feet above the prairie at Fort Garry, and about two miles from it. These ridges are common in the prairies of Red River, and do not necessarily point to an ancient lake margin, as it is probable that most of them were formed under water. They may be traced for many miles, but are sometimes lost in the general rise of the prairie, or blend with higher ridges.

On arriving at St. James's Church we separated into two divisions, Mr. Fleming and Mr. Hime, with the carts and waggon, proceeding to Lane's Post on the Assinniboine, twenty-four miles from Fort Garry, while Mr. Dickinson and myself with two half-breeds, struck in a north-westerly direction across the prairie to Stony Mountain, and thence to the Big Ridge, having arranged to meet at Prairie Portage.

In a wheat field opposite St. James's Church were several pigeon traps, constructed of nets 20 feet long by 15 broad, stretched upon a frame; one side was propped up

* The Rev. W. H. Taylor's Mission. The income of this Mission is derived from the following sources: £100 from the Society for the Propagation of the Gospel, £100 from the Bishop of Rupert's Land.

by a pole 8 feet long, so that when the birds passed under the net to pick up the grain strewed beneath, a man or a boy concealed by a fence or bush, withdrew the prop by means of a string attached to it, and sometimes succeeded in entrapping a score or more of pigeons at one fall. Near the net some dead trees are placed for the pigeons to perch on, and sometimes stuffed birds are used as decoys to attract passing flocks.

In pursuing our course to Stony Mountain we endeavoured to follow the ridge before alluded to, but after tracing it for several miles it became imperceptibly blended with the level prairie. Several ridges were crossed after we lost the first, but in all instances they died away after having preserved their rounded form for two or three miles. Stony Mountain is a limestone island of Silurian age,

Stony Mountain.

having escaped the denuding forces which excavated the Red River valley. It is about four miles in circumference, its highest point is sixty feet above the prairie level. Horizontal layers of limestone, holding very few and obscure fossils, project on its western cliff-like sides. Its eastern exposure is gently sloping, and some ten feet from the summit, the remains of an ancient lake beach are well preserved. Viewed from a distance, Stony Mountain requires little effort of the imagination to recall the time when the shallow waters of a former extension of Lake Winnipeg, washed the beach on its flank, or threw up as they gradually receded, ridge after ridge, over the level floor of the lake, where now are to be found wide and beautiful prairies covered with a rich profusion of long grass.

Leaving Stony Mountain our course lay westerly through a wet prairie to the Big Ridge. Grey cranes, ducks, and plover were numerous on the marshy areas, and in every little bluff * of aspen or willow, the beautiful rice birds (*Dolichonyx oryzivora*) were seen or heard. Where we camped on the edge of a lake near the foot of the Big Ridge, bittern, grackle, and several varieties of duck flew to and fro in alarm at our invasion of their retreats. On the flank of the Big Ridge the cinnamon thrush (*Turdus rufus*) was noticed, but most common of all was the tyrant flycatcher (*Muscicapa tyrannus*), who endeavoured to hold undisputed sway over the bluff he had selected as his home. Near and west of Stony Mountain, many small barren areas occur, covered with a saline efflorescence; they may be traced to the Assinniboine, and beyond that river in a direction nearly due south to La Rivière Sale, and the 49th parallel. These saline deposits are important, as they in all probability serve, as will be shown hereafter, to denote the presence of salt bearing rocks beneath them, similar to those from which the salt springs of Swan River, Manitobah Lake, and La Rivière Sale issue.

Early on the morning of the 17th we ascended the Big Ridge. Its elevation above the prairie is about 60 feet; on its south side it slopes gently to the prairie level, to the north is a plateau, well wooded with aspens, stretching

* The half-breeds call little groves of aspens or willows in the prairies "bluffs."

towards Lake Manitobah; the view from its summit extends far and wide over the Assinniboine prairies; and skirting its base on the south flank are groves of aspen and balsam poplar, with scattered oak trees and willow bushes. The pasturage in the open glades is of excellent quality and very abundant. The ridge is quite level and from 80 to 100 feet broad, devoid of trees, slightly arched and composed of gravel. Here and there it is cut by rivulets draining the marshes in the plateau on its northern side. As it approaches Prairie Portage its apparent elevation diminishes, until at the Portage River it is no longer discernible. We traced it for a distance of seventy miles, and it will be mentioned further on that a similar ridge, but one formed at an earlier period and at a higher level, is seen west of Manitobah Lake, near the Hudson's Bay Company's post, Manitobah House. The older ridge preserves there the same characters of horizontality, uniform outline, gravelly formation, and admirable adaptation to the purposes of a road, which have been already noticed in connection with the Big Ridge north of the Assinniboine and east of Red River. For many miles, ties for a railway might be laid upon both without a pebble being removed, and the only breaks in their continuity occur where streams from the plateau and higher grounds in the rear have forced a passage through them. The older ridge, however, follows the western contour of Lakes Winnipeg and Manitobah, and passes through a country not likely to be first selected by a large body of settlers. The Big Ridge is important in so far that it forms the boundary of land of the first quality, which occupies the low prairie valleys of the Assinniboine and Red River. Soundings subsequently made in Lake Manitobah showed a uniform depth of eighteen feet for a distance exceeding sixty miles along its south-eastern coast, so that if its beds were exposed, it is probable that in process of time it would also become a rich and extensive prairie country, with its present beach distinctly visible as its old boundary. Indeed, the surface of the country between the Big Ridge, the Assinniboine, and Red River, is similar in gently undulating outline to the succession of undrained marshes, ridges, and bogs which exist between the west coast of Lake Manitobah and the older ridge, pointing to a very gradual but constant drainage of this region after a long period of submergence.

We reached Prairie Portage in the evening, where we joined the main party. The Assinniboine at Lane's Post (June 16th) is about 120 feet broad, and its turbid water flows at the rate of 1½ miles per hour. A few miles west of Lane's Post, the saline efflorescence, before noticed as occurring in patches on the prairies and forming small barren areas, is no more to be seen; it consists of chloride of sodium and sulphate of magnesia, with a little chloride of calcium. The first grasshoppers were observed this year at Lane's Post; they were a brood from the eggs deposited by a swarm which alighted on the White Horse Plain in September last.

At Prairie Portage we found an Ojibway encampment in which were some of the refractory personages who had hitherto resisted the humane and unceasing efforts of Archdeacon Cochrane to christianize them. Among the various methods tried by the archdeacon to induce these wanderers to settle and farm,— the first preliminary to the progress of Christianity among wild Indians,—that of presenting the most docile with an ox and plough and teaching them to use it, was the least successful. At the first good opportunity, or during a time of scarcity, the ox and plough would be sold to the highest bidder for very much less than it cost. A promise to add another ox at the end of a year if the first gift was faithfully preserved was of no avail, the charms of the buffalo plains were too tempting, or the seduction of gambling too powerful to be withstood, notwithstanding the most solemn heathen promises. The school, however, gives better hope, and no doubt the rising generation of Indian and half-breed origin at Prairie Portage, will form a thriving, industrious, and Christian community.

Prairie Portage is very delightfully situated sixty miles west of Fort Garry on the banks of the Assinniboine. The prairie here is of the richest description, and towards the north and east boundless to the eye. The river bank is fringed with a narrow belt of fine oak, elm, ash, and ash-leaved maple, but on the south side a forest extends from two to four miles in depth, and then passes into aspen groves; the river abounds in sturgeon and gold eyes, and within eighteen miles there is an excellent fishing station on the coast of Lake Manitobah, where the Portage people take vast numbers of white fish every fall. The old water course of the Assinniboine, near Prairie Portage, is now a long narrow lake, fringed with tall reeds, a favourite haunt of wild fowl and grackle, among which we observed the showy yellow-headed blackbird (*Agelaius xanthocephalus*).

Prairie Portage will eventually become an important settlement, not only on account of the vast extent of fertile country which surrounds it, but because it lies in the track of the buffalo hunters proceeding to the Grand Coteau and the South Branch by way of the Souris River. It is also near to the fertile region drained by White Mud River, and the road to the south-western flanks of the Riding Mountain, passes by the Portage. The current of the Assinniboine is very uniform here, careful leveling showed that it fell one foot two inches in a mile, with a velocity of two miles an hour where the trial was made, but in the middle of the stream the velocity must be considerably greater. The cliff swallow (*Hirundo fulva*) had built its nests in great numbers on the banks of the river, which are about sixteen feet above the level of the water. I counted no less than thirteen groups of nests within a distance of five miles, when drifting down in a canoe. The cliff swallow was afterwards seen in great multitudes on the Little Souris, the South Branch of the Saskatchewan, and the Qu'appelle River.

The first of a series of thunder storms which lasted for some weeks visited us on the 17th; the warm rain fell in torrents and thoroughly wetted all who were exposed. Pigeons were flying in vast numbers across the Assinniboine, and the black tern (*Sterna nigra*) was numerous in the prairies near the settlement. In descending the river for a few miles to inspect its banks, we had occasion to pass by a fish weir, where several Ojibways from the camp near the Portage were watching for sturgeon with spears in their hands. They took no notice of us as we passed, being too busily engaged, but on our return to the encampment we found them waiting with fish to barter for tobacco and tea. We made them a few trifling presents, and, by way of recompense, sustained during the night the loss of a cheese, which after curiously eyeing during supper, they modestly asked for permission to taste. They pronounced it excellent, and in the dead of night (when our half-breeds were sound asleep) they opened the basket in which it had been placed, and quietly abstracted it. In future, when Indians were around, all eatables and articles they might covet were properly secured, and the cheese proved to be our only loss during the exploration.

Leaving Prairie Portage on the morning of the 19th, we took the trail leading to the Bad Woods, a name given to a wooded district about thirty miles long, by the buffalo hunters in 1852, who, in consequence of the floods of that year, could not pass to their crossing place at the Grand Rapids of the Assinniboine by the Plain or Prairie Road. There were four hundred carts in the band, and the hunters were compelled to cut a road through the forest of small aspens which forms the Bad Woods, to enable them to reach the high prairies. This labour occupied them several days, and will be long remembered in the settlements in consequence of the misery entailed by the delay on the children and women.

The trail continued for three miles through a continuation of the low prairies of the Assinniboine, when a sudden ascent of 20 to 25 feet introduced us to a different kind of country, the plateau beyond the Big Ridge, which here crosses the river, and forms the lowest or first step of the Pembina Mountain. The physical features of this boundary to a great table-land will be noticed at length in the sequel. The soil continues poor and sandy for several miles, supporting clumps of aspen interspersed with a few oaks in low places. The view across the Assinniboine reveals in the distance the Blue Hills, and between them and the river is a vast forest, which a subsequent exploration in the autumn showed to consist of oak, elm, ash and aspens, for two to three miles nearest to the Assinniboine; but beyond this limit the forest is almost entirely composed of aspens of small growth.

Grasshoppers were now seen in great numbers, and the first humming-bird was noticed here. The banks of the river showed recent watermarks twelve feet above its present level, willow and other trees overhanging the stream being barked by the action of ice during spring freshets at that elevation. Everywhere rabbits were numerous, and considerable areas occur covered with dead

HALF-WAY BANK, ASSINNIBOINE RIVER

willows and aspens, barked by these animals in the winter, about two feet six inches above the ground. The height of the bank twenty-two miles from Prairie Portage is about 80 feet above the river, denoting a rapid rise in the general level of the country.

On the morning of the 20th we entered the Bad Woods and followed the road cut by the hunters in 1852. The aspens were much disfigured by countless numbers of caterpillars resembling those of the destructive Palmer worm. In the afternoon we arrived at the Sandy Hills; they consist of rounded knolls covered with scrub oak and aspens. Our latitude to-day at the Half Way Bank was ascertained to be 49° 46′ 19″, the height of the prairie 150 feet above the river, the breadth of the valley in which the river flowed 5,680 feet, and the variation of the compass 13° E. After passing the point where the foregoing observations were made, the trail again enters the Bad Woods and continues through them until it strikes the Sandy Hills again. These rounded eminences have all the appearance of sand dunes covered with short grass and very stunted vegetation.

As we emerged from the Bad Woods a noble elk trotted to the top of a hillock and surveyed the surrounding country; a slight breath soon carried our wind as the hunter was endeavouring to approach him, he raised his head, snuffed the air and bounded off. Another terrible thunderstorm came on at sunset, with heavy rain and boisterous wind. The aspect of the country for many miles is that of a plain sloping gently to the east, and studded with innumerable mounds or hillocks of sand, thinly covered with a poor and scanty growth of grass; here and there small lakes or ponds occur fringed with rich verdure, but its general character is that of sterility. From the summit of an imposing sand-hill, formerly a drifting dune, which we ascended on the 21st, the country lay mapped at our feet; as far as the eye could reach, sand-hills, north, east, and west, sometimes bare and ripple-marked, but generally covered with short grass, were exposed to view.

On the afternoon of this day a hail storm of unusual violence caused us to halt. The stones penetrated the bark of our canoes and broke off the gum. The grass-hoppers, which were very numerous just before the storm began, suddenly disappeared, but they might be found quietly clinging to the leaves of grass in anticipation of the storm. After it had passed they re-appeared, apparently in undiminished numbers, although every member of the party, crouching for shelter under the carts and waggon, fully expected the complete annihilation of these destructive and troublesome insects. A wonderful instinct enables them to seek and find refuge from a pitiless hailstorm or a drenching rain. The same evening a thunderstorm of short duration again arrested our progress, but the sun set in gorgeous magnificence, with a brilliant rainbow and vivid flashes of lightning in the east. The cinnamon thrush is not uncommon among the sandy hills; we saw several during the day. The following morning we reached the "pines," for which we had been anxiously looking, but to our disappointment they proved to be nothing more than balsam spruce in scattered clumps.

The heat of the weather began now to be very oppressive, and joined to the incessant attacks of mosquitoes and "bulldogs" proved very exhausting to the horses. At each camping-place we were compelled to make "smokes" to drive away these tormentors, otherwise the persecuted animals would endeavour to approach our camp fires, creating no small confusion in our cooking arrangements. On the 23rd, we traversed a region of sand-hills and ridges, until we arrived at Pine Creek, a distance of eight miles from the preceding camp. Here the sand-hills are absolutely bare, and in fact drifting dunes. Sending the main party in advance, Mr. Dickinson and I set out to examine the valley of the Assinniboine where Pine Creek disembogues. The sand dunes were seen reposing on the prairie level about 170 feet above the river. In crossing the country to regain the carts, our course lay across a broad area of drifting sand beautifully ripple-marked, with here and there numbers of the bleached bones of buffalo protruding from the west side of the dunes, memorials probably of former scenes of slaughter in buffalo pounds, similar to those which we witnessed some weeks afterwards at the Sandy Hills on the South Branch of the Saskatchewan. The progress of the dunes is very marked; old hillocks partially covered with herbage are gradually drifted by the prevailing westerly wind to form new ones. Sometimes the area of pure sand was a mile across, but generally not more than half that distance. The largest expanse we saw was near the mouth of Pine Creek, it is called by the Indians "the Devil's Hills," and a more dreary, parched-looking region could scarcely be imagined.

Opposite the Souris the subjoined section of the north bank of the Assinniboine was measured:—

1. 22 feet pale yellow sand.
2. 1 foot dark blue clay.
3. 12 feet ferruginous gravel and sand.
4. 4 inches dark green hard gravelly clay.
5. 15 feet soft unctuous blue clay to the water's edge.

It is very probable that, of this section, the lowest number represents the weathered shales which were afterwards found exposed a few miles up the Little Souris valley.

We reached the mouth of the Little Souris River on the 24th, and made preparations to cross the Assinniboine at this point. The distance traveled through the Sandy Hills was about forty-eight miles, their breadth does not exceed ten miles. At the mouth of the Souris the grasshoppers were in countless numbers, and so voracious as to attack and destroy every article of clothing left for a few minutes on the grass. Saddles, girths, leather bags, and clothing of every description were devoured without distinction. Ten minutes sufficed them, as our half-breeds found to their cost, to destroy three pairs of woollen trowsers which had been carelessly thrown on the grass. The only way to protect our property from the depredators was to pile it on the waggon and carts out of reach. There were two distinct broods of grasshoppers, one with wings not yet formed, which had been hatched on the spot, the other full-grown, invaders from the prairies south of the Assinniboine. We saw here one of the vast flights of these insects which afterwards were witnessed on a scale of alarming magnitude, giving rise in their passage through the air to optical phenomena of very rare and beautiful descriptions. As we cautiously approached the bank of the river opposite the mouth of the Little Souris, on the look-out for Sioux Indians, some jumping deer and a female elk were observed gamboling in the river. A shot from a Minie rifle dispersed them and started from their lair two wolves who were watching the deer, patiently waiting for an opportunity to surprise them.

The volume of water in both rivers was carefully measured at their point of junction. The Assinniboine was found to be 230 feet broad, with a mean depth of 6 feet, and a current of one mile and a quarter per hour. The Little Souris was 121 feet broad, 2 feet 4

CONFLUENCE OF THE LITTLE SOURIS AND THE ASSINNIBOINE

inches in mean depth, and flowing at the rate of half a mile an hour. Observing numbers of fish rising at grasshoppers in the Souris, we stretched a gill net across the mouth of the river, and succeeded in taking pickerel, gold-eyes, and suckers, the grey and the red. In a second attempt we caught a tartar; a huge sturgeon got entangled in the meshes of the gill net, and before we could land him he succeeded in breaking away and carrying a portion of the net with him.

Signs of Sioux Indians in the neighbourhood led to our keeping watch during the night; and on the morning of the 25th we proceeded cautiously up the valley of the Souris, keeping a sharp look out. On the left bank the Blue Hills of the Souris are visible ten miles from the mouth of the stream, and towards the west, the Moose Head Mountain is seen to approach the Grand Rapids of the Assinniboine. The first rock exposure in the valley was observed about fifteen miles from the mouth of the Souris. It consisted of a very fissile, dark-blue argillaceous shale, holding numerous concretions containing a large percentage of iron, partly in the state of carbonate and partly as the peroxide. Some very obscure fossils were found, with fragments of a large inoceramus. The shale weathers ash white. It is exposed in a cliff about ninety feet high; the upper portion of the cliff consists of yellow sand, superimposed by sandy loam holding limestone boulders and pebbles; the exposure of shale is seventy feet thick, in horizontal layers. The country west of the Souris, so far, is an open, treeless, and undulating prairie. On the east side, the Blue Hills are very picturesque, their flanks and summits are wooded with aspen. Rain as usual, the day closing with a thunderstorm.

On the 26th we arrived at the westerly bend of the

Souris in the midst of a very lovely, undulating country; the river is here fifty feet broad, and in its passage through the Blue Hills it has excavated a valley fully four hundred and fifty feet deep. Rock exposures are of frequent occurrence, the dip being 3° south. Fragments and perfect forms, but very fragile, of a large inoceramus are common. The ferruginous concretions continue disposed in regular layers, and appear to be a characteristic

Valley of the Souris, looking towards the Blue Hills of the Souris.

feature of the cretaceous rocks in this valley. A continuation of the valley of the Souris extends in a direction nearly south-east towards Pembina River, with which it is said by the half-breeds to interlock. Three lakes visible from our camp were stated to be the sources of Pembina River; a little stream issuing from the most westerly of these is called Backfat rivulet, it flows into the Souris. Southwards, Turtle Mountain shows with a faint blue outline on the horizon.

Deer are very numerous at this beautiful bend of the

river; it appears to be a favourite watering-place. The half-breeds of St. Joseph often make it their crossing-place when on a hunting expedition to the Grand Coteau. It is not improbable that it will become a point of importance if ever an emigrant route should be established from Minnesota to the Pacific, *viâ* the south branch of the Saskatchewan; and from the great distance saved by going through St. Joseph instead of Fort Garry, it is not improbable that this may yet be the case.

Valley of the Souris, opposite the Valley of the Backfat Lakes.

On the 27th we succeeded in passing the Blue Hills, and enjoyed on the evening of the same day one of the most sublime and grand spectacles of its kind which it is possible to witness. Before leaving the last ridge of the Blue Hills, we came suddenly upon the borders of a boundless level prairie on the opposite side of the river, one hundred and fifty feet below us, of a rich, dark-green colour, without a tree or shrub to vary its uniform level and yet with one conical hill apparently in its centre. Here we expected to find buffalo, but not a sign of any living creature could be detected with the aid of a good glass. The prairie had been burnt last autumn, and the Buffalo had not arrived from the south or west to people this beautiful level waste. What a magnificent spectacle this vast prairie must have furnished when the fire ran over it before the strong west wind!

From beyond the South Branch of the Saskatchewan to Red River all the prairies were burned last autumn, a vast conflagration extended for one thousand miles in length and several hundreds in breadth. The dry season had so withered the grass that the whole country of the Saskatchewan was in flames. The Rev. Henry Budd, a native missionary at the Nepowewin, on the North Branch of the Saskatchewan, told me that in whatever direction be turned in September last, the country seemed to be in a blaze; we traced the fire from the 49th parallel to the 53rd, and from the 98th to the 108th degree of longitude. It extended, no doubt, to the Rocky Mountains.

A few miles west of the Blue Hills, being anxious to ascertain the dip of a very remarkable exposure of shale with bands of ferruginous concretions, Mr. Dickinson leveled with the utmost care an exposure facing the south, and found it to be horizontal. At the base of the exposure, and on a level with the water's edge we succeeded in finding a layer of rock full of gigantic inoceramus. One specimen measured 8½ inches in diameter, it was very fragile but the peculiar prismatic structure was remarkably well preserved. On attempting to raise it, it separated into thousands of minute prisms so characteristic of this shell.

Vast numbers of pigeons were flying in a north-westerly direction, and our friends the grasshoppers were everywhere abundant. From the Blue Hills to the south bend of the river, rock exposures possessing the characteristics already noticed, occurred at every bend. The first speci-

VALLEY OF THE SOURIS

men of lignite was seen near the mouth of Plum Creek, where we camped on the 29th; it was a water-worn rounded boulder.

After leaving the Blue Hills no trees or shrubs of any description were seen until we arrived at Plum Creek. On low points in the Souris valley some fine oak, elm, balsam poplar, and aspen are found for the first twenty miles. The guelder rose is common on the ravines, wild prairie roses abundant, snowberry and two varieties of cherry of frequent occurrence, as well as woodbine, wild convolvulus, and hop, but for a distance of twelve miles west of the Blue Hills the country is treeless on both sides of the river, and the drift of small depth.

A little beyond Plum or Snake Creek we found numerous pebbles and boulders of lignite; and with a view to ascertain whether the lignite existed *in situ*, an excavation was made in the bank of the river and the stratification for a depth of twenty-five feet exposed. The last outcrop of the cretaceous shales was observed about three miles east of the bank where this trial was made. A few hours' labour revealed five old beaches, probably of a former lake. These beaches were composed of sand and boulders of lignite from the size of a hen's egg to one foot in diameter. No fragment of lignite was found which did not possess a rounded or spheroidal form and a roughly polished or worn surface. An abundant supply was easily obtained for a fire which was soon made on the bank; a strong sulphurous odour was emitted from the iron pyrites in the lignite, and some boulders when broken open exhibited streaks and small particles of a resinous substance like amber.

The excavation exposed the section which is shown in the woodcut on the next page.

The low hills about Plum Creek are sand dunes, and on

their sides an opuntia is very common. The prairie on the west of the Souris as well as on the east is treeless, but the banks of Plum Creek support a thin belt of small forest trees, such as oak and ash, with a few ash-leaved maple. The annual fires prevent the willows and aspens from covering the country, which they would undoubtedly do until replaced by other species, if not destroyed to within a few inches of the root every time the fire sweeps over them. The banks of the Souris here are not more than 40 feet high, with level prairies on either hand, a few miles beyond the Snake Hills. Within four miles of the mouth of Plum Creek, Oak Lake, several miles in diameter, attracted the hunting portion of our party; they brought back some pelicans and a score of duck. Thunder storms as usual to day and yesterday.

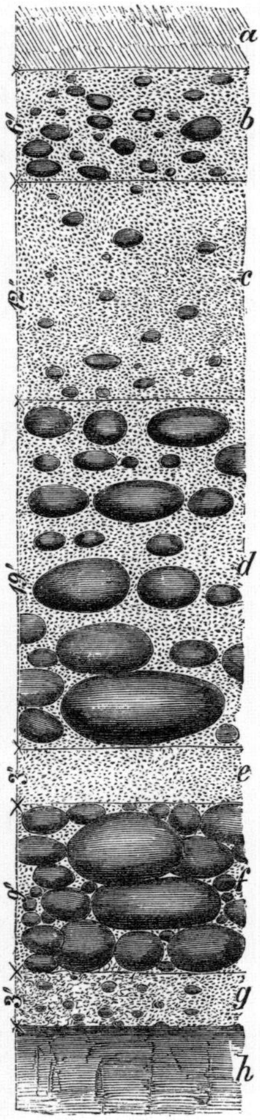

Section on the Little Souris, showing ancient Beaches (*b, c, d, f, g*) with Lignite Boulders.

On the 1st July we arrived at the Souris Sand Hills, and made a section of the river bank where a land slip had produced a fine exposure to the water's edge. The formation consisted of five feet of blue clay above the level of the river, supporting four feet of ferruginous sand and gravel, on which reposed twelve feet of sandy loam and sand to the prairie level. The blue clay, capped by the ferruginous sand, was traced for a distance of $2\frac{1}{2}$ miles, and showed a dip to the south of two feet in the mile, the clay disappearing beneath the water. No organic remains of any description were found, although a careful search was made, but boulders of lignite from 6 to 9 inches in diameter, were frequently seen in the bed of the river. The eggs of the nighthawk (*Chordeiles Virginianus*) were several times found on the bare ground, among the Sand Hills, with no approach to a nest for the helpless young. The parent birds endeavoured to draw us away from their eggs, fluttering as if wounded a short distance from them and uttering cries of distress. The Hudson's Bay Company have a post on the river among the Sand Hills, which is maintained only in winter, during the absence of the Sioux: these savage barbarians being altogether opposed to the approach of civilization near to their hunting-grounds, and entertaining besides a feeling of deadly hostility to the Red River half-breeds.

Near the Company's house we found on the river bank an extensive deposit of bog iron ore, capped by shell marl, and above the marl drifted sand. The banks of the river are here not more than 25 feet high, and on the east side there is a narrow fringe of fine timber. The Bois de Vache (dried buffalo dung) is distributed very abundantly in the prairie and through the Sand Hills and near to the company's post. In fact the buffalo were very numerous during the whole of the winter of 1856 and spring of 1857 on the banks of the Souris, but the great fires during the autumn of last year, have driven them south and north-west, and between the two branches of the Saskatchewan. The country becomes very low after passing the last sand-hills, and over a large extent of prairie south of them, drift timber is distributed, showing the extraordinary rise in the waters of the river during the floods of 1852.

On the low banks which are constantly wearing away, and revealing fresh surfaces at points or bends, some well stratified layers of a brown-coloured deposit, two feet thick, appear about four feet from the present prairie level. They resemble stratified bands of certain varieties of bog iron ore, but a very slight examination is sufficient to show that they are of vegetable origin. They consist of a series of accumulations of buffalo dung, collected doubtless by rains and floods from a considerable area, and deposited in the low part of the Souris valley, which lies to the south of the Sand Hills. A geological formation composed of bois de vache is a novelty not likely to be met with anywhere but on the continent of North America.

On the 2nd of July we observed the grasshoppers in full flight towards the north, the air as far as the eye could penetrate appeared to be filled with them. They commenced their flight about nine in the morning, and continued until half-past three or four o'clock in the afternoon. About that hour they settled around us in countless multitudes, and immediately clung to the leaves of grass and rested after their journey. On subsequent days, when crossing the great prairie from Red Deer's Head River to Fort Ellice, the hosts of grasshoppers were beyond all calculation; they appeared to be infinite in number. Early in the morning they fed upon the prairie grass, being always found most numerous in low, wet places where the grass was long. As soon as the sun had evaporated the dew, they took short flights, and as the hour of nine approached, cloud after cloud would rise from the prairie and pursue their flight in the direction of the wind, which was generally S.S.W. The number in the air seemed to be greatest about noon, and at times they appeared in such infinite swarms as to lessen perceptibly the light of the sun. The whole horizon wore an unearthly ashen hue from the light reflected by their transparent wings. The air was filled as with flakes of snow, and time after time clouds of these insects forming a dense body casting a glimmering silvery light, flew swiftly towards the north-north-east, at altitudes varying from 500 to 1000 feet and upwards.

Some idea of the height of the flight of these insects may be gathered from the opportunity enjoyed by Mr. E. James, who, when standing upon the summit of a peak of the Rocky Mountains, 8500 feet above the level of the Plains in Nebraska territory (14,500 above the sea), saw them above his head as far as their size would render them visible."*

Lying on my back and looking upwards as near to the sun as the light would permit, I saw the sky continually changing colour from blue to silver white, ash grey and lead colour, according to the numbers in the passing clouds of insects. Opposite to the sun the prevailing hue was a silver white, perceptibly flashing. On one occasion the whole heavens towards the south-east and west appeared to radiate a soft grey tinted light with a quivering motion, and the day being calm, the hum produced by the vibration of so many millions of wings was quite indescribable, and more resembled the noise popularly termed "a ringing in one's ears," than any other sound. The aspect of the heavens during the greatest flight we observed was singularly striking. It produced a feeling of uneasiness, amazement, and awe in our minds, as if some terrible, unforeseen calamity were about to happen. It recalled more vividly than words could express the

* Explorations in Nebraska and Dakotah. Preliminary Report of Lieut. G. K. Warren, Top. Eng. U. S. Army, 1858.

devastating ravages of the Egyptian scourges, as it seemed to bring us face to face with one of the most striking and wonderful exhibitions of Almighty power in the creation and sustenance of this infinite army of insects.

In the evening, when the grasshoppers were resting from their long journeys, or in the morning, when feeding on the grass leaves, they rose in clouds around us as we marched through the prairie; if a strong wind blew they became very troublesome, flying with force against our faces, in the nostrils and eyes of the horses, and filling every crevice in the carts. But fortunately, comparatively few flew on a windy day, otherwise it would have been almost impossible to have made headway against such an infinite host in rapid motion before the wind, although composed individually of such insignificant members.

Those portions of the prairie which had been visited by the grasshoppers wore a curious appearance; the grass was cut uniformly to one inch from the ground, and the whole surface was covered with the small, round, green exuviæ of these destructive invaders.

The valley of the Souris, along which we travelled during the day, varies from one quarter to one mile broad; the river is not more than twenty-five feet across, and very shallow. It flows through a rich open meadow, twenty to thirty-five feet below the general level of the prairie, which on either hand is undulating, treeless, covered with short stunted grass, and showing abundance of last year's bois de vache. The first fresh buffalo tracks were seen to-day, and while taking observations for latitude, tracks of a different character and greater significance were discovered by one of the half-breeds; the fresh print of horses' feet, pronounced to be a few hours old, denoting the presence of Sioux or Assinniboines in our neighbourhood.

Before reaching the 49th parallel, the Souris meanders for several miles through a treeless valley, about a mile broad and sixty feet below the prairie level. Turtle Mountain on the east rises nobly from the great plain, the boundary line between British and American territory cutting it. The country west of the Souris is a treeless desert, in dry seasons destitute of water, and without a shrub or bush thicker than a willow twig. We ascertained the breadth of this arid, woodless tract to be at least sixty miles north of the Red Deer's Head River on the 49th parallel. Near the boundary line the Souris expands into a series of large ponds and marshes which are called the Souris Lakes. During periods of high water they form a continuous lake of imposing magnitude, extending many miles south of the 49th parallel, consequently far within the United States territory.

A vast number of gneissoid and limestone boulders are strewed over the hill bank of the Souris, near the 49th parallel, and on a point between a small brook and the river we found a number of conical mounds, and the remains of an intrenchment. Our half-breeds said it was an old Mandan village; the Indians of that tribe having formerly hunted and lived in this part of the Great Prairies. We endeavoured to make an opening into one of the mounds, and penetrated six feet without finding anything to indicate that the mounds were the remains of Mandan lodges. There is a Mandan village near Fort Clark on the Missouri, and in the country drained by the Yellowstone the remains of this once numerous and powerful tribe are now to be found.

The mouth of Red Deer's Head River is within a few

yards of the 49th parallel, Mr. Hime took a photograph of the valley while others of the party made an excursion to the Souris Lakes, within the United States territory, in the hope of finding buffalo to replenish our stores; but although fresh tracks were seen, and skulls and bones in large numbers, the remains of last year's "run," yet no living animal but a "cabri," or prong-horned antelope (*Antilocapra Americana*), was visible.

Turkey buzzards (*Cathartes aura*) were observed hovering at a great height above us, and two young birds were shot in the valley of the Souris. The great scarcity of animal life near the Souris Lakes appeared remarkable, but it might be caused by the desert character of the surrounding country, which was so barren and arid as to be incapable of supporting a scanty growth of herbage on the sandy soil of the prairie.

Having reached the 49th parallel and traced the Souris in search of lignite in position for a distance of a hundred miles, we altered our course to a good camping-ground on Red Deer's Head River, and made preparations for crossing a treeless prairie, at least sixty miles broad, in a direction nearly due north.

The Little Souris or Mouse River rises in British territory, on the flanks of the Grand Coteau de Missouri, near the 105th meridian.* Its valley was reported to us by the Crees of the Sandy Hills on the Qu'appelle to inosculate with Elbow Bone Creek, or the Souris Forks, as this stream is also termed, which flows into the Qu'appelle a few miles west of Long Lake. Captain Palliser indicates † its connexion with the Moose Jaws Forks, also an affluent of the Qu'appelle. It is not in the least degree

* See Capt. Palliser's Map, published in the Blue Book, 1859.

† Ibid.

improbable that connexions exist between both rivers. The Souris crosses the boundary line near the 102nd meridian, and flows at the base of the Grand Coteau, nearly as far south as the 48th parallel, when suddenly turning northward, it re-enters British territory near the 101st meridian.

South of the Souris Lakes it flows in a valley 200 feet below the level of the prairie, with a wooded bottom from one half to two miles wide. The nearest serviceable timber east of the Souris, in the direction of the proposed Pacific railroad near the 47th parallel, is in the valley of Red River, 200 miles distant, and with the exception of cotton wood there is no timber west of the Souris for 400 miles.*

* Governor Stevens's Explorations and Surveys.

CHAP. XIV.

FROM THE BOUNDARY LINE TO THE QU'APPELLE LAKES *viâ* FORT ELLICE.

Indian Signs —Smell of Fire.—The Sioux.—Precautions.—" Something."—Red Deer's Head River.—The Great Prairie, Character of.—Mirage.—Birds.—Grasshoppers.—Limit of Burnt Land.— Pipestone Creek.— Standing Stone.—Country changed. —Forest disappeared.—Approach to the Assinniboine.— Cretaceous Rocks.— Buffalo Bull.— Fort Ellice.— McKay.—Crees.—Hunters.—Provision Trading Posts.—Pemmican.— Dried Meat.—Thunder Storms.—Mammoth Bones.— Ojibway Hunter.— Half-breeds.—En Route for the Qu'appelle Mission.—Grasshoppers.—Thunder Storm.—Trail.— Weed Ridge. — Kinni-Kinnik. — Mode of Manufacturing. —Boulders.—White Crane.— Magpies.— Birds.— Dew.— Aridity of Great Prairie.— Charles Pratt.— Chalk Hills. — Indian Turnip. — Qu'appelle Lakes.— Fresh Arrangements.— Descent and Ascent of the Qu'appelle.—Qu'app elle Mission. — Dimensions of Valley. — Character of Lakes.—White Fish.—Rev. James Settee.—Garden of Mission.—Grasshoppers.—Christian Worship. — Baptism. — "Praying Father" and "Praying Man." —Rum.— Indian Wishes.— Objection to Native Missionaries.—Difficulties arising from the Prejudice of Tribes.—Plain Crees passing away.—En Route.

WHILE engaged in taking observations for latitude at the mouth of Red Deer's Head River, on the night of the 2nd July, John McKay, a Scotch half-breed, observed what he thought to be a wolf, approach the brow of a hill about two hundred yards from us, and after apparently gazing at the encampment for a few minutes it retired beyond view. The night was clear, and our tents being placed in the valley of the river close to its junction with that of the Souris, surrounded by steep hills about one hundred and fifty feet high, an object appearing on the brow of those in our rear could be seen projected against the sky.

McKay took no further notice of the strange visitor than to mention that he saw it and thought it was a wolf; but before we retired to our tents at 2 A.M. we noticed another figure, which he declared to be an Indian, appear near the same spot. Two of the party cautiously approached the foot of the hill, but before they reached it the figure crouched and slowly retired. The horses were gathered near the carts and a watch set, but daylight dawned without the re-appearance of the object of our suspicion. In the morning we endeavoured to discover tracks at the spot where it had appeared, but the hill being composed of gravel, the soil had received no impression which our most sharp-sighted half-breeds could detect. Having verified our observations on Polaris by a solar observation at noon, we started for a new camping-ground about twelve miles up Red Deer's Head River, where we proposed to take in a supply of wood for fuel, before crossing the great prairie to Fort Ellice. On our way thither the old hunter who had joined us at Prairie Portage said he *smelt fire;* we all strained our olfactories to the utmost, but without detecting any odour which might be supposed to proceed from a burning substance; nevertheless the old hunter persisted in the statement that he had "smelt it." We camped at sunset close to the river, and while at supper some of the party distinctly heard the distant neigh of a horse; this of itself would have been considered sufficient warning, but when taken in connexion with the appearance of the object on the hill in the rear of our camp the night before, it was held to be conclusive evidence that we were watched by the Sioux, and that an attempt would be made in the night to steal our horses.

Our camp fires were put out immediately, the carts were placed close together, and a watch organized; the half-breeds did not anticipate an attack until the approach of dawn, but the sudden galloping towards the carts soon after ten o'clock of several horses, who were feeding in the valley about one hundred yards from us, proved that Indians were near us. On hearing the horses approach, the men started up and ran to stop them, which they succeeded in doing before they passed the carts. Each horse was now tethered to a cart or stake, and the half-breeds crawling through the long grass arranged themselves in a half-circle about seventy yards from the carts, each with his gun loaded with buckshot. The night was dark, and perfect silence was maintained in the camp; towards morning one man came in to report: he stated that he had heard "something" cross the river and crawl through the grass within a few yards of him; he waited a few minutes for more to follow before he fired or gave the alarm, and then cautiously crawled through the long grass in track of the "something" which had passed near to him. The track led him to within thirty yards of our tents, and then turned towards the river, and evidently crossed it. Morning soon dawned, and the watchers came in; we examined the tracks described by the half-breed who had first heard the intruders, and they were pronounced to be those of an Indian. Further examination in full daylight showed that we had been surrounded by a band, who, however, perceiving we were on the alert, and that the horses were tethered, made no attempt to steal them. Had it not been for the old hunter's excellent nose, there is little doubt that we should have lost our horses during the night.

On the morning of the 4th, having loaded the carts with wood and taken a supply of water from Red Deer's Head River, which is here a rapid, clear stream, twelve feet broad, we started on a nearly due north course to cross the Great Prairie. The watermarks on the banks of Red Deer's Head River show that it rises fifteen feet during spring freshets, almost filling the low, narrow valley in which it flows. The banks are fringed with small balsam-poplar and aspen; patches of elm and oak occur on the points.

The prairie for many miles north of this river appears to be perfectly horizontal; in passing through it we always seemed to be in the centre of a very shallow depression, with a uniform and well-defined horizon in all directions. Early in the morning the distant outline meeting the clear sky was best defined; as the day wore on refraction magnified the tufts of grass and small willows into bushes and trees, destroying the continuity of the fine horizontal line where sky and earth seemed to meet. Occasionally the effects of mirage were very delusive, beautiful tranquil lakes suddenly appeared in the distance, and as quickly faded from our view. Fortunately the almost daily thunderstorms which had occurred replenished the marshes and small ponds, and gave us an abundant supply of water, but in some seasons the buffalo-hunters suffer much from the want of that necessary of life in crossing this vast treeless desert.

On the afternoon of the 5th we arrived at the northern limit of the burnt prairie, as far as we could judge; south of our point of view, the aspect of the vast level tract was of a dark green hue, with short grass of this year's growth; northwards the colour of the prairie was brown, from the old grass of last year which had not been consumed by the fires. Whenever we approached the old and shallow beds of brooks, boulders became numerous. Some of the little valleys contained ponds, occupying the shallow bed, all of which would probably be united, and form a river in the spring of the year.

Among the birds noticed during this monotonous journey were turkey buzzards, ravens, barking crows and black terns; on the borders of several shallow ponds or marshes, which are often dry in the autumn, ducks were plentiful, and afforded us a grateful supply of fresh food. We saw some herds of cabri, and McKay succeeded in killing a female after a long chase. The grasshoppers were very numerous, and during four days filled the air like flakes of snow; they rose simultaneously when about to take their flight, from areas of two to twenty acres in extent, first perpendicularly to the height of twelve or fourteen feet, then in a slanting direction, until they had attained an elevation of from two to three hundred feet, after which they pursued a horizontal course before the wind. In a light breeze, the noise produced by their wings was like a gentle wind stirring the leaves of a forest.

Our half-breeds informed us that this great prairie west of the Souris continues treeless and arid for a distance of sixty miles, it is then crossed by a river, probably the Moose Mountain Creek, shown on Capt. Palliser's map; beyond this river the prairie continues for eighty miles further without tree or shrub; and as this was the utmost westerly limit to which any of them had journeyed in their buffalo-hunting expeditions, they could afford us no further information respecting its extent. They were most of them familiar with the country south of the Great Prairie, the Grand Coteau de Missouri, where the buffalo range during the summer in vast herds.

On the 6th July we arrived at Pipestone Creek, and found the country swarming with a young brood of grasshoppers, with wings about a quarter of an inch long, showing that their progenitors had arrived in the preceding autumn in time to deposit their eggs in the soil. Innumerable hosts of these insects passed overhead during the day, and on looking up through an excellent marine glass, I could see them flying like scud at an immense height. Had it not been for the thunderstorms which daily refreshed and invigorated the herbage, it is probable that our cattle would have suffered seriously from the devastations of these insects.

Pipestone Creek is 20 feet broad at our crossing-place, with a swift current, and a depth of water varying from 1½ to 3 feet. The valley is narrow, but rich and beautiful in comparison with the desolate prairie lying to the south. Among the trees fringing its banks the ash-leaved maple is most numerous, and the hop, together with the frost grape, is abundant on the edge of the stream. On the hills in its neighbourhood boulders are uniformly distributed, but on the highest a considerable number have been collected together by the Assinniboine Indians, and a rude monument erected in commemoration of a battle fought at a remote period.

The level character of the country disappears after passing Pipestone Creek; the prairie is either undulating and sandy, or varied with low hills of drift, on which boulders are scattered. On the evening of the 6th we camped at Boss Hill Creek, which flows into the Assinniboine through a broad valley among low hills and gentle slopes. From a conical eminence near our camp, Boss Hill, Standing Stone Mountain, and the woods fringing Oak Lake are visible.

The "Standing Stone" is probably the same familiar object in these regions as mentioned by the Rev. John West, who traveled during the winter of 1821 to Brandon House and the mouth of the Qu'appelle on a missionary journey. Mr. West relates that he stopped to breakfast at the Standing Stone, where the Indians had deposited bits of tobacco, small pieces of cloth, and other trifles, in superstitious expectation that it would influence their Manitou to give them buffaloes and a good hunt. During Mr. West's journey, now forty years since, buffalo were very numerous in the winter months on this part of the Assinniboine, and many bands of the race of Indians bearing that name made this part of the country their winter quarters. Mr. West saw an Indian corpse staged about ten feet from the ground, at a short distance from Brandon House, a provision post now abandoned by the Hudson's Bay Company.

We arrived at the Assinniboine about ten miles south-east of the Two Creeks, after passing through a rolling prairie of light sandy soil, and in many places covered with boulders. Small "hummocks"* of aspens, and clumps of partially burnt willows, were the only remaining representatives of an extensive aspen forest which formerly covered the country between Boss Creek and the Assinniboine. So great had been the change during twenty years in the general aspect of this region that our old hunter, who had undertaken to guide us in a straight line across the prairie from Red Deer's Head River, confessed that he did not "know the country" when within ten miles of the Assinniboine; he nevertheless declared his conviction that we should strike the river at the point to which he had promised to lead us. He had not visited it for twenty years, and during that interval the timber, which formerly consisted of aspens and willows, had nearly all disappeared. The old man was correct; the face of the country had changed, the aspen forest had been burnt, and no vestige, beyond the scattered hummocks and burnt willow clumps, remained; his "instinct," as he termed it, and that singular facility which practised prairie wanderers possess,

* A half-breed expression.

of journeying from point to point at great distances apart in a direct line, served him in lieu of memory or compass, for we struck the Assinniboine within two or three miles of the spot to which he had been directed to lead us.

The approach to this river is made by descending a steep slope, which forms the boundary of the prairie two or three miles from its present excavated valley. The plateau thus formed is covered with erratics, consisting of granite, gneiss, limestone, &c. The broad subordinate excavation in which the river flows is about one mile across, and from 200 to 250 feet deep. The narrow plateau covered with boulders points to a former condition of the Assinniboine valley, when a much larger river flowed in a wider and shallower valley 200 feet above its present level. The same remark applies to many other rivers in Rupert's Land, which, although now insignificant streams, yet flow through deep subordinate excavations in a broad but generally shallow and well defined trough with steep margins, erratics being dispersed over both margin and plateau. These records of former physical structure appear to indicate that the water once conveyed by these channels, must have been very largely in excess of the present supply.

On the morning of the 8th we passed through a good grazing country on the high prairie level, but being compelled to descend to the first plateau of the Assinniboine valley for water, we found our progress obstructed by a large number of erratics, which endangered the wheels of the carts. Here, however, we saw the first buffalo bull, and after a chase of half an hour's duration, succeeded in killing him. Although very tough and rather strong flavoured, he was an acceptable addition to our larder. On arriving at the second of the Two Creeks, cretaceous rocks were again recognised. They had the same lithological aspect as those of the Souris; organic remains were scarce, but in sufficient numbers and variety to establish their position. A band of soft yellowish-green substance, resembling some varieties of soap-stone, was observed forming a characteristic feature in the exposure of the shales at this place. An analysis of the material composing the green band is introduced in the Chapter on the Cretaceous Series, Vol. II. In the low valley of the As-

Fort Ellice, Beaver Creek.

sinniboine, to which I descended, similar exposures arising from land-slips were also seen. The section exposed was capped by about ten feet of coarse gravelly drift on the brink of the deep and broad excavation which now forms the valley proper of the river. The shales resemble those on the Souris, but contain fewer fossils, and are perhaps more fissile and less impregnated with oxide of iron.

Three more bulls were seen on the following morning, but being anxious to reach Fort Ellice, and already provided with meat, they were permitted to pass us unmolested. The country in the neighbourhood of Beaver Creek is undulating and attractive, but the soil is sandy, capable only of supporting a short stunted herbage. We arrived at the Fort on the morning of the 10th, and took up our camping-ground on the banks of Beaver Creek, close to the broad and deep valley of the Assinniboine.

Fort Ellice was at one period a post of considerable importance, being the depôt of supplies for the Swan River District, now removed to Fort Pelly. The buildings are of wood, surrounded by a high picket enclosure. Mr. McKay, one of the sub-officers, was in charge at the time of our arrival. Some twenty years ago, before the small-pox and constant wars had reduced the Plain Crees to a sixth or eighth of their former numbers, this post was often the scene of exciting Indian display. Formerly Fort Ellice used to be visited by the Crees alone, now it numbers many Ojibways among the Indians trading with it. The Ojibways have been driven from the woods by the scarcity of game, the large animals, such as moose deer and bear, having greatly diminished in numbers. Many of the wood Indians now keep horses, and enjoy the advantage of making the prairie and the forest tributary to their wants.

On the 11th July, a number of hunters attached to Fort Ellice came in with provisions, such as pemmican and dried buffalo meat, which they had prepared in the prairies a few days before, about thirty miles from the post, where the buffalo were numerous. Fort Ellice, the Qu'appelle post, and the establishment on the Touchwood Hills being situated on the borders of the great Buffalo Plains, are provision trading posts. The Hudson's Bay Company obtain from the Plain Crees, the Assinniboines, and the Ojibways, pemmican and dried meat to supply the brigades of boats in their expeditions to York Factory on Hudson's Bay, and throughout the northern interior. Pemmican is made by pounding or chopping buffalo meat into small pieces and then mixing it with an equal quantity of fat. It is packed in bags made of the hide of the animal, in quantities of about ninety pounds each. Dried meat is the flesh of the buffalo cut into long, broad, and thin pieces about two feet by fifteen inches, which are smoked over a slow fire for a few minutes and then packed into a bale of about sixty pounds. We had many opportunities of seeing the Cree women on the Qu'appelle, cut, prepare, and pack dried meat.

At Fort Ellice (longitude 101° 48′, latitude 50° 24′ 32″, Captain Palliser) the thunder storms were as violent as on the Souris; not a day passed without lightning, thunder, and generally violent rain of half an hour's duration. The grasshoppers at this post had destroyed the crops last year, and, at the time of our visit, the young brood were well advanced, their wings being about one-third of an inch long. Full grown insects from the south were flying overhead or alighting in clouds around us, so that all hopes of obtaining a crop from the garden or potato fields were abandoned for this year. Provisions were very scarce at the post, and had it not been for the fortunate arrival of the hunters with some pemmican and dried meat, we should have been compelled to hunt or kill the ox.

From Mr. McKay I received a particular account of the "Great Bones" on Shell Creek, which had long been a source of wonder and awe to the Indians hunting on the left bank of the Assinniboine, and whose magnificent descriptions led me to suppose they might belong to a cetacean, and were worth a day's journey out of our track to visit and examine. They were seen many years ago protruding from the bank of Shell Creek, 20 feet below the prairie's level. Mr. McKay instructed some of the hunters attached to the post to bring them to him, but no Indians would touch them, and the half-breeds only brought a tooth and collar bone, which were stated by a medical gentleman to whom they were shown to have formerly belonged to a mammoth. Mr. Christie, of Fort Pelly, we were told, went to Shell Creek, with a view to collect more specimens; he obtained some ribs, but in a state of crumbling decay; they were sent to Red River Settlement. The Indians had long regarded these ancient relics as the bones of a Manitou and worthy objects of veneration. An old Indian on Dauphin Lake, to whom reference will be made hereafter, described similar bones in the banks of Valley River leading to Dauphin Lake; but the season was too late when exploring that part of the country to permit of an examination.

On Monday, the 12th, preparations for continuing our journey westward were completed, by engaging an Indian to assist in paddling Mr. Dickinson down the Qu'appelle or Calling River from the Mission to its junction with the Assinniboine. The half of his wages he stipulated to have in advance. Mr. McKay told me he was a bad Indian and not to be trusted, but we could not succeed in getting another. When on the point of starting, a young Ojibway, painted and adorned with feathers, galloped up to the post, entered the room, drew from beneath his moose skin robe two moose tongues and a mouffle, which he quietly handed to Mr. McKay, and, squatting on the floor without speaking a word, lit his pipe. After a few minutes, he informed us that he and his father had killed two moose, thirty miles off, and desired McKay to send for them. Two half-breed hunters also arrived at this moment, in sad plight, hungry and tired, with worn horses and torn clothes. They had come from Fort Union, on the Missouri, having been hunting on the Grand Coteau, where they met a war party of sixty Blackfeet. They fled to the fort, the Blackfeet pursuing them and insisting that the Fort Union people should give them up, a request which was promptly refused.

During the night, the Fort Union people furnished them with a small supply of provisions, and leading them out to the prairies, told them to run for it; they did so, and arrived in safety at Fort Ellice after a harassing journey.

At 4 P.M. on the 12th July, we left Fort Ellice and traveled due west through a pretty country near the banks of the Qu'appelle or Calling River. After breakfast on the following day we arrived at the Cross Woods, which, according to our half-breeds, extend as far as Pipestone Creek; they consist of aspen, with a splendid undergrowth. Here we observed during the morning the grasshoppers descending from a great height perpendicularly, like hail—a sign, our half-breeds stated, of approaching rain. Our route lay through a rolling country, the soil consisting of sandy loam with much vegetable matter in the valleys. Aspen groves are numerous, and many little lakes margined with reeds afford quiet breeding-places for duck. The road is good in summer, but wet and soft in the spring.

The grasshoppers were excellent prognosticators: a violent thunder storm in the afternoon commenced in the east (all preceding storms had come from the west), and was accompanied by exceedingly heavy rain and a very boisterous wind. The storm continued several hours. At 9 in the evening, the air was calm and the heavens clear and bright; at 10, the storm returned from the west, and a more terrific and sublime exhibition of elemental warfare none of us had ever before witnessed. Three times the lightning struck the earth so close to us that there was no perceptible interval between the flash and the shock. It was distinctly heard to *hiss* through the air, and, instead of penetrating the ground at once, it seemed to leap from bush to bush for a distance of 60 or 70 yards. So close did one flash approach that when we had recovered from the shock, and our eyes had regained their power, several of us met each other, groping from cart to cart, to see if any of the party had been struck. It is remarkable that although the wind was blowing violently before and after the two flashes just described occurred, yet, between them, an interval of about three-quarters of a minute, there was a dead calm, and a calm of short duration succeeded each flash in our immediate vicinity.

The trail on the 14th continued through good land for nine miles, with aspen groves on the crown of each undulation, and willow bushes in the hollows, it was succeeded by a prairie three miles across, but of much greater extent longitudinally. Ponds were numerous, abounding with ducks and ducklings. Another rain and thunderstorm on the evening of this day lasted for about an hour. On the following morning we entered a treeless prairie marked at its western extremity by a sandy ridge running N.W. by S.E., known among the Indians as the Weed Ridge. It was covered with the bear-berry from which kinni-kinnik is made. This was the first time we saw this weed since leaving the Sandy Hills of the Assinniboine. The Indians of the prairies generally use the inner bark of the *Cornus sericea*, the red-barked willow as they term it. We also saw them smoke the inner bark of the dog-wood, *Cornus alternifolia*.

The mode in which these barks are prepared is very simple. A few branches about three-quarters of an inch thick and four or five feet long are procured, and the outer bark scraped off after having been warmed over a fire; a knife is then pressed against the inner bark and drawn upwards, for a space of six or eight inches, until the whole of the inner bark is gathered in curly clusters round the stick, it is then thrust in the ground over the embers and roasted until quite dry, when, mixed with tobacco in equal proportions, it forms the favourite kinnikinnik of the North-West Indians. I often saw them smoke bark or the leaves of the bear-berry alone, when their supply of tobacco was exhausted.

The Indian who accompanied us from Fort Ellice to the Qu'appelle Mission, complained of weakness and pain in the chest, he suffered much from cough, and was evidently consumptive; he was treacherous and indolent, and, as will be shown hereafter, soon left us in the lurch.

Beyond the Weed Ridge the country is very undulating; boulders of Silurian limestone and gneiss were strewed on the flanks and summits of the hills. The white or whooping crane (*Grus Americana*) was first seen to-day. This beautiful bird is common in the Qu'appelle Valley and in the Touchwood Hill range. It is a dangerous antagonist when wounded, striking with unerring aim and great force with its powerful bill. When the bird is wounded, the best way to avoid its attacks is to present the muzzle of the gun as it approaches, it will fix its bill in the barrel and may then be destroyed without danger. Instances have been known of this bird driving his bill deep into the bowels of a hunter when not successful in warding off its blow. Magpies are numerous on the Weed Ridge, and the cat bird is heard in every little wooded dell.

On the 15th we passed two streamlets flowing into the Qu'appelle. Their banks were fringed with small timber, and quite lively with birds, which are far more abundant here than on the Souris. On the borders of all the wooded brooks we saw magpies, cat birds, crows, and, occasionally, the solitary thrush; in the wet prairies, the rice bird, black tern, plover, the golden-legged and common, the yellow-headed blackbird, common meadow lark, chipping sparrow, and grackle; on ponds and in marshes, ducks of many species, bittern (*Ardea lentiginosa*), and cranes. In the morning, after a clear night, we always observed heavy dew; this phenomenon was not so frequently noticed on the Souris under similar circumstances. There can be little doubt that the sterility of the Great Prairie between the Qu'appelle and the 49th parallel is owing to the small quantity of dew and rain, and the occurrence of fires. North of the Qu'appelle, the country seemed to be far more humid, and the vegetation far richer and more abundant in many localities than south of that great valley.

Another prairie eight miles broad, bounded by ridges having a N. W. and S. E. direction, succeeded to the one last described, and introduced us on the 16th to a hilly country; the range is called the Indian Head Hills; it contains many beautiful lakes, is well wooded, and forms one of the northern spurs of the Moose Mountain, whose position is given on Captain Palliser's map. The northern slope of the Indian Head Hills is very abrupt, the southern, gentle and undulating. Here we met with Charles Pratt and party going to Red River. Charles Pratt is a half-breed catechist of the Church Missionary Society, well acquainted with the habits of Indians and of buffalo, but apparently scarcely sensible of the importance of his duties and the responsibility of his charge. He gave me a good deal of valuable information respecting the country, and, with characteristic generosity, if not Christian sympathy, told John McKay to take a young heifer belonging to him when we arrived at the Mission and kill it in honour of our arrival. Pratt showed me some specimens of lignite which he had taken from a bed two feet thick at the Wood Hills, about eighty miles south-west of the Hudson's Bay Company's post. He described the hill or range of hills as an island in the prairie, which, like the Stony Mountain near Red River, had escaped denudation, or, it may have been La Roche Percée, visited by Mr. F. W. Lander in 1853*, and by Dr. Hector in 1858.†

An old Indian accompanying Charles Pratt, born in this part of the country, told us that he remembered the time when the whole of the prairie through which we had passed since leaving Fort Ellice was one continuous forest, broken only by two or three narrow intervals of barren ground. They told us that the Plain Crees and the Blackfeet were at war with one another, and that the Crees were hurriedly "pitching east" in order to avoid the Blackfeet. This intelligence had considerable effect upon our old hunter; he made his will and gave it to Charles Pratt to give to Archdeacon Cochrane of Prairie Portage.

The view from the Indian Head range is exceedingly beautiful; it embraces an extensive area of level prairie to the north, bounded by the Aspen Woods on the borders of the Qu'appelle Valley. A portion of the old forest alluded to by the Indian still exists on this range. It consists of aspen of large growth and very thickly set. A few cabri (prong-horned antelope) were seen in the Indian Head range; they used to abound in the country drained by the Qu'appelle.

On Saturday the 17th we entered a very beautiful and fertile prairie at the foot of the Indian Head range, our course leading us in a northerly direction to the Qu'appelle Mission. The common yarrow was very abundant,

* Pacific Railway Explorations.

† Blue Book. Capt. Palliser's Explorations.

and with the hare-bell reminded us of other scenes far away. Six miles from the hills we arrived at a subordinate, shallow, and broad valley, parallel to that of the Qu'appelle. The aspect of its boundary suggested the shore of a lake or bank of a large river. The lower prairie consisted of a sandy loam, in which the Indian turnip was very abundant. We soon came up with a group of squaws and children from the Qu'appelle Lakes who were gathering and drying this root, which the Crees call the *Mis-tas-coos-se-ne-na* or big grass root. The French half-breeds call it the pomme de prairie; the Sioux *Tip-si-na.* It is an important article of food in these regions. The botanical name is *Psoralea esculenta.* Many bushels had been collected by the squaws and children, and when we came to their tents they were employed in peeling the roots, cutting them into shreds and drying them in the sun. I saw many roots as large as the egg of a goose, and among those brought with me to Canada are some of even larger dimensions. The Crees consume this important vegetable in various ways; they eat it uncooked, or they boil it, or roast it in the embers, or dry it, and crush it to powder and make soup of it. Large quantities are stored in buffalo skin bags for winter use. A sort of pudding made of the flour of the root and the mesaskatomina berry is very palatable, and a favourite dish among the Plain Crees.

We reached the Qu'appelle Lakes at 6 P. M., after passing through a magnificent prairie the whole day; in fact the country north of the Indian Head and Chalk Hill ranges is truly beautiful, and will one day become a very important tract. The Chalk Hills are a continuation of the Indian Head range. In the language of the Indians they contain bands of "soft white earth or mud." The half-breeds call them "Chalk Hills." It is a matter of regret that the time at our disposal did not permit us to make an excursion to them, notwithstanding that no indications of rocks in position were seen on the Indian Head range; they were recorded as composed of drift which may or may not conceal rocks in position above the general level of the prairie north of them.

Great was our astonishment on arriving at the Qu'appelle Lakes to find that they were narrow bodies of water, occupying an excavated valley about one mile broad, 250 feet deep, and differing in no important particular from the same valley at its junction with the Assinniboine —120 miles distant by the river, or 134 by the trail. The importance of the Qu'appelle valley began to develope itself when the Crees at the Lakes informed us that it continued through to the Saskatchewan without losing its breadth, and maintained, except for a short distance, a great depth below the prairie level. I determined, therefore, to explore the whole valley from the South Branch of the Saskatchewan to the Assinniboine, and to ascertain the relation it bore to those rivers. With this view the canoes were put in order, the party and supplies divided, and the arrangements detailed in the following paragraph completed.

Mr. Dickinson, with a French Canadian and a Cree half-breed, was instructed to descend the Qu'appelle River from the first Fishing Lake to its mouth. Mr. Fleming and myself were to ascend it from the same starting place to its source, and follow up the valley to the South Branch of the Saskatchewan. Mr. Hime was to explore Long Lake and meet Mr. Dickinson at Fort Pelly. I intended, upon reaching the South Branch, to descend that magnificent river in canoe to the Grand Forks, and then by the main Saskatchewan to Lake Winnipeg and Red River, a distance of about 1000 miles canoe navigation.

FISHING LAKES, QU'APPELLE RIVER

The Qu'appelle Mission is situated between the second and third Fishing Lakes. The situation is beautiful, and the country on all sides of a very novel and peculiar character. Here the Qu'appelle valley is $1\frac{1}{4}$ miles broad and 250 feet deep. On the south a vast level prairie extends to the Indian Head Hills, fertile, inviting, but treeless; towards the north the country is studded with groves of aspen over a light and sometimes gravelly soil. Most beautiful and attractive, however, are the lakes, four in number, which from the rich store of fish they contain, are well named the Fishing Lakes. A belt of timber fringes their sides at the foot of the steep hills they wash, for they fill the entire breadth of the valley. Ancient elm trees with long and drooping branches bend over their waters; the ash-leaved maple acquires dimensions not seen since leaving the Red River, and the Me-sas-ka-to-mi-na (la Poire) (*Amelanchier Canadensis*) is no longer a bush, but a tree eighteen to twenty feet high, and loaded with the most luscious fruit.

The Qu'appelle Mission was established last year (1858). For some time past, however, Charles Pratt, the catechist, has resided where the Mission is situated, and has constructed a comfortable log-house, fenced in a garden, and now possesses six or seven cows and calves. An old half-breed, whose name is obliterated in my note-book, took up his residence with Pratt; he had been engaged for the better part of his life at different fishing stations belonging to the Hudson's Bay Company throughout Rupert's Land, and he declared that in all his experience he had never seen the white fish (*Coregonus albus*) so large, numerous, and well flavoured as in the Qu'appelle Fishing Lakes.

The Rev. James Settee, the missionary, a native of Swampy Cree origin, occupied Pratt's house; he arrived at the Mission last autumn. In the garden where we found him, Indian corn was growing, as well as potatoes, turnips, beans, and other culinary vegetables. The grasshoppers had not yet visited the Mission, but vast flights had passed over it. They were seen passing the Company's post, twenty miles south, on the 8th of the month; they were then flying to the east. They had passed the Mission in 1857, for they visited the Touchwood Hills forty to fifty miles north of it, depositing their eggs in the ground, and during the present summer the young brood, as I learned a few weeks afterwards, destroyed all garden crops at the Touchwood Hills, and on the 28th July took their flight to the south-east.

On Sunday we attended service in Pratt's house; the Rev. Mr. Settee read the prayers in English with great ease and correctness; he preached in Ojibway, and a hymn was sung in the Cree language. Before the sermon the missionary surprised us by waking up a drowsy Indian, who was enjoying a quiet nap in a corner of the room, and leading him to the temporary reading desk, commenced the ceremony of public baptism. My astonishment was not diminished when the reverend gentleman turning to me, without any preliminary notice, said abruptly, "Name this man!" After a moment's reflection I said "John," and without any unnecessary loss of time or words, "John" walked to his bench, and was soon apparently lost, in noisy slumber, to all consciousness of the privileges and blessings of which adult Christian baptism, duly received, had made him the inheritor.

When the Rev. James Settee arrived at the Mission last autumn, the Crees of the Sandy Hills having received intelligence that the bishop had sent a "praying man" to teach them the truths of Christianity, directed messengers to inquire whether "the great praying father had sent plenty of rum; if so, they would soon become followers of the white man's Manitou." The messengers returned with the intelligence that the great praying father had not only omitted to send rum, but he hoped that the Plain Crees would soon abandon the practice of demanding rum in exchange for their pemmican and robes. The messengers were directed to return to the missionary with the announcement, that "if the great praying father did not intend to send any rum, the sooner he took his praying man away from the Qu'appelle Lakes the better for him."

Encampment in the Qu'appelle Valley.

There are very few tents about the Mission at present, the Indians being in the plains engaged in hunting the buffalo. Mr. Settee speaks English very fluently, the field for his labour is extensive, but not at present promising. When conversing with the Crees of the Sandy Hills, many of them expressed a wish to have their children taught by white men, but they did not appear to like the idea of their being taught by a native of a different origin. This is an important point to be observed in the selection of native missionaries. The school, however, appears here, as elsewhere among Indian tribes, to be the only sure ground for establishing the true faith among them. "Teach my children for two or three years, but let me follow the ways of my fathers," said the son of the chief of the Sandy Hills to me. Many expressed a wish that their little ones should know the white man's cunning, and learn to cultivate the soil, but they would stipulate to remain themselves still the wild prairie Indians, hunting the buffalo, and occasionally tasting the savage excitement of war.

It is a wrong policy to send a Swampy Cree among the Plain Crees, or an Ojibway amongst the Crees, as a teacher and minister of religion. These highly sensitive and jealous people do not willingly accept gifts or favours which involve any recognition of mental superiority in the donor from one not of their own kindred, language, and blood; although he may be of their own race. An Ojibway remains always an Ojibway, and a Swampy Cree a Swampy Cree, in the eyes of the haughty and independent children of the prairies, and they will never acknowledge or respect them as teachers of the "white man's religion."

Several of my half-breeds appeared to think that Mr. Settee would have troublesome times, and that he would not be able to make much impression among the Plain Crees. What has to be done must be done soon, the time is short and the race is fast passing away; in another generation we shall probably lament the disappearance of a tribe which twenty years ago could muster one thousand mounted warriors, and who, in all the pride and barbarous pomp of Indian display, were accustomed to approach the Hudson's Bay Company's posts singing their Hi-he-ah, Hi-he-ah, or war song, in savage unison.

On the 20th July the canoes were launched on the Third Fishing Lake, and having seen Mr. Hime *en route* for Long Lake, my carts and horses on the way to the Grand Forks of the Qu'appelle, and Mr. Dickinson started for the mouth of the river, I embarked with Mr. Fleming, an Ojibway, and a Cree half-breed, and paddled up stream with a view to trace out the valley to its junction with the South Branch of the Saskatchewan. The succeeding Chapter contains a narrative of this exploration, which is followed by Mr. Dickinson's description of his canoe voyage to the Assinniboine. We arranged to meet at Fort Ellice forty-three days after our simultaneous departure from the Third Fishing Lake.

CHAP. XV.

THE QU'APPELLE VALLEY. — FROM THE MISSION TO SAND HILL LAKE.

Depth of the Fishing Lakes.—Cross-Sections.—Confervæ.—Lower Lakes 66 Feet deep.—Birds.—Vegetation.—Water-mark.—Third and Fourth Fishing Lakes.—Fish.—Soundings in Fourth Lake.—Fishing Lakes probably once united.—Geese.—Pelicans.—Fourth Lake.—Water-mark.—Aspect of Valley in 1852.—Qu'appelle River.—Prairie.—Depth of Valley.—White Cranes.—Section of Alluvial Flats.—Temperature.—Character of Prairie.—Birds.—Shrubs.—Antelope.—Hare.—Roses.—Grand Forks.—Plain Crees.—Temperature of River.—Ice Marks.—Buffalo Tracks.—Character of Stream.—Willow Bushes.—Fetid Air.—Drift Clay.—Erratics.—Freemen's Houses.—Prairie.—Want of Timber.—Thunder Storms.—Touchwood Hills.—Indians.—Tolls.—Diplomacy.—Indian Resolve.—The Grand Forks.—Long Lake.—Souris Forks.—Souris of Qu'appelle and Assinniboine.—Dimensions of Valley.—The Grand Coteau.—Prairie Fires.—Indian Signs.—A Prairie on Fire.—Buffalo.—Consequence of Prairie Fires.—Reclamation of sterile Areas.—Indian Telegraph.—Scarcity of Wood.—Ancient Indian Encampment.—The Plain Crees.—Cree Tents.—Provisions.—Buffalo Pound Hill Lake.—Indians.—Aspect of Country.—Coteau de Missouri.—Last Mountain.—Treeless Plain.—The Grand Coteau.—Buffalo.—Birds.—Plain Crees, Camp of.—The Qu'appelle Valley.—Marrow.—Precautions.—Sand Hill Lake.—Crees.—Bois de Vache.—Salt Lake.—Dimensions of Valley.—Erratics.—Cross the Qu'appelle Valley.—Camp at Sand Hill Lake.

THREE quarters of a mile from the mouth of the little stream joining the second and third Fishing Lakes, the lead showed 44 feet of water. This great depth surprised us, as we had been paddling, since leaving the Mission, in shallows not exceeding four and five feet in depth. Cross sections subsequently made, showed that the lakes were generally deep on the north and shallow on the south side. An abundant growth of green confervæ covered the surface, which, in its aggregations and general distribution, reminded me of a similar profusion on the Lake of Woods during August, in 1857. The hill sides of the valley are deeply ravined and wooded, but the hills they separate are bare; we soon noticed too that the north side began to show far less timber than the south, and of more stunted growth. The snowberry was seen in every hollow. Ash-leaved maple and elm were numerous on the south side of the lake, together with the mesaskatomina.

Two excellent photographs, taken near the Mission, of the lakes and hills, display the chief characteristic of the valley with the fidelity which can only be attained by that wonderful art.

Soundings near the middle of the lake showed 56 feet, which when added to 249 feet, the depth of the valley below the prairie as ascertained by trigonometrical measurement, make the total excavation 305 feet. Another sounding 200 yards from the N.W. point, gave 57 feet of water. This was the greatest depth we obtained, but Mr. Dickinson found the lower lakes to be 66 feet deep. The shores of gravel are strewed with blocks of drift limestone and the unfossiliferous rocks. Gulls are numerous about these remote lakes, and a pair of eagles have had their eyrie for many years in a fine elm tree, near the west end of the Third Fishing Lake. The hop grows very luxuriantly in the thin belt of woods on the south side, and the frost grape hangs in beautiful festoons from the drooping branches of the elm. The water mark shows that this lake rises from six to seven feet above its present level.

A low plateau, inundated every spring, separates the Third from the Fourth Lake. It is the delta of two ravines which in the spring and autumn, bring down a large quantity of water from the prairie above. Third Fishing Lake is connected with Fourth Fishing Lake by a rapid stream flowing through the plateau, about 100 feet broad. At its mouth we saw a large number of fish rising at the grasshoppers, which dropped from flights of these insects passing over at the time. In the same stream were many large fish, and among them several individuals of a species to which further reference will be made. Soundings in the Fourth Lake showed 54 feet; this depth was maintained for a long distance with great regularity. In fact, these lakes appear to be nearly uniformly deep, and point to an excavating force, or peculiarity of rock formation deserving of further inquiry. The deltas at the mouth of the ravines, coming in from the prairie at right angles to the general course of the valley, give a clue to the mode in which the lakes were separated one from the other. It is very probable that they were once all united.

Geese appeared in large numbers in the Fourth Lake, and at its western end we saw a splendid flock of pelicans containing thirty-five individuals; as we approached they sailed majestically round and round, but took flight before we arrived within gunshot. Magpies are very numerous in the thin woods fringing the lakes; so also are grackles, the cat bird, and many smaller birds. The Fourth Lake is very shallow at its western extremity, six feet being the greatest depth recorded. The hills on the north side are quite bare, and trees on the south side are found only in the ravines. The lake is full of weeds and its water emits a very disagreeable odour, but the watermarks show, that during spring freshets, its level is eight feet higher than in the summer season. This is an important fact when taken in connection with the alleged

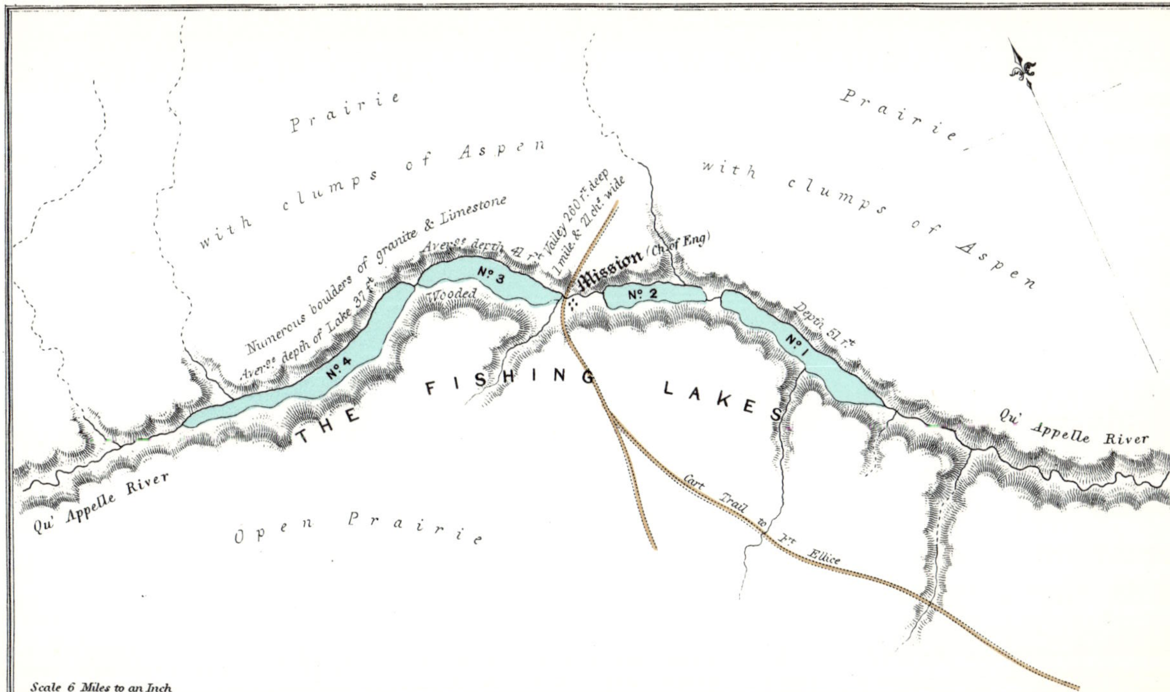

Engraved by Edwd. Weller

Plan of the Fishing Lakes, Qu' Appelle River.

appearance of the whole valley during wet springs; it is then said to resemble a broad river from a few miles east of the Saskatchewan to the Assinniboine. In 1852, a year memorable in Rupert's Land for the great floods which covered an immense tract of country, the Indians represent the Qu'appelle Valley as filled with a mighty river throughout its entire length, flowing with a swift current from the lakelets at the Height of Land, soon to be described, to the Assinniboine, and as a mountain torrent through the short distance of twelve miles, which separates them from the South Branch of the Saskatchewan.

After leaving the Fourth Lake and the marshes at its west extremity, we paddled, sailed, or tracked up a narrow swift stream, four or five feet deep, seventy feet broad, and winding through an alluvial flat in a valley of undiminished breadth and depth. The hill sides were now absolutely bare, not a tree or shrub was to be seen. We had reached the point where timber ceases to grow in the valleys of the rivers except in peculiar situations; the altitude of the banks could not be less than 280 or 300 feet. The prairie on either side is also treeless and arid. On the 21st, after spending a restless night owing to the attacks of multitudes of mosquitoes, we left the canoe in the hands of our half-breeds to track up the stream, and, ascending to the prairie, walked for some miles on the brink of this great excavation. We waited five hours for the canoe to reach us, the windings of the stream involving a course three times as long as a straight line up the valley. The hill sides here began to acquire a more imposing altitude, and probably exceeded 300 feet. White cranes appeared in flocks of four and seven together, but they were so wary that it was impossible to approach them.

The river was often seen to draw near to either side of

the Great Valley, having excavated a channel 10 to 12 feet deep in the alluvial flats through which it pursued its tortuous course. Its banks revealed the following section: —

6 inches light vegetable mould with sand,
4 inches yellow clay,
10 inches light vegetable mould (former surface),
9 feet yellow clay,
2 to 3 inches ferruginous sand to the level of the river.

The last layer was hard, compact, and very coarse-grained. The river is here 60 feet broad, and flows at a rate of a mile and a half an hour. At the mouth of Long Creek, an insignificant affluent, the hills are covered with limestone and granitic boulders; the north side is treeless like the vast prairie beyond it, the south side has aspens in the ravines and aspen groves on the prairie. The width of the valley remains uniform, never exceeding one mile and a quarter or being less than one mile. The pasturage in the flats is superb, the grass being long and very thickly set. Robins, magpies, and yellow birds enliven small aspen groves on the south side, or the thickets of cherry, mesaskatomina, dogwood, and snow-berry, which fill the hollows and ravines; the cat bird is also common, and the tyrant flycatcher everywhere. In the river were vast numbers of ducks and geese; the young birds frequently made us an excellent meal; but no four-footed animals were seen, with the exception of one prong-horned antelope and one prairie hare.

In the afternoon of this day we made many miles by sailing before a strong east wind, notwithstanding a heavy rain and thunder-storm; we were glad to be able to push on through this seemingly interminable and now monotonous valley, as the air from the marshes on either side of the river was fetid and oppressive. A scramble to the summit of the steep hill bank, 300 feet high, though very fatiguing, was amply repaid by the cool, pure and delightful breeze blowing over the desolate prairies around us. Roses of three different varieties, red, white, and variegated, were numerous on the upland; and in the morning, when the dew was on them, or at night, when it was falling, the fresh air from above came down in puffs into our deep, hot valley with delicious and invigorating fragrance. On the fourth day after our departure from the lakes we sighted the Grand Forks; leaving the canoe, I hastened on to a point where the men with the carts and horses were to await our arrival, and found them safely encamped on a beautiful meadow, anxiously looking for us. An empty cart and a couple of horses were despatched for the canoe, still some miles below us, and in the evening we were joined by Mr. Fleming and the two voyageurs.

During the day the temperature of the River was found to be 74°. At the mouth of a dry bed of a stream which we called Maple Creek, some very old trees of the ash-leaved species were observed. Many of them showed marks where they had been tapped. The willows which fringed the banks of the Qu'appelle were barked by ice eight feet above the surface of the water. Numerous buffalo tracks began to appear before we reached the Forks, and where these animals had crossed the river, they had cut deep roads to the water's edge, and lanes through the willow bushes. The bones of many a young bull and cow were seen sticking out of the banks where they had been mired.

The tortuous character of the stream before we took the canoe out of the water, may be imagined from the fact that eleven hours constant, steady tracking enabled us to progress only five miles in a straight line through the valley, and not less than 200 courses and distances were recorded in the canoe. Some little time was lost in crossing from one side to the other in order to avoid the willow bushes, which only grew on the inside of a bend, rarely or never on the outside or longest curve. The breadth of the river where we left it was forty feet, and the speed of its current one mile and a quarter an hour. The fetid air from the marshes made most of the party feel unwell, and I therefore determined to carry the canoe in a cart on the immediate edge of the prairie, keeping the valley in constant view, and occasionally descending into it and crossing it, to ascertain by levelling and measurement its leading dimensions.

No rock exposure was anywhere to be seen; drift appears to cover the country to a great depth. Where land slips have occurred and exposed an almost perpendicular section, yellow gravelly clay alone is visible. Some of the limestone erratics strewed over the sides of the ravines resemble those frequently seen on the south-east side of Lake Winnipeg.

Near our camp, on the 23rd, were six or seven log-houses, occasionally inhabited during the winter months by *freemen*, that is, men no longer in the service of the Company. The prairie above the freemen's houses slopes gently to the edge of the valley from the distant horizon on both sides. Clumps of aspen vary its monotonous aspect, and though clothed with green herbage, due to the late abundant rains, the soil is light and poor. Some distance back from the valley it is of better quality, the finer particles not having been washed out of it; the grass there is longer and more abundant, but the greatest drawback is the want of timber.

Since we have been on the Qu'appelle we have frequently noticed thunderstorms towards the north-west and north, in the neighbourhood of the Touchwood Hill range, which did not reach us ; the day before yesterday (21st July) a very violent thunderstorm in the Qu'appelle valley, which delayed us for several hours, did not wet the carts ten miles to the south. Rain clouds appear to follow the Touchwood Hill range ; the frequency of storms in that region is proverbial, and the richness of the vegetation there proves that an abundant supply of rain falls during the hot summer months.

Soon after sunset our camp received an unexpected addition of six " Bungays," * who were on their way to Fort Ellice with dried buffalo meat and pemmican.

They had been hunting between the two branches of the Saskatchewan, and represented the season as very dry and the buffalo scarce. We passed a quiet and friendly night with them, and on the following morning made them a small present and pursued our way to the Grand Forks.

I happened to be about 100 yards in advance of the carts, after we had traveled for about a quarter of an hour, when hearing a loud clatter of horses' feet behind me, I looked round and saw the six Indians approaching at a gallop. One of them, who had represented himself as a chief, seized my bridle, drew the horse's head round, and motioned me to dismount. I replied by jerking my bridle out of the Indian's hand. My people came up at this moment and asked in Cree what this interference meant. " We wanted to have a little more talk," said the chief ; " we are anxious to know the reason why you are traveling through our country." It turned out after a

* Crees and Ojibways of mixed origin.

little more "talk" that they wished to establish a sort of toll of tobacco and tea for permission to pass through their country, threatening that if it were not given they would gather their friends in advance of us, and stop us by force. We knew that we should have to pass through about 100 tents, so there was some little meaning in the threat. The old hunter, however, knowing Indian habits and diplomacy well, at once remarked that we were taking a large present to the chief of the Sandy Hills, and we did not intend to distribute any tobacco or tea until we had seen him, according to Indian custom. They tried a few more threats, but I closed the parley by unslinging a double-barrelled gun from the cart, and instructing the men to show quietly that they had theirs in readiness. Wishing the rascals good day, we rode on; they sat on the ground, silently watching us, but made no sign. In the evening one of them passed near us at full gallop, towards some tents which we saw in the distance as we ascended the hill at the Grand Forks.

One rather significant statement they made proved to be correct, namely, that the Plain Crees, in council assembled, had last year "determined that in consequence of promises often made and broken by the white men and half-breeds, and the rapid destruction by them of the buffalo they fed on, they would not permit either white men or half-breeds to hunt in their country or travel through it, except for the purpose of trading for their dried meat, pemmican, skins, and robes."

We crossed to the north side of the Qu'appelle when we arrived at the Grand Forks, and ascended the hill bank to the prairie. The Grand Forks consist of the junction of two deep and broad valleys bearing a great resemblance to each other; the south valley is that in which the Qu'appelle river flows, the other is occupied by Long Lake, or Last Mountain Lake, forty miles in length, and from one-half to two miles broad, being in fact an exact counterpart of the Qu'appelle Valley and Lakes. It is narrow, deep, filled throughout with water, and is said to inosculate with the South Branch of the Saskatchewan, some miles below the Elbow. In its general aspect Last Mountain Lake is similar to the Fishing Lakes. A rapid, winding stream, 30 feet broad, runs from it into the Qu'appelle. Both valleys are of uniform breadth and depth, and very little narrower than, when united, they form the main valley of the Qu'appelle river. From the Grand Forks to the Souris Forks (Elbow Bone Creek) the country is treeless, slightly undulating and poor. The Indians say that the Souris River of the Qu'appelle, coming from the Grand Coteau de Missouri, inosculates with an arm of the Souris of the Assinniboine before described, and that a canoe in high water might pass from one river to the other without a portage. If this be the case, the diversion of the waters of the South Branch down the Qu'appelle valley would acquire additional importance, and give value to an immense extent of territory, now comparatively inaccessible, and very insufficiently watered.

A few miles west of the Souris Forks the Qu'appelle is 19 feet wide and 1½ feet deep, but the great valley is still a mile broad and 200 feet deep. Here on the 25th we caught a glimpse of the blue outline of the Grand Coteau, with a treeless plain between us and the nearest part which is called the "Dancing Point of the Grand Coteau;" and has long been distinguished for the "medicine ceremonies" which are celebrated there. After passing these forks, the country is more undulating, small hills begin to show themselves; the general character of the soil is light and poor; the herbage consists of short tufted buffalo grass, and plants common in dry arid plains. This afternoon we saw three fires spring up between us and the Grand Coteau. They were Indian signs, but whether they referred to the presence of buffalo, or whether they were designed to intimate to distant bands the arrival of suspicious strangers, we could not then tell, and not knowing whether they originated from Crees, Assinniboines, or Blackfeet, we became cautious. In a few days we ascertained that the fire had been put out* by Crees, to inform their friends that they had found buffalo.

The grandeur of the prairie on fire belongs to itself. It is like a volcano in full activity, you cannot imitate it, because it is impossible to obtain those gigantic elements from which it derives its awful splendour. Fortunately, in the present instance the wind was from the west, and drove the fires in the opposite direction, and being south of us we could contemplate the magnificent spectacle without anxiety. One object in burning the prairie at this time, was to turn the buffalo; they had crossed the Saskatchewan in great numbers near the Elbow, and were advancing towards us, and crossing the Qu'appelle not far from the Height of Land; by burning the prairie east of their course, they would be diverted to the south, and feed for a time on the Grand Coteau before they pursued their way to the Little Souris, in the country of the Sioux, south of the 49th parallel.

Putting out fire in the prairies is a telegraphic mode of communication frequently resorted to by Indians. Its consequences are seen in the destruction of the forests which once covered an immense area south of the Qu'appelle and Assinniboine. The aridity of those vast prairies is partly due to this cause. The soil, though light, derives much of its apparent sterility from the

* A native expression; "put out fire," signifies to set the prairie on fire.

annual fires. In low places and in shallow depressions where marshes are formed in spring, the soil is rich, much mixed with vegetable matter, and supports a very luxuriant growth of grass. If willows and aspens were permitted to grow over the prairies, they would soon be converted into humid tracts in which vegetable matter would accumulate and a soil adapted to forest trees be formed. If a portion of prairie escapes fire for two or three years the result is seen in the growth of willows and aspens, first in patches, then in large areas, which in a short time become united and cover the country, thus retarding evaporation and permitting the accumulation of vegetable matter in the soil. A fire comes, destroys the young forest growth and establishes a prairie once more. The reclamation of immense areas is not beyond human power; the extension of the prairies is evidently due to fires, and the fires are caused by Indians, chiefly for the purpose of telegraphic communication, or to divert the buffalo from the course they may be taking. These operations will cease as the Indians and buffalo diminish, events which are taking place with great rapidity.

The extension of the prairie country must have exercised a powerful influence upon the Indian population of Rupert's Land. By the progressive limitation of their hunting grounds during the winter season, hostile tribes would possess greater opportunities of destroying one another than when spread over the prairies. Migratory bands of Indians dependent upon wild animals for their support must diminish or increase with the area over which their sustenance extends, and it is apparent that the extension of absolutely treeless prairies and of sterile soil, the formation of "Plains," in a word*, is un-

* See the succeeding chapter for the distinction between Prairie and Plain.

favourable to the increase of the buffalo, the elk, the moose, the antelope and the bear,—animals which always seek the protection of "woods" during the terribly inclement winters of the north-western part of the American continent.

Wood began to be a great treasure in the prairie after passing the Moose Jaws Forks; we were compelled to go supperless to bed on the night of the 25th, because we had neglected to take a supply at the last aspen grove we passed, thinking that the bois de vache (dried buffalo dung) would be found in abundance, but the fires had burnt it also, and not even a fragment was to be procured. No tree or shrub, or even willow twig could be seen in any direction from our camp on the morning of the 25th. Our customary breakfast of tea and buffalo meat was impossible. We had to content ourselves with uncooked pemmican and water from a marsh.

Immediately on the banks of the Qu'appelle Valley near the "Round Hill" opposite Moose Jaws Forks, are the remains of ancient encampments, where the Plain Crees, in the day of their power and pride, had erected large skin tents, and strengthened them with rings of stones placed round the base. These circular remains were twenty-five feet in diameter, the stones or boulders being about one foot in circumference. They wore the aspect of great antiquity, being partially covered with soil and grass. When this camp ground was occupied by the Crees, timber no doubt grew in the valley below, or on the prairie and ravines in detached groves, for their permanent camping grounds are always placed near a supply of fuel.

Making an early start in search of wood, we came suddenly upon four Cree tents, whose inmates were still fast asleep; about three hundred yards west of them we found ten more tents, with over fifty or sixty Indians in all.

They were preparing to cross the valley in the direction of the Grand Coteau, following the buffalo. Their provisions for trade, such as dried meat and pemmican, were drawn by dogs, each bag of pemmican being supported upon two long poles, which are shaft, body and wheels in one. Buffalo Pound Hill Lake, sixteen miles long, begins near the Moose Jaws Forks, and on the opposite or south side of this long sheet of water, we saw eighteen tents and a large number of horses. The women in those we visited on our side of the valley and lake, had collected a great quantity of the mesaskatomina berry which they were drying. In gathering the mesaskatomina, which the Indians represented to be scarce in the valley of the Qu'appelle, they break off the branches of the trees loaded with fruit in order that they may collect the juicy berries with greater ease to themselves, never thinking that this practice continued from year to year must diminish and ultimately extirpate the shrub which they prize so highly, and which forms an important part of their summer food. They announced the cheering intelligence that the Chief Mis-tick-oos, with some thirty tents, was at the Sandy Hills impounding buffalo. Leaving the hospitable Crees after an excellent breakfast on pounded meat and marrow fat, we arrived at Buffalo Pound Hill at noon. The whole country here assumed a different appearance; it now bore resemblance to a stormy sea suddenly become rigid; the hills were of gravel and very abrupt, but none exceeded 100 feet in height. The Coteau de Missouri, particularly the "Dancing Point," is clearly seen from Buffalo Pound Hill towards the south, while north-easterly the last mountain of the Touchwood Hill Range looms grey or blue in the distance. Between these distant ranges a treeless plain intervenes.

Ponds and lakes are numerous on the Grand Coteau

side, and it is probably on this account that the buffalo cross the Qu'appelle valley near the Moose Jaws Fork and west of Buffalo Pound Hill Lake; in the winter they keep towards the Touchwood Hills for the sake of shelter, and the excellent herbage which grows in the beautiful meadows between the aspen clumps. The prairies there too are not so often burned as south of the Qu'appelle, the valley of that river serving as a great barrier to prevent the onward progress of the devastating fires.

We now began to find the fresh bones of buffalo very numerous on the ground, and here and there startled a pack of wolves feeding on a carcass which had been deprived of its tongue and hump only by the careless, thriftless Crees. On the high banks of the valley the remains of ancient encampments in the form of rings of stones to hold down the skin tents are everywhere visible, and testify to the former numbers of the Plain Crees, affording a sad evidence of the ancient power of the people who once held undisputed sway from the Missouri to the Saskatchewan. The remains of a race fast passing away give more than a transient interest to Buffalo Pound Hill Lake. The largest ancient encampment we saw lies near a shallow lake in the prairie about a mile from the Qu'appelle valley. It is surrounded by a few low sandy and gravelly hills, and is quite screened from observation. It may have been a camping ground for centuries, as some circles of stones are partially covered with grass and embedded in the soil.

At noon on the 26th we rested for a few hours opposite to a large camp of Crees on the other side of the lake; our sudden appearance at the edge of the prairie threw them into a state of the greatest excitement as evinced by their haste in collecting their horses and gathering in groups in the valley below. A few of them set out to ride round the lake but in the wrong direction, so that the chance of their overtaking us was highly doubtful. This magnificent sheet of water, never less than half a mile broad and sixteen miles long, shadowed forth what the Qu'appelle valley might become if a river like the Saskatchewan could be made to flow through it. As we neared the height of land, the physical structure of this great valley became a deeply interesting and almost exciting subject of inquiry. So far it had preserved its breadth and depth with astonishing uniformity all the way from the Mission, and we were within forty miles of the South Branch of the Saskatchewan. The hill banks of the river now became clothed with shrubs again, and the ash-leaved maple and elm appeared in the ravines, sustained, no doubt, by the presence of so large a body of water as Buffalo Pound Hill Lake.

Towards evening we arrived at another Cree encampment, where we were again hospitably treated to beaten buffalo meat and marrow fat. Birch bark dishes full of that nutritious but not very tempting food were placed on the ground before us and we were requested to partake of it. The Indians took a piece of the pounded meat in their fingers and dipped it into the soft marrow. A hunting knife which I employed for the same purpose excited their admiration, and after allowing them to examine it, I placed it as I thought securely in my coat pocket; on the following morning, however, the knife was not to be found, nor did I ever see it again. They were delighted to receive a small present of tea and tobacco, and while I was engaged in the tent with the men, the girls, children and old women assembled round the carts asking if we had any rum, and snuffed the boxes and bags containing provisions, in search of that odoriferous stimulant. We left our hospitable friends in the evening and camped about three miles from the last Cree tent. The chief of the band, an old man, accompanied us for some distance, expressing very amicable sentiments, and hinting that it would be as well to keep a watch over our horses during the night, for there were some young scamps among his band who would think it an honour to steal a white man's horse. Visitors came during the evening, and from their actions we thought it advisable to keep watch and tether the horses; observing these precautions they retired at an early hour after a friendly smoke. On the following morning when looking for my hunting knife I was very suspiciously reminded of the old chief's caution; it flashed upon me that the cunning fellow had himself secretly abstracted the knife while pressing his friendly advice.

At dawn we were *en route* again, and towards noon approached the Sandy Hills, the valley continuing about 140 feet deep and maintaining its width. Two days before our arrival the Indians had been running buffalo, and many carcases of these animals were scattered over the arid, treeless plain through which our route lay. Several herds of buffalo were visible wending their way in single file to the Grand Coteau de Missouri distinctly looming south of the Qu'appelle valley. After travelling through a dry, barren region, strewed with erratics, until 2 P.M., we arrived at the lake of the Sandy Hills, and on the opposite side of the valley saw a number of tents, with many horses feeding in the flats. When within a mile of the lake a buffalo bull suddenly appeared upon the brow of a little hill on our right. A finer sight of its kind could hardly be imagined. The animal was in his prime and a magnificent specimen of the buffalo. He gazed at us through the long hair which hung over his eyes in thick profusion, pawed the ground, tossed his head and snorted with proud disdain. He was not more than fifty yards from us, and while we were admiring his splendid proportions he set off at a gallop towards some low hills we had just passed over.

Our appearance on the brink of the valley opposite the tents surprised the Indians, they quickly caught their horses and about twenty galloped across the valley, here quite dry, and in a quarter of an hour were seated in friendly chat with the half-breeds. We kindled a fire with bois de vache, of which there was a vast quantity strewn over the plain, but no wood was near at hand. When the men were going to the lake for water to make some tea, the Indians told us it was salt, and that the only fresh water within a distance of some miles was close to their camp on the opposite side of the valley. We were therefore constrained to cross to the other side and erect our tents near to the spring. Advantage was taken of our passage across the valley to make an instrumental measurement of its leading dimensions. It was found to be 140 feet deep, estimating from the abrupt edge of the bank, and one mile five chains broad. The depth below the general level of the prairie is considerably greater, for there was a descent of fifty or sixty feet by a gentle slope not included in the foregoing measurement. A vast number of erratics strewed this slope, indeed it was with great difficulty that we steered the carts through the formidable accumulation of boulders which beset our path. The bed of the Qu'appelle is quite visible in the valley, but on account of the porous nature of the soil, the overflow from Sand Hill Lake penetrates it in dry weather, and reappears about half a mile below in the form of a little stream about ten feet broad, issuing from a marshy tract occupying the entire breadth of the valley. When crossing it the carts and horses sank deeply in the soft grassy bottom, already much cut up by the passage of a large number of buffalo during the week preceding our arrival.

Although still early in the afternoon, the difficulty of obtaining water and fuel, as well as a desire to procure a guide from the Indians, induced us to camp at the east end of Sand Hill Lake with the Crees by whom we were surrounded. Advantage was taken of the time at our disposal, and the opportunity offered at a large camp, to obtain some information respecting their habits and mode of life.

Scarcely had we made a distribution of tobacco and tea, when a buffalo bull appearing on the opposite side of the valley near where we had passed in the morning, afforded one of the young Indians an opportunity of showing his skill and bravery in attacking this formidable animal single handed and on foot, a conflict which is briefly described in the following chapter as seen through a good telescope from our camp on the south side of the valley.

CHAP. XVI.

SAND HILL LAKE TO THE SOUTH BRANCH OF THE SASKATCHEWAN.

Encounter with a Buffalo Bull.—Interior of Tents.—Barter.—Watchers.—Dogs.—Eyebrow Hill.—Prairies.—Prairies and Plains.—Difference between Prairies and Plains.—Limits of the Prairie Country in the United States.—Growth of Timber in River Bottoms.—Plains in Rupert's Land.—Origin of Prairies.—Grand Coteau de Missouri.—Extent and Boundaries of.—Character of the Grand Coteau.—Elevation of.—Vegetation of.—Eyebrow Hill.—Source of Qu'appelle.—Buffalo.—Character of Qu'appelle Valley.—Water-marks.—Sandy Hills.—Distribution of Boulders.—Section.—Rock Exposure.—Mis-tick-oos.—Sand-Dunes.—South Branch.—The Qu'appelle Valley.—Cree Camp.—Height of Land.—Section of Valley.—Levels.—Buffalo Pound.—Camp Moving.—"Dead Men."—Old Buffalo Pound.—Horrible Spectacle.—New Pound.—Bringing in Buffalo.—Slaughter in Pound.—"A Talk."—Objections to Half-breeds.—To the H. B. Co.—Demeanour of the Indians.—The Wants of Mis-tick-oos.—His Tent.—His Wives.—Rock Exposure.—Boulders in Valley.—Character of the South Branch.

Armed with his bow from the bois d'arc, his arrows from the mesaskatomina, neatly feathered with the plumes of the wild duck, and headed with a barb fashioned from a bit of iron hoop, the young Plain Cree threw off his leather hunting shirt, jumped on a horse, and hurried across the valley. Dismounting at the foot of the bank, he rapidly ascended its steep sides, and just before reaching the top, cautiously approached a large boulder which lay on the brink, and crouched behind it.

The buffalo was within forty yards of the spot where the Indian crouched, and slowly approaching the valley as he leisurely cropped the tufts of parched herbage which the sterile soil was capable of supporting. When within twenty yards of the Indian the bull raised his head, snuffed the air, and began to paw the ground. Lying at full length, the Indian sent an arrow into the side of his huge antagonist. The bull shook his head and mane, planted his fore feet firmly in front of him, and looked from side to side in search of his unseen foe, who after driving the arrow, had again crouched behind the boulder. Soon, however, observing the fixed attitude of the bull, a sure sign he was severely wounded, he stepped on one side and showed himself. The bull instantly charged, but when within five yards of his nimble enemy, the Indian sprang lightly behind the boulder, and the bull plunged headlong down the hill, receiving after he had passed the Indian, a second arrow in his flanks. As soon as he reached the bottom, he fell on his knees, and looked over his shoulder at his wary antagonist, who, however, speedily followed, and observing the bull's helpless condition, sat on the ground within a few yards of him and waited for the death gasp. After one or two efforts to rise, the huge animal drooped his head and gave up the strife. The Indian was at his side without a moment's pause, cut out his tongue, caught his horse, — an excited spectator of the conflict,—and galloping across the valley, handed me the trophy of his success.

We made ourselves acceptable to the Indians by offering them a present of powder, shot, tea, and tobacco, and in return they invited us to partake of pounded meat, marrowfat and berries. The chief of the band assured us that his young men were honest and trustworthy; and in compliance with his instructions, property would be perfectly safe.

I visited the interior of most of their tents, and found the squaws almost exclusively engaged in drying buffalo meat. A couple of table-spoonfuls of tea, and a small plug of tobacco, always ensured a hearty welcome, and in return they generally presented me with a choice piece of buffalo meat from a fat cow, or a small skin of marrowfat. One of the young men took a fancy to a checked flannel shirt I was wearing at the time, and offered his saddle for it; declining the bargain, he added his bow and a quiver of arrows; I told him to bring the bow and arrows to my tent at night, and I would give him a new shirt for them. He said he should prefer a white one, and then the buffalo would not mind him; and when he came to complete his bargain, he selected a white jersey in preference to a showy coloured check. From time to time, scouts would come in and go out towards the Grand Coteau, on the look out for Blackfeet, and as nightfall approached, the wandering horses were gathered closer to the camp. The dogs, however, are their great protection; it is almost impossible for any stranger to approach a camp without arousing the whole canine population; and the passage of bands of buffalo during the night-time is signalised by a prolonged baying, which, however suggestive of sport and good cheer, is most wearisome to those who are anxious to rest. During the night a heavy rain filled the hollows with water, and gave us promise of an abundant supply until we arrived at the Sandy Hills, where the main body of Plain Crees were encamped. On the following day, the 28th, I rode to the Eye-brow Hill range, a prolongation of the Grand Coteau, and distant from the Qu'appelle Valley about four miles. It was there that the Indians told me I should find one of the sources of the Qu'appelle river. After an hour's ride I reached the hills, and quickly came upon a deep ravine at the bottom of which bubbled a little stream about three feet broad. I subsequently followed its course until it entered the prairie leading to the great valley, and traced it to its junction with the main excavation, through a deep narrow gully.

The Eye-brow Hill range is about 150 feet above the prairie, and forms the flank of a tableland stretching to the Grand Coteau, of which it is the northern extension. The recent tracks of buffalo were countless on the hill sides, and in the distance several herds could be seen feeding on the treeless plateau to the south. On the flanks of the Grand Coteau the true prairies may be said to terminate, and the plains to commence. It is doubtful whether the term "Plain" is not now applicable to a large portion of the country west and south of the Qu'appelle Mission. The destruction of "woods" by fires has converted into sterile areas an immense tract of country which does not appear necessarily sterile from aridity, or poverty of soil.

The Plains and Prairies of America occupy regions differing widely from one another in physical characteristics. The phraseology of the half-breeds tends to mislead a traveler not familiar with the precise application of the words they use. Such terms as "woods," "prairies," and "plains" are illustrations of this apparent want of precision, which if employed without explanation in a written narrative, would very probably cause considerable misapprehension, and lead to deductions wholly at variance with fact.* A tract of country may be described as a "wooded country," conveying the idea that timber covers the surface and is capable of affording a supply of that indispensable material for building purposes and fuel; but in Rupert's Land, west of the Low Lake Region, and south of the 53rd parallel, the "woods" consist generally of

* Vide Col. Emory's Remarks quoted in the Chapter on the Climate of the Southern Part of Rupert's Land, Vol. II.

small aspens very rarely exceeding six inches in diameter or twenty-five in altitude, and most frequently distributed in detached groves, " bluffs," or belts. The same remark applies to the use of the word " prairie," and to prairie country; prairies may be level, rich and dry, sustaining luxuriant grasses and affording splendid pasturage; they may be marshy and wet, or undulating—" stony," " sandy," or " salt." Such indefinite terms as " open prairie," " rolling prairie," " alluvial prairie," not unfrequently employed in describing without limit as to space, the vast unpeopled wastes,—often beautiful and rich, often desolate and barren,—of the Prairies and *Plains* of America, are sometimes both physically and geologically wrong, and serve to convey the impression that the large areas to which they are applied possess, if not a fertile, at least not an unkindly soil, or an arid climate, rendering husbandry hopeless.

The difference between " Prairies " and " Plains " will be best shown by describing their limits in the United States and Rupert's Land. In the United States the true *Prairie* region extends " over the eastern part of Ohio, Indiana, the southern portion of Michigan, the southern part of Wisconsin, nearly the whole of the States of Illinois and Iowa, and the northern portion of Missouri, gradually passing, in the territories of Kansas and Nebraska, into the *Plains*, or the arid and desert region which lies at the base of the Rocky Mountains. This passage takes place in the region between the meridians of 97° and 100°, west of which belt the country becomes too barren to be inhabited and worthless for cultivation. The passage from the heavily wooded region of the north and east into the treeless plains of the west is a gradual one, and the disappearance of the underwood and the predominance of " oak openings," or groves of oak and other forest trees, not crowded together, but scattered over the surface at a considerable distance from one another, without any low shrubs or underbrush between them, is the characteristic of the border of the Prairie region."*

The growth of timber in the river bottoms in the United States disappears altogether at the borders of the Plains about the 98° of longitude.† In Rupert's Land, south of the 52nd parallel, trees cease in the river bottoms except near the Upper Qu'appelle Lakes, in longitude 104°; but they occur only on the *northern* aspect of the south side of the deep valley west of longitude 102°, or a few miles west of Fort Ellice.

The true limit of the Plains in Rupert's Land, east of the South Branch, is well shown by the Grand Coteau de Missouri. The country east of that natural boundary may be classed as Prairie country, over the greater portion of which forests of aspen would grow if annual fires did not arrest their progress. The plateau of the Grand Coteau forms the true Plains of Rupert's Land, where both soil and climate unite in establishing a sterile region. Mr. J. D. Whitney considers the absence of forest on the rich prairies of the United States to be mainly due to the physical nature of the soil; "the extreme firmness of the particles of which the prairie soil is composed, is probably the principal reason why it is better adapted to the growth of its peculiar vegetation than to the development of forests."‡ The origin of the groves scattered over the Prairies is traced to ridges of coarse material, apparently deposits of drift, on which, from some local cause, there

* Report on the Geological Survey of the State of Iowa, by James Hall, State Geologist, and J. D. Whitney, Chemist and Mineralogist.

† Ibid.

‡ Geology of Iowa.

has never been an accumulation of fine sediment. The origin of the prairies and plains of Rupert's Land, as well as the present distribution of timber, will be discussed in another chapter. The Grand Coteau de Missouri, distinctly visible from the Eye-brow Hill, begins in latitude 45°, about sixty miles south-west of the head of the Coteau des Prairies in latitude 45° 55′; the intervening valley is occupied by James's River, an affluent of the Missouri. Its boundary pursues a course nearly due north, under the 99th meridian as far as the 47th parallel, when, turning north-westerly, it enters British territory near the 104th meridian, and still preserving a north-westerly direction comes on the South Branch of the Saskatchewan, a few miles from the Elbow, in longitude 108°.* The region east of the Grand Coteau belongs to the prairie region, the Grand Coteau itself and its prolongation towards Battle River, from its eastern boundary to the foot of the Rocky Mountains, constitutes the "Plains" properly so called of the north-western territories of the United States and of British America. From the character of its soil and the aridity of its climate, the Grand Coteau is permanently sterile and unfit for the abode of civilised man. The length of its abrupt flank, not including small bays or indentations, is about 650 miles from its commencement in the valley of James River to the Elbow of the South Branch of the Saskatchewan. For a distance of 380 miles the course of the Missouri is approximately parallel to the flanks of the Grand Coteau, and at an average distance of 50 miles from it. This vast tableland rises from 400 to 800 feet above the Missouri, and bears the same

* Vide Map accompanying Capt. Palliser's Reports; also Military Map of Nebraska and Dakotah, by Lieut. G. K. Warren, Top. Eng. U. S. Army.

altitude in relation to the high prairies through which the Shayenne meanders to the Red River of the north, on the one hand, and James River to the Missouri on the other.

The vegetation on the Grand Coteau is very scanty, the Indian turnip is common, so also is a species of cactus; no tree or shrub is seen, and it is only in the bottoms and marshes that rank herbage is found.*

In the afternoon I bade farewell to our Cree friends, and riding west joined the carts on the south side of Sand Hill Lake, on the brink of which we travelled until we arrived at the gully through which the stream from the Eye-brow Hill range enters the Qu'appelle valley. It was here nine feet broad, and three deep, having received accessions in a short course through the prairie from the hills where I had observed it scarcely three feet broad. We camped in the valley, and employed the evening in taking levels.

About four miles west of us we saw the Sandy Hills, and could discern the great valley passing through them, and containing, as the Indians had alleged, ponds which sent water both to the South Branch and the Assinniboine; an important physical fact which we afterwards verified instrumentally and by optical proof. We found the streamlet from the Eye-brow Hill range strike the Qu'appelle Valley eight and a half miles west of Sand Hill Lake, and four miles from the height of land where the ponds lie. The fall between the ponds and our camp was about five feet, and the valley 150 feet deep, and one mile seventy chains broad. The Eye-brow Hill stream had excavated a channel nine feet deep in the

* Vide Explorations and Surveys for a Railroad to the Pacific, by Governor Stevens.

bottom of the great valley, and was joined, by a sluggish brook coming from the ponds above, a few yards from our camp. Water marks on the hill banks showed that the entire breadth of the valley is flooded during spring freshets.

The Sandy Hills commence on the north side about two miles west of Sand Hill Lake as it appears in summer. They are drifting dunes, and many of them present a clear ripple-marked surface without any vegetation, not even a blade of grass. They have invaded the great valley, and materially lessened its depth. One feature in its banks is worthy of special notice. Many boulders or erratics are distributed over the western extremities of small hills or ridges into which the steep banks are broken, 70 to 120 feet above the level of the flats. These ridges have the form of long, narrow islands, their longitudinal axes being parallel to the sides of the valley, and the erratics are deposited and arranged on the top of each ridge, and at the west extremity. The form of these ridges is also peculiar, they are sharp at the west end where the erratics lie, and rounded at the east end. The slope is gentle at the west end, abrupt at the east end. This peculiarity is a constant feature of all the ridges seen on the sides of the banks of the valley. They vary in height from 10 to 30 feet, and in length from 60 to 140 feet, and in breadth from 20 to 80 feet. They have evidently some relation to the excavating force which has produced this great valley, and cannot be attributed to the long continued action of a small stream; however competent running water may be to produce deep and long depressions in loose drift, or a soft friable rock.

A section of the bank of the Eye-brow Hill stream in its course through the flats, showed fine clay brought by recent rains from the hill banks, sand blown from the dunes, and loam produced by the blending of the two. Where it leaves the prairie the little river has exposed a section of a drift hill round the base of which it sweeps. Gravelly drift is seen to repose upon an ochreous stratified rock, seamed with veins of selenite. It exhibits a stratum of yellow and red ferruginous clay about six feet thick, and below it a hard greenish sandstone, in which gigantic concretionary masses are numerous. Veins of selenite penetrate the greenish coloured rock, but are most abundant in the ferruginous clay. This is the first rock seen in position above the Mission. Subsequent comparison with the rocks on the South Branch showed it to belong to the uppermost member of the cretaceous series.

On the morning of the 29th, we prepared to visit the main body of the Crees at the Sandy Hills, and with a view to secure a favourable reception sent a messenger to announce our arrival, and to express a wish to see Mis-tick-oos, their chief. Soon after breakfast we crossed the valley and threaded our way between sand dunes; one dune was found to be seventy feet high, quite steep on one side, beautifully ripple-marked by the wind, and crescent-shaped; from its summit we saw the woods and hills beyond the South Branch of the Saskatchewan, and what was more delightful to us, traced with the eye the Qu'appelle valley apparently with undiminished depth and breadth through the Sandy Hills, until it was lost as it dipped towards the South Branch.

At eight o'clock A.M. we came in sight of the Cree camp, and soon afterwards messengers arrived from Mis-tick-oos, in reply to the announcement we had transmitted to him of our approach, expressing a hope that we would delay our visit until they had moved their camp half-a-mile further west, where the odour of the putrid buffalo would be less annoying. We employed the time in ascertaining the exact position of the height of land, and soon found a pond from which we observed water flowing to the Saskatchewan and the Assinniboine. The pond was fed by a number of springs and small streams, a foot or two broad, issuing from the Sandy Hills at right angles to the valley.

Sand hills or dunes cover the country for a considerable distance on both sides. We selected this spot to level across it, and found the depth to be 110 feet below

Transverse Section of the Valley of the Qu'appelle at the height of land.

the first plateau, its breadth, although partially invaded by sand dunes, seventy-three chains, or nearly one mile. On the south side it rises much more gently than on the north, and shows two terraces, both of which are covered with drifting sand. Here we commenced taking the levels to the South Branch, twelve miles distant from us, an operation which we soon found necessary to close for the present, in consequence of the arrival of about sixty Cree horsemen, many of them naked with exception of the breech cloth, and belt. They were accompanied by the chief's son, who informed us that in an hour's time they would escort us to the camp.

They were about constructing a new pound, having literally filled the present one with buffalo, and being compelled to abandon it on account of the stench which arose from the putrifying bodies. We sat on the ground and smoked, until they thought it time for us to accompany them to their encampment. Mis-tick-oos had hurried away to make preparations for "bringing in the buffalo,"* the new pound being nearly ready. He expressed, through his son, a wish that we should see them entrap the buffalo in this pound, a rare opportunity few would be willing to lose.

We passed through the camp to a place which the chief's son pointed out, and there erected our tents. The women were still employed in moving the camp, being assisted in the operation by large numbers of dogs, each dog having two poles harnessed to him, on which his little load of meat, pemmican, or camp furniture was laid. After another smoke, the chief's son asked me, through the interpreter, if I would like to see the old buffalo pound, in which they had been entrapping buffalo during the past week. With a ready compliance I accompanied the guide to a little valley between sand hills, through a lane of branches of trees, which are called "dead men" to the gate or trap of the pound. A sight most horrible and disgusting broke upon us as we ascended a sand dune overhanging the little dell in which the pound was built. Within a circular fence 120 feet broad, constructed of the trunks of trees, laced with withes together, and braced by outside supports, lay tossed in every conceivable position over two hundred dead buffalo. From old bulls to calves of three months old, animals of every age were huddled together in all the forced attitudes of violent death. Some lay on their backs, with eyes starting from their heads, and tongue thrust out through clotted gore. Others were impaled on the horns of the

* A half-breed expression signifying "to drive the buffalo from the Prairie towards and into the pound," or Ponds, as they are termed in half-breed phraseology.

old and strong bulls. Others again, which had been tossed, were lying with broken backs two and three deep. One little calf hung suspended on the horns of a bull which had impaled it in the wild race round and round the pound.

The Indians looked upon the dreadful and sickening scene with evident delight, and told how such and such a bull or cow had exhibited feats of wonderful strength in the death-struggle. The flesh of many of the cows had been taken from them, and was drying in the sun on stages near the tents. It is needless to say that the odour was overpowering, and millions of large blue flesh flies, humming and buzzing over the putrefying bodies was not the least disgusting part of the spectacle. At my request the chief's son jumped into the pound, and with a small axe knocked off half a dozen pair of horns, which I wished to preserve in memory of this terrible slaughter. "To-morrow," said my companion, "you shall see us bring in the buffalo to the new pound."

After the first "run," ten days before our arrival, the Indians had driven about 200 buffalo into the enclosure, and were still urging on the remainder of the herd, when one wary old bull, espying a narrow crevice which had not been closed by the robes of those on the outside, whose duty it was to conceal every orifice, made a dash and broke the fence, the whole body then ran helter skelter through the gap, and dispersing among the sand dunes escaped, with the exception of eight who were speared or shot with arrows as they passed in their mad career. In all, 240 animals had been killed in the pound, and it was its offensive condition which led the reckless and wasteful savages to construct a new one. This was formed in a pretty dell between sand hills, about half a mile from the first, and leading from it in two diverging rows, the bushes they designate "dead men," and which serve to guide the buffalo when at full speed, were arranged. The "dead men" extended a distance of four miles into the prairie, west of and beyond the Sand Hills. They were placed about 50 feet apart, and between the extremity of the rows might be a distance of from one and a half to two miles.

When the skilled hunters are about to bring in a herd of buffalo from the prairie, they direct the course of the gallop of the alarmed animals by confederates stationed in hollows or small depressions, who, when the buffalo appear inclined to take a direction leading from the space marked out by the "dead men," show themselves for a moment and wave their robes, immediately hiding again. This serves to turn the buffalo slightly in another direction, and when the animals, having arrived between the rows of "dead men," endeavour to pass through them, Indians here and there stationed behind a "dead man," go through the same operation, and thus keep the animals within the narrowing limits of the converging lines. At the entrance to the pound there is a strong trunk of a tree placed about one foot from the ground, and on the inner side an excavation is made sufficiently deep to prevent the buffalo from leaping back when once in the pound. As soon as the animals have taken the fatal spring they begin to gallop round and round the ring fence looking for a chance of escape, but with the utmost silence women and children on the outside hold their robes before every orifice until the whole herd is brought in; they then climb to the top of the fence, and, with the hunters who have followed closely in the rear of the buffalo, spear or shoot with bows and arrows or fire-arms at the bewildered animals, rapidly becoming frantic with rage and terror, within the narrow limits of the pound.

PLAIN CREES DRIVING BUFFALOES INTO A POUND

A dreadful scene of confusion and slaughter then begins, the oldest and strongest animals crush and toss the weaker; the shouts and screams of the excited Indians rise above the roaring of the bulls, the bellowing of the cows, and the piteous moaning of the calves. The dying struggles of so many huge and powerful animals crowded together, create a revolting and terrible scene, dreadful from the excess of its cruelty and waste of life, but with occasional displays of wonderful brute strength and rage; while man in his savage, untutored, and heathen state shows both in deed and expression how little he is superior to the noble beasts he so wantonly and cruelly destroys.

Mis-tick-oos, or "Shortstick," is about fifty years old, of low stature, but very powerfully built. His arms and breast were deeply marked with scars and gashes, records of grief and mourning for departed friends. His son's body was painted with blue bars across the chest and arms. The only clothing they wore consisted of a robe of dressed elk or buffalo hide, and the breech cloth; the robe was often cast off the shoulders and drawn over the knees when in a sitting posture; they wore no covering on the head, their long hair was plaited or tied in knots, or hung loose over their shoulders and back. The forms of some of the young men were faultless, of the middle-aged men bony and wiry, and of the aged men, in one instance at least, a living skeleton. I inquired the age of an extremely old fellow who asked me for medicine to cure a pain in his chest; he replied he was a strong man when the two Companies (the Hudson's Bay and the North West) were trading with his tribe very many summers ago. He remembers the time "when his people were as numerous as the buffalo are now, and the buffalo thick as trees in the forest." The half-breeds thought he was more than 100 years old.

When Mis-tick-oos was ready to receive me, I proceeded to the spot where he was sitting surrounded by the elders of his tribe, and as a preliminary, rarely known to fail in its good effect upon Indians, I instructed one of my men to hand him a basin of tea and a dish of preserved vegetables, biscuit, and fresh buffalo steaks. He had not eaten since an early hour in the morning, and evidently enjoyed his dinner. Hunger, that great enemy to charity and comfort, being appeased, I presented him with a pipe and a canister of tobacco, begging him to help himself and hand the remainder to the Indians around us. The presents were then brought and laid at his feet. They consisted of tea, tobacco, bullets, powder, and blankets, all which he examined and accepted with marked satisfaction. After a while he expressed a wish to know the object of our visit; and having at my request adjourned the meeting to my tent in order to avoid sitting in the hot sun, we held a "talk," during which Mis-tick-oos expressed himself freely on various subjects, and listened with the utmost attention and apparent respect to the speeches of the Indians he had summoned to attend the "Council."

All speakers objected strongly to the half-breeds' hunting buffalo during the winter in the Plain Cree country. They had no objection to trade with them or with white people, but they insisted that all strangers should purchase dried meat or pemmican, and not hunt for themselves.

They urged strong objections against the Hudson's Bay Company encroaching upon the prairies and driving away the buffalo. They would be glad to see them establish as many posts as they chose on the edge of the prairie country, but they did not like to see the prairies and plains invaded. During the existence of the two companies, all went well with the Indians, they obtained excellent pay, and could always sell their meat, skins, robes, and pemmican. Since the union of the companies they had not fared half so well, had received bad pay for their provisions, and were growing poorer, weaker, and more miserable year by year. The buffalo were fast disappearing before the encroachments of white men, and although they acknowledged the value of fire-arms, they thought they were better off in olden times, when they had only bows and spears, and wild animals were numerous. They generally commenced with the creation, giving a short history of that event in most general terms, and after a few flourishes about equality of origin, descended suddenly to buffalo, half-breeds, the Hudson's Bay company, tobacco, and rum. I asked Mis-tick-oos to name the articles he would wish me to bring if I came into his country again. He asked for tea, a horse of English breed, a cart, a gun, a supply of powder and ball, knives, tobacco, a medal with a chain, a flag, a suit of fine clothes, and rum. The "talk" lasted between six and seven hours, the greater portion of the time being taken up in interpreting sentence by sentence, the speeches of each man in turn.

During the whole time we were engaged in "Council" the pipe was passed from mouth to mouth, each man taking a few whiffs and then handing it to his neighbour. It was a black stone pipe, which Mis-tick-oos had received as a present from a chief of the Blackfeet at the Eagle Hills a few weeks before.* When the pipe came round to me I usually replenished it, and taking a box of "vespers" from my pocket, lit it with a match. This operation was observed with a subdued curiosity, each Indian watching me with-

* See Chapter on Indian customs, superstitions, &c., Vol. II.

out moving his head, turning only his eyes in the direction of the pipe. No outward sign of wonder or curiosity escaped them during the "talk." On one occasion the pipe was out when passed to the Indian sitting next to me; without turning his head he gently touched my arm, imitated the action of lighting the match by friction against the bottom of the box, and pointed with one finger to the pipe. They generally sat with their eyes fixed on the ground when one of them was speaking, giving every outward sign of respectful attention, and occasionally expressing their approval by a low gurgling sound. When the talk was over, I went with Mis-tick-oos to his tent; he then asked me to reproduce the match-box, and show its wonders to his four wives. One of them was evidently sceptical, and did not think it was "real fire" until she had ignited some chips of wood from the lighted match I presented to her. I gave a bundle to Mis-tick-oos, who wrapped them carefully in a piece of deerskin, and said he should keep them safely,—they were "good medicine." I made his wives happy and merry by distributing about a pound of tea amongst them, and their hilarity knew no bounds when the incident narrated in the second volume* took place.

Mis-tick-oos apologised for the smallness of his tent, remarking that he had collected twenty skins for a new one, and when he had obtained two more he should get his wives to make them up. I asked how it was that a chief with four wives should have a smaller tent than many of his young men around us. No reply was given to this question, and the half-breed interpreter said that there was a mystery attached to the tent formerly possessed by Mis-tick-oos, which he would endeavour to solve, but

* Chapter on Indian customs and superstitions, &c., Vol. II.

none of the people present would talk about it, so he surmised that there might be some disappointment, or perhaps the death of a relative connected with it. From information afterwards obtained, and verified at the Hudson's Bay Company's post at the Touchwood Hills, it appears that a favourite son of Mis-tick-oos was killed while hunting for the post some months before, and that in accordance with the customs of the Crees, Mis-tick-oos had presented the officer in charge of the post with his tent of twenty-two skins as a "grief offering."

Early on the morning of the 30th I retraced my steps to examine an exposure of cretaceous rock forming part of the bank at the summit level of the Qu'appelle valley, while Mr. Fleming continued taking the levels to the South Branch. The rock is a sandstone, dipping very slightly to the south-west. The length of the exposure, east and west, is about fifty yards; it is covered with drifting sand. Near the summit the layers are highly fossiliferous, and almost wholly composed of *Avicula Linguæformis* (Evans and Shumard); above and below the fossiliferous portion there is a coarse greenish-coloured sand, interstratified with brown ferruginous layers. The thickness visible is about twelve feet. The rock occurs near the bend of the valley at its summit level; the exposure is perpendicular, and about sixty feet above the bottom of the valley. Some of the beds, those which are unfossiliferous are very soft and friable, easily disintegrating, and may, farther west, be the origin of some of the sand dunes distributed over so wide an area in this part of the country. In descending the slope from the summit level to the Saskatchewan, the boulders on the ridges in the valley were found to be generally deposited upon the west side. The inclination of the boulders was towards the east, those forming the upper stratum were inclined against or superimposed upon the west side of those beneath, leading to the inference that the current which directed the course of the ice which bore them, came here, as on the other side of the summit level in the valley, *from the west.*

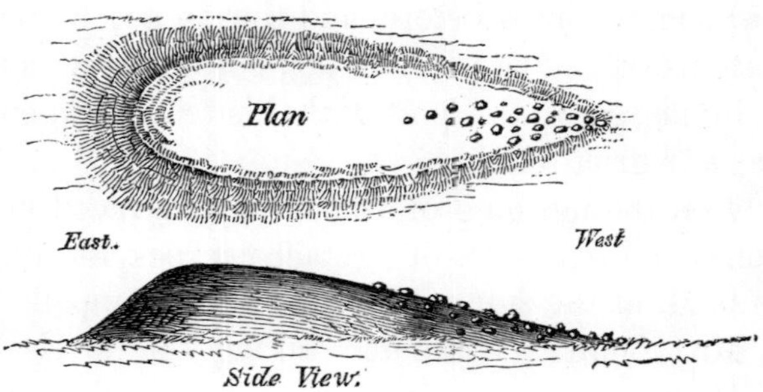

Ridges, with Boulders, on the East and West side of the height of land in the Qu'appelle Valley.

About fourteen miles from the South Branch there is a gigantic erratic of unfossiliferous rock on the south side of the valley. It is seventy-nine feet in horizontal circumference, three feet from the ground ; and a tape stretched across the exposed portion, from side to side over the highest point, measured forty-six feet. The Indians place on it offerings to Manitou, and at the time of our visit it contained beads, bits of tobacco, fragments of cloth, and other trifles.

At noon I bade farewell to Mis-tick-oos, and joining the carts we wended our way by the side of "the River that Turns," occupying the continuation of the Qu'appelle valley, to the South Branch of the Saskatchewan. The carts were accompanied by several Indians, who watched with much curiosity the progress of taking the levels, and were very anxious to know what "medicine" I was searching for when sketching the position of the erratics in the valley.

Now and then a fine buffalo bull would appear at the brow of the hill forming the boundary of the prairie, gaze at us for a few minutes and gallop off. The buffalo were crossing the South Branch a few miles below us in great numbers, and at night, by putting the ear to the ground, we could hear them bellowing. Towards evening we all arrived at the South Branch, built a fire, gummed the canoe, which had been sadly damaged by a journey of 700 miles across the prairies, and hastened to make a distribution of the supplies for a canoe voyage down that splendid river. We were not anxious to camp at the mouth of "the River that Turns," in consequence of a war party of Blackfeet who were said to be in the neighbourhood of the Cree camp, watching for an opportunity to steal horses, and if possible to "lift a scalp."

The Indians who had accompanied us hastened to join their friends as soon as they saw we were ready to embark, and just as the sun set, the canoe containing Mr. Fleming and myself, with two half-breeds, pushed off from the shore; the remainder of the party in charge of the old hunter, retired from the river with the carts and horses to camp in the open prairie, where they would be able to guard against a surprise by the Blackfeet, or the thieving propensities of treacherous Crees. Great precautions were undoubtedly necessary, as sure signs had been observed within three miles of the Sandy Hills, proving that a war party of Blackfeet were skulking about. The Crees, always accustomed when on the South Branch to their attacks, merely adopted the precaution of posting watchers on the highest dunes, about a mile from their camp, but in accordance with the friendly advice of Mis-tick-oos, we embarked at this late hour in the evening with a view to avoid surprise and mislead any watchful eyes that might have taken note of our movements. We drifted a mile or two down the river until we came to a precipitous cliff showing a fine exposure of rock, which proved a temptation too great to be resisted, so we drew the canoe on the bank and camped for the night on the east side of the river, making arrangements to watch in turns.

The first view of the South Branch of the Saskatchewan, fully 600 miles from the point where the main river disembogues into Lake Winnipeg, filled me with astonishment and admiration. We stood on the banks of a river of the first class, nearly half a mile broad, and flowing with a swift current, not more than 350 miles from the Rocky Mountains, where it takes its rise. We had reached this river by tracing for a distance of 270 miles, a narrow deep excavation continuous from the valley of one great river to that of another, and exhibiting in many features evidences of an excavating force far greater than the little Qu'appelle which meanders through it, was at the first blush, thought capable of creating. How were the deep lakes hollowed out? lakes filling the breadth of the valley, but during the lapse of ages not having increased its breadth, preserving too, for many miles, such remarkable depths, and although in some instances far removed from one another, yet maintaining those depths with striking uniformity. What could be the nature of the eroding force which dug out narrow basins 54 to 66 feet deep at the bottom of a valley already 300 feet below the slightly undulating prairies, and rarely exceeding one mile in breadth? It was easy to understand how a small river like the Qu'appelle could gradually excavate a valley a mile broad and 300 feet deep. The vast prairies of the north-

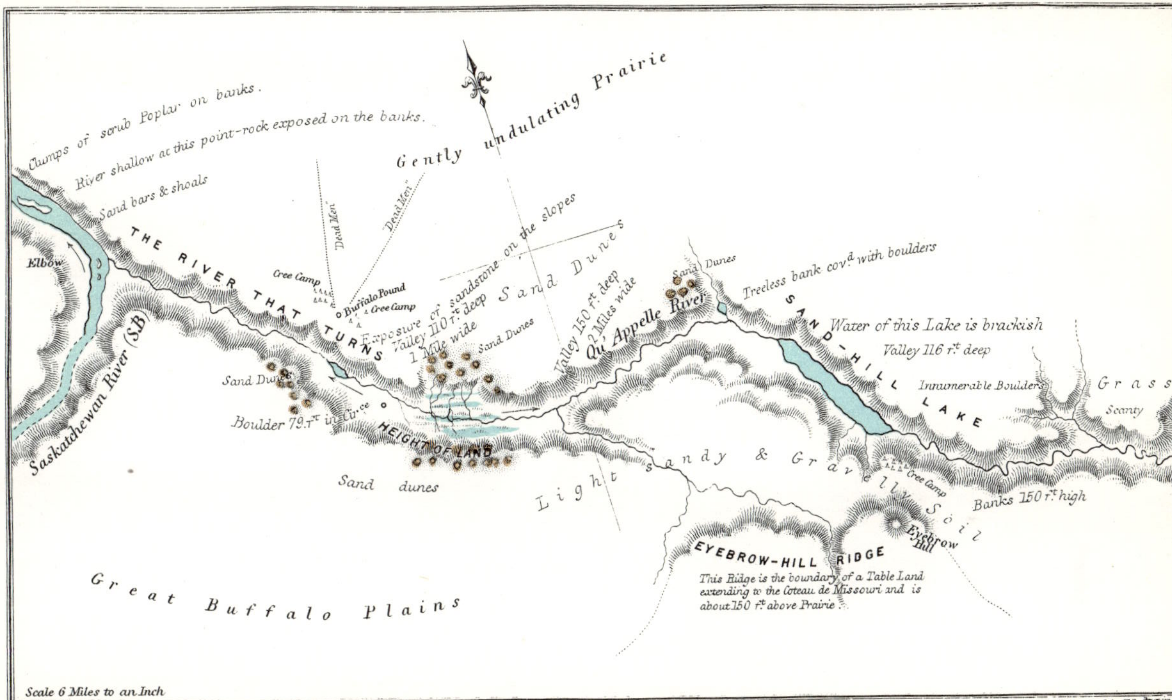

Engraved by Edw.d Weller

Plan shewing the Junction of the Qu'Appelle Valley with the Saskatchewan.

west offer many such instances; the Little Souris River, for example, in passing through the Blue Hills; the Assinniboine, for a 150 miles, flows through a broad deep valley, evidently excavated by its waters; the rivers in western Canada often flow in deep eroded valleys; but in no instance to my knowledge are deep and long lakes known to occupy a river valley, where nearly horizontal and very soft rocks preclude the assumption that they may have been occasioned by falls, without bearing some traces of the force which excavated their basins. They seem to point to the former existence of a much deeper valley now broken into detached lakes by the partial filling up of intervening distances. It was certainly with mingled feelings of anxiety and pleasurable anticipation that we embarked on the broad Saskatchewan, hoping during our long journey down its swift stream to find some clue to the origin of the curious inosculating valley of the Qu'appelle we had traced from one water-shed to another.

CHAP. XVII.

FROM THE QU'APPELLE MISSION TO FORT ELLICE, DOWN THE QU'APPELLE RIVER.

The Second Fishing Lake — Depth of.—Indian Map. — Origin of name Qu'appelle, or "Who calls River."—The First Lake, or Pakitawiwin.—Great Depth of First Lake. — Fish.— Confervæ. — Depth of Valley.—Width of River.— High-water Mark.— Valley flooded. — Affluents.—Depth of Valley.—Crooked Lake, or Ka-wa-wa-ki-ka-mac—Dimensions of.—Effects of Fires.— Trees in Valley.— Boulders.— Character of the Country.—Indian Surprise.—Indians.—Summer Berry Creek.— Dimensions of Valley.—Valley and Prairie Scene.—Camp Scene.—Character of Valley. — Ka-wah-wi-ya-ka-mac, or Round Lake — Dimensions of. — Stony Basin.—Granite Boulders.—Little Cut-Arm Creek.—The Scissors Creek.— Rock Exposure. -- Grasshoppers.— Big Cut-Arm Creek.— Dimensions of Qu'appelle. — Flooding of Valley.—Timber.—Undergrowth —Birds. — Minks. — Deer. — Uniformity of Qu'appelle Valley.— Table showing the dimensions of the Qu'appelle Valley and Qu'appelle Lakes.

MR. DICKINSON'S NARRATIVE.

DEAR SIR,—Soon after parting from you on the morning of July 20th, at the Church of England Mission in the Qu'appelle Valley, my instruments for surveying, with watch, a magnetic compass, a log line and sounding line, all arranged for ready use, and a cargo of kettles, pans, pemmican and blankets stowed away, our little canoe commenced its voyage down the river. In half an hour we reached the lake, which is generally called the second of the Fishing Lakes. Before venturing to go down it we were obliged to stop for the purpose of gumming the canoe, as it was leaking more than was desirable. To save time we took breakfast here. The distance between this lake and the one at the Mission is a mile and a half, while the actual length of the river is upwards of two miles. Its width averages eighty feet, and its depth three feet; the rate of current, which is nearly uniform throughout its length, is one mile per hour. The difference of level between these two lakes, obtained instrumentally on a previous day, is 1·50 feet. These measurements, not valuable in themselves, are taken for the purposes explained in the "Rules for conducting the Exploring Survey," namely, as the means for calculating approximately the total fall in the river. I may mention, that at every opportunity similar measurements and observations were made, with the assistance of Mr. John Fleming, from which we were able to deduce some general rules for guiding us in estimating the fall in rivers, and also, that the log line was found to be most invaluable in ascertaining the rate of the canoe on the rivers as well as on the lakes, being a much more accurate way than that of estimating it by the eye.

The canoe being now declared to be sea-worthy, we started on our way again. The lake is three miles and a quarter long and three-quarters of a mile in breadth, extending between the slopes of the valley, and appearing to be merely an expansion of the river, but on trial found to be something more than that. For some distance out from the mouth of the river it is only from three to four feet deep, but on trying it, when we were about half a mile distant, with a sounding line thirty feet long, to my great surprise, I could find no bottom; having added more line, the depth proved to be forty-two feet. About the middle of the lake the depth is forty-eight feet.

A stream a quarter of a mile in length, flowing sluggishly through a marsh, connects this lake with the next, the first of the Fishing Lakes, or as it is in Cree, *Pakitawiwin.* All the Indian names of the lakes and tributaries of the

Qu'appelle I got afterwards on my arrival at Fort Ellice, from an old Indian seventy years of age, who had been once upon a time a great hunter and warrior, now in peace and comfort spending his remaining days at the hospitable Fort. With a piece of charred wood he drew on the floor a map of the Qu'appelle Valley from the Fishing Lakes to the Assinniboine, showing every little creek so accurately that I easily recognised them. Mr. McKay, who was then in charge of the Fort, kindly acted as interpreter on the occasion. The Cree name of the Qu'appelle river is *Katapaywie sepe*, and this is the origin of the name as told me by the Indian :—A solitary Indian was coming down the river in his canoe many summers ago, when one day he heard a loud voice calling to him; he stopped and listened and again heard the same voice as before. He shouted in reply, but there was no answer. He searched everywhere around, but could not find the tracks of any one. So from that time forth it was named the "Who Calls River."

Pakitawiwin is six miles long and half a mile wide, and is most wonderfully deep. In one place, by means of putting together various pieces of cord, sashes, &c., the sounding line being too short, the depth was found to be about sixty-six feet. The mean of several depths is fifty-two feet. It is famous for the quantity and quality of its fish. For three miles we passed through a dense decaying mass of confervæ, which an east wind had driven to the upper parts of the lake. The smell of it was most unpleasant; the men pushed through it as hard as they could, no easy matter, as it impeded the progress of the canoe considerably. The valley here is about the same depth as it is at the Mission, but the slopes are not so precipitous; one of them, that on the south side, has been the whole way covered with a dense growth of young aspens, and the other has been bare of trees except in some of its many hollows and ravines.

Leaving the lake we now descend the river at an average speed of four miles an hour, the rate of current being generally about one mile and a quarter per hour. Paddling was easy work, but the steering by no means so, for the bends of the river are innumerable and very sharp, and the waters sweep round them with great velocity; oftentimes, but for the strong and dexterous arm of the steersman, the canoe would have been dashed against the bank; as it was he could not avoid sometimes getting entangled among the overhanging branches of the willows. The width varies from one chain to one and a half, and the depth from four and a half to two feet. The bed, for the most part, consists of soft mud, and is quite free from boulders, as is the case the whole way to the mouth, excepting in one place to be mentioned hereafter. The high water mark, very apparent on the willows growing along the banks, was eight feet over the present level of the water; the whole bottom of the valley, I was told, is often flooded to a depth of three feet.

Nineteen small creeks flow into this portion of the river, two only of them having names, the first and second Pheasant Creeks, called in Cree *Akiskoowi sepesis;* named after a hill which lies to the north some miles away, from near which they both take their rise. I took a cross section of the valley here, and found it to be 320 feet deep and seventy-eight chains wide; it is, I think, the deepest part of it. At noon, on July 23, we reached Crooked Lake, called in Cree *Kawawak-kamac*, the most picturesque of the Qu'appelle Lakes. Several streams draining the prairies on both sides have excavated deep and wide gorges opening into the main valley, which here sweeps in graceful curves, so that Crooked Lake seems to be embosomed amongst hills, and thus differs from the others which have very much the appearance of a gigantic canal. It is a little more than six miles in length, and its mean width is three-quarters of a mile. The greatest depth I found was thirty-six feet, and the mean of several soundings was thirty-one feet. The south slope, as before, is clothed with a dense foliage of young aspens, willows, and dogwood; a great contrast to the opposite side, on which only grows short and scanty grass, leaving the granite boulders which lie scattered over it, exposed to view; only in the ravines and the deep hollows are seen patches of young aspens and straggling oaks which have escaped the devastating fires.

For some time I could not understand why one side should be covered with trees and the other quite bare, the soil on both being exactly similar, until I discovered unmistakeable evidences of fire, which may be the cause of it. On inquiry afterwards I found that Indians often travel along the valley on the north of the river, which accounts for the fires being on that side.

Between the gravelly beach and the first of the slopes a fringe of willows runs all round the lake, and several points of low land jut out on both sides, on which grow oak, elm, and ash; not very large trees certainly, but healthy and thriving looking, and giving additional beauty to the landscape.

I ascended a bluff on the north side by a well-worn deer path, on which there were many foot-marks quite fresh, for the purpose of taking some observations connected with the survey and seeing the nature of the surrounding country. A gently undulating prairie, dotted with clumps of small poplars and willows stretched away on every side, and as far as I could see, the soil was a light sandy and gravelly loam, and in many parts strewed with boulders. I rather think that such is the character of a considerable extent of this section of the country.

As I stood upon the summit of the bluff, looking down upon the glittering lake 300 feet below, and across the boundless plains, no living thing in view, no sound of life anywhere, I thought of the time to come when will be seen passing swiftly along the distant horizon the white cloud of the locomotive on its way from the Atlantic to the Pacific, and when the valley will resound with the merry voices of those who have come from the busy city on the banks of Red River to see the beautiful lakes of the Qu'appelle. The view down the valley, where the river, after issuing from the lake, commences again its strange contortions, was doubtless very pretty, but it showed too the trouble that was before me, that there would be no rest for eye or finger, such as I had when taking long straight courses on the lake.

Again re-seated in the canoe we soon passed out of the lake into the river, the current of which for some distance is very strong and rapid, about two and a half miles per hour according to the log-line, and the width averages 70 feet, and the depth 3 ft. 6 in. A little way down it, as we swiftly and noiselessly glided round a sudden bend, we were borne by the current very close indeed to a group of Indian women who were enjoying the pleasures of a bath, quite as much to our astonishment as to theirs. First a loud chorus of screams arose, and then there was a rushing about for blankets and other apparel, which they adjusted with most wonderful rapidity, and then away they scampered to their wigwams laughing heartily as they went. Presently men and boys came trooping down to us simply arrayed in blankets, some worn in rather a *negligé* fashion, for the day was very hot. The chief man of the party, which consisted of six families, invited me in the most polite and hospitable manner to go to his lodge and have something to eat; but I had to decline as he had told me previously, in answer to a question as to how many days' journey it was to Fort Ellice, that we would have to sleep four or five times before we reached it, and this was now our fourth day from the Mission; and, moreover, I thought that the interior of a wigwam would not be a very agreeable place on such a hot day.

While we were speaking, the young ladies, whom we had so unintentionally disturbed, came down one by one to see us. Although their toilets were quite completed, so very modest were they that they remained behind the bushes and peeped at us through the branches. Having given the men some tobacco, and received in return a large supply of Pembina berries (High-bush cranberries), we wished them good-bye and resumed our journey. We went at the average rate of four miles an hour for two hours and a half, and camped before sunset at the foot of a bluff on the south side of the valley, of which I had taken a bearing from the end of the lake, and close to a creek about ten feet wide called *Nipimenan sepesis*, or Summer Berry Creek.

The valley is here of the same breadth as heretofore, that is, about one mile, and its depth is from 250 to 300 feet. The bottom is covered with willows interspersed with young sugar maples, with here and there an open patch of long luxuriant grass. With some difficulty I made my way to the level of the prairie through a dense and tangled mass of aspens and underwood of willows, dog-wood, and rose trees; but the beauty of the glorious sunset, and the cool refreshing breeze that came across the plains, more than repaid the trouble. I need not try to describe the exceeding beauty of the scene, for I could not; I will merely state what the components of the picture were. The sun just merged from behind a bank of crimson clouds reflected in the waters of Crooked Lake; part of the valley in deep shade and part brightly illuminated. The vivid green of the young poplars on one side, and on the other large granite boulders lying on the bare and rugged surface of the slope. The blue smoke of the wigwams rising up high and straight from the bottom of the valley. The river, with its complicated coils, gliding among the willow bushes. To the south the great prairie, ocean-like, with its many islands of aspens and single trees, looking in the distance, and by twilight, like becalmed ships. As this view just dissolves away, another rises very pleasant to see,—our camp fire is now burning brightly below, and over it swings a kettle, and passing round and about it are my two men, one busily engaged in preparing supper, the other in spreading out the blankets on the ground between the fire and the canoe.

Next morning (24th) we started as soon as it was daylight, glad to escape from our insatiated tormentors the mosquitoes and black flies, that would not let us rest or sleep all night. While at breakfast at eight o'clock a great thunder-storm from the south-west came upon us. Having thrown an end of the tarpaulin over the canoe, and resting the other end on the paddles stuck into the ground, we got beneath it and very soon fell fast asleep, and slept till one o'clock, when I was awoke by the sudden calm, for the storm had apparently only just then ceased.

The valley and river still retain their old character and dimensions till we come to the lowest of the lakes, called *Ka-wa-wi-ga-kamac*, or Round Lake, which varies from one mile to half a mile in width, and is nearly five miles long.

The name is by no means an appropriate one, as it is far from being round. The mean of some soundings I took was twenty-eight feet, the greatest being thirty feet. On the sand banks which are at the head of the lake were myriads of duck, and large numbers of geese were swimming about in every direction, and a few great northern divers or loons. We camped at a place about two and a half miles down the river, called the Stony Barrier, the Cree of which is *Asinni-pichigakan.* For about 100 yards in length the river is full of large and small granite boulders, rendering it quite impassable for the smallest canoe when the water is low; at this time the water was just high enough to admit of us passing over it.

Two miles down the river from this spot a little stream brings in its gatherings from the prairies on the south, rejoicing in the name *Isquawistequannak Kaastaki*, which means, "where the heads of the women lie." A long time ago two women, one a Cree and the other an Ojibway, were killed by the Mandans on the banks of this stream; their bodies were left unburied, and their skulls are still lying there, from which circumstance the stream derives its name. This was all my informant at Fort Ellice knew of the story. The next creek which is dignified with a name is the "Little Cut-arm," or *Kiskipittonawe sepesis*, the origin of which I could not find out; it flows in from the north.

A few miles further down, another creek, ten feet wide and very rapid, joins the Qu'appelle on the other side; its name is *Pesquanamawe sepesis*, which may be rendered into English "the Scissors Creek;" it is not a very literal translation, but is the best that can be given. The incident to which it owes its name exhibits a peculiar habit of the Indian, but is one that cannot be told. Near this spot there is an exposure of rock on the north slope of the valley, which on examination proved to be a shale similar to that on the Little Souris, but so decomposed that the amount or direction of its dip could not be ascertained. There are several extensive patches where the surface of the rock has been reconverted into soft mud, very much cracked, and on which no grass grows. On digging into it I found the mud to be three inches thick, then fragments very small and soft, and gradually increasing in size and hardness to a depth of about two feet, where the rock is perfectly hard but very much shattered. About fifteen miles to the east of this the rock is again to be seen on the south slope of the valley, also much broken.

On the 26th vast clouds of grasshoppers, flying towards the east, passed high over our heads, without intermission, for nearly two hours. It was the last large flight I saw.

Big Cut-arm Creek, or *Kichekiskapettonano sepesis*, the last to be noted, joins the Qu'appelle about twenty miles from its mouth, and is the largest of its affluents. It is twenty-five feet wide and three feet deep where it issues from a wide ravine on the north side. The Qu'appelle thence to its mouth is from eight to twelve feet deep, and varies in width from seventy to ninety feet, and the rate of current is a mile and a half per hour.

There is much good land in the valley from the Fishing lakes to the Assinniboine, but as it is flooded every spring, it is questionable whether it will ever be of much importance. For ten miles up it there is an abundance of timber, consisting of aspens, balsam-poplars, elm, black ash, oak, birch, and sugar maple; none, however, exceeding one foot six inches in diameter, and few so large. The underwood is chiefly composed of dog-wood, roses, cherries, and pembinas, intertwined with convolvuli and vetches. In this wooded part the birds are innumerable. Kingfishers, blue jays, and Canada jays, cat-birds, and American magpies, flitted from tree to tree uttering their discordant notes. Cherry-birds and pigeons were calmly and listlessly perched on the dense trees, having eaten plentifully of their favourite fruits, while the tyrant flycatcher, when alone or with some companions, chased and worried the crows, ravens, hawks, and eagles, who tried in vain to escape from them. The beautiful white-bellied swallow swiftly skimming the surface of the river, helped in addition to enliven the valley. Ducks and geese crowded the river for several miles; there were enough of them, I should think, to supply all the markets in Canada. Minks were perpetually crossing and re-crossing the river in front of the canoe. I was told that deer are sometimes very numerous in the valley, but I was only fortunate enough to see two jumping deer who were coming down to the river to drink, but the moment they got a glimpse of us away they bounded up the slope. The only other animal we saw was a little prairie wolf, Togany, as he is called by the Indians, that was standing by the edge of the river, and who was so much astonished at our sudden appearance that he never thought of running away, but stood staring at us incapable of motion.

The wonderful uniformity of the valley, or that part of it which I have described, necessarily causes a great deal of repetition in the description of it; so similar is its character throughout, that my two men, half-breeds, well accustomed to mark any peculiarities in the features of a country, said, that though they might pass up and down it several times, they thought they would often be at a loss to know in what part of it they were. The length of the valley from the second Fishing Lakes to its junction with the valley of the Assinniboine is 110 miles, while the river itself is about 270 miles long, which will give an idea of its tortuous course. We arrived at its termination on the evening of July 27th, and having hauled up the canoe on the bank, walked across to Fort Ellice, distant about three miles, where I was kindly received by Mr. McKay.

Very truly yours,

J. A. DICKINSON.

H. Y. Hind, Esq., &c. &c. &c.

CHAP. XVIII.

FROM THE ELBOW OF THE SOUTH BRANCH OF THE SASKATCHEWAN TO THE NEPOWEWIN MISSION ON THE MAIN SASKATCHEWAN.

Rocks on the South Branch.—Cretaceous.—Altitude of Exposure.—Character of. — Selenite. — Fossils.— Concretions.— Mesaskatomia Berry.—Character of River.—Drift.—Rock Exposures.—Fibrous Lignite.—Treeless Prairie.—Cree Camp.—Mud Flats.—Rock Exposure.—Concretions.—Treeless Banks and Prairie.— Low Country. — Driftwood. — Ripple Marks.—Dimensions of the South Branch.—The Moose Woods.—Water and Ice Marks.—Forest Timber.—Character of River.—Treeless Prairie.—Boulders.—Soundings.—Buffalo.—"The Woods." — Rate of Current.—Boulders, Arrangement of.—Artificial Pavement.—Tiers of Boulders.—Temperature.—Balsam Spruce.—Former Aspen Forest.—Good Country.—Water Marks.—Soundings.—Stratified Mud.— Fall of River.— Character of River.—Colour and Temperature of North and South Branch.—The North Branch.—Absence of Indians.—Absence of Animal Life.—Grizzly Bear.— Aridity of Country through which the South Branch flows.—Current of North Branch.— Coals Falls.— Boulders.— Trees.—The Grand Forks.—The Main Saskatchewan.—Fort à la Corne.— The Rev. Henry Budd.—The Nepowewin Mission.—Cubic Feet of Water in North and South Branch and Main Saskatchewan, or Ki-sis-kah-che-wun.—Opening and Closing of the River.

THE first rock in position on the South Branch below the Qu'appelle valley is a cretaceous sandstone, exposed on the river bank, for some miles. The altitude of the highest part of the exposure is sixty feet above the level of the water. It is capped by about seven feet of drift (*a*), which rests on twenty feet of soft and easily disintegrated sandstone of a pale yellowish-grey colour (*b*), containing a large number of small, pale yellow, spheroidal bodies, varying from one-tenth of an inch to one inch and a half in diameter, and composed of sand. Below this soft stratum there occurs a layer of sandstone (*c*) about three feet six inches thick, which is broken into an irregular projecting outline by the protrusion of a series of immense concretions of a flat spheroidal form, like that of a lemon slightly compressed parallel to its longest diameter. The concretions vary from three feet to six feet in horizontal dimensions. They are very hard in the centre, and show concentric rings for at least six inches from their outer casing, which is a shell of gypsum, often passing into selenite. Selenite is found in this and lower strata in

Section on the South Branch of the Saskatchewan, showing layers holding Avicula Linguæformis and Avicula Nebrascana.

veins and fragments. Some of the concretions thrust out their rounded forms from the face of the cliff, others have been broken off and show their internal structure. A grey sandstone (*d*) with a slight tinge of green, soft and friable, then occurs for a space of four feet; it is succeeded by five feet of hard sandstone (*e*) containing a vast number of cylindrical forms, slightly conical, and showing traces of organisation. Below this stratum a layer of sandstone six feet thick occurs (*f*), holding spheroidal forms, which vary in size from six inches to two feet in diameter; they are formed of yellow sand containing a hard central calcareous nucleus often four inches to one foot in thickness, and composed almost altogether of an aggregation of *Avicula Nebrascana* (Evans and Shumard). The stratum in which they are embedded holds *Avicula Linguæformis* (Evans and Shumard).

A second layer (*g*) of large concretions follows, similar in external aspect to those already described. Below them there is a persistent layer of hard calcareous sandstone about four feet thick (*h*), containing *Avicula Linguæformis.*

The lowest stratum exposed is a soft sandstone (*i*) about six feet above the river, and passing beneath its level. This rock is worn into caves by the action of water. The formation is nearly horizontal, with a slight north-westerly dip. For several miles this upper cretaceous* rock continues to form the river bank. The concretionary masses are persistent, bold, and prominent, and about three miles in a north-westerly direction from the point where they were first observed, those of the lower stratum are nearly on the same level as the water, thus showing a northwesterly dip of about three feet in the mile.

The banks of the river slope gently from the prairie on the south-west side to an altitude of about 250 feet above it, they then assume the form of steep declivities. On the north-west side the sandstone cliff rises abruptly from the river to a height varying from thirty to sixty feet, when it meets the foot of an undulating slope which extends to the prairie level. Trees, consisting chiefly of aspen and the mesaskatomina, are found in patches on both sides. The river continues for many miles about 700 yards

* See Chapter on the Cretaceous Series, Vol. II.

broad, with numerous sand-bars and low alluvial islands. The drift above the sandstone is gravelly, and many small sand dunes occur on the hill bank sloping to the prairie, into which they have progressed to a considerable distance. A treeless prairie, boundless and green, except where the patches of drifting sand occur, is visible on either hand from the top of the bank; below, the river glides with a strong current, two, and two and a half miles an hour, filling the broad trench it has eroded. The mesaskatomina is very abundant; shrubs or trees eighteen to twenty feet high, loaded with fruit perfectly ripe and of excellent flavour are numerous in every grove, the berries being of the size of large black currants, and very juicy, sweet, and wholesome.

During the morning of the 31st three Crees from a camp on the east bank came to the river; they shouted to us, asking us to land, an invitation we declined. They were "pitching eastward" to avoid the Blackfeet. About twelve miles below the Qu'appelle the river becomes narrower, being not more than a quarter of a mile broad, but full of mud flats and shoals. The banks are more sloping, and frequently broken into two terraces, the upper one being the prairie. The lower terrace is studded with small groves, the intervals consisting of pretty grassy areas, smooth as a lawn.

About fifteen miles from the Qu'appelle Valley the drift is occasionally exposed in cliffs, which disclose its structure twenty to thirty feet above the river. It consists of coarse sand stratified in curves, and often containing beds of gravel; it is also frequently capped by the same material enclosing small boulders. The dip of the rocks to the north-west, and the aspect of the drift, appear to indicate a depression, which may have been the seat of a large lake during earlier periods.

Some exposures of sandstone are visible on the river at intervals lower down, and the drift above them is stratified, containing layers of boulders of the same character as the sandstone below, and so regularly placed as to lead, when viewed from a small distance, to the belief that they are part of rock in position. Thirty miles from the Qu'appelle the rock appears on the south-west side, and consists of a white sandstone, with impressions of fragments of leaves, and some brown, fibrous lignite.

A treeless prairie, with a few sand dunes forms the country on either side for a distance of thirty-eight miles, which comprised the extent of our voyage during the day. As evening began to close upon us we came to a camp of Crees just after they had crossed the river. They numbered nineteen tents, and in order to avoid them we drifted several miles further down, and built our fire close to the river at the mouth of a small gully leading from the prairie, 100 feet above us. Mud flats and sandbars continue as before, but the river is not more than a third of a mile broad.

A narrative of a canoe voyage down a river flowing through a prairie country must necessarily involve numerous descriptive repetitions, which will appear perhaps less tedious and more readable in the form in which they were registered at the time in my note book, than if I were to attempt a connected narrative. I shall therefore strictly follow the daily record of what we observed, at the risk of its being nothing more than a dry enumeration of not very interesting facts.

August 1st.—Morning revealed a fine exposure of rock on the river bank where we camped last night. There is a change in the aspect of some of the strata, they occur massive, in rusty red and greenish-gray sandstone layers, with the concretionary bands as before described. A belt of sandstone twelve feet from the river level is capped by brown and red argillaceous layers, forty feet thick in the aggregate. Drift sand, ten feet thick to the prairie level, succeeds. The upper portion of the drift is hard and reddish coloured; as it approaches the clays below, it partakes of an argillaceous character. The upper stratum of the sandstone weathers reddish brown, with bands of deep red and purple. Below this a greenish-grey stratum occurs enveloping concretions of a reddish-brown colour. The concretions are hard and argillaceous. The greenish-grey matrix is soft when weathered, otherwise hard, and may be split without difficulty into thin layers. The concretions occur in the sandstone in forms easily detached, and often contain abundance of *Avicula Linguæformis.* If the clays above the sandstone are rock in position, the exposure has an altitude of about sixty feet. Fragments of fibrous lignite, dark-brown and sometimes approaching to black in colour, occur in the sandstone. The attitude of the rocks is nearly horizontal, and the greenish-grey sandstone is identical with the formation seen on the south bend of the Qu'appelle above Sand Hill Lake; the red layers are similar lithologically to those observed at the height of land in the same valley, holding the same species of shells. Sometimes layers of grey sandstone occur which are easily split; they contain the impressions and remains of plants. The position of these rocks is about forty-five miles from the Qu'appelle valley.

The river banks and the whole country is now much lower; this subsidence began about four miles south of our camp. The banks at our second camp are not more than 100 feet in altitude, and are becoming lower as we proceed north. They are treeless areas, and so is the prairie on either side, with few detached exceptions.

The river is about half a mile broad, with a current in the lead fully two miles and a half an hour. Large drifted trees are sometimes seen on the beach, and one pine was noticed this morning. They have probably traveled from the flanks of the Rocky Mountains.

About 60 miles from the Elbow, small forests of aspen begin to show themselves on the banks, after passing through a low country, which is an expansion of the river valley. Ripple marks are numerous on the fresh mud, the furrows lying parallel to the course of the stream, they are quite recent and similar to those observed on Red River in the spring. The ash-leaved maple begins to show itself, the aspen being still the prevailing tree, but the " woods are not continuous, and the prairie on either side of the river remains bare; it is fast regaining its former altitude. Sand hills are visible in the distance from the top of the bank, and on its side groves of the mesaskatomina are very abundant and the fruit fine flavoured. The exposed cliffs consist of reddish loam, but rock in position is no longer seen below them. At a point fifty-three miles from the Elbow we made a careful section of the river, and found its breadth to be nearly one-third of a mile (28 chains); its greatest depth was ten feet on the east side, but on the west side there is another channel with nine feet of water.

Approaching the Moose Woods we passed for several hours between a series of low alluvial islands from ten to twelve feet above the water. They sustain some fine elm, balsam-poplar, ash, ash-leaved maple, and a vast profusion of the mesaskatomina. The river valley is bounded by low hills leading to the prairie plateau four to eight miles back. The country here furnishes an excellent district for the establishment of a settlement. The spot where we encamped for the night is an extensive, open, undulating meadow, with long rich grass, and on the low elevations rose-bushes in bloom grow in the greatest profusion. It is only ten feet from the water, yet it does not appear to be flooded in the spring; water-marks and ice-marks are nowhere seen above four feet from the present level of the broad river.

August 2nd.—The region called the Moose Woods, which we entered last evening, is a dilatation of the Saskatchewan flowing through an extensive alluvial flat six miles in breadth, and cut into numerous islands by the changing course of the stream. This flat is bounded by sand hills, some of which are nothing more than shifting dunes. The woods are in patches, and in the low land consist of balsam-poplar, white wood, and aspen. Small aspen clumps cover the hills, but no living timber of importance has been seen as yet, although many fine dead trunks are visible, probably destroyed by fire. The river continues to flow through a broad alluvial flat for about twenty-five miles. Its water is very turbid, like that of the Mississippi, holding much solid matter in mechanical suspension.

Beyond the Moose Woods the banks close upon the river, and have an altitude not exceeding sixty feet. The breadth of the stream contracts to 250 yards, with a current fully three miles an hour. On the east bank the prairie is occasionally wooded with clumps of aspen, on the west side it is treeless, and shows many sand hills. During the afternoon we landed frequently to survey the surrounding country. Nothing but a treeless, slightly undulating prairie was visible; many large fragments of limestone not much water-worn lie on the hill banks of the river, which are about 100 feet in altitude. The river continues very swift, and maintains a breadth of 250 yards. Frequent soundings during the day showed a depth of ten to twelve feet. A little timber displays itself occasionally on the east bank below the level of the prairie. The dead bodies of buffalo are seen floating down the stream, or lodged on sand-bars in shallow water. The banks expose occasionally yellow drift clay with numerous boulders; the soil of the prairie appears to improve as we progress northwards, and the grass is no longer stunted and withered. Little rapids occur at the bends of the river, but there is always deep water on the other side. A heavy thunder-storm compelled us to camp two hours before sunset.

August 3rd.—The river at our camp is not more than 200 yards broad, but deep and swift; the volume of water it carries here, about eighty miles from the Grand Forks, is much less than at the Elbow, where it is half a mile broad. No doubt evaporation during its course through arid plains is competent to occasion a large diminution. Recent water-marks show a rise of five and eight feet, but near the top of the lowest bank stranded timber occurs twenty-five feet above the present level of the river. On both sides a treeless prairie is alone visible. There is a remarkable absence of animal life, no deer or bear have been seen, the tracks of buffalo are everywhere, but they have already passed to the east. The nights are cold but fine, and dew is very abundant. The prairie level is not more than eighty feet above the river.

At 8 A.M. we arrived at a part of the river where it showed an increase in breadth, it is now about a quarter of a mile broad, still flowing through a treeless prairie, in which only one low hill is visible. This character continues for many miles, the hill banks then begin to increase in altitude, and are about 100 feet high, but the river still flows through a dreary prairie for thirty miles from our camp, after which "the Woods," as they are termed, begin; they consist of a few clumps of aspen on the hill flanks of the deep valley. The face of the country is changing fast, and is becoming more undulating, patches of aspens showing themselves on the prairie ; here and there, however, the remains of a heavier growth are visible in clusters of blackened trunks ten to fourteen inches in diameter. During the afternoon we anchored to measure the rate of the current. The river is 200 yards broad, and it flows three miles and a half an hour ; its average depth is seven and a half feet.

Some remarkable exposures of drift, consisting of clay, enclosing horizontal tiers of boulders, (*c*, *f*,) often occur after entering the wooded parts of the South Branch of the Saskatchewan. The drift is exposed in cliffs at the bends

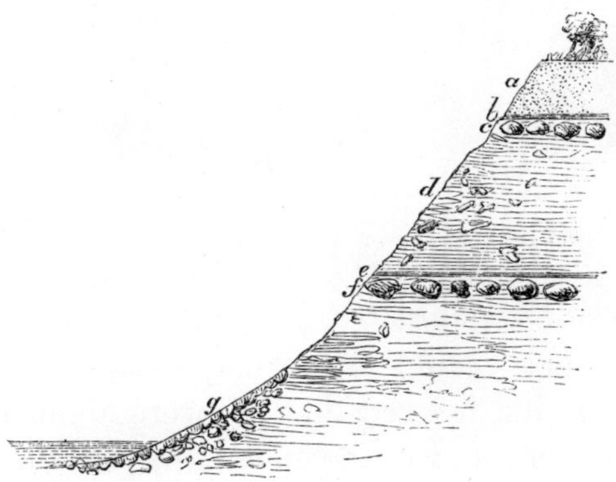

Horizontal tiers of Boulders in drift on the South Branch, with polished Boulder Pavement at the edge of the River.

of the river, from fifty or eighty feet in altitude. The fragments of shale, slabs of limestone, and small boulders imbedded in the clay (*d*) are not arranged according to the position they would assume if dropped by floating ice ; some of them stand in the drift with their longest axis vertical, others slanting, and some are placed as it were upon their edges. They have the same forced arrangement and position as the shale, &c., in the blue clay at Toronto.

In many places close to the water's edge and rising from it in a slope for a space of twenty-five to thirty feet, the fallen boulders are packed like stones in an artificial pavement (*g*), and are often ground down to a uniform level by the action of ice, exhibiting ice grooves and scratches in the direction of the current. This pavement is visible for many miles in aggregate length at the bends of the river. Sometimes it resembles fine mosaic work, at other times it is rugged, as where granite boulders have long resisted the wear of the ice and protected those of softer materials lying less exposed.

Polished and Grooved Pavement of Boulders on the South Branch.

Two tiers of boulders (*c*, *f*) separated by an interval of twenty feet, are visible in the clay cliffs lower down the river. When first noticed they (*f*) were about fifteen feet above the water; as we descended the stream they were seen to rise above its level, preserving evidently a nearly horizontal position. The lower tier (*f*) consists of very large fragments of water-worn limestone, granite, and gneissoid boulders; above them is an indurated sand containing pebbles; this is superimposed by an extremely fine stratified clay (*e*), breaking up into excessively thin layers, which envelope detached particles of sand, small pebbles, and aggregations of particles of sand. Above the fine stratified clay, yellow clay and unstratified sand occur. The fine clay must have been deposited in very quiet water, a microscopic examination subsequently made, failed to reveal any diatomaceæ. The polished pavement at the foot of the cliff was observed this afternoon inclined at a high angle, so much so indeed, that it was difficult to walk upon it.

Towards evening the country began to improve, and the timber to include a few elm and birch. In the prairie are clumps of aspen; on the flats, which occur regularly at the inside of each bend of the river opposite steep clay cliffs on the outside of the curve, fine aspens are common and the herbage is very luxuriant. Seventy-five miles from the Grand Forks we leveled the most rapid part of the river seen during the voyage thus far, and found the fall to be 2·85 feet per mile.

August 4th. — Temperature of air at 8 A.M. 61°, of the South Branch 67°. The balsam-spruce begins to appear in groves. The river winds between high wooded banks, with low points and wooded bottoms on one side, high cliffs also wooded with aspen and spruce groves on the opposite bank. The flats are covered with a rich profusion of vetches, grasses, and rose bushes. There are traces everywhere of a former fine aspen forest, with clumps of elm and ash; the dead trunks of these trees, eighteen inches in diameter, being frequently concealed by the undergrowth, offer a rude and stubborn obstacle to progress on foot through the tangled mass of vegetation which covers the rich flats. A view obtained from a low hill coming down to the banks of the river, continues to show a deep valley about three-quarters of a mile broad, through which the river winds from side to side in magnificent curves. The polished pavement on the banks was frequently seen during the day scarred with ice furrows and scratches. During the whole afternoon we passed swiftly through a good country, well fitted for settlement, as far as we could judge from soil and vegetation. Low islands are numerous in the river, and extensive alluvial flats spread out in an expansion of the valley, but water-marks are well preserved seven and nine feet above the present level. The banks of loose clay, when not protected by the pavement before described, being slowly undermined, fall bit by bit into the river. A violent thunder-storm at 5 P.M. compelled us to camp. Soundings during the day showed ten to fourteen feet water in the channel; the current maintaining its speed of three to three miles and a half an hour.

August 5th.—The early part of the morning was employed in examining the surrounding country, which gave evidence of an excellent soil, and timber sufficient for the first purposes of settlers. Much of the timber, however, has been burnt, and the country is fast becoming an open prairie land.

The stratified layers of fine mud (*b, e*) before described, were found this morning forty feet from the water's edge, above the horizontal layer of boulders (*c, b*) which has again made its appearance. The small aggregations of sand are still distributed between the thin layers of fine clay. A great change is coming over the character of the stream; its fall, as ascertained by leveling, exceeds at some of the bends two feet in the mile, with a very rapid current, sometimes six miles an hour. Large boulders are numerous in the bed of the river, but there is always a passage from fifty to sixty yards broad, often, however, very tumultuous, and for a small heavily freighted canoe very rough, and at times hazardous. The hill banks are getting higher as we approach the North Branch, and the balsam-spruce appears in patches and strips. The river sweeps in grand curves at the foot of high bluffs, in whic fine exposures of the drift may be seen, while on the opposite side are low alluvial points covered with aspens, thick and impenetrable. Six miles from the Grand Forks yellow clay cliffs, 120 feet high, appear at the outside curve of the bends, and where the adjoining flat begins, balsam-spruce two feet in diameter is not uncommon.

At half-past two P.M. we arrived at the North Branch, coming upon it suddenly and finding ourselves in its waters almost before we were aware of its proximity. The temperature of the South Branch was 67°, of the North Branch 62°; an important difference at this season of the year. It is perhaps a fair standard by which to estimate the climatic character of the regions of country through which these rivers flow. The water of the South Branch is yellowish-brown in colour, and turbid; of the North Branch a shade lighter and clearer; the one more resembled the waters of the Mississippi, the other those of the St. Lawrence. The South Branch is the larger river of the two at the Grand Forks.

Throughout the entire length of our voyage we have been surprised at the extraordinary absence of animal life. Of quadrupeds, we have seen half a dozen wolves, two or three badgers, several beaver, skunks, minks, foxes, and a number of dead buffalo; of birds, eagles, geese, a few ducks, kingfishers, cliff martins, pigeons, crows, cranes, plover, hawks, and a few of the smaller birds; but no deer, or bear, or live buffalo; and if we had been compelled to depend altogether upon our guns for a supply of provisions, it is probable that our voyage of 250 miles down the South Branch would have been attended with some inconvenience and delay. Early in spring and late in the autumn game is more abundant, but during the summer season the smaller rivers in the prairies, the ponds and lakes which abound throughout the country north of the Touchwood Hills, to be afterwards described, are the haunts of vast numbers of aquatic birds and of the larger four-footed animals which now form the small remnant of the earlier representatives of animal life in these wilds, before the fur trade led to their destruction, either for the sake of their flesh or skins.

With the exception of the Cree encampments passed during the first and second days of our voyage, we did not meet with a single Indian or half-breed. Once or twice "smokes," which, from their being soon answered in another quarter, we presumed to be signals, and might be raised by Blackfeet in the distant prairies, appeared on the west side of the river.

Once only were we disturbed in camp, and this may have been a false alarm. Both of our half-breeds came into the tent some time after we had retired to rest, and in a low tone whispered "a grizzly bear," at the same time seizing a rifle and a double-barrelled gun which were purposely placed at the foot of the tent ready for any unwelcome intruder upon our repose. The night was dark and the fire nearly out. Our men declared they had seen a large animal within ten yards of us, and pronounced it to be a grizzly bear; the alarm they testified was the only proof of the presence of that terrible animal, for the patient watching of the whole party during the greater part of the night, and a careful search for tracks next morning failed to satisfy me that we had been disturbed by this deservedly dreaded monster of the western plains.

That the grizzly bear is sometimes found far down the South Branch is a well known fact, and he is such a daring and formidable antagonist, that proper precautions are always advisable. A large camp fire often fails to deter this animal from making an attack, and when a large fire might attract the attention of wandering parties of Blackfeet which were known to be following the Crees, who had crossed the river some distance above us, it would not have been wise to have availed ourselves of this doubtful security. Our camp was at the edge of a cliff, we therefore were sure of not being attacked in our rear, and the greater part of the night was passed in quietly watching the open space in front of us. It was the steady determination of the half-breeds to watch after a fatiguing day, that led me to suppose they had really seen a grizzly bear, for under ordinary circumstances no people are so unwilling to deprive themselves of sleep during the night in the prairies as those who have lived the greater part of their lives in them, without they have the best reasons for keeping themselves awake.

The very small number of tributaries received by the South Branch between the Elbow and the Grand Forks is a remarkable proof of the aridity of the region through which it flows. For nearly 200 miles the South Branch receives but one small affluent from the east, and on the west side, where the water-shed is of much greater breadth, but where we should expect to find a more arid climate, it receives eight insignificant brooks. From the Lumpy Hill to the Grand Forks, a distance of about sixty miles by the course of the river valley, four streamlets cut its eastern bank. The water-shed on the east side has not an average breadth exceeding twelve miles, and two of the tributaries proceed from ponds in valleys cutting the low dividing ridge, which, like those of the Qu'appelle, are tributary to Long Lake or the main Saskatchewan, as described in the succeeding chapter.

After resting for some time at the junction of these mighty rivers, the South Branch being about 180 yards, the North Branch 140 yards broad, their currents meeting one another at the rate of three and a half miles an hour, we turned our canoe up stream, and attempted to stem the tide of the North Branch of the Saskatchewan in search of the " Coal Falls."

During the afternoon of the 6th and the morning of the 7th of August we occupied ourselves in dragging the canoe up the North Branch. Paddling was quite out of the question, the current being from six to seven miles an hour a few hundred yards above the Forks, and continuing rapid for a distance of seven miles, that being the furthest limit of our exploration up the north Saskatchewan. This rapid current is maintained for eighteen miles above the Grand Forks ; the valley of the river, as far as we saw it, resembles in almost all particulars the last ten miles of the South Branch ; but the river channel is much more obstructed by boulders, and the depth and volume of water considerably less. It is doubtful whether in its present condition a steamer drawing more than two feet of water could ascend it, and in dry seasons the boulders and rapids would probably present an insuperable obstacle. The river was high at the time of our visit, nevertheless in descending we had a few narrow escapes from striking against huge boulders just concealed by the water. If some of these were removed, the chief difficulties during low summer levels to steamers of shallow draft and great power would vanish.

The character of the Coal Falls, above the point we reached, is described by the people at Fort à la Corne to be similar to the part we saw. The hill banks expose drift in which large masses of cretaceous rock, containing fish scales, are imbedded. Fragments of lignite are numerous, but no rock was seen in position. The breadth of the valley is about half a mile, and 150 feet deep, through which the river winds from side to side like the South

Branch. The low points are covered with aspen, the hill banks with white spruce, aspen, Banksian pine, and poplar. Just below the junction of the two branches, after they unite to form the main Saskatchewan at the Grand Forks, there is an extensive flat, on which the remains of an old post of the Hudson's Bay Company is situated.

The main Saskatchewan, or Ki-sis-kah-che-wun, as the natives call it, is a noble river, sweeping in magnificent curves through a valley about one mile broad, and from 150 to 200 feet deep. We paddled rapidly round eight points, making a distance of sixteen miles in three hours, and towards evening sighted Fort à la Corne, with the Nepowewin Mission on the opposite or north side of the river. As the description of the Saskatchewan and the valley in which it flows at Fort à la Corne applies equally to the river between it and the Grand Forks, it is unnecessary to incur the risk of needless repetition by enumerating the features of each of the eight points or bends we passed, and of the valley through which the river flows. At Fort à la Corne we made measurements of its leading dimensions, a section of the bed of the river, ascertained its rate of current, examined the cliffs, points, and flats, which are so curiously reproduced at every bend, which will be amply sufficient to illustrate the most interesting and important features of this noble stream between the Grand Forks and a short distance below Fort à la Corne; from that point the country begins to assume a different aspect, and will require an independent notice.

The Saskatchewan opposite Fort à la Corne is 320 yards broad, twenty feet deep in the channel, and flows at the rate of three miles an hour. The mean depth of the river is fourteen feet, but it is in the memory of those living at the fort, when it was crossed on horseback during a very dry season.

An approximate estimate of the number of cubic feet of water passing down the South Branch, North Branch, and main Saskatchewan, gives the following numbers :—

	Cubic feet per hour.
South Branch	123,425,616
North Branch	91,011,360
Main Saskatchewan, at Fort à la Corne . .	214,441,290
Main Saskatchewan near Tearing River . .	206,975,000

The following table will show the comparative magnitnde of the Saskatchewan *:—

NAMES.	Area of drainage in square miles.	Length in miles.	Discharge in cubic feet per second.		
			Low water.	Mean.	High water.
Amazon	2,400,000	4,000	. . . .	. . . . . .	1,700,000
Mississippi . . .	1,226,000	4,400	447,200	. . . . . .	1,270,000
St. Lawrence . .	565,000	2,600	. . . .	900,000	
Niagara	237,300	. . .	370,589	389,000	406,000
Ganges.	432,000	1,680	36,300	207,000	494,200
Main Saskatchewan	240,000	. . .	. . . .	59,289	
South Branch . .	. . . . . .	. . .	. . . .	34,284	
Nile	520,200	2,240	23,100	220,000	
Ohio, at Wheeling .	25,000	. . .	1,400	. . . . . .	260,277
Thames	5,000	215	1,330	. . . . . .	7,900
Rhone	38,000	560	7,000	21,000	204,000
Rhine	88,000	700	13,400	33,700	164,000
Ottawa (Grenville)	80,000	700	35,000	85,000	150,000
French River . .	4,700	. . .	9,500		

Subjoined is the period of opening and closing of the Saskatchewan for several years at Fort à la Corne.

River Closed.		*River Opened.*	
1851 . .	November 6th.	1852 . . .	April 12th.
1852 . .	„ 11th.	1853 . . .	„ 14th.
1853 . .	„ 13th.	1854 . . .	„ 14th.
		1856 . . .	„ 9th.
		1857 . . .	„ 16th.
		1858 . . .	„ 20th.

---

* This table, with the exception of the Main Saskatchewan and the South Branch, is from the Report of the Ottawa Survey, by T. C. Clarke, Esq., C.E.

The Nepowewin Mission is situated on the north bank of the Saskatchewan, opposite to Fort à la Corne.

Nepowewin Mission and Fort à la Corne.

The Rev. Henry Budd, the native resident missionary, was ordained priest in 1853 at Cumberland Station, near the mouth of the Pasquia river. He had long laboured at this missionary outpost as a catechist. So early as 1840 he set out from Red River Settlement to make preparations for erecting a church and establishing a station at the mouth of the Pasquia. In 1852 Mr. Budd started from Christchurch, Cumberland station, for the Nepowewin, where he arrived on the 8th of September, and on the 13th of the same month commenced to clear a small piece of ground on the river bank, opposite to Fort à la Corne, for the erection of missionary buildings. The name Nepowewin is derived from an Indian expression signifying "the standing place," where the natives are accustomed to await the arrival of the Hudson's Bay Company's boats, as they are tracked up the north side of the river. Mr. Budd's house, garden, and little farm is a pattern of neatness, order, and comfort, yet it is difficult to say whether the prospects of this mission are favourable or otherwise. I cannot resist the impression that the selection of a permanent missionary station should be determined more by the agricultural capabilities of the locality, than for any special advantages which it may possess as a fishing station, or as lying on a main line of communication. The area of fertile land at the Nepowewin is limited to the points of the river, and perhaps does not exceed 400 or 500 acres at each point, these areas being separated from one another by the river, or by a bend which sweeps the foot of the banks of the deep valley, and which involve the ascent and descent of the bank, perhaps 250 feet high, in order to effect a communication between those which lie on the same side.

The valley of Long Creek, five miles south of the Nepowewin, appears to furnish a very large area of land of the best quality, and will probably yet become the seat of a thriving community, while the Nepowewin will remain a mere fishing station or landing-place. But when these events take place, the wild Indians will have passed away, and the white race occupy the soil, yet it is to be hoped that the descendants of some of those heathen wanderers who have here the opportunity of hearing of Christ and His kingdom, may find a permanent home near the Nepowewin, so long distinguished for the medicine feasts which are celebrated in the pine woods crowning the banks of the Saskatchewan, and whose remains in the form of painted idol posts, I saw almost within sight of the mission station, on the opposite side of the swift flowing river.

CHAP. XIX.

FROM THE NEPOWEWIN MISSION ACROSS THE COUNTRY TO FORT ELLICE.

Sandy Strip on the Saskatchewan.—Banksian Pine.—Indian Idols.—Medicine Feasts.—Rev. Henry Budd.—His Journal.—Fine Country.—Long Creek.—Old Forest.—Fires, Extent of.— Extension of the Prairies. — Former Extent of wooded Country.— Effect of Fires.— Long Creek.— Hay Ground.—Moles.—Humidity of Climate.—A Bear.—Source of Long Creek.—The Birch Hills.—Flowers.—Aspect of Country.—Carrot River. — The Lumpy Hill of the Woods. — Lakes. — The wooded Country.— Former Extent of.—Limits of good Land —Raspberries.—Mosquitoes.— The Height of Land.—Continuation of the Eye-brow Hill Range.—Valley inosculating with the South Branch and Main Saskatchewan. — Grasshoppers.—Character of the Country.—Birds.—Destruction of Forests.— The Big Hill.—Boulders. — Limit of wooded Country.—Belts of Wood. —Great Prairie.—Character of the Country.—Salt Lakes.—The Touchwood Hills.—Beautiful Country.—Excellent Soil.—The Quill Lakes.— Flowers.—White Cranes.—The Heart Hill.—The Last Mountain.—The Little Touchwood Hills.—Lakes.—Touchwood Hill Fort.—Ka-ou-ta-at-tin-ak.—Touchwood Hills Range.—Long Lake.—Devil's Lake.—Garden at the Fort.—White-Fish in Long Lake.—Burnt Forest.—Grasshoppers. —Winter Forage for Horses.—White-Fish.—Buffalo.—Climate of Touchwood Hills— Humidity of. — Medicine Man. — "Wampum." —Trail to Fort Ellice.—Marshes.—Little Touchwood Hills.—Character of Country Changes.—Depressions.—Pheasant Mountain.—File Hill.—Character of the Country.—Heavy Dews.—Cut-Arm Creek.—Willow Prairie.—Little Cut-Arm Creek.—Rolling Prairie.— Attractive Country.— Spy Hill.— Boulders.—Aspen Groves increasing.—Sand Hills.—The Assinniboine.

THE trail from Fort à la Corne to the old track leading from Fort Ellice to Carlton House ascends the hills forming the banks of the deep eroded valley of the Saskatchewan in the rear of the fort. It passes through a thick forest of small aspens until near the summit, when a sandy soil begins, covered with Banksian pine and a few small oak.

This sandy area occupies a narrow strip on the banks of the river, varying from half a mile to four miles broad. South of the sandy strip the soil changes to a rich black mould distributed over a gently undulating country where the pine gives place to aspen and willows in groves, the aspens occupying the crest of the undulations, the willows the lowest portion of the intervening valleys. On the slopes the grass is long and luxuriant, affording fine pasturage. The general aspect of the country is highly favourable for agriculture, the soil deep and uniformly rich, rivaling the low prairies of Red River and the Assinniboine. Our course lay along the banks of Long Creek, which flows in a small depression parallel to the South Branch of the Saskatchewan, and enters the main river near Fort à la Corne.

The large poles of a great medicine tent, erected in the spring to celebrate the annual goose dance, were standing on the top of the hill sides of the valley which the Saskatchewan has excavated. Four painted posts, about five feet high, remained, two on the outside and two on the inside of the ring of the medicine tent. These were the images of Manitou the Indians invoked during the celebration of important ceremonies. The features of a man were roughly carved on each post, and smeared with patches of vermilion and green-coloured paint over the cheeks, nose, and eyebrows. When decorated with fresh paint, feathers, strips of leather, and a painted robe of elk, moose, or buffalo skin, these idols inspire the most superstitious awe among the untutored savages who carve and ornament them. But the awe of many becomes terror, and the superstition absolute idolatry, when illumined by fires at night, and invoked as the representatives of all-powerful Manitou, the whole assemblage jumping in time to the wild song and monotonous drum of the conjurors, circle round their idols, and join in chants to the praises of the spirits they represent.

The Rev. Henry Budd thus records in his journal the progress and continuances of these annual idolatrous ceremonies :—

"*March* 31*st.*—The Thickwood Crees—so named to distinguish them from the Plain Crees—having formed a considerable party, are preparing to commence their spring feasts, &c. ; they have brought some of the best of their last winter's hunt for the purpose. They will be feasting and dancing for several days and nights together.

"*April* 1*st.*—The Indians are busy this morning putting up a large tent, where they intend to keep their feast and dance. The first feast to be kept up is in honour of the God Pahkuk, for having preserved, as they believe, the Indians the whole of last winter, and given them plenty of animals of all kinds to live upon. I hear there is to be no Mitawin kept up here this spring, for what reason I have not yet learnt. Whether it is because we are here, I do not know, but it is the first spring for a long time that that ceremony is not to be kept up here. Old Mahnsuk has arrived to-day from his spring hunt : he has been away nearly one month ; but the river is so dangerous to cross now that we shall not be able to go and see him. The drum is going the whole of the evening in preparation for to-morrow.

"*April* 2*nd.*—The feast has commenced betimes this morning, and the drum has had no rest the whole of last night. The dance does not commence until there have been some long speeches put forth, and the feast over.

"*April* 3*rd : Lord's-day.*—The Indians have been dancing and drumming the whole of yesterday and last night, and this very likely will continue for some time yet.

"*April 4th.*—The Indians are still carrying on their dance and feasting: they are preparing some more places for dancing in. Their great dance, the goose dance, is not yet commenced. This dance is repeated every spring and fall, in honour to the gods for preserving the Indians.

"*April 6th.*—The Indians are now at the height of the goose dance: that over, there will be several ceremonies of less importance to be performed, before the Indians are considered to be in a proper state for enjoying their summer.

"*April 12th.*—We could see some of the Indian tents stripped of their covering, nothing but the bare poles standing, which intimates that those are going away."

During the whole of the afternoon of the 10th we passed through a good farming country. The remains of aspen forests, in which trees of large growth are numerous, are still to be seen standing in groves, or with blackened trunks lie hidden in the long luxuriant herbage until rudely encountered by the carts and horses as they push their way through the rank and tangled grass. Raspberries were abundant in patches but not yet ripe; they were fully ripe a fortnight since on the Qu'appelle, 200 miles south.

Some of the small aspens had been nipped at the extremities of the branches by frost when in full leaf; the tops of many were black and drooping.

About four miles from Long Creek, and perhaps ten from the South Branch, a low range of hills running north-east and south-west, are still covered with an aspen forest of the same age as the blackened poles which stand in clumps on all sides. These poles are from nine to twelve inches thick; the young aspens are from four to six inches in diameter. "The fire" was here last year, and we have now traced the extent of that vast conflagration from Red River to the South Branch, and over four degrees of latitude. The Rev. Henry Budd states that in the autumn of 1857, north, south-east, and west of the Nepowewin Mission the country appeared to be in a blaze. The immediate banks of Long Creek, with the exception of a narrow strip in the prairie south of the Qu'appelle, is the only part of the country in which we have not yet recognised traces of last year's fire. The annual extension of the prairie from this cause is very remarkable. The limits of the wooded country are becoming less year by year, and from the almost universal prevalence of small aspen woods it appears that in former times the wooded country extended beyond the Qu'appelle, or three or four degrees of latitude south of its present limit. It must however be borne in mind that the term wooded country south of the Saskatchewan is applied to a region in which prairie or grassy areas predominate over the parts occupied by young aspen woods. The southern limit of the wooded country is some distance north of the Touchwood Hills range, but there are areas north and south of the Qu'appelle where the remains of aspen forests of large dimensions exist, and young forests are in rapid process of formation perhaps soon to be destroyed by fire.

This lamentable destruction of forests is a great drawback to the country, and a serious obstacle to its future progress. It appears to be beyond human power to arrest the annual conflagrations as long as the Indians hold the prairies and plains as their hunting grounds. Their pretexts for "putting out fire" are so numerous, and their characteristic indifference to the results which may follow a conflagration in driving away or destroying the wild animals, so thoroughly a part of their nature, that the annual burning of the prairie may be looked for as a matter of course as long as wild Indians live in the country. A fire lit on the South Branch of the Saskatchewan may extend in a few weeks, or even days, to Red River, according to the season and the direction and force of the wind.

Long Creek maintains a breadth of six feet, flowing clear but sluggishly through a broad shallow depression, where wild hay is as abundant as if the whole valley were one continuous beaver meadow. The burrows of moles are very numerous, indeed wherever the soil is very rich these little animals are to be found in large numbers; they form excellent indicators of the fertility of a soil, being never seen where the land is poor and sterile. Ponds and lakes are abundant, this extensive distribution of water pointing to a much more humid climate than that of the country south of the Qu'appelle. In the morning I killed a black bear which was leisurely cropping the willows on the banks of Long Creek.

Our trail on the 11th lay through an equally fertile country. The burrows of foxes and badgers have only twice shown a light gravelly substratum on low ridges, otherwise the black mould is everywhere distributed. A chain of lakes, lying westerly from our course, give rise to Long Creek. The Lakes are from 200 yards to a third of a mile broad, and form a continuous series connected by a small rivulet for a distance of ten miles. A hill range, called the Birch Hills, whose western flanks we turned, is said by Indians to extend to the rear of Fort Pelly. The Birch Hills form the dividing ridge between the water which flows into the main Saskatchewan and the Assinniboine, or Red Deer and Swan Rivers. The remarkable profusion of flowers gives extraordinary beauty to large open areas; they generally occur in parterres of several acres in extent occupied by one species, here the yarrow, there the fire weed, then a field of a species of helianthus, followed by *Liatris scariosa.* When viewed from an eminence, the country appeared to be clothed with pink, white, yellow, and blue, in singular contrast to the uniform tint which prevails on the great prairies of the Little Souris. The valley of Long Creek offers by far the most attractive features for settlement of any part of the country through which we have passed since leaving Prairie Portage.

Our course now followed the windings of a shallow brook which runs into the South Branch. It meanders through a fine broad and rich valley with hills on its south-eastern side gently sloping towards it, and covered with the dead yet standing trunks of burnt aspens. The soil of this valley differs in no particular from that of Long Creek; the flowers are equally numerous and showy, consisting of the same varieties, and distributed in large patches occupied by a single species.

We passed near to the source of a river which flows into the main Saskatchewan at the Pas, about 140 miles distant from us. It is called Carrot River or Root River, and rising within twelve miles of the South Branch, drains an extensive area of wooded country, passing also in its course through numerous lakes. The head waters of Root River being within ten or twelve miles of the South Branch show that the height of land between the two water-sheds maintains the same distance as on the Qu'appelle, and at the north fork of that valley near the Moose Woods.

The valley through which the small tributary of the South Branch flows, separates the Lumpy Hill of the Woods from the west flank of the Birch Hills; it is rich in alluvial meadows, ponds and lakes. A view from the

Lumpy Hill, which I ascended on the evening of the 11th is very extensive. The altitude of this eminence is about 400 feet above the general level, and from its summit an undulating open country, dotted with lakes and flanked by the Birch Hills is visible towards the east; south and south-west is a lake region, also north and north-east. These lakes are numerous and large, often three miles long and two broad. Seventeen large lakes can be counted from the Lumpy Hill; low ranges of hills can also be discerned in several directions. The most important of these are the Bloody Hills, the Woody Hills, far in the prairie west of the South Branch, and the chain of Birch Hills running from the Lumpy Hill easterly. The view extends to the borders of the wooded land; beyond is a treeless prairie. The so-called wooded land now consists of widely separated groves of small aspens, with willows in the low places. The Cree Indian guide we took from the Lake of the Sand Hills states that formerly the woods extended in one unbroken range to the borders of the prairie, which may be twenty-five miles south-east of the Lumpy Hill.

Much of the soil on the south and east of the Lumpy Hill is sandy and poor, in fact we have reached the limit of the good land, and are about to enter a comparatively sterile country. Low hills and long ridges running north-east by east, and south-west by south, diversify the general level character of the prairies, as seen from the Lumpy Hill. This eminence consists of drift sand and clay as far as my opportunities of observation enabled me to judge, with boulders on its summit; the western side is very steep, and partially covered with a burnt forest of birch. Raspberries of large size abound on the west side, but the mosquitoes start from the bushes in such countless myriads that it is next to impossible to linger five minutes to pick the delicious fruit. I offered the Cree guide a piece of tobacco for a tin cup full of raspberries, he tried to win it, but after a short struggle with these terrible insects he rushed from the hill side and buried his face in the smoke of the fire we had lit in the hope of expelling them from the neighbourhood of our camp; the horses became quite frantic under the attacks of their tormentors, holding their heads over the smoke, and crowding together in a vain endeavour to avoid the clouds of insatiable insects which surrounded us. Both man and beast passed a miserable, restless, and sleepless night.

The early part of the morning of the 12th was spent on the summit of the Lumpy Hill. A strong breeze drove the mosquitoes away, and permitted me to enjoy a quiet view of the country, which lay mapped about 400 feet below. After breakfast the trail, taking a direction nearly due east, passed over a series of hills and through intervening valleys, constituting a height of land. This range may be from thirteen to fifteen miles from the South Branch; it appears to be a continuation of the Eye-brow Hill range on the Qu'appelle, before described, receiving in its easterly prolongation the name of the Birch Hills, which limit the valley of the north Saskatchewan, as far as the rear of Fort Pelly. As soon as we had passed the crest of this range and entered a small prairie east of the hills, a valley through the range became apparent to our right. From lakes in this shallow depression water passes during spring freshets, to the South Branch and also to the North Branch by a tributary of Carrot River.

Grasshoppers were seen during the day flying to the north-east. They are the first that have been observed since leaving the Fishing Lakes on the Qu'appelle. The vegetation still continues luxuriant, lakes are numerous, aspen groves scattered here and there, and flowers abundant. Wild-fowl are found on all the lakes; and brown and white cranes, together with smaller waders of many species in the marshes. As we approach the great prairie the country becomes more undulating, and the soil light-coloured and poor; the aspens, which cap some of the low hills, are still large, although many are nothing more than dead trunks. The "wooded" country through which we passed during the day is only so called in remembrance of former forest growth. If the devastating fires continue for a few more years it will become a treeless prairie to the Lumpy Hill, and the aspen and birch woods will then be limited to the country between that eminence and the North and South Branches of the Saskatchewan. A young brood of grasshoppers has been seen to-day, showing that these destroyers reached this part of the country last autumn.

After traversing a very undulating country, in which are low ranges of hills and conical mounds with limestone boulders on their summits, we arrived about noon on the 13th at the Big Hill, a point of some interest, for south and south-east of it lies a boundless, undulating prairie. The summit of the Big Hill is covered with huge granite or gneissoid and limestone boulders, indeed on all the hills which surround the Big Hill boulders are very generally distributed. The limit of the so-called "wooded country," is about seventy miles from the North Branch in an air line, and thirty miles from the South Branch.

From the summit of the Big Hill the "Buffalo Cart Plain," and "Lake where the moose died," are visible; both noted localities in the wild history of these regions. South-east of the Big Hill the trail winds through a dreary labyrinth of dome-shaped hills, many of them covered with boulders. One or two small streams flowing into Ashes' Lake, a large sheet of water to the north of the trail, meander through this dreary prairie; timber is only found in the form of small aspens, on low ridges or near ponds.

The 14th brought us to a better country, still undulating, yet containing many beautiful lakelets fringed with aspens. The soil is light and the herbage scanty, a fit introduction to the "Carry Wood Plain," which lies at the foot of the Touchwood Hills. From a low gravelly ridge forming the north-western boundary of the Carry Wood Plain, the Indian guide pointed out a hill towards the south-west, which he said was close to Long or Last Mountain Lake, already stated in a former chapter to join with the Qu'appelle at the Grand Forks.

August 15*th*.—In journeying from the Lumpy Hill we crossed three "belts of woods," as the Indian guide termed them, before arriving at the great prairie west of the Touchwood Hills. These belts, which consist of groves of small aspen following a low gravelly ridge about a mile broad, and having a north-east and south-west direction, are separated by prairie valleys which sustain in their lowest parts a good soil and fine pasturage. Each belt diminished to a point some ten or fifteen miles south-west of our track. The points or termination of these belts are visible from the summit of mounds on our trail, not more than fifty feet high; beyond them is a treeless prairie, stretching away to the South Branch. The "belts of wood" become broader in a north-easterly direction until they merge into the wooded country between the Birch Hills and the Saskatchewan. There are many delightful spots in the belts, the herbage is clean as a well shaven lawn, the clumps of aspen are neatly rounded as if by art, and, where little lakes alive with water-fowl abound, the scenery is very charming, and appears to be the result of taste and skill, rather than the natural features of a wild and almost uninhabited country.

In the prairie valleys, and often when surrounded by conical hills, the ponds are fringed with boulders, while water-marks show that during the spring a large area is flooded. This is particularly the case at the foot of the Touchwood Hills. This great extent of pond and marsh affords food and shelter to vast numbers of aquatic birds. Grey geese were seen here for the first time; the Canada goose is very abundant; and duck, teal, cranes, and bittern are numerous. The lakes and marshes all contain salt or brackish water, which we found to our discomfort was not suitable for culinary purposes, or for slaking thirst. Tea made from it had a nauseous taste, and possessed the medicinal effect which might be supposed to result from preparing that beverage with a weak solution of Epsom Salts. The Touchwood Hills, as seen from this open "salt prairie," present a bold outline gently rising from the vast level, and maintaining a course nearly due east and west for ten or twelve miles.

In the afternoon we began the ascent of a gradual slope at the foot of the Touchwood Hills, following for some distance against the stream the course of a small brook which comes from the summit of the range, bright, cool, rapid, and sweet. At 6 P.M. we reached the summit plateau, and then passed through a very beautiful undulating country, diversified with many picturesque lakes and aspen groves, possessing land of the best quality, and covered with the most luxuriant herbage. From the west side of the summit plateau the Quill Lakes are seen to the north-west; these bodies of water have long been celebrated for the large numbers of goose quills which were occasionally collected there by Indians and brought to the fort for exportation. There is no timber visible on the west side of the range, with the exception of small aspen and burnt willow bushes. All the wild flowers, so numerous and beautiful in the valley of Long Creek, are met with on the summit plateau of the Touchwood Hills, of even larger growth and in greater profusion. Little prairie openings fringed with aspen occur here and there, through which the trail passes; we then come suddenly on to the banks of a romantic lakelet, in which ducks with their young broods are swimming, and flocks of white cranes start from their secluded haunts at the unexpected intrusion. The breadth of this beautiful plateau is about four miles, its level above the "salt prairie" to the west may be abcut 500 feet. Our course lay diagonally across it, so that we had to pass through seven miles of this delightful country. The Heart Hill, with others not seen before, come into view as we approach the eastern limit and begin a descent to Touchwood Hills Fort. The Last Mountain is visible in the west, but blue in the distance; the Little Touchwood Hills lie before us, the trail to Fort Ellice stretching towards their eastern flank. The country between the two ranges is dotted with lakes and groves of aspen. From a small hill near the fort I counted forty-seven lakes.

Touchwood Hills Fort, August 16th.—Arrived at the fort after sunset last evening. It is situated on the south-east flank of the range near the foot of a hill from which an extensive view of the country is obtained. Heart Hill, or *Ka-ou-ta-at-tin-ak*, is about 700 feet above the general level of the plain, and seven miles in an air line nearly due north (true) of the post. The general direction of the range is N. 45° E. (true). It appears to consist of a series of drift hills, many of which rise in rounded dome-shaped forms from the summit plateau. The Last Mountain bears S. 44° W., about twenty-five miles distant from the post, and the end of Long Lake, as it was pointed out to me by the guide, bears S. 70° W., distant from the fort a good day's journey, or about thirty miles. The Little Touchwood Hills bear S. 27° E., and have a general direction parallel to the main range. At the foot of the Heart Hill and on its northern flank is a lake about five miles long, running east and west close to its foot, and is said to contain white fish. Devil's Lake, which is connected with Last Mountain Lake, lies about forty miles due west of the post.

The guide who accompanied Mr. Hime to Long Lake from the Qu'appelle Mission* describes Long Lake to be broader at its northern extremity than elsewhere, but preserving throughout its length of forty to fifty miles a breadth of one mile on an average.

The garden, or rather the remains of a garden, in the rear of the fort, produces every variety of vegetable grown in Canada, but the efforts to cultivate it are almost abandoned in consequence of the depredations committed by Indians from the prairies, when they arrive in autumn with their provisions for trade, such as buffalo meat and pemmican. A few of the lakes near the fort are known to contain fish, and it is probable that all of the large fresh water lakes in this beautiful region also abound in them. The officer in temporary charge of the post stated that the people here had only known of the existence of white-fish in the Last Mountain Lake for three years; they are now taken there in the fall, and it is probable that the fishery recently established will become of great importance to this part of the country. The Plain Crees are not fisher-

* Page 325.

men like the Ojibways, they did not know how to catch fish when the attention of the people at the Touchwood Hills Fort was first directed to the treasures of Last Mountain Lake. Mr. Hoover, the officer in charge at the time of my visit, told me that he had first observed white-fish under the ice in November of 1854, and since that period they have established a fishery which provides the fort with an ample supply for winter consumption. The white-fish weigh on an average 7 lbs., but 10 lbs. each is not uncommon.

The timber on the Touchwood Hills is nearly all small and of recent growth, fires years ago having destroyed the valuable forest of aspen which once covered it. The remains of the forest are still seen in the forms of blackened poles, either standing erect or lying hidden in the rich covering of herbage which is found everywhere on the south-west flank of the range. So luxuriant and abundant is the vegetation here, that horses remain in the open glades all the winter, and always find plenty of forage to keep them in good condition. The cows are supplied with hay, the horses are worked during the winter, either journeying to Fort Pelly or to the Last Mountain Lake to fetch fish. Buffalo sometimes congregate during the winter in the beautiful prairie south of the fort in vast numbers.

On page 319 mention is made of a descent into a lower prairie before reaching the Qu'appelle Valley, whose boundary bore the aspect of a lake shore. On the north side of the Qu'appelle, and distant from it about fifteen miles, Mr. Hime saw a low ridge during the whole day's journey west from the Fishing Lakes towards Last Mountain Lake. This ridge was traced parallel to the Qu'appelle Valley for a distance of twenty miles in a westerly direction from the Fishing Lakes, how far east it extends is unknown.

These opposite ridges seem to show that a former lake or arm of the sea, from thirty to forty miles broad, has left a record behind.

Last year the grasshoppers visited the Touchwood Hills and deposited their eggs; this year the new brood consumed every green leaf in the garden, and made local ravages in the surrounding country. They took their flight on the 28th July for the south-east, and during the period of my visit but few were to be seen.

Snow falls on the Touchwood Hills to the depth of two feet and a half in the woods, and in the prairie where aspen groves are numerous it is not unfrequently found one foot and a half deep. In the great treeless prairie to the south where the herbage is short, the snow is drifted off by winds. The climate of the Touchwood Hills is evidently very humid; thunder storms appear to travel in the direction of this range and occasion a copious precipitation as they pass over it. Not only are lakes very numerous and well supplied with water, but there are several living streams flowing from the range. Indeed the whole country from the Touchwood Hills to the Riding Mountain, including the country about the head waters of the Assinniboine is dotted with innumerable lakes, annually replenished by summer rains.

The humidity of the climate of this part of British America contrasts in a remarkable manner with the variable character of that of the region under the same meridian lying six or seven degrees further south, in the territory of the United States.

When Lieutenant Warren reached Fort Pierre in latitude 44° 24′, longitude 100° 30′, in 1855, he was informed that there had been no rain or snow there for *more than a year*. The appearance of the vegetation confirmed the statement, as there was scarcely a green spot anywhere to be seen. In the summer of 1855 there was abundance of rain. At Fort Union in July, 1856, no rain had fallen during that year, and in many places there was a great scarcity of grass; after the 15th July rain was abundant. Again, in 1856, very little rain fell at Fort Pierre, so that on the 5th of October the grass had all dried up, although at the same period of the previous year it was everywhere green.*

During the two nights we remained at this post we were disturbed by a noted conjuror who was performing his ceremonies over the suffering form of a complaining woman who lay in his medicine tent near to the fort. His drum and song were heard nearly the whole of the night, and his incantations are described in another chapter as well as the remedy for the sickness of the poor squaw, which the conjuror suggested as infallible.

While at the Touchwood Hills, and indeed whenever I thought there was a chance of obtaining a "specimen," or information respecting it, I made inquiries respecting the celebrated "wampum." I was often told that a real Indian wampum belt is now extremely rare in Rupert's Land, and no one with whom I conversed on the subject had been fortunate enough even to see one.

The name "*wampum*" is applied to little cylinders made from sea-shells perforated through their longest diameter, and generally strung upon leather or sinews. The shells, which the Indians inhabiting the interior formerly obtained by traffic from the tribes on the Atlantic coast, were broken into fragments and ground by friction into the necessary cylindrical form. The little cylinders were strung into bracelets, belts, earrings, and

* Explorations in Nebraska and Dakotah.

ornaments of different kinds, and they were also used as money, as tokens of peace and friendship, and as records of important transactions.*

* "Les *Colliers* et les *Branches de porcelaine* étaient un agent universel en usage de tout tems chez les sauvages dans presque toute l'Amérique. Ils les employaient comme monnaie dans les transactions commerciales, comme ornement et parure dans les fêtes, comme annales pour l'histoire, comme gage et sanction dans les traités, comme satisfaction dans la réparation d'une injure ou d'un crime.

Les grains qui les composaient et qui portaient bien improprement le nom de *porcelaine,* provenaient de certains coquillages marins connus sous différens noms. On les a appelés *vignols, escargot de mer, concha venerea,* et chez les Italiens *porcella.* C'est de ce nom, dit le P. Lafiteau (t. ii., p. 200), qu'on a fait *porcelaine.*

Les sauvages les brisaient en morceaux, et en les frottant sur des pierres, ils leur donnaient la forme de petits cylindres applatis ou allongés. On en trouvait de blancs et de violets; ceux-ci étaient les plus estimés. Les sauvages les perçaient par l'axe du cylindre, et les enfilaient sur des lanières de cuir. Dans cet état on les appelait *Branches de porcelaine.* Les *Colliers,* sous la forme d'une ceinture de deux pieds de long environ, étaient composés de plusieurs branches, dont les grains étaient liés entre eux comme dans un tissu, et disposés avec art, de telle sorte que le mélange des couleurs produisait des dessins variés. Les colliers ordinaires avaient douze rangs de 180 grains chacun. Les sauvages en fesaient des ceintures, des bracelets, des pendants d'oreille, et quelquefois des plaques qu'ils suspendaient sur la poitrine et sur le dos. (Sagard.)

Les grains ainsi travaillés recevaient généralement le nom de *Wampum.* Les coquillages se tiraient surtout des côtes de la Nouvelle-Angleterre, et de la Virginie.

Les Andastoes étaient célèbres pour ce genre de commerce. Champlain mentionne cette spécialité dans sa carte.

On en recueillait aussi beaucoup sur la côte de Long-Island. Les Hollandais, habitans de ces parages, se livraient à cette spéculation. Van Tienhoven la présentait à ses compatriotes en 1650, comme un motif pour eux de venir coloniser ces rivages, *dont elle était,* dit-il, *la richesse.* On voit en effet que jusqu'en 1673, il y avait peu de monnaie en circulation, même parmi les colons hollandais. Six grains de Wampum blanc, et trois noirs valaient deux sols. En 1683, le maître d'école de Flutsburg recevait encore son traitement dans cette monnaie.

Jacques Quartier, qui précéda Champlain de plus d'un demi-siècle sur les rives du St. Laurent, fut curieux de connaître comment les peuples de ces contrées, qui paraissaient si peu développés sous le rapport des arts, pouvaient se procurer cet ornement, à qui il donne le nom d'*Esurgny* et

In reply to a speech delivered by Lieut.-Colonel Mackey to the Indian tribes assembled at the island of St. Joseph, Lake Huron, in 1829, reference was frequently made by the speakers to "wampum." Minutes of the speeches were made at the time*, and the following extracts will show the different application of this significant token. *A Minominie chief:* "I beg of our father to view this Wampum as a pledge of our being faithful children." "With this Wampum we also make a road to his newly kindled fire." "This Wampum I expect to see next year" (delivering the Wampum and pipe). *A Chippewa chief:* "The great Master of Life gave us pipes and Wampum for the purpose of conveying our ideas from man to man." "This Wampum reaches to Penetanguishene" (delivering the Wampum). *A Chippewa chief*

quelquefois d'*Esurguy.* "C'est leur richesse," dit-il, "et la chose qu'ils estiment être la plus précieuse, comme nous faisons de l'or." (Troisième Voyage, c. i.)

Il recueillit la tradition, et voici ce qu'il raconte de cette pêche merveilleuse: "Quand un homme a deservi la mort, ou qu'ils ont pris aucun ennemi à la guerre, ils le tuent, puis l'incisent sur les cuisses et par les jambes, bras et épaules à grandes taillades; puis ès-lieux où est le dit Esurgui, avalent le dit corps au fond de l'eau, et le laissent 10 ou 12 heures, puis le retirent à mont et trouvent dedans les dites taillades et incisions les dits cornibots, desquels ils font des pâtesnostres et de ce usent comme nous faisons d'or et d'argent, et le tiennent la plus précieuse chose du monde." (IIIe. Voy. c. vii.)

On est tenté de regarder cette histoire comme fabuleuse. Peut-être les sauvages auront-ils voulu abuser de la crédulité d'un étranger, ou cacher leur secret, en lui racontant cette pêche étrange. C'est sur son autorité que beaucoup d'autres historiens ont adopté le même récit sans nouvelle recherche. Cependant quand Lescarbot publia son *Histoire de la Nouvelle-France,* en 1609, cette pêche ne se faisait plus dans le St. Laurent, et les peuples de la contrée n'avaient plus ces coquillages. "Peut-être," ajoute Lescarbot, "ils en avaient perdu le métier; car ils se servent fort de Matachaiz (grains de rassade) qu'on leur porte de France."—*Relation abrégée de quelques Missions des Pères de la Compagnie de Jésus dans la Nouvelle-France, par le R. P. F. J. Bressany, de la même Compagnie. Traduite de l'Italien, par le R. P. F. Martin, de la même Compagnie. Montreal,* 1852.

* Sessional Papers. Legislative Council. Canada, 1847.

from Lake Superior: "With Wampum like this my ancestors made a road to Montreal many years since; one end of this string is tied to my village at Sha-qua-me-cong and the other I wish you to tie at Penetanguishene to your new fire" (delivering the Wampum). *An Ottawa chief:* "As a proof of our determination we make a road with this Wampum, the end of which we expect to see to-morrow* (meaning next year) at Penetanguishene (where the presents were to be delivered), and trust it will continue clear for generations to come."

On the 17th August we left the Touchwood Hills, and followed the trail leading to the lesser range bearing the same name.

A chain of hills joins the Greater and Lesser Touchwood Hills, having a course nearly north-west and south-east, or at right angles to those of the main ranges. In this subordinate range there are many conical elevations, some of them well wooded up to their summits, but the forest trees are small. The trail to Fort Ellice winds round the base of dome-shaped hills, past small lakes and aspen bluffs, through luxuriant herbage, and over an excellent soil. About nine miles from the Fort it begins to ascend the eastern flank of the Little Touchwood range, and gently winding up it for several miles, it finally reaches an extensive marsh which occupies a portion of the summit plateau. The marsh is but the introduction to numerous lakes, which continue to diversify the country in all directions.

On the following day we entered a region differing in many points from the rich tract we had left. Gravelly hills and areas of coarse drift sand form the surface of the country for a few miles, and apparently continue in

* "To-day" signifies, this year; "to-morrow," next year; "night," winter; "morning," spring; "play," war; "milk," rum. In the minutes of the speeches the metaphorical allusions are curious and interesting.

a south-west direction to the Qu'appelle. They are succeeded by a number of curious depressions or hollows, circular or oval in form, and varying from one quarter to one mile in diameter, often with a lake in the centre, but without visible outlet. The land is hilly in which they occur, and the elevations form a ridge running nearly north-west and south-east, towards File Hill, like the general direction of the hill ranges before described, but the country is so undulating that it is difficult to ascertain the true character of the surface until we arrive at the summit plateau. Here boulders are seen, the sand is coarse and mixed with a little clay, so as to resemble a gravelly loam on the ridges and hills, as well as on their flanks, but in the hollows and valleys the soil is excellent, and the herbage very luxuriant.

In returning from Last Mountain Lake, in an easterly direction, Mr. Hime crossed a ridge supporting clumps of poplar, and then struck into an open prairie country, which soon became a series of high gravelly knolls, with numerous boulders on them. About fifteen miles east of Last Mountain Lake, he ascended a high range of gravelly knolls, running from north to south, and then came to a valley 150 feet deep; in the bottom of this valley was a small creek, expanding into a chain of ponds, each about one quarter of a mile long, and three chains broad. Ascending the opposite bank, another ridge of gravelly knolls was passed, and a descent made into the prairie, which continues rolling, and interspersed with willow and aspen clumps and gravelly ridges, until File Hill is approached, when a more humid tract begins, dotted with marshes and ponds, in which innumerable hosts of duck find shelter and breeding places. On nearing File Hill, the soil improves in character, and the country becomes very picturesque and attractive.

August 19th.—The view this morning from the summit of a mound revealed a rolling, treeless prairie, stretching on all sides, and bounded only by the horizon. The wooded range of Pheasant Mountain appears low in the south-west, serving only to destroy the uniformity of the general outline. Numerous lakes, ponds, and marshes, covered with wild fowl, are visible in every direction. The soil in low situations is good, supporting long grass which afforded fine pasturage for our cattle. The ridges and mounds are gravelly, and a few boulders of the unfossiliferous rocks are seen here and there. It is remarkable that east of the Touchwood Hills no limestone boulders have been yet noticed, but limestone gravel is common.

The Pheasant Mountain runs north-east and south-west, and may be from fifteen to twenty miles long. Like its western companion, File Hill, it is wooded with aspen, and full of ponds and lakelets. At its foot the half-breeds report a lake ten to fifteen miles in length, on the south-east side, which we thought we could see from our point of view. The Greater and Lesser Touchwood Hills, the Pheasant Hill, and the File Hill, all appear to be rich, humid tracts, which will become important centres when civilisation in conjunction with population reaches these solitudes. North of the Carlton trail, and in the direction of Fort Pelly, the country is marshy, and abounds in ponds and wet places, which emit a very disagreeable odour when disturbed in passing through them. Beaver Mountain, a continuation of the Touchwood Hills, is seen from this wet prairie. The wet grass reminded me that the dews in the Touchwood Hills are very heavy and abundant at this season of the year. Last night dew was deposited a few minutes after the setting of the sun, although the sky was cloudy, and prevented direct radiation. This phenomenon has been noticed several times; the setting of the sun appears to admit of the cooling of the air sufficiently to allow the dew point to be quickly attained on the surface of vegetables, notwithstanding the screen of clouds which must necessarily obstruct radiation into space, but it would also appear to show that the temperature of the clouds must be very low. With the thermometer at 65° in the air, ten minutes after sunset, and under a cloudy sky, I have three times observed dew form since leaving Fort à la Corne. On clear nights dew has always been deposited during the summer, often so copiously as to wet the tents.

On the 20th we crossed a rapid stream with a swift current, ten feet broad and one and a half deep, flowing towards the Qu'appelle. It was thought to be Cut Arm Creek; it meanders through a prairie covered with low willows, and named the Willow Prairie, which embraces an extensive area of excellent land, sustaining fine pasturage. Limestone boulders were seen again to-day, but the country preserves a uniform and level character, with a few gravelly ridges and mounds; neither lakes nor marshes are numerous, and wood for fuel is very scarce. Little Cut Arm Creek, which we crossed during the morning, flows in a ravine about 80 feet deep and 400 broad. The next day lakes began to appear again, the prairies to become more rolling and intersected by ridges, which preserve a certain amount of parallelism, generally from north-east to south-west. The aspen replaces the willow in small clumps, and after passing Big Cut Arm Creek, the country is decidedly undulating, attractive, and very well watered. Large hills appear near the Big Cut Arm, which flows in a valley 1200 feet broad and 180 feet deep, resembling that of the Qu'appelle, from which we are not now far distant. We camped in the evening near to Spy Hill, called also *Ka-pa-kam-a-ou*, or "Some one knocked;" from a tragic incident which occurred during the early history of the fur trade. A war party of Assinniboines were following a body of Crees, and had despatched a spy to ascertain the nature of their camping ground. The spy placed himself on the summit of "Spy Hill" during the night time, in order to examine the camp of the Crees at the early dawn. A Cree warrior started at nightfall to Spy Hill for the purpose of watching the Assinniboines. When the morning dawned the Cree observed that he was lying within a few yards of the Assinniboine spy; he approached him stealthily, and succeeded in dealing a fatal blow before his enemy was aware of his presence, hence the name *Ka-pa-kam-a-ou*, or "Some one knocked."

Before reaching Spy Hill we saw a tent in the distance, prettily pitched on the banks of a small stream. Riding thither with an interpreter, I found an Ojibway family, consisting of one man, two wives, and several children. The Indian was a celebrated hunter, and showed me the produce of his summer hunt, which he was taking for barter to Fort Ellice. It consisted of twenty-two dressed moose skins, some bundles of sinews, and one or two red deer skins. He had killed the moose and deer on the west flanks of the Riding and Duck Mountains, and appeared to be proud of his success, which was certainly extraordinary at the present day. The money value of the twenty-two moose skins would be about eleven pounds sterling at Fort Ellice, and the Indian would receive in trade articles not exceeding in their aggregate value one-third of that sum. When one Indian during the summer kills twenty-two large animals like the moose, which would be at least three times as much as he would require to feed his family, there can be no room for astonishment that the influence of the fur trade has been mainly instrumental in reducing many parts of the country, once very thickly stocked with wild animals of the deer tribe, to a comparative desert, scarcely able to support the few wandering savages who depend upon the chase for their subsistence.

August 22nd.—The Blue Hills across the Assinniboine are visible from Spy Hill, so also are those on the Qu'appelle. Spy Hill is a gravelly eminence about 120 feet above the prairie ; near to it boulders of the unfossiliferous rocks are very numerous, and of large dimensions. One of gneiss measured thirteen feet in diameter. Our old hunter remarked that the aspen groves were much more numerous west of Spy Hill at the present time, than when he first remembered the country, forty-three years ago. After crossing a sandy prairie flanked on our left by numerous bare sand hills, we reached the Assinniboine at the mouth of the Qu'appelle early in the afternoon, and having forded that river in preference to the Qu'appelle, we had the pleasure on the following day of meeting Mr. Dickinson within a mile of the ferry, on his way to Fort Ellice, our place of rendezvous. The distance from Fort à la Corne to Fort Ellice by the route we followed is about 330 miles.

CHAP. XX.

THE QU'APPELLE VALLEY.—FORT PELLY TO THE SETTLEMENTS ON RED RIVER.

THE QU'APPELLE VALLEY.—Leading Dimensions.—Character of the Great Plain it intersects.— Elevation above the South Branch. — Lakes in the Qu'appelle Valley— Depths of.— Timber on.—Valley flooded.— Effects of a Dam across the South Branch.—Diversion of the Waters of the Saskatchewan.—Table showing leading Dimensions of the Qu'appelle River, Lakes, and Valley.—THE ASSINNIBOINE.—The West Bank.—Fort Pelly.—White Sand River.—Manitou Lake.—Little White Mud River.—Leech Lake.—Character of Country. — Crops at Fort Pelly.— Swan River. — Snake Creek.—Fertility of Swan River.—Mr. Dawson's Description.—Thunder Mountain.—Porcupine and Duck Mountains.—Dividing Ridge between the Swan and Assinniboine Rivers.—Miry Creek.—Riding Mountain.—Shell River.—River Terraces.—Indian Graves.—Little Saskatchewan.— Cretaceous Shales on Birds-tail Creek.—On Rapid River. —Termination of Riding Mountain.— White Mud River. — Ancient Beach. — Beauty of White Mud River.—Rat River.—Prairie Portage.—The Settlements.

THE QU'APPELLE VALLEY.

THE valley of the Qu'appelle River joins the Assinniboine about five miles above Fort Ellice. It is 269 miles long, and appears to be a former continuation of the South Branch, in a direction nearly due east, to the low regions now occupied by Lakes Manitobah and Winnipeg. Its western extremity issues from the South Branch at the Elbow, or the point where that river, from a south-easterly course, suddenly takes and preserves for 250 miles a north-easterly course, until it joins with the North Branch.

The narrowest breadth of the bottom of the Qu'appelle valley is half a mile; its greatest breadth about one mile and a half. Its shallowest part is about 120 feet below the level of the prairie, and its greatest depth is between 350 and 400 feet. It cuts a gently sloping plain extending from the South Branch to the Assinniboine. The surface of this plain is slightly undulating, and in a few localities broken by elevations which have a general direction from the north-east to the south-west; the north-western sides being abrupt and steep, the south-eastern descents gradual and undulating. The Touchwood Hills, Lumpy Hill, the Pheasant Mountain, the File Hill, &c., are among the most prominent of these elevations. So gradual is the general slope of this great plain, and so extensive is its surface that Elbow Bone Creek or the Souris Forks* inosculates with the Little Souris River, which after a course leading it sixty miles south of the boundary line, returns north and joins the Assinniboine about 115 miles in a south-easterly direction from Fort Ellice.

The highest part of the bottom of the Qu'appelle valley is only 85 feet above the South Branch at its summer level, and from 75 to 78 feet above it during the spring elevation of its waters. This occurs at a point distant 11½ miles from the junction, where a lake is found, which discharges itself both into the Saskatchewan and Assinniboine. Before connecting with the Assinniboine, it falls about 280 feet in 256 miles, or 1 ft. 1 in. per mile. The difference of level between the South Branch at one end of the Qu'appelle valley and the Assinniboine at the other, does not exceed, according to our estimate, 200 feet.

In its long, deep, and narrow course there are eight

* On the map accompanying Captain Palliser's Reports, Moose Jaws Forks is shown by a dotted line to inosculate with the Little Souris, and Elbow Bone Creek is named "Many Bone Creek." This difference in names does not affect the remarkable fact of an inosculation of the Little Souris with the Qu'appelle.

lakes, having an aggregate length of fifty-three miles. Most of these lakes abound in white-fish of great size and the finest quality. They are connected with Last Mountain Lake, occupying another valley running north-westerly, a counterpart of that of the Qu'appelle, inosculating with it at the Grand Forks, and, as reported by Indians, with the South Branch some thirty miles north of the Elbow. Numerous soundings of the Qu'appelle Lakes, showed them to hold from forty to sixty-six feet of water, which depths are maintained with great regularity. Timber ceases in the valley about 168 miles from the Assinniboine. It appears again at the Grand Forks and Moose Jaws Forks, 194 miles from the Assinniboine, and occurs also in small quantities at the Sandy Hills, near the height of land. Moose Jaws Forks is well wooded for a considerable distance: it comes from the flanks of the Grand Coteau de Missouri, whose blue outlines are distinctly visible from this point of the Qu'appelle valley.

It is stated elsewhere, that we frequently found water-marks eight feet above the level of the Qu'appelle River in August, 1858. When the snow melts in the spring there is a continuous water communication from Fort Garry to the Sandy Hills of the Qu'appelle, down which large bateaux might drift without necessarily touching land. According to the testimony of the Crees who hunt on this river, the whole valley from the Sandy Hills to the Assinniboine was converted into a lake in 1852, a year memorable in Rupert's Land for its extraordinary humidity.

The construction of a dam 85 feet high and 800 yards long would send the waters of the South Branch down the Qu'appelle valley and the Assinniboine into Red River, thence past Fort Garry into Lake Winnipeg.

The same result would be produced if a cutting were made through the height of land in the Qu'appelle valley to the depth of forty or fifty feet, and a dam some thirty or forty feet high thrown across the South Branch. A second low dam below the mouth of the Souris Forks would send these waters through the valley of that river into the Little Souris, thence into Lake Winnipeg by the Assinniboine and Red River. The time may yet arrive when the future population of Rupert's Land and Dakotah territory will find it advantageous to construct these or similar works, even if they should be for the purposes of irrigation or inland navigation.*

LEADING DIMENSIONS OF THE QU'APPELLE, OR CALLING, RIVER VALLEY, AND OF THE LAKES WHICH OCCUPY IT.

Table showing the length, breadth and depth, of the Qu'appelle Valley at different points.

			Miles.	Chains.
Length of Valley from the South Branch of the Saskatchewan to the Assinniboine			269	0
Breadth of valley	70 miles	from the Assinniboine	0	78†
Do.	120	do.	1	21
Do.	177	do.	1	30
Do.	239	do.	1	5
Do.	253	do.	1	70
Do.	258	do.	0	73‡

At its junction with the Assinniboine its breadth exceeds one mile.

At its junction with the Saskatchewan its breadth exceeds one mile and a quarter.

			Feet.
Depth of the valley	70 miles	from the Assinniboine	320
Do.	120	do.	265
Do.	177	do.	250
Do.	239	do.	220
Do.	253	do.	140
Do.	258	do.	110
At its junction with the Assinniboine the prairie slopes to the valley of that river, and its depth here is			240
At its junction with the Saskatchewan the prairie also slopes to the valley of the Saskatchewan, and its depth was estimated to be			140

* Vide "On the Qu'appelle or Calling River and the diversion of the waters of the South Branch of the Saskatchewan down its valley, with a view to the construction of a steamboat communication from Fort Garry, Red River, to near the foot of the Rocky Mountains;" Report on the Assinniboine and Saskatchewan Exploring Expedition by the Author.

† One mile less 44 yards. ‡ One mile less 154 yards.

Table showing the length, breadth, mean depth, greatest depth, and distance from the Assinniboine, of the Lakes in the Qu'appelle Valley.

Name of Lake.	Length.	Breadth.	Depth.		Distance from Mouth.
			Mean.	Gr.	
	m. ch.	Chains.	Feet.	Feet	m. ch.
Round Lake, or Ka-wa-wi-ya-ka-mac	4 56	60	28	30	41 20
Crooked Lake, or Ka-wa-wa-ki-ka-mac	6 10	60	31	42	56 0
Fishing Lake, No. 1, or Pa-ki-ta-wi-win	6 0	40	52	66	108 0
" " No. 2	3 25	40	32	48	114 20
" " No. 3	4 30	60	41	57	119 20
" " No. 4	8 50	60	37	54	124 12
Long Lake	. . . .	60 as far as seen.	. . .	. .	168 0
Buffalo Pound-hill Lake	16 0	40	. . .	. .	194 20
Sandhill Lake	4 50	45	. . .	. .	239 50
Total length of the Lakes . .	53 61				

NOTE.—The breadths and depths are the *means* of several measurements. The distances are taken along the centre of the valley.

FORT PELLY TO RED RIVER.

We spent two days in the valley of the Assinniboine near Fort Ellice, being occupied in making a section of the valley. We found its breadth to be one mile and thirty chains, and its depth 240 feet below the level of the prairie on either hand. The river is 135 feet broad, with a greatest depth of 11·9 feet, a mean depth of eight feet, and a current flowing at the rate of one mile and three-quarters per hour.

After drifting down the Qu'appelle from the Mission in canoe, Mr. Dickinson crossed the country to Fort Pelly, passing for the first fifteen miles through a very sterile region, the soil being a light sandy clay, and in many places consisting of pure sand, covered principally with a low growing creeper, bearing berries like the juniper; the grass is very short and scanty, and the aspens, which are the only trees, are very small. North of Wolverine

Creek the country improves very much as to its soil and vegetation, but it abounds in marshes, swamps, and ponds of various sizes, around which grow willows and young aspens, and this character continues for about sixty miles.

Thence to Fort Pelly the country is densely covered with aspens from five to fifteen feet high, and with willows of different kinds, the trail winding through the Beaver Hills as far as White Sand River. There are open spaces to be seen now and then, where the luxuriance of the vegetation is remarkable. Lakes and ponds are very numerous throughout, encircled with large aspens and balsam-poplars.

Several rivers and creeks flow into the Assinniboine from the west, into which many of the marshes and swamps might be easily drained. White Sand River, which is the largest of them, is seventy feet wide, four feet deep, and very rapid.

The Indians say that White Sand River rises in a small lake in the Touchwood Hills, named "Manitou Lake;" so called, it is alleged, in consequence of a whirlpool it contains which carries the water of the lake round four times in twenty-four hours. During the winter season this whirling motion is attended with noise and commotion beneath the ice, which forms first round the edges of the lake, and then slowly narrowing the area of open water, finally closes it, the whirling motion still continuing below the surface.

Little White Mud River reveals a curious feature in the topography of this region. It rises in Leech Lake,—a marshy sheet of water,—half way between Round Lake, in the Qu'appelle valley, and Fort Pelly. From the east end of Leech Lake, a tributary of Big Cut Arm Creek runs into the Qu'appelle, and from its west end Little White Mud River rises, which joins the Assinniboine through White Sand River, near Fort Pelly. Near

Leech Lake, the Indians who hunt in this part of the country (Ojibways and Swampy Crees) cultivate small patches of potatoes.

The country drained by the lower portion of White Sand River is very low and wet; swamps and bogs are numerous, together with innumerable small marshy lakes. At the crossing place, on the trail from the Qu'appelle Mission to Fort Pelly, the bank is high and very steep, being about fifty feet above the water. The river runs at a speed of five miles an hour, and was about four and a half feet deep at the end of July, 1858. An exposure of shale ten feet thick occurs at the crossing; it is capped by forty feet of yellow clay. The shale is probably cretaceous, and of the same age as that on the Qu'appelle and the Riding Mountain, to be described in the proper place. On the north side of the river, which is low and alluvial, an abundance of white sand has given a name to this tributary of the Assinniboine. Grasshoppers were numerous on the north side of the White Sand River.

The crops at Fort Pelly had been beautiful at the beginning of the season, but were all, excepting the potato, completely devoured by the grasshoppers in July. After a short stay at Fort Pelly, Mr. Dickinson visited Swan River, by the valley of Snake Creek, with Mr. Macdonald, the gentleman in charge at Fort Pelly, and Mr. Hime. This beautiful valley contains all the requirements necessary for a settlement. The timber is very plentiful and of a good size; the balsam-spruce is abundant, and averages two feet in diameter five feet from the ground. There is also some fine tamarack, varying from 1 ft. 6 in. to 2 feet in diameter. The balsam and aspen-poplar grow to a large size, and are everywhere to be seen. The land, for the most part, is good sandy loam, and is watered by numerous streams.

Snake Creek is about 13 feet wide, and 1 ft. 6 in. deep ; it yields plenty of fish, so also do one or two small brooks running into it. Swan River is from 90 to 100 feet wide, and 14 feet deep, its current is very rapid, being about three miles an hour. The valley, which is from 80 to 100 feet below the general level of the country, is most rich and fertile, but almost altogether filled up with trees, such as poplar, balsam-spruce, and willows.

Mr. Dawson, who in the spring of 1858 ascended Swan River in canoe, thus describes the country through which he passed :—

"From Winnipegoos Lake to Swan Lake the distance is about six miles. The stream which connects them, here appropriately enough called Shoal River, varies in width from 150 to 300 feet. It is shallow, and has a very swift course.

"About Swan Lake the country is highly interesting. Numerous islands appear in the lake: to the north an apparently level and well wooded country extends to the base of the Porcupine Range, while to the south the blue outline of the Duck Mountain is seen on the verge of the horizon.

"Ascending from Swan Lake for two miles or so, the banks of Swan River are rather low. In the succeeding ten miles they gradually become higher, until they attain a height of nearly 100 feet above the river. The current is here remarkably swift, and the channel much embarrassed by round boulders of granite mixed with fragments of limestone, which latter is the rock proper to the country, although it does not crop out, so far as we could see, on any part of Swan River. Landslips occur in many places where the banks are high, exposing an alluvial soil of great depth resting on drift clay or shale, of a slightly bituminous appearance.

"About thirty miles above Swan Lake the prairie region fairly commences. There the river winds about in a fine valley, the banks of which rise to the height of eighty or one hundred feet. Beyond these an apparently unbroken level extends on one side for a distance of fifteen or twenty miles to the Porcupine Hills, and for an equal distance on the other, to the high table-land called the Duck Mountain. From this south-westward to Thunder Mountain the country is the finest I have ever seen in a state of nature. The prospect is bounded by the blue outline of the hills just named, while, in the plain, alternate wood and prairie present an appearance more pleasing than if either entirely prevailed.

"On the 10th of June, the time at which we passed, the trees were in full foliage, and the prairie openings presented a vast expanse of green sward.

"On approaching Thunder Mountain, which seems to be a connecting link between the Porcupine range and the Duck Mountain, the country becomes more uneven. Some of the ridges on the shoulder of the Thunder Mountain even show sand, but there are wide valleys between them.

"On leaving Swan River to cross to Fort Pelly, the land rises rapidly to a plateau elevated about 250 feet above the level of the stream. The road then follows for some distance a tributary of Swan River, which runs in a beautiful valley, with alternate slopes of wood-land and prairie. Numbers of horses were quietly feeding on the rich pasture of this valley when we passed, and what with the clumps of trees on the rising grounds, and the stream winding among green meadows, it seemed as if it wanted but the presence of human habitations to give it the appearance of a highly cultivated country. The Hudson's Bay Company keep a guard here to take care of the numerous horses attached to their establishment of Fort Pelly."

A very low and narrow dividing ridge separates the waters of the Assinniboine from those of Swan River, affording another instance of the remarkable character of the water-partings and of the uniform inclination of the great prairie-plains of Rupert's Land. Miry Creek which flows into Snake Creek, an affluent of Swan River, is not more than three miles from the Assinniboine. The one river running far south along the edge of the escarpment which forms the abrupt north-eastern boundary of these table lands, the other breaking through it at right angles and reaching the Winnipeg by a course which curiously enough, has the same direction and turnings on the east side as its counterpart has on the west. A bateau or canoe may descend the Assinniboine from Fort Pelly and reach Lake Winnipeg by Red River; a bateau or canoe may also descend Miry Creek in the spring and passing down Swan River reach Lake Winnipeg by the Little Saskatchewan or Dauphin River.

On the 26th August we set out on our return to the settlements, taking the trail on the east bank of the Assinniboine. Our route lay on the flanks of the Duck and Riding Mountains, and through a country admirably adapted for farming purposes. On the morning of the 27th the herbage was covered with hoar frost, but without any injury to vegetation. Ponds and lakes are very numerous on the flanks of the Riding Mountain, but as far as our opportunities enabled us to judge, the whole country, with the exception of narrow ridges, possesses a rich black fertile mould, supporting very luxuriant herbage, and on the mountain an ample supply of timber, consisting chiefly of aspen of large dimensions. The Riding and Duck Mountains consist of a succession of slopes and terraces on their south-western sides, the ascent being almost imperceptible to the thick impenetrable forest which covers the highest plateau. On Birdstail Creek cretaceous shales identical with those on the Assinniboine crop out in different places, but organic remains are scarce and indistinct.

Shell River separates the Duck from the Riding Mountain, and its valley affords a very interesting illustration of river terraces. Between the present bed of the river and the prairie plateau it cuts near the crossing place, it discloses three distinct terraces, visible on both sides of the river, but more distinctly marked on the left bank than on the right. On the prairie banks of Shell River are several gravelly ridges resembling in most particulars the Big Ridge; the elevation is thirty feet above the level on which they are formed. The length of the slope to the summit varies from 100 to 180 feet; the course is nearly north and south, or nearly at right angles to the point where they touch the bank of the stream. The ridges die away in the general rise of the prairie as they recede from Shell River. They are thickly covered with boulders.

Although several landslips have revealed the structure of the banks of Shell River, yet no rock in position was observed. The banks consist, according to Mr. Dickinson, of clay, sand, and gravel mixed with a few boulders. Near the confluence of Shell River, or A-se-sepee as it is termed by the natives, with the Assinniboine, some Indian graves are placed on the edge of the bank, sad memorials not unfrequently met with in travelling through these beautiful wastes.

On Saturday, 28th August, we arrived at the Little Saskatchewan or Rapid River, which Mr. Dickinson had explored for a distance of one hundred miles from its source. The valley of this river is extremely beautiful and fertile until within a few miles of its junction with the Assinniboine; it offers probably the most attractive and desirable place for settlement in any part of the country west of Red River. The stream abounds in fish, the flats in the valley are covered with the richest herbage; timber, consisting of aspen, poplar, and oak, is abundant; the prairies on either side are clothed with the greatest luxuriance of vegetation, the scenery is very attractive, and the river navigable down stream for canoes and bateaux to the Assinniboine. Where the Rapid River enters the Riding Mountain balsam and white spruce appear, and our explorations on the east flank of the range showed that large birch, spruce, poplar, and aspen flourished on the summit plateau.

Fires here as elsewhere have damaged the forest which once covered the country. Vast numbers of young oak and aspen are springing up in all directions on the prairie fringing the river near the trail. Birds are very numerous throughout this region, and every lake contained duck with their young. The aspen groves and willow clumps were alive with grackle and yellow birds congregating in flocks. Humming-birds were also observed as well as the American cuckoo and the solitary thrush. While in the marshes, herons, cranes, and bitterns were disturbed in groups as we cautiously approached in search of duck.

In a brook emptying into Rapid River, I found an exposure of the cretaceous shales before described as occurring on the Assinniboine and the Little Souris. The rock was very fragile, and contained a few fossils in an imperfect state of preservation.

On the 29th we reached the south-eastern termination of the Riding Mountain, and obtained a fine view of the successive steps of which it is composed. These were three in number, each step being separated by a gently sloping terrace; subsequent observation showed that we were encamped on a fourth terrace which is continuous with Pembina Mountain. The entire mountain appeared, from our point of view, to be densely covered with forest trees, and effectually resisted all attempts to reach the summit on the west side with horses on account of the fallen timber and thick growth of aspens. The country through which we passed during the day was very wet and swampy in many places, but on the ridges the soil is dry and gravelly; we were in fact, descending the Pembina Mountain, which being here extended over a great breadth, is not easily recognised. In the afternoon we arrived at a beautiful ridge, running N. 12° W. and S. 12° E. One side of this ridge is partly excavated by the White Mud River, and exhibits finely stratified gravel, consisting almost altogether of small limestone pebbles, with a few belonging to the unfossiliferous rocks. The ridge is gently sloping towards the east, and precipitous towards the west, having on either hand a level country, higher on the west side than on the east. I have no doubt this ridge is a continuation of a former lake boundary at a higher level than the Big Ridge of the Assinniboine. Some fine oak grows on the banks of White Mud River near the ridge, and ash-leaved maple begins to show itself again.

Our course on the 30th lay through the prairies drained by White Mud River. This tract of country is second only in beauty and fertility to the valley of Rapid River. Not only is the herbage of surprising luxuriance, but the trees in the river bottoms are of very large dimensions, and consist of oak, elm, ash, maple, aspen, and poplar. Near the crossing place there is a fish weir, where large quantities of pike, suckers, gold-eyes, and other species, are taken by the people of Prairie Portage, who have established a fishing station here, as well as one at Lake Manitobah, some miles further east.

The woods fringing the river at the crossing place are very important. The oak and elm are of the largest size, being often found 2 ft. to 2 ft. 6 in. in diameter, with tall, clean trunks. The hop and vine twine around the underbrush, and give a very attractive appearance to the belt of woods which fringe White Mud River

Wishing to ascertain the character of this stream to its outlet, we gummed the canoe, and once more launched it for a short voyage down the White Mud River, to the fishing station on Lake Manitobah. Mr. Dickinson proceeded down the river, the carts, with Mr. Hime, journeyed on towards Prairie Portage, while I rode to the fishing station, in company with a half-breed who was familiar with the history and progress of the station since its commencement.

We soon arrived at Rat River, a stream of much interest in connection with the floods of the Assinniboine. Down its valley the water of that river flows into Lake Manitobah during freshets, and by making a very shallow cut, a permanent communication in time of high water, could always be maintained. The fishing station at the mouth of White Mud River consists of about half a dozen houses, which are only tenanted during the fishing season. Very large quantities of white-fish are caught here, and no doubt when the demand requires it the station at the mouth of White Mud River will become an important source of supply. The Assinniboine prairies extend to the banks of Manitobah Lake, and their elevation as seen here and at

Oak Point is not twelve feet above the level of that extensive but shallow sheet of water.

We camped on the banks of Rat River, and the following day made a nearly due south course through a rich but treeless prairie to Prairie Portage on the Assinniboine. In making this traverse we passed the shallow, winding, but dry bed of a brook several times, a tributary of Portage River. In wet seasons this bed is occupied with drainage water from the Bad Woods, while Rat River rises within three miles of the Assinniboine in the same locality. The valley of Rat River and of the dry watercourse may yet become of vast importance if it should ever happen that the commercial inducements for effecting a communication with the South Branch by way of the Qu'appelle valley, should lead to the construction of works for that purpose.

On the 31st of August we arrived at Prairie Portage, and reached the settlements at Red River on the 4th of September, after an absence of nearly three months. Our course from Prairie Portage lay through the prairies already described.*

* Chapter XIII.

CHAP. XXI.

FROM FORT À LA CORNE, DOWN THE SASKATCHEWAN, TO THE GRAND RAPID AND LAKE WINNIPEG.

Departure from Fort à la Corne.—Object of the Expedition.—Equipment.—"Bull-Boats."—Birch-bark Canoes.—General Direction, Current, and Breadth of the Saskatchewan. — Character of its Valley. — Country through which the River flows well adapted for Settlement.—Pemmican Portage.—Cumberland House.—Description of Cumberland.—The Saskatchewan and surrounding Country between Cumberland and the Pas. — Indian Hunter.— Sturgeon. — The Pas.— Christ Church.—Gradual Depression of the Country bordering the River.—Alluvial Flats.—Marshes.—Delta.—Muddy Lake.—Rock Exposure.—Marshes and Mud Flats.—Cedar Lake : its Situation and Dimensions.—Surrounding Country.—The Saskatchewan between Cedar Lake and Lake Winnipeg.—Cross Lake Rapid : its Dimensions.— Enter Cross Lake.— Meet a Brigade of Boats.—Cross Lake : its Dimensions and Altitude.—Surrounding Country.—The Saskatchewan east of Cross Lake.—Rapids : their Dimensions.—Smooth Reach. — Drift Clay Banks. — The Grand Rapid : Portage ; Running the Rapid ; its Dimensions ; Character of its Excavated Bed ; Magnificence of the Upper Portion of the Cataract ; Mode of Ascending it ; Remarks in Relation to surmounting this Barrier and making the Saskatchewan available for Steam Navigation.—Indian Encampment.—Lake Winnipeg.

MR. FLEMING'S NARRATIVE.

Dear Sir,—On the 9th of August, 1858, we set out from Fort à la Corne to continue the canoe voyage down the main Saskatchewan to Lake Winnipeg, thence to coast along the western shore of that lake to the Red River. The principal object of this journey was to complete the track-survey and reconnaissance of the Saskatchewan, which had been began at the Elbow of the South Branch, and to survey and examine the west coast of Lake Winnipeg with a view to describe the topography and general character of the region bordering the line of exploration.

The birch-rind canoe in which we embarked was eighteen feet long, two feet six inches wide, with a round bottom, and drew about a foot of water, a depth sometimes too great in places where the river abounds in mud-shallows and sand-bars. This small and light craft was one of those which we had hauled overland from Selkirk settlement, and although new at starting, it had become battered and worn during our summer's campaign on the prairies. The crew consisted of two half-breeds, one of Ojibway and the other of Blackfoot origin. Their experience and skill in canoeing, "woodcraft," and hunting rendered them well adapted for the service to be performed.

A canoe-voyage on these north-western waters is generally monotonous and not often accompanied by that excitement and adventure which is still not unfrequently met with in journeying across the great prairie-plains of the North American Continent. A relation of incidents, and a description of objects to which observation was directed, on this voyage of exploration, may therefore possess but little interest compared with a portrayal of prairie life and travel, buffalo hunts, and scenes among the savage and often hostile tribes distributed over the great plains of the far West.

Bark canoes are not often seen so high upon the Saskatchewan, there being a scarcity of birch-bark in the region through which the north and south forks flow. These great prairie-rivers are generally crossed and often descended in "bull-boats" or "parchment canoes" by the Indians, for great distances. These bull-boats are made of one or two buffalo skins, stretched on a light frame, stitched together, and the seams covered with tallow and ashes. Hunters and trappers frequently set out from Fort à la Corne, on horseback or on foot, to the

Moose Woods or the great prairies on the south Saskatchewan, and return in bull-boats laden with dried-meat and skins, both craft and cargo being the proceeds of their hunt. Bark canoes, although more durable than bull-boats, are nevertheless very fragile and require to be handled with great care; the seams and cracks in the bark require constant "gumming" and attention. Our canoe being leaky, owing to injuries it had sustained in crossing the plains from Red River to the Elbow of the South Branch, was the source of much trouble until we reached Cumberland House, where, through the courtesy of the gentleman in charge, we were enabled to procure a new canoe and some other necessaries.

The general direction of the Saskatchewan from Fort à la Corne to Cumberland House is north-easterly. The current continues strong for a considerable distance below Fort à la Corne, where the average rate was found to be three miles an hour. In some places the mean velocity of the current exceeds this, as ascertained by repeated trials; and at the points or alluvial promontories at the great bends a small rapid is frequently seen, generally caused by a submerged spit or reef of boulders and gravel protruding into the river; but the water is turbulent in its passage over these shoals only, which are always on one side of the river: in the bay opposite them it is quite smooth and deep, averaging in the channel nineteen feet.

At Fort à la Corne the breadth of the Saskatchewan according to trigonometrical measurement is 965 feet, and its immediate banks are high; the sides of the valley, which are much higher, being no great distance from the river. The breadth of the river continues very uniform, but its banks become gradually lower, the hill sides of the valley at the same time diverging. About twenty miles below Fort à la Corne the banks of the river are low, and the general character of the adjacent country considerably changed. The high cliffs before seen at the great bends give place to rich alluvial flats, supporting a forest of fair-sized balsam-spruce and poplar, and the valley becomes so broad that the high banks are nowhere observed.

Having been occupied some time near Fort à la Corne in making a transverse section of the river, ascertaining its fall by leveling, and measuring its rate of current by log-line (adopting the mean of a series of observations), it was at a late hour when we got fairly under-weigh, and we did not accomplish more than twenty-three miles the first day.

As the day began to wane we drew up our canoe on a low boulder-promontory of this fast-flowing river, and were soon reclining upon the polished and rounded pavement, beside the ruddy and cheerful blaze of a fire of driftwood. The stillness of night gradually crept on, until nothing was heard but the rippling and surging of the water over the smooth boulder-stones at our feet. The Saskatchewan or "the river that runs swift" is truly well named, for even upon the smoothest and deepest parts of the river, long lines of bubbles and foam, ever speeding swiftly but noiselessly by, serve to indicate the velocity with which this mighty artery courses unceasingly onward, swelling as it goes, with the gatherings of its many wide-spreading tributaries, to mingle its restless and muddy waters in the Arctic seas.

The second day of our journey (August 10th) we embarked at 6 A.M., and passed during the day the "Big Birch Islands," and many others; they are all alluvial deposits, and some of them are overflowed in spring. The banks of the river are now quite low, and the country on either side is very flat; but it still continues well adapted for agricultural purposes and settlement, the soil being a rich alluvial loam of a considerable depth, well watered and drained by many fine creeks, and clothed with abundance of timber for fuel, fencing, and building. In some places stony points, projecting into the river, contract it to a width of five or six chains; stretching out from these points there are shoals over which, as already observed, the current is very strong and rough. Among the islands the river attains a width of a quarter to half a mile, but where it is broad its depth is diminished in many places by mud flats. We stopped to camp for the night about half-past six P.M., nearly fifty-three miles from where we started in the morning.

August 11*th.*—We left our last night's resting-place at daybreak this morning, and passed through an excellent tract of country all day; the soil on both sides of the river consisting of a very rich alluvial deposit, ten feet in thickness, above the surface of the water, well wooded with large poplar, balsam-spruce, and birch, some of the poplars measuring two and a half feet in diameter; and, as far as I was enabled to ascertain, the land continues good for a great distance on either side, but more especially on the south side of the river. In many places the river is studded with large alluvial islands supporting a most luxuriant growth of poplar and willows. Among these islands the channel is sometimes intricate, being occasionally interrupted by sand-bars and snags. We encamped about 6 P.M., having attained a distance of about forty-seven miles to-day. We passed a sleepless night; a terrific thunder-storm coming on after dark, and having no tent to protect ourselves from the driving rain, we were drenched to the skin. Our constant tormentors, the mosquitoes, were also excessively annoying.

The general character of the country we passed during the next day is excellent, the soil being rich and the timber of fair quality. The depth and breadth of the river is variable; in one or two places it is impeded by mud-flats and shoals, sometimes holding snags and sawyers. About noon we came to the mouth of a tributary stream 100 feet broad, flowing into the Saskatchewan from the north, which we supposed to lead to Cumberland, as it corresponded to the description given to us at the Nepowewin, but being desirous of keeping the main river, we went on until reaching an old carrying place, called "Pemmican Portage," leading to the fort, where we discharged and hauled up the canoe. I despatched the Blackfoot half-breed to the fort, and he returned in the evening reporting the road very wet and marshy. We came to-day nearly twenty-nine miles, so that the distance between Fort à la Corne and Cumberland House, by the windings of the river, is upwards of 150 miles.

August 13*th.*—Owing to the thick growth of rushes and the shallowness of the water in many parts of the marsh between the Saskatchewan and Pine Island Lake, we had to go over to Cumberland this morning in the empty canoe, pushing it through the marsh until we reached a strip of dry ground, about half a mile wide behind the fort. Mr. Edward M'Gillivray, the gentleman in charge *pro tem.*, received us very hospitably. I obtained from him some pemmican and flour, and got him to procure for me a new canoe, for which I had to wait, as it was not quite finished. In the forenoon a brigade of boats from the McKenzie River arrived and departed *en route* to York Factory. One of the boats contained Mr. Anderson, chief factor, who was going direct to Red River and Canada. Although Mr. Anderson left Cumberland three days before us, in a boat of four or five tons burthen, well manned and equipped, and infinitely better adapted for encountering the boisterous gales of Lake Winnipeg, than our little canoe, we reached the mouth of Red River only twenty-four hours after him.

On Saturday the 14th August we were aroused at daybreak, by the singing of the voyageurs of another brigade of boats just arriving. It proved to be a detachment from York Factory, bringing Mr. J. G. Stewart, chief trader, in charge of Cumberland. Our canoe was not finished till late in the afternoon, when I should have started had I been supplied with a guide for Cedar Lake and the Grand Rapid; but the only man that was competent and willing to go, being one of Mr. Stewart's boatmen, and they having received their usual holiday allowance of rum on reaching their destination, no arrangement could be made with him. I was consequently compelled to remain till Monday. During the day, Mr. Stewart, from whom I received the most kind and hospitable attention, opened some packs and enabled me to get one or two articles of clothing, of which I stood greatly in need.

Cumberland House, the chief depôt or fort of the Cumberland District of the Hon. Hudson's Bay Company, is situated on the south shore of Cumberland or Pine-Island Lake; in latitude 53° 57′ N., and in longitude 102° 20′ W. of Greenwich (according to Sir John Richardson). It is about two miles in an air line north of the Saskatchewan, on the north side of what is called "Pine Island," a tract of land of considerable extent, between the Saskatchewan and Pine-Island Lake, isolated by two branch rivers connecting the lake with the Saskatchewan. The stream we passed before reaching Pemmican Portage is the western connection, and bears the name of Big-Stone River; it is about six miles long by its windings, and about two chains wide. When the water of the Saskatchewan is high, it passes through this channel or canal into Pine-Island Lake, and when low, the water from the lake flows into the Saskatchewan. At the time my survey was made (16th August, 1858), Big-Stone River was flowing into the Saskatchewan, at the rate of one and a half miles an hour. The eastern connection is about the same size as Big-Stone River, and joins the Saskatchewan some distance below Pemmican Portage; it is called "Tearing River," and is the route followed by the McKenzie River boats. The Saskatchewan boats go by these rivers when they require to call at Cumberland.

The country around Cumberland is low and flat; the soil in some places is a stiff clay, but in general it consists of a gravelly loam a few feet in thickness, covering a horizontal bed of white limestone, and supporting a light growth of poplar and birch. Occasional groves of spruce (the so-called pine of Rupert's Land, from which Pine Island derives its name) are seen here and there. The land being so little raised above the lake and river, a great deal of it is submerged during the spring floods, and some portions upon which the water remains become marshes and swamps, but many of them could be drained and improved without much difficulty.

There are about ten acres of ground enclosed and under cultivation at Cumberland. I observed a field of barley, and another of potatoes, both looking well, within the fort palings; and there is an excellent garden adjoining the chief factor's house; the soil appeared rich and fertile, bearing an exuberant growth of rhubarb, cabbage, peas, carrots, and other vegetables.

Cumberland House being at the junction of two great lines of water communication, one leading from the Pacific, and the other from the Arctic Seas to the Winnipeg basin, is a place of importance, and was formerly one of the Company's principal depôts. Within the fort there are a number of buildings, one of them (the store-house) is a very large edifice, containing extensive machinery and appliances for pressing and packing furs, and making pemmican. Cumberland has been visited by several celebrated Arctic explorers. In the garden there is a sun-dial which was brought from England, and erected by Sir John Richardson, and Sir John Franklin remained here a portion of the winter of 1819, while on his first overland expedition to the Polar Seas *viâ* the McKenzie River.

Cumberland House, Pine Island Lake.

The weather was very fine during our short sojourn at Cumberland, and although enjoying the hospitality of the gentleman in charge, I regretted the unavoidable delay as the season was already far advanced. Mr. Anderson very kindly offered me a passage in his boat to Red River, which of course I could not accept as his route lay along the east side of the lake.

The gentlemen at the fort appeared astonished that a bark canoe had come such a distance down the south

Saskatchewan, and that it had been dragged thither over the prairies from Red River; they looked upon it as a novelty in traveling in the north-west.

Messrs. Stewart and Anderson went in search of Sir John Franklin in 1855, under the auspices of the Hon. Hudson's Bay Company. They descended Back's Great Fish River to the Arctic Seas in bark canoes, and obtained some relics near Point Ogle; they also met Esquimaux in the vicinity of Montreal Island who had actually seen the whites, and who confirmed the account which had been previously given to Dr. Rae.

Sunday, August 15th.—A beautiful day. Another brigade from Methy Portage came in and left about noon to-day; bound for York Factory under the pilotage of the veteran guide, L'Espérance. The guide's boat contained a gentleman in the Company's service, from the McKenzie River district, who was proceeding to Montreal by the round-about way of York Factory and England.

August 16th.—We left Cumberland this morning in our new craft, a three-fathom birch-bark canoe. Not being so deep nor of the same beam as the old one, our load of baggage, instruments, and provisions, sank it to within a few inches of the gunwale, rendering it rather unsafe in a heavy sea. I succeeded in getting an Indian guide, through the kindness of Mr. Stewart, but could not prevail upon him to accompany us farther than the Grand Rapid, which ultimately proved fortunate for us, as, had he continued with our party, the pemmican, upon which we had now solely to depend till we reached Red River, would have been exhausted much sooner than it was. We returned to the Saskatchewan *viâ* Big Stone River; and passed the mouth of Tearing River about fourteen miles further down. Between the mouths of these rivers, the Saskatchewan flows occasionally among low alluvial islands, wooded with small poplar and willows, and in many places its depth is lessened by mud-flats and sand-bars; its banks are now low, alluvial flats, only two to three feet above the water, covered with grey willows and sapling poplar. The current in this part of the river is slacker than before, the average rate, as measured by the log, being two miles an hour. We camped about a quarter to seven, P.M., and made a section of the river, which gradually increases in breadth and volume of water; a number of soundings, taken at intervals across the river with the hand-lead, showed a mean depth of twenty feet; the width of the river at this point being 980 feet. I leveled about three quarters of a mile along the bank of the river here, to ascertain its fall.

To-day we met an Indian hunter and his family in a small canoe, the first we had seen since we set out on our journey; indeed the only signs of animal life yet observed in this solitary region were a young black fox, that came down to the brink of the river to slake his thirst, but scampered away as soon as a shot was fired at him, and a beaver in the vicinity of some trees, felled by these industrious animals. The Indian had been catching sturgeon and drying them for future use. This excellent fish abounds in many parts of the Saskatchewan, and it is one of the chief articles of food in the country. The Indians, as well as those in charge of the posts, have frequently nothing else to live upon for months, and the failure of the sturgeon fishing is often the cause of much distress and starvation. The sturgeon sounds are collected at the forts, and form an important article of export.

August 17*th.*—We embarked at 4 A. M., and observed no material change in the general character of the river and adjacent country during the day. The banks of the river are similar to those already described, being low alluvial flats not exceeding two feet above the water, and covered with willows and patches of balsam-poplar. The tract of country back from the river is rather low and wet; and the Indians make portages in one or two places from the river to small lakes north of it. The current is now much slacker than before, being only one to one and a half miles an hour.

About thirteen miles below Tearing River, Fishing-Weir Creek falls into the Saskatchewan; by which, during high water, boats sometimes go to Cumberland. About fourteen miles farther down, at what is called the Big Bend, the general direction of the Saskatchewan changes from a north-easterly course, which it has maintained from the Grand Forks, to a south-easterly one. This Big Bend is the most northerly point on the river, being very near the 54th parallel of latitude. The Pas or Cumberland missionary station, where we arrived about sunset, is nearly twenty-two miles below the Big Bend. About three miles above, or west of the Pas, the Saskatchewan makes an abrupt semi-circular curve, (called by the Indians "The Round Turn,") causing eddies and whirlpools, the river being at the same time diminished in width. The depth of the river was here found to be thirty-three feet, and its breadth about ten chains. Near the Round Turn there is a wooded ridge, upwards of fifty feet high, about half a mile from the north bank of the river. About three-quarters of a mile above the Pas, Root River, a long affluent with a width at its mouth of two chains, empties into the Saskatchewan.

The Pas, or Cumberland Station, is a missionary post of the Church of England, situated at the confluence of the Saskatchewan and the Basquia River, a tributary about three chains wide at its mouth.

Christ Church, as will be seen in the sketch I made of the Pas, is a neat and rather imposing edifice; and it seemed like getting back to civilization again, after all our wayfaring, when, on rounding one of the majestic sweeps of the river, the pretty white church, surrounded by farmhouses and fields of waving grain, burst unexpectedly upon our view. It was on a calm summer's evening, and the spire was mirrored in the gliding river, and gilt by the last rays of the setting sun.

The Pas, or Cumberland Mission.

The church is situated on the right or south bank of the river; near it is the parsonage, a large and commodious building, occupied by the Rev. E. A. Watkins, the present incumbent. Adjoining the church there is a neat school-house with several dwelling houses; and on the opposite side of the river there are six or seven houses, but they seemed to be uninhabited and in a dilapidated condition; the Indians, for whom they were erected, disliking a settled life devoted solely to the pursuit of agriculture;

and preferring the wandering and precarious life of a hunter in their native wilds. The river banks at the Pas are ten to twelve feet high, composed of light-coloured drift clay holding boulders and pebbles of limestone, and the surface soil is a dark gravelly mould, well adapted for cultivation ; but the surrounding country is said to be low and swampy with marshy lakes. Barley and other crops growing here looked well and were just ripening. Mr. Watkins' garden also looked well, and he kindly supplied us with some onions to make our pemmican more palatable. Mr. Watkins had just arrived from Red River in a freighter's boat. He had been twenty-five days on the route, having encountered much stormy weather on Lake Winnipeg.

August 18th.—Having to make some observations this morning, and Mr. Watkins wishing to send some letters with me, we did not leave the Pas till about 9 A.M. From the Pas, the Saskatchewan flows in a north-easterly direction through a low flat country, wooded with scrub poplar, and balsam-spruce, for about eight miles ; when again turning suddenly it resumes its south-easterly course, forming a great bend or elbow. About a mile below the mission, a branch, three chains wide, leaves the Saskatchewan, and cutting across the tongue of land embraced by this elbow, affords a navigable passage about three miles shorter than by the main river ; although it is the route generally followed by the boats, had I availed myself of it I must have left a considerable portion of the Saskatchewan proper unsurveyed. It was with the greatest difficulty that our Indian guide could be prevented from taking us by this short cut instead of the main river. When remonstrated with, and requested to return, after proceeding some distance down the smaller branch, he said, shaking his head, "you will find yet that the river forks off in many a branch."

About six miles from where this branch or canal rejoins the Saskatchewan, another branch, leading from Moose Lake and House, falls in; before uniting with the great river it separates into two branches forming a Y, the distance between the mouths being about half a mile. From the Pas to this point the character of the country bordering the river gradually deteriorates, the banks becoming lower and lower, and the timber more scrubby and scanty. The alluvial flats are in many places only one to two feet above the water, and they are at some points covered with drift wood, showing that they are flooded at certain seasons.

We stopped to cook dinner opposite the Moose Lake branch, where, by ascending a tree, I succeeded in getting a view of the surrounding country. The banks are here three feet above the river, supporting a thin strip of grey willows along the water's edge; and about half a chain back from the river there commences an extensive marsh or swamp with rank reeds and rushes, interspersed with ponds of open water, and dotted with clumps or islands of balsam-spruce and willows as far as the eye can reach. From Moose Lake Fork to where we camped, about sixteen miles further down, a slight improvement is observed on the immediate banks of the river; occasional groves of young ash, elm, and ash-leaved sugar maple are seen, but the flats behind are generally very low, and covered only with grey willows and sapling poplar.

We started on Thursday, August 19th, at break of day, with wet baggage and blankets, a thunder-storm with heavy rain having come on during the night. About four miles below our camping place, one or two branches leave the main river and flow to the north into a marshy expanse of water, about one mile broad and two to three miles long, called "Marshy Lake." Between this point and Cedar Lake are seen all the characteristics of a great alluvial delta. The Saskatchewan ramifies into many different channels, some of them return to the parent stream forming large islands, and several flow into Muddy Lake and other expansions of the main river, before finally emptying into Cedar Lake.

The country bordering the Saskatchewan from Marshy Lake towards Muddy Lake and Cedar Lake, consists of low mud-flats not exceeding eighteen inches above water, supporting along the river's edge a belt of willows, alder, dogwood, and long rank grass; in the rear is an extensive marsh with occasional islands of small poplar and spruce. These flats, being so little above water, are flooded every spring after the ice breaks up, and no camping place can then be found for a considerable distance up the river. A very rich mud is deposited during these floods, raising and extending the flats every year.

Muddy Lake, near which we were compelled to remain for some time, owing to a boisterous head wind, is apparently a dilatation of the Saskatchewan in a northerly direction; it is about two miles wide, and extends to the north for about four miles. We effected a landing on a point of the river four to five feet above the level of the water, where we found an exposure of light-coloured limestone in horizontal beds along the water's edge, and several large detached masses adjacent. This was the first outcrop of rock *in situ* we met with on the main Saskatchewan, and I made a very careful search for fossils, but, being unsuccessful, had to content myself with some specimens of the rock. On examining the point it was discovered to be an island eight chains long and four broad, with the river on one side, and on the other a vast reedy marsh interspersed with large ponds. This island is a favourite camping and fishing place of the Swampy

Crees, there being on it a clump of good sized poplar, the only timber fit for fuel for miles around; and here they hold their great councils, dog feasts, and medicine dances. Its name in Swampy is *Kash-ke-bu-jes-pu-qua-ne-shing*, signifying, "Tying the mouth of a drum."

Between Muddy Lake and Cedar Lake the Saskatchewan meanders through an immense marsh with tall reeds and rushes. It is now no longer an integral stream, but is divided into a maze of reticulating branches. According to our Indian guide, land is being formed here very fast; and what is now marsh and mud-flats was, within his recollection, open navigable water for a considerable distance back from where the Saskatchewan at present debouches into Cedar Lake through its numerous mouths. In one or two places we saw the trunks and branches of stranded trees sticking above water, where alluvial flats or shoals of mud and drift timber are in course of formation.

The Indians informed me that beyond these extensive alluvial flats and shallow marshes there is not to their knowledge anything but "muskeg" or boggy swamps for a very great distance on either side. I could see no high ground of any kind, and the character of the country bordering the Saskatchewan, as above described, may be said to continue back from the river for many miles.

We entered Cedar Lake on the morning of the 20th August, and coasted along the north shore till about noon, when we ran into a fine little harbour to eat dinner, after making a long traverse. In the afternoon, while crossing a wide and deep bay or sound stretching far to the north (the extremity being below the horizon), a stiff breeze sprang up, soon raising a very heavy sea, in which our canoe became almost unmanageable, pitching tremendously and shipping a great deal of water. On the 21st we breakfasted at the Rabbit Point, and entered the portion of the Saskatchewan issuing from the east end of the lake about noon.

Cedar Lake (so called from the occasional groves of cedar — a tree rarely seen in Rupert's Land — growing on its shores, particularly at its western extremity) is an expanse of water of considerable extent, in which the turbid waters of the Saskatchewan are allowed to disseminate and settle before re-uniting into one great river and rushing down the Grand Rapid into Lake Winnipeg. It is situated in about 53° 15′ N. latitude, and 100° W. longitude; and is nearly thirty miles long, with a breadth at its widest part of about twenty-five miles; its coast line embracing an area of water of about 312 square miles. Cedar Lake being more than sixty feet higher than Lake Winnipeg, is consequently upwards of 688 feet above the sea level. The only tributary it has of any size, besides its principal feeder the Saskatchewan, is a branch leading from Moose Lake and House, which enters it from the north. I was unable to obtain soundings of the lake in consequence of the high winds and stormy weather that prevailed during our voyage through it, but so far as I could learn it has sufficient depth of water for the largest craft, except at the west end, where the Saskatchewan is rapidly filling it up.

The northern coast of Cedar Lake is deeply indented and very low, and the country continues flat for a long distance back. At some of the points and on many of the islands along the coast, there are exposures of limestone in horizontal beds, the top of the strata being a few feet above the surface of the lake. It is to be regretted that, owing to the stormy weather, and the rate at which we were obliged to travel, no opportunity was afforded for collecting specimens. The mainland and islands being well wooded with balsam-spruce, birch, poplar, tamarack, cedar, and Banksian pine, could furnish an abundant supply of fuel ; thus offering, like the Saskatchewan, facilities to steam navigation ; but a considerable portion of the land is reported to be swampy and unavailable for agricultural purposes.

Cedar Lake is separated from Lake Winnipego-sis (the Little Winnipeg) on the south, by a low isthmus called the Mossy Portage, by which the Hudson's Bay Company formerly sent their supplies to the Swan River district. (They are now sent *viâ* the Little Saskatchewan.) The distance between the lakes at this point is about four miles, and Lake Winnipego-sis is about four feet higher than Cedar Lake, according to Mr. Dawson's measurements.

The portion of the Saskatchewan between Cedar Lake and Lake Winnipeg is nearly twenty miles in length, and its general direction is easterly. Through this channel, the great volume of water brought down for many hundred miles by the main river, and its north and south branches, together with that collected by many tributaries through a wide extent of country, is disembogued by one grand mouth into Lake Winnipeg.

Where the Saskatchewan issues from Cedar Lake the bed of the river is divided for a short distance into two channels by an island. We entered the smaller or south channel, and found it only two or three chains wide, for a distance of about a quarter of a mile. At its narrowest part, near the beginning, the Indians have a fishing station, and white fish and sturgeon are caught there in abundance. Along the side of this watercourse there is an outcrop of horizontal limestone, three to four feet in thickness, above the water, covered with a thin coating of vegetable mould, supporting small poplar, willow, and dogwood. I brought away some specimens of the rock, but could find no fossils. The current in this channel, as in most places where the river is narrower than usual, is strong, measuring two and a half to three miles an hour.

About half a mile below Cedar Lake, on the right or west bank of the river, which is now more than half a mile in width, is situated Cedar Lake House, a winter trading-post of the Hon. Hudson's Bay Company, lately established with a view to compete with the "Freemen," who come annually from Red River to trade with the Indians in this locality.

Between Cedar Lake and Cross-Lake Rapid, a little below which the Saskatchewan expands into Cross Lake, the river is very broad, and widens here and there into deep bays and funnel-shaped indentations. It grows narrower again a little above the rapid, where a projecting point of limestone, obstructing the current, causes a small smooth rapid on the south side, with a fall of about eight inches. The Cross-Lake Rapid is occasioned by a band of limestone intersecting the bed of the Saskatchewan nearly at right angles; and this is the first interruption of any magnitude to the even flow of the river. The Saskatchewan is let down by this rapid about five and a half feet in a short distance. There is a large island near the south side of the river, extending the length of the rapid, and dividing it into two channels. The broadest or northern channel is that which came under my observation. It is about thirty chains wide, and is the route followed by the Hon. Hudson's Bay Company's boats. In order to ascend the rapid, the Company's boats of four to five tons burden have to be "tracked" or dragged up with half cargo, and the other half of their load has to be carried over the portage, a distance of 230 yards. The fall from the west to the east end of the portage (obtained by leveling) is 4·08 feet, and from the east end of the portage to the quiet water below, about one and a half feet, making a total fall of 5·58 feet. Loaded boats run the rapid without difficulty, and if the channel were cleared of boulders and improved, it might be ascended by a powerful steamer.

Having spent some time in making observations at Cross-Lake Rapid, it was late in the afternoon when we entered Cross Lake; where our Indian guide left us, although he had agreed to pilot us down the Grand Rapid. He expressed himself anxious to return to his family at Moose Lake, and could not be induced to go farther. During the return journey, upon which he set out in a little canoe that he picked up coming down the river, he would have several days' hard paddling against a swift current. His departure was not objectionable, however, inasmuch as, had he continued longer with us, our slender stock of pemmican must soon have given out. We could spare him but a handful to carry him to his destination.

At the east end of Cross Lake we met Mr. Christie (a gentleman in the service of the Hon. Hudson's Bay Company, who had recently been appointed to the charge of Edmonton House,) in command of a brigade of boats, *en route* from York Factory to Edmonton and the Rocky Mountain district. Mr. Christie's heavily laden boats (fourteen in number) were manned by a motley group of Indians, half-breeds, Orkney-men, Norwegians, and negroes; they had just made the laborious ascent of the Grand Rapid, and thus far their progress had been very slow. Mr. Christie represented the many difficulties which had to be contended with in a boat voyage; the detentions on the lakes by contrary winds; the strong currents and rapids that had to be encountered in ascending the rivers; and the difficulty of procuring men suitable for the work; (each boat requiring six to eight experienced voyageurs;) and he expressed a hope that the long talked of steamers would soon make their appearance on Lake Winnipeg, to replace the present tedious, toilsome, and expensive mode of conveyance.

In reply as to whether there would be sufficient business to warrant the placing of steam vessels on these north-western waters, (irrespective of the establishment of a continental route to the Pacific, through British territory;) I was informed that there would be plenty of freight to carry for the present requirements and traffic of Rupert's Land; as during the year (1858) no fewer than 167 freight boats of the largest class, (four to five tons each) belonging to private traders and merchants, as well as the Hon. Hudson's Bay Company, (many of them loaded with valuable furs,) had passed Norway House, at the northern outlet of Lake Winnipeg, *en route* to York Factory; and returned with heavy cargoes of merchandise brought by sea to York, consisting chiefly of the usual supplies for Selkirk settlement, ammunition, and a variety of goods for the prosecution of the Indian trade both by the Company and "Freemen." The aggregate quantity of freight transported by this fleet of boats from the seaboard to Lake Winnipeg, and thence distributed along its principal feeders would be upwards of 800 tons. It is well known that there are large quantities of goods imported by other lines of communication — chiefly through the United States territory at present; and as the York Factory route is to be partially abandoned, a large portion of the importations of Rupert's Land will have henceforth to enter the Winnipeg Basin from the south, so that there will doubtless be sufficient commerce in view of the great water facilities afforded by the country, to encourage the initiation of steam navigation.

After remaining at Mr. Christie's encampment about

an hour, we set off again in the hope of reaching the Grand Rapid before dark. We soon entered a rapid by which we were lowered about two and a half feet in a distance of ten chains, followed, after an interval of smooth water, by another about a mile long, but with an easy inclination, the descent in that distance not being above seven and a half feet; it being nearly dark when the foot of the latter was reached, we camped for the night (August 21).

Cross Lake doubtless derives its name from its shape and the peculiar position it bears in relation to the Saskatchewan, of which it is evidently a dilatation. It is an oblong sheet of water, upwards of eight miles in length, having its longitudinal diameter at right angles to the general trend of the river; three miles is its greatest transverse diameter, and this breadth is about the distance between the termination and beginning of the bed of the river on either side of the lake. The altitude of Cross Lake in relation to Cedar Lake and Lake Winnipeg, acquired by leveling the rapids and measuring the currents in the river, would make its approximate elevation above the sea about 680 feet. It is reported to be deeper than Cedar Lake; and its banks on the east and west side are more abrupt and rocky, but its northern and southern shores are very low. Along the coast there are some fine groves of balsam-spruce, and aspen, but the land back from the lake is very flat and poorly wooded, a great portion of the original forest having been destroyed by fire; large tracts of burnt and dead timber are seen here and there; the blackened trunks of poplar and spruce indicating the ridges or dry areas over which the conflagration extended, and the lifeless tamaracks revealing the swamps or flooded land. The lake extends so far to the north, its extremity in that direction is not seen from the traverse line, being below the horizon of the spectator. In the northern arm of the lake there are several wooded islands, but as they were some distance from our track I was unable to ascertain the nature of their formation.

There being two rapids between Cross Lake and the Grand Rapid, the Saskatchewan may be said to descend by four distinct steps from Cedar Lake to Lake Winnipeg; the first one east of Cross Lake, having a length of about ten chains with an estimated fall of two and a half feet, occurs half a mile below the recommencement of the channel of the river, and appears to be attributable to a low and nearly level belt of limestone, through which the river has gradually excavated its way by three separate channels. The middle channel, by which we descended the rapid is only three to four chains wide, and could apparently be ascended by a steamer without difficulty, as it is deep and appears to be free from boulders. The other channels might even be more favourable for steam navigation, being broader as far as could be observed, and containing a greater volume of water; they are, however, a little out of the direct course, and for this reason are not followed by the boats. The smooth portions of the river are broad here; the width above the two islands formed by these three channels being more than half a mile, and below them upwards of three-quarters of a mile. About a mile below the foot of the first rapid the second one begins. Its length is fully a mile, and its approximate fall is not more than seven and a half feet. It has a long gradual slope, with a deep channel of rolling, but comparatively unbroken water in the middle; the water is more turbulent at the sides, where the current is interrupted by points of limestone rock, boulders, and débris. The exposures of limestone on the points, are four to six feet in thickness above the water, with a horizontal stratification. The loaded boats of the Hon. Hudson's Bay Company descend this rapid easily, and as they are generally "tracked" up with the whole of their lading, a lightened steamer, with powerful engines, might surmount it by taking the best channels and other precautions.

It is about four miles from the foot of this last rapid to the beginning or summit of the Grand Rapid. In that distance the river is smooth and deep, but has a very swift current, especially where its bed is contracted. The width of the river in this interval is much diminished, varying from nine chains to a quarter of a mile, and the rate of current is from three to three and a half miles an hour. There are one or two large boulders in the bed of the river here, over and around which the water boils and bubbles like a cauldron; and now and then shoals on the north side of the channel are indicated by the rippling water and ground-swell occasioned by the current in passing over them. The land between Cross Lake and the Grand Rapid is generally low and flat, but thickly timbered with balsam-spruce, poplar, tamarack, and birch. At the second rapid, east of Cross Lake, the banks on the north side of the river are eight to ten feet above the surface of the water, and are composed of a light-coloured drift clay. These clay banks gradually increase in height towards the Grand Rapid, where they attain an elevation of upwards of twenty feet; but it is probable that the surface of the country is nearly level, and that it is the descent in the river which causes the apparent rise in its banks.

August 22nd.—This being Sunday we did not proceed on our journey till after breakfast (about 8 A.M.). However desirable it might have been, under other circumstances, to have remained inactive on this day, in the position in which we were placed, like a ship at sea, with a limited supply of provisions, and a long and hazardous voyage before us, it would have been altogether out of the question; indeed, the loss of a day or even an hour might have compromised the safety of the whole party.

In about an hour we reached the beginning or west end of the portage at the head of the Grand Rapid, where my various instrumental observations and measurements in relation to the rapid began. In order to commence operations we disembarked and made the portage, which of course is never done by boats in descending the river. Yet, notwithstanding that boats invariably "run" the whole of the rapid, it would be extremely perilous to descend the upper portion of it in a small heavily laden canoe without a guide.

So much having to be done with so few hands, our little party exhibited a scene of unusual activity and exertion from the time we landed at the top of the rapid until we camped in the twilight on the coast of Lake Winnipeg. The first thing to be accomplished was the transportation of the canoe and the heavier articles of luggage to the east end of the portage; to effect this the united energies of the party were required, and owing to the length of the portage it occupied some time. Whilst the Ojibway was carrying the remainder of the lading, I was engaged with the Blackfoot in making a survey of the portage and rapid, chaining across in one direction and leveling back in another, and so forth.

About 4 P.M. the various observations were completed, and everything had arrived at the east end of the portage. The different operations involved the crossing of the portage (more than a mile in length) many times during the day. While dinner was preparing I occupied myself in making a sketch of the cataract and examining the character of the perpendicular limestone cliffs at its side.

After eating a hasty meal we re-embarked to run the lower portion of the rapid.* The voyageurs wished me to walk through the woods to the foot of the rapid (probably to lighten the canoe), but as the day was already far advanced, and being anxious to reach Lake Winnipeg, as well as for other reasons, I deemed it expedient to go down "in canoe."

In running the rapid we followed as closely as possible the instructions given to us by our old guide on the Plains (John Spence), who had often piloted the old North-West Company's *North* canoes down its entire length. In attempting, according to his directions, to cross from the north to the south side of the rapid in order to get into what was reported to be the best channel for a small canoe, such was the fierceness of the current, and the turbulence of the great surges and breakers in the middle, that we were nearly engulfed; and although every nerve was strained we were swept down with impetuous velocity, and did not get near the other side till we were about three quarters of a mile below our starting point. We were then impelled with astonishing swiftness along the south side of the torrent, often in dangerous proximity to the rugged wall of rocks bounding the channel, and now and then whizzing past—almost grazing—sharp rocky points jutting out into the river, against which the thundering waters seethed and foamed in their fury. During the descent the voyageurs exerted themselves to the utmost of their strength, and evinced an admirable degree of coolness and dexterity.

The Grand Rapid is acknowledged by those who have witnessed it, and who have had opportunities of traversing

* The part here designated as the *lower portion,* although the Grand Rapid is one continuous torrent from beginning to end, is that below the east end of the portage, and is more than one mile and a half in length.

the great river systems of the continent, to be unsurpassed (as a rapid) in magnificence and extent, as well as in volume of water. It is certainly a formidable barrier to the navigation of the Saskatchewan.

The following are the dimensions of the leading features of the Grand Rapid :—

1. *Its Length.*—The portage path is nearly straight, with a magnetic course from the upper to the lower end, of S. 60° E.; it is 87 chains 40 links in length; the

The Grand Rapid of Saskatchewan.

distance between its extremes by the river is a little more than this, as the river describes an arc of which the portage is the chord, but as the head of the rapid is a little below the west end of the portage, this distance may be adopted as the length of the upper or most precipitous portion of the rapid. The distance from the east end of the portage to the foot of the rapid by our track is 129 chains. This would make the whole length of the rapid 216 chains 40 links, or nearly $2\frac{3}{4}$ miles.

2. *Its Descent.*—By leveling carefully along the por-

tage path, I ascertained the fall between the smooth water at the head of the rapid to the general level of the water at the east end of the portage to be 28·58 feet; and after observing instrumentally the descent in the lower portion of the rapid as far as the nature of the country would allow, I closed my levels on a bench-mark at the surface of a pond of still water, fed by an eddy at the lower end of the portage. The fall in the lower portion of the rapid, acquired by leveling and by careful estimation, is about 15 feet; this would give 43½ feet as the total descent of the rapid.

3. *Its Breadth and Depth.*—The width of the river, at the upper end of the portage, is about 20 chains; at the head of the rapid, about 7 chains further down, where there is an island in the bed of the river, it is about 30 chains; and at the lower end of the portage, where the rapid emerges from the highest limestone plateau, its width is about 10 chains. From thence it gradually widens towards the foot of the rapid, where it attains a width of 25 chains. I was unable to obtain soundings of the rapid, but, from the depth and volume of water above and below it, where the river is much broader, it is undoubtedly deep.

The Grand Rapid, throughout almost its entire length, washes the bases of perpendicular escarpments of rock. It passes through two plateaux of brittle buff-coloured limestone, with a horizontal stratification; the top of the first, or upper plateau, being nearly on a level with the surface of the water at the head of the rapid, and underlying a stratum of light-coloured clay, twenty-three feet in thickness, in which are embedded boulders and pebbles of limestone; the whole overlaid by about eight inches of vegetable mould, and clothed by a forest of balsam-spruce, tamarack, and poplar. The surface of this plateau continues nearly level as far as the lower end of the portage, where the top of the rock is 25·36 feet above the surface of the water, and about the same height above the lower plateau. The lower plateau continues some distance further down, but is soon hidden by drift clay-banks, which, at the foot of the rapid, have an altitude of twenty to thirty feet above the water.

It is not improbable that the Grand Rapid is the result of the eroding influence of the great body of water in the river, upon the rock through which it flows—the limestone being of a crumbling and yielding nature. At a remote period, the water of the Saskatchewan was perhaps lowered from the top of this rock formation, by a perpendicular cataract; the precipitous leap most probably began at the lower end of the portage, or at the eastern limit of the highest limestone plateau, from whence the river gradually wore away the rock, at the same time diminishing the height of the fall, until it became a foaming rapid from beginning to end.

The upper portion of the Grand Rapid—of which I succeeded in getting a sketch—presents a scene that strikes the beholder with wonder and admiration. The great body of water that has been stealing along, swiftly but silently, for many miles, appears to be suddenly imbued with life—the rippling of the river becoming gradually more turbulent, until the tumultuous surges grow into huge, rolling billows, crested with foam, like waves in a tempestuous sea. The great rollers and breakers seem, to the spectator, to be continually changing in shape and appearance, owing to the lines of surf and the peculiar colour of the water; but although the mighty cataract thus appears to be for ever changing, it really rolls on for ever the same.

The ascent of the Grand Rapid is one of the most

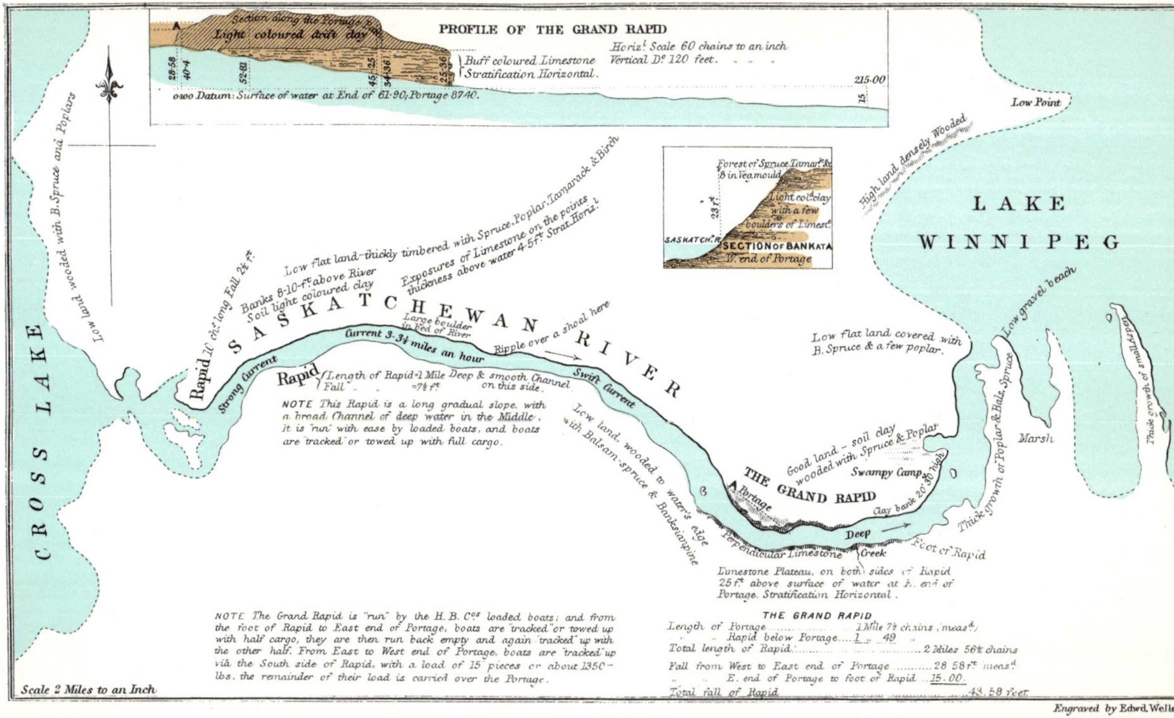

Plan shewing the Grand Rapid of the Saskatchewan

laborious duties that has to be performed on a boat voyage from Lake Winnipeg to the Saskatchewan district. The Hon. Hudson's Bay Company's brigades surmount this fearful interruption to the upward navigation of the Saskatchewan in the following way: On arriving at the foot of the rapid, every boat discharges one-half of its cargo of four to five tons. Thus lightened, they are then "tracked" (towed) up to the beginning of the portage—the whole of the crew of six or eight voyageurs, with the exception of the bowsman and steersman who remain in the boat, being engaged in the labour of tracking. Each man is attached to the tracking-line by a leather belt, or portage-strap, passing round his body; and harnessed in this manner they drag the boat along, running and scrambling barefooted over the slippery and jagged rocks at the sides of the cataract. When the lower end of the portage is reached, the boat is emptied, and "run" back again to the foot of the rapid, and from thence hauled up as before, with the remainder of its load. The whole of the lading is then carried over the portage, exclusive of fifteen pieces, or about 1,350 lbs., which is left in the boat. With this ballast, the boat is pulled across to the south side of the rapid, to be tracked up, as the towing-path is better there than on the north side. In consequence of the rapidity and violence with which the upper portion of the rapid flows, in ascending it, it is necessary to employ the "main line,"—a much thicker and stronger rope than is generally used for tracking To this line the crews of one or two boats are lashed, and thus they run along the top of the cliffs of limestone,—there being no footing at the bottom of these walls of rock—hauling the heavy craft up the surging cascades. The utmost strength of the bowsman, with his pole, and the steersman, with his long sweep oar, is required, to prevent the boat from being dashed to pieces among the rocks.

Small brigades, feebly manned, often haul their boats over the portage. The portage road bears evidence of this, as it is deeply scored and furrowed by the keels of boats from beginning to end.

The boats used by the Hudson's Bay Company are built at the principal posts of the various districts. They are heavy, and strongly built, but last only about three years, having to undergo such severe strains and hard usage in the rapids and on the portages on the routes followed. They are of the whale-boat build, about thirty feet long and six broad, sharp fore and aft, with flat floors, which make them very leewardly.

Although the Grand Rapid is the most serious obstacle that the Company's boats have to encounter, it is not the only difficulty they meet with on the Saskatchewan. The whole ascent of the river is one of labour and fatigue. The current is so swift—as the name of the river is well known to imply—that the voyageurs would track nearly all the way to the Rocky Mountains, if the banks of the river would allow; but where the river passes through marshes and swamps, they have no alternative but to pull against the current, however strong it may happen to be.

Before finally determining upon any works or measures for overcoming the Grand Rapid, in order to render the whole of the Saskatchewan navigable for steam vessels from Lake Winnipeg, without interruption; it would be necessary to make a more extensive and elaborate survey; but probably sufficient information and data have been acquired during this reconnaissance from which schemes might be devised, and suggestions offered, for surmounting the difficulty. To navigate the Saskatchewan at present, a steamer would evidently have either to be built above the rapid, hauled over the portage, or "warped" up the rapid itself. Seeing that the Company's large bateaux are hauled up the rapid by manual labour, it does not seem impracticable for an empty steamboat, with engines of great power, to ascend it, by the aid of hawsers and guy-ropes stretched from the steamer to the land, using, along with capstans, the motive power of the steamer as far as available. But in any case, unless a canal were constructed, a transhipment of cargo bound upwards would have to take place, whether there were steamers plying above and below the rapid, or whether steamers were forced up the rapid ; so that it would be necessary to construct a good road or tramway on the present line of portage. The features of the country in the vicinity of the Grand Rapid are very favourable for a road, and even for a settlement, as the banks of the river are high, with a considerable depth of good soil, from the second rapid east of Cross Lake to near Lake Winnipeg. There is also abundance of timber for fuel and building.

From the foot of the Grand Rapid, the Saskatchewan flows with a pretty strong current, in a northerly direction till it enters Lake Winnipeg. Its mouth has a width of about twenty-eight chains, and is a little over two miles below the lower end of the rapid. On the coast of Lake Winnipeg, immediately east of the mouth of the Saskatchewan, there are several deep and narrow bays, or estuaries, marshy at their inner extremities, and separated by narrow points or spits of gravel, by which it seems not improbable the Saskatchewan entered the lake at some period of its existence, and that north-easterly gales and shoves of ice have driven up these barriers, and caused the river to excavate new outlets.

We visited an Indian encampment on the north bank of the river, a little below the foot of the rapid, in the expectation of procuring some sturgeon, but were unsuccessful—the fishery carried on here by the Indians having failed this year. This encampment of two lodges was the only one we saw on the main Saskatchewan. It had been a larger camp, but eight families had just left it, previous to our arrival, for their winter quarters at the Little Saskatchewan. They are Swampy Indians, and generally winter at Fairford, whence they proceed in summer to the Grand Rapid; where, by assisting in dragging the boats and *portaging*, they get a small recompence in the shape of tea, tobacco, or pemmican. They occupy the time between the arrivals of the different brigades of boats, in catching and drying fish, and generally leave after the last fleet has passed up in the autumn.

We entered Lake Winnipeg at sunset, and camped not far from the mouth of the Saskatchewan, upon a narrow spit of gravel, separated from the wooded shores by a marsh. The night was clear and beautiful, and the lake wonderfully calm. From our bivouac, where we lay with cramped limbs outstretched on the shingle-beach, could be seen the great headland "Kitchi-nashi," vanishing away to the south-east in the far distant horizon. A view very extensive and beautiful, but which betokened many hours of paddling and tracking out of the direct course to Red River. To the east and north the only limit to our gaze was the dim horizon of the great lake which lay tranquilly outspread before us like an unruffled sea.

CHAP. XXII.

FROM THE GRAND RAPID OF THE SASKATCHEWAN TO THE RED RIVER SETTLEMENTS *viâ* THE WEST COAST OF LAKE WINNIPEG.

Enter Lake Winnipeg.—Cape Kitchi-nashi.—Storms.—Detained on an Island.—Windbound on the Mainland.—Tempest.—Repulsed by the Wind.—Character of the Coast: the Sand Beaches and Swamps.—War Path River.—Verifying Rate of Canoe.—Indians.—Tracking.—Limestone Point.—Encountering a head Wind and Storm.—Lightening Canoe.—Starving Indians.—The Little Saskatchewan.—The Prominent Features of the Coast.—Formation of Cape Kitchi-nashi.—Limestone Exposures.—Tributary Streams.—General Character of the Country.—Indians and Fishery at Little Saskatchewan.—Indian Chart.—Inaccuracy of the Maps of the Lake.—Depart from the Little Saskatchewan.—Windbound again for three Days.—Provisions exhausted.—Contrary Winds.—Driven back and stopped.—The Cat Head.—Windbound again by a Hurricane.—Barrier of Boulders.—Eagle.—Stopped by foul Winds again at the Wicked Point.—Pike Head and River.—Indian Fishing-Weir.—Opportune Supply of Fish.—Wide Traverse to Grindstone Point.—Grassy Narrows.—Sandy Bar.—Arrive at the Settlements.—Conclusion.

MR. FLEMING'S NARRATIVE — (*continued*).

It was on the evening of Sunday the 22nd August, 1858, that we glided from the mouth of the Saskatchewan into Lake Winnipeg, but our voyage through this great inland sea was not fairly begun until the following morning, when we embarked at an early hour (4.20 A.M.) in our little canoe. Being favoured with a light breeze for a few hours we reached the neck of the great promontory, *Cape Kitchi-nashi*, about noon. From the mouth of the Saskatchewan to this point the coast trends to the south-east, and is indented in a remarkable manner by a series of deep bays of every shape and size. As it would require unlimited time and resources to penetrate into every sinuosity of the coast, we generally steered straight from point to point, although in doing so some long traverses had to be made.

The northern coast-line of the promontory being nearly straight with fine sand-beaches, affording tolerably good footing, we tracked along the shore for the remainder of the day; although this was hard enough work, the men were glad to avail themselves of it, as a change or relief from paddling. By working fifteen hours to-day we were enabled to camp at the extreme point of the headland, where, the night being favourable, the magnetic variation of 15° E. was observed. The Ojibways call this cape "Kitchi-nashi," and the Swampys "Missineo," both names signifying "Big Point." By some it is called "The Détour."

August 24th.—A fine morning, the lake quite calm. After doubling the cape we overtook eight small canoes containing the band of Indians (Swampy Crees) who left the Grand Rapid on Sunday, 22nd. In a short time a light breeze sprang up, and by hoisting a blanket we sailed at a pretty good rate for some hours. About 2 P. M. the wind began to increase in strength and turned suddenly against us, so that we had to run in behind a low point of sand and gravel for shelter. Although the wind still continued high we started again and made a traverse to a small sand island on which we were obliged to remain, being then over two miles from the main land, and the storm having increased in violence. A storm of wind soon raises a very heavy sea on Lake Winnipeg on account of its little depth of water.

The island on which we were detained is one of the Gull-egg group, which, with a point of sand protruding from the main land, forms a pretty good harbour on the south side of the neck of the great promontory. The Indians were nearly destitute of provisions, and followed us to the island, where they fortunately got a plentiful supply of eggs and young gulls; but having little ammunition, they brought down only a few old ones, although they hovered in countless numbers over the island, screaming at the wholesale destruction of their young brood.

August 25*th.*—The storm raged all night, and this morning we found ourselves surrounded by a foaming sea on a low island of sand about 100 yards in length, and so narrow that the spray from the breakers dashed completely over it. The gale blew hard from the east till about noon, when it began to subside; I then determined upon starting on our course, but, seeing a thunder-storm approaching, decided upon taking dinner before making the attempt. It was well that we did so, because just as we were hastily swallowing our meal of pemmican, the thunder-storm, accompanied by strong wind and heavy rain, burst upon us with great violence. Some of the Indians were endeavouring to reach the next island in the line of traverse, but had to abandon the attempt and drive before the gale to the main-land, three miles off.

The storm soon abating again, we crossed to the next island and thence to the main-shore; and after coasting along for some miles, encamped on a sandy point, where we found a small clump of poplar and spruce.

August 26*th.*—Last night the Northern Lights or Aurora Borealis were unusually brilliant, darting and playing about with extraordinary rapidity in all directions, sometimes extending to the zenith and sometimes to the south of it. The voyageurs said they portended a coming storm, and their prognostications proved correct. The night was clear with a bright moon till about midnight, when a cold north-westerly wind arose, followed in a very short time by a stormy sea. The gale soon veered round to the north increasing to a perfect hurricane, and during the day the lake was white in all directions with breakers and foam. A heavy surf breaking along the coast and tearing away large portions of the bank on which we were camped, warned us to move our canoe and lading back from the shore; yet, notwithstanding every precaution, some of our paddles and poles were swept away during the night. A large marsh being in our rear we could retire but a few yards from the raging lake to wait for the abatement of the storm.

August 27th.—After midnight the wind began to decrease gradually, and by daybreak it had so far subsided as to permit us to continue our voyage. By breakfasting at a point where we found an outcrop of limestone I was enabled to procure some fossils. This, the first rock exposure observed since leaving the Saskatchewan, is apparently the termination of a ridge running at right angles to the coast-line, and bounded on either side by marsh and swamp. The top of the rock is ten feet above the surface of the lake, and is covered by a stratum of boulders and drift two feet in thickness, supporting small poplar, tamarack, spruce, birch, and Banksian pine; there are only six feet of the limestone exhibited, the remaining four feet being concealed by a talus of boulders and débris. The high water-mark of the lake reaches to the top of the talus.

A contrary wind arising about noon detained us four hours at the mouth of a creek, which we ascended a short distance. The entrance, or where the creek cuts through the sand-beach enclosing a marsh, is one chain wide; within the sand-beach the creek expands into a deep pond thirty chains in diameter, surrounded by a marsh; this pond is fed by the inner portion of the creek, a broad and sluggish stream five feet deep, meandering through a tamarack swamp. It is reported by the Indians to have its source a long distance inland. As there is but one and a half to two feet of water over the bar this could only be used as a harbour for boats. It is about half-way between the Gull Islands and War-Path River.

We set off again after the wind had moderated a little, but were compelled to camp in an hour and a half in the lee of a point, on the weather side of which an adverse wind was blowing hard, driving before it a heavy sea. Being thus repulsed again by the wind, I directed my attention to the character of the coast in the vicinity of our bivouac. Along the shore there extends a long straight sand-beach, sixty feet wide and arched like a road-way; on the inner side of this beach there is a tamarack and black spruce swamp, with a bottom of black muck and moss two feet in thickness, covered with water. This "muskeg," is said to continue for a great distance back. By leveling I found the surface of the water in the swamp to be only eight inches higher than the lake; and as the crown of the sand beach is only four and a half feet above the level of the water, and is covered with drift-wood, it is evident that the lake washes into the marsh during high water.

Leaving camp at 4.30 A.M., August 28th, we reached the mouth of War-Path River at 1 P.M. The Indians say this river rises in lakes, and draining a great extent of swampy country, is very large in spring. There are three feet of water over the bar at its mouth; the channel at the entrance is contracted in summer by the sand to a width of forty feet, with an average depth of four feet; within the entrance there is a basin thirty chains broad, forming a boat harbour of easy access.

After tracking for several hours along straight sand-beaches, which separate marshes from the lake, we camped nearly opposite Caribou Island, on a coast similar to that which we left in the morning. The Indians came up with us, and erected their lodges in our neighbourhood. They were now quite friendly, but at first their chief and principal men looked upon us with suspicion, especially while noting or making any observations. They seemed to be under the impression that there was something going on which would ultimately deprive them of their country. The failure of their summer fishery at the Grand Rapid had rendered them badly off for provisions, and the only food they could get on this inhospitable coast was an occasional gull. Whoever had the good fortune to obtain one of these sea-birds was soon surrounded by a group of Indian children, watching with anxious eyes for the intestines of the animal, which they threw upon a fire for a few minutes, and then ate apparently with great relish.

August 29th.—Embarking this morning at daylight, we reached Limestone Point about eleven o'clock, after making a traverse of three miles against a strong head wind. On this point there is a very fine exposure of light-coloured limestone, containing numerous fossils, some of which I succeeded in procuring. The outcrop on the point is fourteen feet in thickness above the lake, in massive horizontal layers, overlaid by two and a half feet of drift and fragments of limestone that have evidently been broken up by ice. This headland is the abrupt termination of a narrow ridge of limestone clothed with aspen, spruce, and birch; it is about two miles long, running nearly north and south. On the west side of it is Portage Bay, so called by the Indians, as they sometimes make a portage from the foot of it, across the neck of the point.

After remaining here about two hours we proceeded on our journey. On rounding the point we found the wind on the east side of it blowing directly in our teeth, and it required the utmost exertions for two hours to force the canoe against a high gale and stormy sea, until we got into the lee of a small island, it being impossible to land on the main shore. The canoe leaked and shipped so much water during this traverse, that in order to lighten her we were compelled to throw overboard some of the heavier of our geological specimens. It was with great regret I saw one of them, a very large and fine *orthoceratite*, consigned to the deep.

On the island we found part of the Indian band, but the greater portion were hurrying on to the Little Saskatchewan to get fish, as they had nothing to eat. We saw them in the distance, battling against the wind and sea, their little canoes like specks tossing among the swells and breakers. The occupants of some of the canoes were only women and children; one canoe was paddled by two Indian women and two little boys eight or nine years of age; they, especially, would have hard work in reaching the mouth of the river. The Indians remaining on the island were chiefly the more feeble of the party, and being ravenously hungry they were all in the marshes busily engaged in pulling up and eating the roots of bulrushes. The storm increased towards evening, and we were obliged to camp on the island ourselves.

August 30th.—Although the unfavourable wind had diminished but little this morning, we plied our paddles so well, and made such good headway against it, that we entered the mouth of the Little Saskatchewan or Dauphin River about 11 A.M. We tracked up the river to the Indian encampment, about four miles from its mouth, for the purpose of procuring fish, and found the Indians at the rapids scooping large numbers of excellent white-fish from the eddies.

It would perhaps be as well to give here a short recapitulation of the character and general topography of the west coast of the lake between the main Saskatchewan and the Little Saskatchewan.

The distance from the mouth of the Main to the mouth of the Little Saskatchewan, by our track along the coast, or by the course that canoes or row-boats would be likely to pursue is about 140 miles; but the distance by the coast-line embracing every sinuosity of the shore is much greater.

The most prominent feature in the line of coast is the great headland, Cape Kitchi-nashi. This immense promontory begins to stretch out into the lake in a direction a few degrees north of east, about fifteen miles south of the Saskatchewan. Its extreme point is about twenty-four miles in an air line from the general line of the coast, and its width varies from three to six miles and upwards; its neck is indented by several deep bays, some of which might be available as harbours or roadsteads. The formation of the cape is peculiar; it is very low and flat on the north side, while on its southern boundary the coast is comparatively high and abrupt. Its northern side consists of a series of marshes separated from the lake by a narrow sand-beach, and these marshes gradually blend into a tamarack and spruce swamp. Along the south side of the cape there is a continuous escarpment of light-coloured clay, twenty-five to forty feet high, yet even on the top of these high banks the character of the land is of the poorest description, being nothing but a "muskeg" or trembling swamp, containing a thin growth of very scrubby tamarack and spruce, covered with drooping moss.

The extremity or apex of the promontory is a very low and broad sand-beach covered with water-worn boulders; the lake is also dotted with boulders a long way out from the shore, there being a sand-bar or continuation of the point under water, on which they rest. From the size and position of the cape, and the dangerous shoals extending out from it, if beacons or lighthouses are ever required on the lake for the safety and convenience of shipping, no more suitable place could be selected for the erection of one than here.

The coast north-west of the cape, as already stated, is very low, and much broken by deep and narrow bays.

From Cape Kitchi-nashi to the Little Saskatchewan the coast trends generally to the south-east. Between these points limestone is exposed in six places. The exposures are the precipitous extremities of ridges, forming points at intervals along the coast. The stratification in every instance is horizontal, but the escarpments vary in height above the lake; they increase in altitude from four to fourteen feet towards the south. These ridges are generally wooded with aspen and other deciduous trees; and the swamps intervening are timbered with tamarack and spruce; some of the spruce near the coast are pretty large. Between the ridges, low sand-beaches extend along the coast. These beaches separate ponds and open marshes, averaging from a quarter to one mile wide, from the lake; in the rear of the marshes is the great tamarack and spruce swamp, or "muskeg."

The tributary streams in this part of the coast are not numerous, and they are generally of no great size. The chief are the Gull-Egg Rivers or the Two Rivers, the War-Path River, Jumping River, and one or two others without name; they are not in themselves large, but their estuaries might be available as harbours for boats.

The character of the country exhibited on the coast extends almost an unlimited distance back; indeed the Indians report the whole of the country between Lake Winnipeg and Lake Winnipego-sis as one vast "*muskeg*" —the great moose hunting-grounds of the Swampys.

Although the country here described is quite unfit for agricultural purposes, it is not altogether valueless; there are large areas of good timber along the coast, available for fuel, and the limestone cropping out at the various points is well adapted for building.

A person ascending or descending the Little Saskatchewan would be greatly deceived in relation to the character of the country through which this rapid river flows, if he were to judge by the appearance of its banks without penetrating beyond. I made a section of the river and its excavated bed at the Indian encampment, about three and a half miles from its mouth. At this place the river was 360 feet broad, four and a half feet deep, and ran over a stony bottom. Its banks were from twenty-five to thirty-five feet high, composed of light-coloured clay, holding boulders and pebbles of limestone. Along the margin of these banks there was a zone of tall poplars, extending about 300 feet back from the face of the bank. This narrow strip of good timber and land soon blended into a great muskeg or tamarack and spruce swamp. This swamp consisted of dead and dwarfish tamaracks covered with pendant moss, and struggling through several feet of wet and trembling moss and black mould, through which a pole could be shoved for several feet.

The Indians were catching very fine white-fish in large quantities at this point with little trouble. The river seemed to be crowded with them. At various places along the brink of the stream, enclosures of stones were constructed, beside which an Indian stood with a large scoop-net attached to a pole, filling the stone enclosure from time to time as he scooped them from the eddies in his vicinity. The excellence and abundance of these white-fish, and the proximity of this locality to the salt region, together with the facilities for water communication, may lead to the establishment of a fishery here on an extensive scale. The Indians were curing the fish without salt, by splitting them very thin and drying them in the sun.

We were invited into one of the tents to partake of some fish which they had cooked for us. A large birch-bark dish of fish was set before us, the whole of which we had to eat, according to Indian custom, not to give offence to our entertainers. As an extra mark of favour, an old fellow in a very dirty blanket strewed a few grains of black-looking salt over my portion. This salt he carried beneath his belt in a piece of birch-bark.

Being without a guide I got one of the Little Saskatchewan Indians to draw me a map of the lake between Bushkega Islands and Grassy Narrows, showing the traverses and route to be taken between the islands in order to cross the great arms of the lake, Fisher Bay and Washow Bay. This Indian chart was of great service to us; the best and most recent maps of the lake to which I had access being so incorrect: on them the general contour of the coast north of the Little Saskatchewan is tolerably well delineated, but to the coast north and south of the Dog's Head Straits they bear very little resemblance; the large islands are omitted altogether, and the Great Black Island is represented as forming the extremity of a promontory on the mainland between two bays.

From the beginning, our canoe was very weak, the bark being of the poorest description and badly put together, and having now become quite frail I tried to barter with one of the Indians for a new and stronger one; but, taking advantage of our situation, he placed upon it a much greater value than I felt inclined to give.

Having made sections of the river, and examined the country bordering the Little Saskatchewan, we left it on the 31st of August, but were detained the greater part of the day on a point only a few miles from the mouth of the river, by unfavourable wind and in consequence of the sickness of Louis, our steersman, who, being a pretty old man, was disabled from over exertion in the storm on Sunday.

On the 1st of September, while sailing with a side wind across the mouth of a deep bay, in which there was rather a heavy sea rolling, a large swell broke over us, throwing in a great deal of water, and from the evening of the 1st of September until the morning of the 5th we were windbound on a low marshy point on the north-east side of the great bay into which the Little Saskatchewan empties. The spot on which we were imprisoned is very much circumscribed, being a narrow sand-beach, about a chain in length, and bounded on three sides by an extensive marsh. During the three days that the storm lasted the wind blew a hurricane from the N.N.W., raising a tremendous sea on the lake, and the surf beating along the shore, washed away several yards of the sand-beach on which we were encamped. The weather was clear the first day, but on the second and third days it rained almost incessantly, and it was then for the first time on our voyage that we really felt the want of a sufficiency of food, as our stock of provisions was reduced to a few pounds of rather mouldy pemmican, which I determined to eke out as long as possible, being still a great distance from Red River (upwards of 170 miles by the canoe route), and with that object in view we made it a rule to eat only one meal a day while we were windbound, unless we were fortunate enough to procure some additional food, in the shape of wild fowl or other animals. We succeeded in getting a grey gull on the second day, on which we made an excellent repast.

On the morning of the 5th, just before we started, an Indian and family from the Dog's Head came to us, they had been windbound seven days on an island not far from where we were, they said they never saw such a continu ous succession of winds and storms on the lake before, and informed us that a freeman's boat which passed during the night had been thirty days between Red River and the Saskatchewan, a distance that has been accomplished by a boat, with a favourable wind, in three days. After bartering with this Indian for a small "rogan" of fish pemmican (dried fish pounded and mixed with sturgeon oil), we proceeded on our journey, glad to get away from the dreary spot. Although there was still a heavy retarding ground-swell on the lake, we paddled many miles before halting. On stopping to cook breakfast we were greatly disappointed to find that the fish pemmican which we were so thankful to get, was nearly all rotten, there being only a small portion on the top that could be eaten, the remainder had to be thrown away.

A contrary wind freshened up again about noon, but we continued struggling against it, until in attempting to round a point we were completely driven back, and narrowly escaped foundering among the huge swells and breakers that dashed high over the boulders extending out from the beach; we saved the canoe by jumping into the surf and throwing the lading rapidly ashore. As soon as we got everything out of the reach of the waves that were dashing their spray over the dripping shingle beach into the swamp behind, I sent the Ojibway off into the marshes to try to procure us some food. Not making his appearance at night-fall I despatched the Blackfoot in search of him; they both returned very late, having wandered many miles along the coast, but brought nothing with them. The Blackfoot half-breed (who had received missionary instruction at Red River) attributed the Ojibway's want of success to the fact of his hunting on Sunday.

Embarking at daylight on the 6th we reached the Cat Head at 2 P.M., after a hard paddle against an adverse wind and rough sea. This bold headland consists of a perpendicular escarpment of buff-coloured limestone in

The Cat Head, Lake Winnipeg.

massive horizontal layers, the top strata overhanging the base; the summit of the rock is thirty to thirty-five feet above the lake, and is covered with drift and boulders to the depth of three feet, on which grow scrubby poplar, spruce, and tamarack. The water is quite deep up to the foot of the cliff, and as no landing can therefore be effected, I was unable to make a minute examination of the rock. There is a series of low, arched caverns in the base of the cliff, in which the waves and swells washing to and fro make a singular hollow noise, and for this reason the Indians think it is the abode of a *manitou*.

Some of the Swampys say Cat Head is so named because an Indian hunter was killed there by falling over the precipice while chasing a wild-cat or lynx. The profile of the upper, or overhanging portion of the cliff, bears a singular resemblance to the "cat-head" of a ship.

The wind becoming more foul we were compelled to camp on a point about a mile and half south-east of the Cat Head, at the extremity of the north-western side of Kinwow (Long) Bay.

During the next day (7th September) the wind blew hard from the east, and the waves on the lake rolled mountains high, so that we could not venture out, having a long traverse before us. The narrow point or peninsula upon which we were detained, is of a peculiar character, consisting of a straight barrier or ridge of boulders about three-quarters of a mile long, running at right angles to the coast, and connecting it with a small area or island of limestone a few feet high; this barrier resembles very much a railway embankment, or a rip-rap breakwater; although it is twenty to twenty-five feet high, the waves wash over it during the great storms on the lake in the fall of the year. A spruce tree growing on this peninsula has been trimmed into a "lopstick," by Angus Macbeth, from which the locality has derived the name of Macbeth's point. There are also two high cairns of stones on the point, but whether they were erected by the Indians, or by half-breeds I did not ascertain.

The morning of the 8th dawned, but there still seemed to be little chance of our getting off, and our prospects now began to look cheerless enough; we had but a handful of pemmican and one charge of ammunition left; while deliberating whether to eat the last remnant of our food, a bald-headed eagle came wheeling in great circles over us; he poised himself for an instant as if about to descend upon his prey, when he was fortunately brought down with our last charge of shot. He proved to be a large bird with magnificent plumage; a Cree or Blackfoot would have given a good horse for his wings or tail. By eating nearly every portion of the animal, except his feathers, we managed to make him serve for two or three meals.

The wind moderated sufficiently at last to permit us to resume our journey, but we had a fatiguing paddle for two hours in crossing Kinwow Bay. The extremity of this long arm of the lake was below our horizon, and the wind came sweeping out of it in great squalls. The wind veered round to the east and stopped us again about noon at the Wicked Point, where we spent the afternoon in drying our clothes and blankets, and gathering sand cherries.

September 10*th.*—The wind fell and allowed us to reach Pike Head yesterday morning. We at once ascended the Pike or Jack-Fish River to the "basket" or weir erected across it by the Indians, about half a mile from its mouth, for the purpose of procuring fish. The basket was much broken, and when we arrived, was covered with turkey-buzzards waiting to pounce on any fish that might get entangled in its meshes. By repairing the basket and watching it all night we caught an abundance of fish of four species, viz. gold-eyes, wall-eyed pike, suckers, and pike. It rained without intermission during the day, and as the wind continued unfavourable we remained at the basket, gutting fish to take with us.

We generally boiled our fish, making use of the liquor in which they were cooked as a substitute for tea; and having succeeded in capturing a small badger, by pouring water into his burrow, we got sufficient fat or oil to enable us to have fried fish occasionally.

The Indians catch fish by means of these weirs or traps in large numbers, and this mode of capture is practised by most of the North American tribes; the traps varying in construction according to the locality or the ingenuity of those by whom they are erected. The one by which we were fortunately enabled to procure a supply of fish at the Pike River consisted of a fence of poles, stretching from one side of the river to the other; they were sloping in the direction of the current, like the inside of a mill-dam, and allowed the water to pass through but not the fish. Near the river bank, on one side, there was an opening in the weir about a yard in width to allow the fish descending the river to pass into a rectangular box, with a grated bottom sloping upwards, through which the water flowed and left the fish dry. The fish very seldom entered this pound in daylight, but during the night they poured in in great numbers. In order to secure all that come into the trap when it is prepared for catching fish, an Indian sits beside it all night with a wooden mallet in his hand, with which he strikes the larger fish on the head to prevent them jumping out. He is kept busily employed pitching them out on the bank, and in the morning there is a large heap for the women to clean and cut up. The fish came into the trap almost as fast as we could pitch them out, and we caught in a short time 111 gold eyes, 44 wall-eyed pike (called perch by the half-breeds), 16 sucking carp (or suckers), and 11 pike, making 182 altogether.

The average width of the Pike River is about a chain, and its depth about five feet, with a moderate current; its banks half a mile from its mouth are of light-coloured clay five to ten feet high, and covered with a rich dark mould, supporting a thick growth of aspen, spruce, tamarack, birch, and balsam. Near the basket there is an old log-house, formerly a missionary station, but now abandoned. When the Indians come to fish here, they cut up the flooring and timber of this house for fuel, instead of availing themselves of its shelter.

September 11*th*.—Having stowed away as many fish as we could find room for in the canoe, we left the Jack-Fish River in the morning, and being favoured with a fair wind, sailed without stopping till dark, when we camped on a small island in the entrance to Fisher Bay. On Sunday the 12th we had to encounter a brisk contrary wind from the south, but, by working sixteen hours against it, and making some wide traverses between the islands, we succeeded in reaching the point opposite Dog's Head, at the beginning of the narrows, before night set in.

By starting at daylight and sailing along the east coast of the lake on the 13th we got in sight of the Grindstone Point about 2 P.M., when we set out on a longer and more dangerous traverse than any we had yet accomplished. We had to cross from the east coast of the lake to the Grindstone Point on the west coast, a distance of about twelve miles. From the shape of the lake with its many deep and broad bays this great traverse is unavoidable. When we started from the east side of the lake, the high escarpment of rock forming the point seemed quite low and blue in the distance. By spreading a blanket we were assisted for a while by a side wind; but the wind soon changed and freshened, so that we had to lower sail and ply our paddles with all our strength until reaching the point, nearly four hours from the time we left the east shore. Taking advantage of a little moonlight, which enabled us to coast along a straight shore after dark, we did not stop to camp till arriving at the Little Grindstone Point. The east coast of the lake from the Dog's Head to where we left it to cross to Grindstone Point, consists of a succession of knolls or low domes of granite and gneiss rising generally eight to ten feet above the water, and clothed on their flanks with a scrubby growth of timber, chiefly Banksian pine, spruce, and a few aspen; there are, generally, ponds and swamps between the granite knolls, and the coast line is much broken by deep inlets and small well-sheltered bays, forming excellent harbours and coves for boats. The east coast, north and south of the straits, is described as being similar to this; abounding in harbours, and for this reason it is the route by which boats invariably go to York Factory, and generally to the Saskatchewan. Opposite the mouth of Great Washow (Deep) Bay there is an inlet or passage called Loon's Straits, formerly a canoe route of the old North-West Company. There is always more or less current through the narrows of the lake at the Dog's Head. This current is sometimes flowing north and sometimes south, the direction depending upon the prevailing direction of the wind.

By making an early start on the 14th, and creeping along in the shelter of the land, we were enabled to dine at Grassy Narrows. Although our fish had not improved any since leaving Pike River, we always possessed keen appetites, and were now by no means fastidious. Sailing from Grassy Narrows across a bay into which White-Mud River empties, we arrived at the Sandy Bar a little after dark and camped.

September 15th.—The wind and weather being favourable to-day, by working fifteen and a half hours, we reached the marsh near the mouth of Red River about dark. We found an Indian encamped on the sand-beach hunting the ducks which are in countless numbers in these marshes at this season. He had killed 100 "stock" ducks during the day, and generously gave us a liberal supply.

We reached the Stone Fort about dark on the 16th September, where I succeeded in procuring a horse, and left the men to track the canoe up the river. A ride of twelve miles brought me to the middle of Selkirk Settlement, and by 11 P.M. I was in our old quarters, after a canoe voyage of forty-eight days in all; nine of which were occupied in descending from the Elbow of the South Branch of the Saskatchewan to Fort à la Corne, fourteen from thence to the mouth of the Saskatchewan, and twenty-five days in traversing Lake Winnipeg.

The whole distance traveled and explored in canoe is over 940 miles; 600 of which being down the Saskatchewan, and 340 miles open lake navigation. In performing this latter part of the journey with a little frail canoe, heavily laden, we were completely windbound for twelve days, and had to contend nearly all the time we were moving with boisterous head winds, foul weather, and a hand to mouth sustenance. This will, in some measure, account for the slow rate of progress we unwillingly made through Lake Winnipeg. I must take this opportunity of bearing testimony to the unwearied labour, patient endurance, and unflinching devotion of my two voyageurs. Their conduct while they were my companions, for nearly two months was beyond all praise; and they sustained hardships and risks of no ordinary description without a murmur.

Very truly yours,

JOHN FLEMING.

H. Y. Hind, Esq. &c. &c. &c.

END OF THE FIRST VOLUME.

AN OJIBWAY SQUAW, WITH PAPOOSE

see p. 123, vol. 2

NARRATIVE

OF

THE CANADIAN RED RIVER

EXPLORING EXPEDITION OF 1857

AND OF THE

ASSINNIBOINE AND SASKATCHEWAN

EXPLORING EXPEDITION OF 1858

BY

HENRY YOULE HIND, M.A. F.R.G.S.

PROFESSOR OF CHEMISTRY AND GEOLOGY IN THE UNIVERSITY OF TRINITY COLLEGE, TORONTO

In Charge of the Assinniboine and Saskatchewan Expedition

VOLUME II

CONTENTS

OF

THE SECOND VOLUME.

THE

ASSINNIBOINE AND SASKATCHEWAN EXPLORING EXPEDITION

OF 1858 (CONTINUED).

CHAPTER XXIII.

LAKE WINNIPEG.—RED RIVER TO THE LITTLE SASKATCHEWAN.

CHAP. XXIV.

THE MOUTH OF THE LITTLE SASKATCHEWAN, OR DAUPHIN RIVER, TO THE SALT SPRINGS ON WINNIPEGO-SIS LAKE.

CHAP. XXVII.

WINTER JOURNEY WITH DOGS FROM FORT GARRY TO CROW WING.

CHAP. XXVIII.

INDIAN WEALTH. — THE BUFFALO. — THE HORSE AND THE DOG.

CHAP. XXIX.

INDIAN CUSTOMS AND SUPERSTITIONS.

CHAP. XXX.

INDIAN POPULATION OF BRITISH AMERICA.

CHAP. XXXI.

INDIAN TITLE TO RED RIVER.

CHAP. XXXII.

MISSIONARY LABOUR AND ITS RESULTS.

CHAP. XXXIII.

THE HUDSON'S BAY COMPANY.

CHAP. XXXIV.

THE COMMUNICATION BETWEEN CANADA AND RED RIVÈR.

CHAP. XXXV.

THE COMMUNICATION BETWEEN RED RIVER AND THE PACIFIC. — THE IMPORTANCE OF THE BASIN OF LAKE WINNIPEG.

GEOLOGY OF THE BASIN OF LAKE WINNIPEG.

CHAP. XXXVI.

SURFACE GEOLOGY.

CHAP. XXXVII.

THE LAURENTIAN AND HURONIAN SERIES.

CHAP. XXXVIII.

THE SILURIAN AND DEVONIAN SERIES.

CHAP. XXXIX.

THE CARBONIFEROUS SERIES. — JURASSIC FORMATION.

CHAP. XL.

THE CRETACEOUS SERIES. — TERTIARY FORMATIONS.

CLIMATE OF THE SOUTHERN PART OF RUPERT'S LAND.

CHAP. XLI.

CHAP. XLII.

THE LOCUSTS AND THE FLOODS.

LIST OF ILLUSTRATIONS

IN

THE SECOND VOLUME.

*** *The following Illustrations are from Photographs taken by Mr.* HUMPHREY LLOYD HIME, *Photographer to the Assinniboine and Saskatchewan Expedition, or from Sketches by Mr.* JOHN FLEMING, *Assistant Surveyor and Draughtsman.*

CHROMOXYLOGRAPHS.

WOODCUTS.

MAPS AND PLANS.

THE
ASSINNIBOINE & SASKATCHEWAN EXPLORING EXPEDITION
OF 1858

CONTINUED

THE

ASSINNIBOINE AND SASKATCHEWAN EXPLORING EXPEDITION OF 1858.

CHAPTER XXIII.

LAKE WINNIPEG.—RED RIVER TO THE LITTLE SASKATCHEWAN.

Mouth of Red River.—Aurora.—Weather Signs.—Channel at the Mouth of Red River.—Storm.—Character of the South Coast of Lake Winnipeg.—Fresh-water Shells.—New Land.—West Coast.—Confervæ.—The Willow Islands.—Clay Cliffs.—Good Land.—Drunken River.—Aurora.—Rock Exposure.—Deer Island.—Section on Deer Island. Equivalent of the Chazy Formation.—Fishing Ground.—Miskena.—Grindstone Point.—Rev. Mr. Brooking.—Rocks at Grindstone Point.—The Little Grindstone Point.—East Coast of Lake Winnipeg.—Punk Island.—Yellow Ochre.—Coast near Dog's Head.—Limestone Cave Point.—Fissured Rocks.—Jack Fish River.—Fisher Bay.—The Cat Head.—Little Saskatchewan Bay.—East Coast of Lake Winnipeg.—Dimensions of Lake Winnipeg.—Elevation and Distance of the Lake Superior Watershed.—Elevation of the Mississippi Watershed.

The early period of the year during which the arrival of winter may be expected to close the navigation of the lakes and rivers of Rupert's Land, makes every autumnal day valuable for continuing an exploration in canoes or batteaux. A fortnight, however, after our return from the prairies was necessarily occupied at what may be appropriately termed Selkirk Settlement, in writing reports and making preparations for a voyage through Lake Winnipeg, the Little Saskatchewan River, and Lake Manitobah to the Salt Region on the shores of Winnipego-sis Lake. Mr. Dickinson prepared for an exploration of the country between the Lake of the Woods and Red River, and between the Assinniboine and the 49th parallel. Both parties were ready by the 18th, and at noon started on their respective routes.

Freighter's Boat.

In a Red River freighter's boat of four tons' burden, with a crew of seven men, and accompanied by Mr. Fleming, I reached a point about seven miles below the Indian Settlement, being aided by a fair wind, and camped at dusk. On the following morning, the temperature of the air at sunrise was 63°, of the river, 59°. We arrived at the mouth of the river at 10 A.M., and hastened to avail ourselves of a south-east wind just beginning to rise. Last night the aurora was very beautiful, and extended far beyond the zenith, leading the voyageurs to predict a windy day. The notion prevails with them that when the aurora is low, the following day will be calm; when high, stormy. The temperature of the mouth of the river was 59°, and of the open lake, $1\frac{1}{2}$ mile from shore, $58\frac{1}{2}$°. Rain commenced as soon as we were fairly in Lake Winnipeg *, the wind suddenly chopped round to the north, driving a dense fog before it, and in a few minutes enveloped us in a misty shower. The steersman instantly turned about and made for the mouth of the river, there being no harbour nearer than the Willow Islands, at least fifteen miles distant. The breeze rapidly increased to a gale as we regained calm water inside the bar at the mouth of Red River.

The wind subsided about 2 P.M., and a shot heard from a direction due south of where we lay, induced some of the voyageurs to exclaim, that the wind would soon come from that direction, according to an impression common among these excellent observers and interpreters of "signs," that a shot heard against the wind is a good omen. But our steersman placed more faith in the aurora, and thought we had not "taken all the wind out of it yet." The sky having a threatening appearance, we determined to camp.

There are six mouths to Red River, winding through extensive marshes; the channel through which we passed was the main outlet; its breadth varies from twenty to twenty-eight feet, and on either side shelves rapidly from four to eighteen feet of water. At 3 P.M., when just on

* Winnipeg, from the Ojibway—*We*, dirty; and *nepe*, water.

the point of starting, one of the voyageurs suggested that we should wait for a few minutes longer as he had observed the water of the lake coming in at the mouth of the river, and thought that the wind would soon blow strong from the north, although at the time the sky was clear and a calm prevailed. In less than half an hour a fresh northerly breeze sprang up, scud appeared drifting before it, and the waters of the lake flowed rapidly up the river into the vast marshes which extend for many miles inland at the southern extremity of Lake Winnipeg. The weather at this season of the year is very changeable, and renders boat navigation of the lake rather hazardous. In anticipation of a storm, we made ourselves as comfortable as circumstances would permit on a low spit of sand, with the lake before us, the river on our left hand, and interminable marshes east and south of us.

Soon after sunset, the breeze from the north rose into a gale; the water of the lake ran like a rapid up the river channel into the swamps, and a terrific swell soon set in from the lake, breaking upon the sandy beach with a stunning noise. The water rose to within six inches of the level of the spit on which our tent was pitched and threatened every instant to submerge it. At 10 P.M., the gale was at its height, and as we sat upon a stranded trunk of a tree, looking out upon the lake, a truly magnificent scene lay before us. Huge crested breakers covered the lake as far as we could see through the gloom, lighting up the coast with long glistening streaks of white foam. The noise was so overpowering that we had great difficulty in hearing one another speak; the waves broke over the narrow spit which formed the low bank of the river where our boat was moored and the tent pitched; our camp ground was reduced to a strip of sand eight yards broad and seven inches above the river on one side, with overflowing swamps on the other; if the storm had continued half an hour longer we should have been compelled to take to the boat and drift into the reeds, at the risk of being stranded when the gale subsided and the water retired from the marshes into the lake.

Beach of Lake Winnipeg, near the Mouths of Red River.

For many miles the south coast of Lake Winnipeg consists of alternate strips of sand sustaining willows, and narrow, reedy marshes running parallel to the coast line. Some of these sand strips show many years of duration when well protected by drift timber, others are of recent origin, clean and bare, enclosing ponds in which rushes are only just beginning to show themselves. They are the records of the progress made by new land in its invasion of the lake at and near the mouths of Red River. A northerly gale throws up a bar or beach about one hundred yards from the main shore; on the new beach drifted timber accumulates, and in process of time becomes consolidated by the gravel and sand which is washed between the logs. Willows soon grow on the new soil thus formed, and bind the whole into a firm beach with a marsh in the rear. A heavy gale may sweep the new land away or throw up another beach about one hundred yards in advance of it, on which the process of consolidation is renewed. For ages past this work of construction and destruction has been greatly in favour of the former. Hence it arises that, with the exception of the newly formed spit at the mouth of the river, there is no accessible camping ground for several miles up the stream; marshes surround the spits or old beaches on which the willows grow, and extend in all directions as far as the eye can reach.

The beach and marshes contain an infinite number of fresh-water shells belonging to the genera *Helix*, *Bulimus*, *Succinea*, *Pupa*, *Planorbis*, *Lymneus*, &c. &c. For many hundred yards together the beach is covered with perfect or disintegrated forms of these shells, thrown up by the waves upon the sand.

We employed ourselves during our unexpected detention in examining the coast, sounding the river, and in shooting and fishing. Our sporting brought us only six duck, three plover, and three large pike. The flesh of the pike was of a delicate salmon colour, more like that of the salmon trout of the Canadian lakes than of the common pike.

Sept. 21st.—Rising at 4 A.M. in half an hour we were *en route*, the morning just beginning to dawn; temperature of the air at sunrise, 51°, of lake 59°. The west coast for a few miles is elevated from five to six feet above the lake; here and there a low beach of limestone gravel, sand, and a few granite boulders, is fringed with a belt of tall aspens which grow within twenty feet of the water's edge. Behind the belt of aspens is a marsh, then another belt of aspens also followed by a marsh. This succession continues for a distance of about three miles before good land supporting heavy aspens is to be found in large areas. Near to the spot where we breakfasted, an excellent illustration of the prevailing character of the west coast, thus far, occurs. A sandy beach covered with shingle had separated a former bay from the main body of the lake. On this beach, which was not twenty feet broad, or more than five above the lake level, willows, dogwood, and grasses were growing; a large pond lay inside, fringed with rushes; it was tenanted by hosts of duck. In the rear of this pond a narrow strip of land clothed with aspen, separated a marsh from it, which had doubtless once been a bay of the lake, then a pond, and finally a marsh.

At 11 A.M., a vast quantity of confervæ appeared in clusters on the surface of the lake, resembling in every particular a similar organism noticed in extraordinary profusion on the Lake of the Woods in August, 1857. The sudden appearance of this "weed," indicated a calm, according to the experience of our voyageurs. A calm did occur for a short time, soon, however, followed by rain in the north, which fortunately did not reach us.

Inland ponds cut off from the lake by low beaches appear as far as the Willow Islands, where we arrived in the afternoon. These islands were found to consist of a few small sandy areas and one long narrow strip of sand

and gravel, stretching into the lake in an easterly direction, and separated from the shore by a narrow channel. They are fast wearing away, and in the memory of some of the voyageurs, were covered ten years since with willows, poplar, and a few spruce. They have probably afforded much of the material for the formation of the beaches which have cut off portions of the lake on the south-west coast, the sand and shingle being drifted along the shore by the long waves which every breeze from the north or a northerly direction creates. The depth of water near the coast is very small; soundings showed twenty-nine feet of water one mile north of Willow Island, the deepest part yet observed. Near Willow Island we met an Indian in a canoe with his wife and two children: he was going to Red River. I gave him some tobacco and his squaw a small quantity of tea; in return he unrolled a piece of birch bark and handed me the moufle of a moose, at the same time remarking that he was a conjuror and would "make us a fair wind." The steersman replied that a fair wind for us would be adverse to him. "Ah," said the conjuror, "but I will make one for you and two for myself."

In the afternoon I landed to examine some cliffs of clay which appear about twenty-three miles from the mouth of the river. They were sixteen feet in altitude, and exposed a clean surface of stratified marl, reposing on a brownish-black clay. The stratification was in thin horizontal layers, easily detached one from the other. The brownish-black clay showed a very tenacious character, so much so, that it was very difficult to break off with the hand masses larger than ten or twelve cubic inches, in any other direction than that of the plane of stratification. It was worn by the action of the waves into a great variety of forms, and on the beach lay large numbers of egg-shaped and spherical bodies, varying from one foot in length and three inches in diameter, to small round pellets of the size of peas. They were covered with minute pebbles, or with sand, and when broken, showed a nucleus of the tough clay which had assumed its regular form by constant rolling on the beach. No organic remains were found, and the impression conveyed by the aspect of the clay and the marl by which it was capped, satisfied me that it was of the same age as the clay and marly substratum of the Red River and Assinniboine Prairies.

The timber in the forest consisted of aspens and birch, with a few oak, elm, and ash. Our steersman, who knew the country well, informed me that good land on which large timber grew, did not extend more than one mile from the lake. It is succeeded by spruce and tamarac (*Larix Americana*) marshes, the trees being of dwarfish dimensions. The afternoon was calm and warm, so far verifying the predictions of our voyageurs, which they had based on the sudden appearance of the "weed," in the morning.

Sept. 22nd.—Last night was cold, calm, and beautiful, the thermometer fell to 36° at 10 P.M., and to the freezing point before daybreak. Donati's comet shone a fine celestial object; and notwithstanding the brightness of the moon, then nearly full, a splendid aurora was distinctly visible, and the heavens presented a peculiarly beautiful spectacle. We camped near the mouth of Drunken River, a small stream which would make an excellent boat harbour, if widened at its outlet. The clay cliffs and marl disappeared before we arrived at our camping place; the shore again consisting of a beach, with a swamp or marsh, fringed with small spruce and tamarac in the rear. I aroused the men at 4 A.M. The aurora at that hour was a splendid object, and appeared in the form of sudden flashes of low arcs of light, complete from east to west, rising in vast waves from one constant luminous base, a few degrees above the horizon. The undulations of pale light followed one another with great rapidity and regularity for many minutes together.

A strong westerly breeze early this morning, soon enabled us to reach the Sandy Bar, fourteen miles from Drunken River, and then the Grassy Narrows, a distance of seven miles. Both of these points are low, sandy, and gravelly peninsulas stretching out into the lake opposite to Big Black Island. The first exposure of limestone was seen on a small island opposite Big Black Island, which we named Guano Island. It dipped very slightly to the south-west; a search for fossils was fruitless, but on Big Black Island, and those adjacent to it, near the Little Grindstone Point, limestone of Lower Silurian age appears in the form of low mural cliffs on the west shores, which alone were seen. This limestone is a continuation of a fine exposure afterwards found on Deer Island, where we arrived at 1 P.M.

The following section occurs on Deer Island:—

Shingle Beach (Limestone):

No. 1. Four feet of dark green argillo-arenaceous shale, with thin layers of sandstone of uneven thickness—Fucoids very abundant in the sandstone. The weathered sandstone is reddish brown; fresh surfaces are white or grey. White iron pyrites assimilating the forms of disks, spheroids, and shells occurs in the sandstone.

No. 2. In many respects like the former; the sandstone layers are from one to four inches in thickness and predominate over the shaly portions. Its thickness is six feet. The character of these formations (1 and 2) is very variable; the green argillaceous portion sometimes predominates, and occasionally the sandstone.

No. 3. Ten feet of sandstone with green bands of a soft argillaceous rock, from one quarter to four inches in thickness. The sandstone often white, but generally red. A persistent green band, a few inches thick, filled with

Exposure on Deer Island, near Grindstone Point.

obscure forms resembling fucoids, is very characteristic. The red-coloured sandstone is often soft and friable, the white frequently embodied in the red. Both red and white contain obscure organic forms. The green patches which are found throughout the sandstone contain impressions of fucoids; an *Orthoceratite* was found in the sandstone. In some parts of the exposure on Deer Island the sandstone layers are much harder, although partaking of the characters already described. When thus hard, the white portion is extremely brilliant, of a pure white, and very siliceous; it would form an excellent material for the manufacture of glass. Forms, coloured brown, often pervade the white sandstone, and appear to resemble fucoids and corals replaced by brown ochreous sand.

No. 4. Eighteen feet of limestone, perfectly horizontal, very hard, and breaking off the cliff, where the soft sandstone has been weathered away, in huge rhomboidal slabs, eight to twenty-five feet in diameter, and four to ten inches thick.

The surface of the limestone shows silicified shells and corals: among the shells an *Orthoceras* nine inches in diameter was seen, with fossils belonging to the genera *Rhynchonella* and *Tetradium*. This formation is equivalent to the *Chazy* of New York and Canada, and consequently lies near the base of the Lower Silurian Series.

In the shingle immediately below the cliff, many fine *Orthoceratites* were found, with a large *Maclurea*, and *Catenipora escharoides*.*

Limestone forms the west coast for some miles south of Big Grindstone Point, where we arrived in the evening. This part of Lake Winnipeg is very beautiful, resembling, in many pleasing particulars, the scenery on the justly celebrated Lake Simcoe, Canada West, near the Narrows, where wooded islands rising from the lake in clusters and rows, are suggestive of tranquil summer retreats. Between Grindstone Point and Deer Island, the lead showed sixty feet of water. It is the great fishing-ground of some of the bands of Indians who make this part of the lake their

* For the determination of the fossils from this and other localities in the region about Lake Winnipeg, Manitobah, &c., I am indebted to E. Billings, Esq., F.G.S., Palæontologist to the Canadian Geological Survey.

wintering place. White fish are very abundant, and caught by the Indians in large numbers; their flavour is not so fine as those of Lake Manitobah, or of the Qu'appelle Lakes. Sturgeon are also numerous, and, according to the belief of the miserable natives who fish here during the winter, the deep part of the lake is their great place of resort at that period of the year, where they lie with *Mis-ke-na*, "the chief of the fishes," in the southern portion of Lake Winnipeg.

Longfellow alludes to the same superstition held by Lake Superior Indians, in the song of "Hiawatha," when he makes his hero go—

> "Forth upon the Getche Gumee,
> On the shining Big-Sea-Water,
> With his fishing-line of cedar—
> Of the twisted bark of cedar—
> Forth to catch the sturgeon Nahma,
> Nishe-Nahma, King of Fishes,
> In his birch canoe exulting:
> All alone went Hiawatha."

We approached Grindstone Point after dark, and observed a camp-fire on the beach, with a freighter's boat close in shore. It belonged to the Rev. Mr. Brooking and his family, who were returning to Rossville from Red River. Mr. Brooking is a Wesleyan missionary, for some years a resident in Rupert's Land, and engaged in the unthankful labour of attempting to Christianize the Indians. He had traveled from the head of Lake Winnipeg to Red River Settlement, to obtain medical advice for Mrs. Brooking, who was very unwell. Our interview was short, the voyageurs in Mr. Brooking's boat being anxious to take advantage of a fair wind which had just arisen. As soon as supper was ended they embarked and proceeded by moonlight on their lonely journey. He was twenty days in coming from Norway House to Red

River, having been kept back by contrary winds. His prospects of traversing the lake rapidly were now more favourable, as the south wind which prevailed would soon drive a freighter's boat to Norway House.

Sept. 23*rd.* — The rocks at Grindstone Point, about six miles north of Deer Island, are similar to those already described in the previous section. Being further north, the exposure is higher, and the sandstone bands more

Grindstone Point.

fully shown. Beneath No. 1 of Deer Island, a hard, yellow, compact sandstone appears, and is exposed for a space of four feet above the level of the water. Strata No. 1 and No. 2 of Deer Island appear in a slightly different form here: the sandstone bands are thicker; the green shaly portion more distinct as a separate band, and two feet thick; while above the hard yellow sandstone, the base of No. 1 appears in the form of a purple band of very soft sandstone, about one foot in thickness, containing a vast number of *stains*, which seem to have been occasioned by fucoids.

At Little Grindstone Point, the limestone No. 4 of Deer Island comes to the water's edge. The sandstone No. 3 is just below its level. Little Grindstone Point is a little more than seven miles south-west of Big Grindstone Point, and the altitude of the limestone, where it touches the sandstone at the last-named place, is about twenty-five feet, which would give an inclination of a sectional exposure in a south-westerly direction of about three feet in the mile. It appeared, however, to have a slight westerly dip, showing the true dip to be a few degrees more to the west than south-west, as was afterwards ascertained. In the limestone, turbinated shells are numerous, with *Orthoceras* of large dimensions. The scenery is attractive, and, in a geological point of view, eminently interesting. The opposite coast is formed of the unfossiliferous rocks belonging to the great Laurentian Series, which extends from Labrador to the Arctic Ocean. Within three miles of Grindstone Point, islands of this important formation occur a short distance in advance of the east coast, which is wholly composed of it.

The depth of Lake Winnipeg immediately opposite Grindstone Point is forty-eight feet. A storm afforded us another opportunity of examining the fossiliferous rocks of this locality, for no sooner had we started in the direction of the "Granite Islands," opposite the point, than the wind turned round to the north and compelled us to seek shelter in a bay of Punk Island, three miles southeast of the Grindstone Point.

On Punk Island, strata 1, 2, 3, and 4 of Deer Island were recognised in a bay, with some lithological differences. In Nos. 1 and 2 here, which could scarcely be distinguished from one another, *Modiolopis parviuscula* (N.S.) *, is very numerous.

On the north-east side of Punk Island, above the purple sandstone mentioned as occurring at Big Grindstone Point, a thin stratum of buff coloured limestone occurs, possessing some peculiarities. On raising slabs there is seen between each stratum a soft and very pure ochre of a beautiful yellow colour; it is found filling small depressions in the limestone, and in layers, from one-eighth to half an inch in thickness. The ochre, when moist and fresh, is easily worked by the fingers, quite destitute of gritty or hard particles, of an uniform pale yellow colour, and when burned, of a beautiful cinnabar red. It is used by the Indians in both states as a pigment; the limestone in which it occurs is extremely porous and often honey-combed.

Sept. 24th. — At half-past 2 A.M., the wind being fair, and the sky clear, we prepared to start. There was a sharp frost during the night, and the thermometer registered 28°. We made the traverse of Great Washow Bay, thirteen miles across, and breakfasted at a point half-way between Bull's Head and Dog's Head. The limestone cliffs here were about thirty feet high, and occupy the coast from Bull's Head to Whiteway's Post, opposite the Dog's Head. Where seen at breakfast, the coast is fringed with broken masses, which lie piled one on the other in picturesque confusion. Ascending the cliff, I found large portions detached from the main body, forming deep clefts or cracks. Some of these fissures were twelve feet wide and twenty feet deep, others three feet wide and of greater depth. Sometimes the fissures were roofed with masses which had slipped forward, forming

* See Chap. on the Silurian Series.

long narrow caves lined with moss. One cave was more than sixty feet long, and, with the exception of a small aperture, closed at one end and roofed throughout. We named the spot Limestone Cave Point. From the description given by one of the voyageurs who had wintered near this place and knew the country well, the rock along the coast, from the Bull's Head to Pike Head, is fissured in the manner described. Very roomy caverns can be found, which are easily converted into excellent wintering houses for trappers. The sides of the fissures are perpendicular, and the fracture is so even as to form chambers of a rhomboidal shape. The passages between them are beautifully covered with moss, while, gracefully drooping overhead, the birch and white spruce obstruct the rays of the sun, giving to these lonely cells a gloomy and desolate aspect. The limestone is similar to that which has already been described as No. 4 of Deer Island. At the Narrows, or Dog's Head, the limestone and unfossiliferous rocks are in close proximity; the east side of the strait being composed of the Laurentian Series, on the west side, of lower Silurian limestone.

The wind being favourable we sailed during the whole day, and at 4 P.M. reached the mouth of Jack Fish River, making a traverse across Fisher Bay, a very deep indentation whose western limit could not be seen from the canoe. In Fisher Bay islands are numerous, and some of them of large area, such as Great Moose Island, in the mouth of the bay, and Juniper Island, four miles to the north. Due west of the Dog's Head, Black Bear Island contains an excellent boat harbour, a feature worthy of note, as it occurs near the beginning of the great traverse across Fisher Bay. Jack Fish River issues from a marsh separated from the lake by a belt of sand and shingle about 100 yards broad. The river runs in a westerly direction from a series of small lakes and swamps, through a level, low country abounding in fine spruce and tamarac forests, broken by gravelly ridges supporting poplar and birch. The breadth of the river at its mouth is thirty feet, but where it passes through the swamp it is broad and deep, and so continues for some distance into the country. Jack Fish River is a favourite fishing station of a tribe of Swampys, and was once the seat of a missionary establishment.

It has been mentioned in a subsequent chapter that this tribe were deterred during the winter of 1858 from wintering here, by a threat from a noted conjuror of the Grand Rapid, illustrating the abject position in which superstition frequently places these unfortunate people.

Leaving Jack Fish River, or the Pike Head, as it is also termed, from a promontory bearing that name near to the mouth of the stream, we coasted under sail past Wicked Point across the traverse of Kinwow Bay, rounded Macbeth Point, and camped at Point Turn-again, beyond the Cat Head. The coast at the Cat Head is very precipitous; the limestone cliffs rise about thirty-five feet from the water, without any intervening beach, so that boats cannot land, and must necessarily push on until a narrow beach is found a few miles beyond it. Limestone cliffs, similar in all respects to those of the Cave Point, occupy the coast at intervals as far as the Cat Head, and probably fringe the Mantagao-sebe Bay, as they are seen near the mouth of the Little Saskatchewan, and on the north point of the great bay which derives its name from that river. Taking advantage of a fair wind and fine night we carried on across Lynx Bay, and camped at half-past eleven, P. M.

At half-past four on the following morning a westerly

wind enabled us to pass the Bushkega and Sturgeon Islands, and make the traverse across the Little Saskatchewan Bay to the mouth of the river. In making the traverse we could not see the extremity of this deep indentation in a south-easterly direction, where the Mantagao-sebe debouches. The greater portion of the bay was coasted by Mr. Fleming, during his voyage from the Grand Rapid to the mouth of Red River. The temperature of the Little Saskatchewan was found to be $52\frac{1}{2}°$, of Lake Winnipeg one degree higher.

The description of the west coast of Lake Winnipeg from the mouth of the Little Saskatchewan to the Great Saskatchewan is given in Mr. Fleming's narrative, Chap. XXII., Vol. I. In order to complete a description of the entire coast line of this lake, I append the following extract from Sir John Richardson's "Journal of a Boat Voyage through Rupert's Land and the Arctic Sea."

"The eastern coast-line of Lake Winnipeg is, in general, swampy, with granite knolls rising through the soil, but not to such a height as to render the scenery hilly. The pine-forest skirts the shore at the distance of two or three miles, covering gently rising lands, and the breadth of continuous lake surface seems to be in process of diminution, in the following way. A bank of sand is first drifted up, in the line of a chain of rocks which may happen to lie across the mouth of an inlet or deep bay. Carices, balsam-poplars, and willows speedily take root therein, and the basin which lies behind, cut off from the parent lake, is gradually converted into a marsh by the luxuriant growth of aquatic plants. The sweet gale next appears on its borders, and drift-wood, much of it rotten and comminuted, is thrown up on the exterior bank, together with some roots and stems of larger trees. The first spring storm covers these with sand, and in a few weeks the vigorous vegetation of a short but active summer binds the whole together by a network of the roots of bents and willows. Quantities of drift-sand pass before the high winds into the swamp behind, and, weighing down the flags and willow-branches, prepare a fit soil for succeeding crops. During the winter of this climate, all remains fixed as the summer left it; and as the next season is far advanced before the bank thaws, little of it washes back into the water, but, on the contrary, every gale blowing from the lake brings a fresh supply of sand from the shoals which are continually forming along the shore. The floods raised by melted snows cut narrow channels through the frozen beach, by which the ponds behind are drained of their superfluous waters. As the soil gradually acquires depth, the balsam-poplars and aspens overpower the willows, which, however, continue to form a line of demarcation between the lake and the encroaching forest.

"Considerable sheets of water are also cut off on the north-west side of the lake, where the bird's-eye limestone forms the whole of the coast. Very recently this corner was deeply indented by narrow, branching bays, whose outer points were limestone cliffs. Under the action of frost, the thin horizontal beds of this stone split up, crevices are formed perpendicularly, large blocks are detached, and the cliff is rapidly overthrown, soon becoming masked by its own ruins. In a season or two the slabs break into small fragments, which are tossed up by the waves across the neck of the bay into the form of narrow ridge-like beaches, from twenty to thirty feet high. Mud and vegetable matter gradually fill up the pieces of water thus secluded; a willow swamp is formed; and when the ground is somewhat consolidated, the willows are replaced by a grove of aspens.* Near the first and second Rocky Points †, the various stages of this process may be inspected, from the rich alluvial flat covered with trees and bounded by cliffs that once overhung the water, to the pond recently cut off by a naked barrier of limestone, pebbles, and slabs, discharging its spring floods into the lake by a narrow though rapid stream. In some exposed places the pressure of the ice, or power of the waves in heavy gales, has forced the limestone fragments into the woods, and heaped them round the stems of trees, some of which are dying a lingering death; while others, that have been dead for many years, testify to their former vitality, and the mode in which they have perished, by their upright stems, crowned by the decorticated and lichen-covered branches which protrude from the stony bank. The analogy between the entombment of living trees, in their erect position, to the stems of *Sigillariæ*, which rise through different layers in the coal-measures, is obvious." ‡

The following are the approximate leading dimensions of Lake Winnipeg:

Area of Lake	8500 square miles.
Length, not including Play Green Lake	280 statute miles.
Greatest breadth	57 " "
Length of coast line	930 " "
Approximate height above the sea . .	628 feet.

* The fact of the formation of these detached ponds, marshes, and alluvial flats, points either to a gradual elevation of the district, or to an enlargement of the outlet of the lake, producing a subsidence of its waters.

† The strata at these points contain many gigantic Orthoceratites, some of which have been described by Mr. Stokes in the Geological Transactions.

‡ If one of the spruce firs included in the limestone debris, had its top broken off, and a layer of mud were deposited over all, we should have the counterpart of a sketch of Sir Henry de la Bêche's Manual (p. 407). The thick and fleshy rhizomata of the *Calla palustris*, marked with the cicatrices of fallen leaves, and which are abundant in these waters, bear no very distant resemblance to *stigmariæ*.

This estimate of the altitude of Lake Winnipeg above the sea level, was deduced in 1857 from the levels taken across the portages along the line of the canoe communi cation between Fort William on Lake Superior, and Fort Alexander on Lake Winnipeg. The height of the dividing ridge which separates these lakes from one another, is 1485 feet above the level of the sea; and distant, by the canoe route, 104 miles from Fort William, and 510 miles from Fort Alexander.

Major Long found the sources of the St. Peter and Red Rivers to be 830 feet above the ocean, and Lake Winnipeg 630 feet above the same level—a difference of only two feet in excess of the estimate we made in 1857.

When it is considered that the St. Peter River is an affluent of the Mississippi flowing into the Gulf of Mexico, and that Red River communicates with Lake Winnipeg, which sends its surplus water to Hudson's Bay by Nelson River, the extraordinary lowness of the southern watershed becomes apparent.

Lake Winnipeg freezes every winter, ice frequently forming to the thickness of five feet, and sometimes not leaving the upper end of the lake before the 10th of June.

CHAP. XXIV.

THE MOUTH OF THE LITTLE SASKATCHEWAN, OR DAUPHIN RIVER, TO THE SALT SPRINGS ON WINNIPEGO-SIS LAKE.

The Little Saskatchewan.—Height of Bank.—Country in Rear.—Tracking.—Swamps.—Banks of River.—Ojibway Camp.—White-fish.—Character of Country.—Canoe Fleet.—Spruce.—Boulders.—Marsh.—St. Martin Lake.—"Money."—Pounded Fish.—War-path River.—War Roads.—Ojibway, Sioux, Swampy, Cree, Blackfoot, and Crow.—Wavys.—Fine Land.—The Narrows.—Boulder Barriers.—Sugar Island.—Indians.—Gneissoid Islands.—St. Martin Rocks.—Beach Barriers.—Depth of St. Martin Lake.—Thunder Island.—Thunder-storm.—Partridge Crop River.—Rushes.—Old Mission.—Low Country.—Indian Farmer.—Wide-spread Marsh.—Fairford.—The Character of the Country.—The Mission.—Evening Service.—Rev. Mr. Stagg.—The Farm.—Hudson's Bay Company's Post.—Rum.—Lake Manitobah.—Progress of the Season.—Rocks.—Fossils.—The Coast.—Steep Rock Point.—Devonian Rocks.—Indian Superstition.—Water-hen River.—Eagles.—Character of Water-hen River.—Pelicans.—Indians.—Wood and Prairie Indians.—Barter.—Winnipego-sis Lake.—Ermine Point.—Elms.—Salt Spring.—Snake Islands.—Duck Mountain.—Snake Island Fossils.—Arrive at Salt Springs.

A FEW hundred yards above the mouth of the river, horizontal Lower Silurian limestone shows itself on both sides, and it is through this rock that the Little Saskatchewan has excavated its bed. The limestone contains fossils in abundance, but in very bad state of preservation in many of the layers. They are similar to those found on Lake Winnipeg at Cave Point, and in its lithological aspect there is no appreciable difference between the exposures in either locality. The Little Saskatchewan, as its name implies, has a very rapid current, varying from one to four miles an hour. The banks are not more than twenty to twenty-five feet above its level near the mouth, and diminish in altitude in ascending the stream. They are fringed with aspen, poplar, spruce and tamarac. In the rear, swamps occur, often covered with deep moss, and containing islands of tamarac and a few balsam spruce of fair dimensions, but scarcely suitable for any other purposes than those which a limited settlement might occasion.

The river proving too rapid for using the sweeps, we were compelled to track up, a difficult and tedious labour to the men, but offering an excellent opportunity for making traverses into the country, which, however, were never deep, the swamps soon arresting progress inland. The general aspect of the river for the first four miles is very attractive, resembling in many particulars Rainy River. About three miles from the lake the limestone disappears, being covered with drift or alluvial clay. The banks rise gently with the stream, which is rapid and shallow. The yellow autumnal foliage of the aspens contrasts beautifully at this season of the year with the spruce and tamarac, and gives a charming appearance to the river banks. Towards evening we arrived at a camp of Swampys, containing four tents. They had an abundance of white-fish, and told me the river was "full of them." Anxious to test the statement I intimated a wish to purchase a score of fresh fish, and offered an Indian some tea and tobacco if he would catch them immediately. He accepted the offer, entered his canoe, and crossing over to a well-known eddy, in fifteen minutes brought back twenty white-fish, weighing on an average three pounds each. We camped close to the Swampys, on the edge of a cranberry marsh, as we knew that if we tracked a mile or so up the stream they would follow us, and our party might be increased by others in advance of

SUSAN, A SWAMPY HALF-BREED

them. As it was, the guns they fired upon our arrival had been heard, so that at sunset several canoes came swiftly down the stream, filled with men and women to "learn the news." The whole body camped close to us, and what with talking, shouting, screaming of children and howling of dogs, we enjoyed no rest until late in the night.

Rising at day-break on the following morning, Sept. 26th, a few hours were occupied in examining the country in the rear of the camp. The banks of the river are here about twenty feet above the present level of the river, but the country is very marshy, and clothed with tamarac and spruce behind the belt of aspens which fringe the river banks. After breakfast, the wind being fair, we hoisted sail, and in company with our Swampy friends proceeded up the river. A little fleet of twenty-three canoes, each with a birch bark sail, glided quickly ahead of us, but the breeze freshening, we soon caught and passed them one by one. About nine miles from the mouth, the banks of the river are not more than ten feet above its present level, but are rarely flooded. They consist of alluvial clay, and sustain many groves of fine spruce and aspen. At some of the bends there is a large accumulation of boulders, consisting chiefly of the unfossiliferous rocks. The colouring of the trees, at this season of the year, was unusually delicate and beautiful, nearly all the aspens in front being yellow, while those in the rear, protected in some measure from the night frosts, still retained their green.

About five miles from St. Martin Lake, a wide-spreading marsh begins, on the edge of which we camped, our Indian friends soon closing with us; when the wind failed, the squaws towed their canoes with lines made of twisted cedar bark. Some of the old men were anxious to show me some specimens of "Money" they had carefully folded in bits of cloth or birch bark. The "Money," respecting which they have no distinct idea except that it is "white," according to information they have obtained from half-breeds, consisted of fragments of selenite, iron pyrites, and silver mica. They profess to know where a large quantity of this "Money" is to be found, and demand tea and tobacco for the intelligence. These Indians had been making their autumnal fishing hunt, and possessed large birch bark vessels filled with pounded white-fish, previously dried and smoked, a miserable substitute for pemmican. They had also sturgeon bladders filled with white-fish oil. The pounded fish and the oil form part of their winter stores; some samples which were submitted to me for inspection, with a view to barter, were the reverse of inviting.

The chief pointed out a portage path between the Little Saskatchewan or Dauphin River, and the War-path River, which forms the war-road of the Ojibways and Swampys of Lake Winnipeg when they proceed on their periodical excursions against the Sioux. This war road was much used in the earlier history of the natives of the Low Country, but on account of the great diminution in their numbers, which has taken place during the present century, war is no longer a pursuit or pastime with them, as with the Lac la Pluie Ojibways and the Sioux.

The selection of certain tracts of country for the "War-path" is probably determined by the facilities presented for communication and concealment combined. The following are celebrated "war-paths," where hunting is generally disallowed, although game from that circumstance is usually most abundant.

1. "The War-path River" and war road of the Lac la Pluie Ojibways, and the Sioux, from Rainy River to

Red Lake River, thence across the prairies in the Valley of Red Lake River to Miniwaken or Devil's Lake, in Dakotah Territory.

2. "War-path River," from the south-west corner of the Lake of the Woods to Roseau River, thence to the prairies west of Red River — the same tribes.

3. "War-path River," from Lake Winnipeg to the Little Saskatchewan, thence to the prairies south of Manitobah Lake — the old war-path of the Swampy Crees, the Assinniboines and Sioux, also of the Swampy Crees and the Lake Winnipeg Ojibways.

4. The "War-road," near the Elbow of the South Branch of the Saskatchewan, on the flanks of the Grand Coteau, of the Blackfeet and Plain Crees.

5. The "War-road" of the Sioux, Blackfeet and Crows, in the valley of the Yellowstone.

September 27*th.* — A stormy uncomfortable night. Wavys, as the half-breeds term them, Wa-wa in Cree, (*Anser hyperboreus*), flying to the south early this morning in large flocks, were regarded as a sure sign of approaching winter. The Indians say there is some fine land with large trees in the rear of this part of the river, but from our camp to St. Martin Lake, about thirteen miles in an air-line from Lake Winnipeg, the banks are low and marshy. St. Martin Lake once reached, small eminences, which in this flat country almost deserve the name of hills, appear on the south and north coasts, before entering the Narrows. In general the shores are very low, particularly to the south-east. The Narrows are caused by a remarkable barrier of boulders, chiefly consisting of the unfossiliferous rocks, about six feet above the lake and twenty feet broad. On the west side of the barrier there is an extensive wide-spreading marsh, but the water of the lake is clear, as in most limestone regions.

We arrived at this isolated body of water, soon after noon, and camped on a beach or barrier thrown up in the form of semicircular ridges, about half a mile across the arc, and connected in the form of the letter S. In the formation of these ridges granite or gneissoid boulders are first pushed by ice upon a limestone gravel bar, aspens and willows grow on the ridges rapidly formed by sand and gravel washed up in the rear of the boulders, and the space partly enclosed or sheltered by the curve is soon filled with reeds, thus forming extensive marshes at the eastern extremity of St. Martin Lake. Near the channel which separates this maze from the main body of the Lake, a new beach is now in process of formation, and consists at present of a long semicircular line of stranded boulders, over which the sea washes in easterly and westerly gales. Round about the boulders, limestone gravel is accumulating, and thus, in this direction at least, the lake is slowly diminishing in size, the materials being in great part supplied from the wearing away of islands, and the adjoining coast.

We succeeded in passing the Narrows before breakfast on the morning of the 28th, and made our way into the main lake through a channel varying from three to nine feet in depth, kept open no doubt by the Partridge Crop River, which takes the name of the Little Saskatchewan or Dauphin River, after it has passed through St. Martin Lake. Having landed on Sugar Island, we were followed by a little fleet of eighteen canoes, whose owners appeared determined to reach Fairford before us, if possible. The chief with some ostentation informed me that the island belonged to him, but he had no objection to my exploring it. He further stated, that as chief of the band he claimed the whole country from Fisher River, on Lake Winnipeg to the mouth of Partridge Crop River

Sugar Island is a favourite camping ground of the Dauphin River Swampys, who now occupy this part of the country. We found some graves near to a garden in which potatoes were planted. A few pieces of tobacco procured us a small supply of this vegetable, so rare and valuable in these regions. Sugar Island is so named from a grove of the ash-leaved maple, the trees of which bore old marks of tapping.

Sugar Island.

I found here what appeared to be partially metamorphosed sandstone rock, tilted at an angle of 50°, with a S. 30° W. strike. At one extremity of the island it approached the character of gneiss, at the other extremity it presented the appearance of impure sandstone layers tilted at a high angle. Sugar Island is about a mile from the Narrows, and lies S. 75° E. from three small islands, which upon examination were found to consist of gneiss intersected with quartz veins. The rock on Sugar Island is exposed on one side in the form of a precipitous cliff 20 feet high. On the opposite side it slopes gradually to the water's edge.

We went out of our course to visit the gneissoid islands before referred to. The first island bore nearly due east of Sugar Island. It consists of gneiss with rose coloured felspathic veins, pursuing a general direction of S. 40° E. The axis of the island is also S. 40° E., and the gneiss is intersected by fissures nearly at right angles to one another, one set bearing S. 20°—40 E. The surface of the gneiss on the highest point, which may be 23 feet above the lake, is polished and furrowed in a direction S. 55° E. The south-east shore is precipitous, the opposite sloping.

The second island consists of a dome-shaped mass of gneiss with large quartz veins meandering through it. The third island, within a few yards of the first and second, shows far less metamorphic action, and with a strike S. 15° W., has a dip 75° from the vertical. It is precipitous to the N.W. and slopes to the S.E.

Proceeding along the south-west coast we found a barrier of beaches along the shore about 300 yards distant from it, on which boulders of the partially metamorphosed sandstone and gneiss were piled up, farther on were worn and large unworn fragments of a siliceous limestone, which, however, was nowhere found in position. The occurrence of these gneissoid islands in a flat limestone country is very interesting; the metamorphosed sandstone shows that the epoch of their elevation must have been before the deposition of the limestone found on Thunder Island, to which we next proceeded, and after the deposition of the sandstone on Sugar Island. The three gneissoid islands, having no name, we called

St. Martin Rocks. It is not improbable that the epoch of their elevation was simultaneous with outbursts which have been observed in other parts of the continent. At noon we arrived at a semicircular island of beaches similar to those at the east end of the lake. They are due to the great shallowness of St. Martin Lake, which, with an area of over three hundred square miles, was in no part of our course found to be more than eighteen feet deep, and often only five and six feet, for long distances.

In the afternoon we landed on an island where stratified limestone, in horizontal layers, was exposed. The limestone possessed some singular peculiarities. Numerous cup-shaped forms, of very large dimensions, were visible in projecting masses over the whole of the surface exposed. Many of these cups were fully thirteen inches in diameter at the surface, and would hold at least one quart of water. They consisted of concentric rings, or cups, regularly arranged, and from ten to fifty or more in number. The thickness of each cup varied from one-tenth to one-quarter of an inch. A single specimen resembled a gigantic onion which had been cut in half, with a few of the inner layers extracted, leaving a cavity or depression. Many square yards of surface were variegated with this structure. The colour of the limestone is a buff-yellow, its fracture is uneven, and masses are difficult to separate, being extremely hard and siliceous. The height of the exposure is sixteen feet, and so nearly horizontal, that no inclination could be detected. The island having no name, and being remarkable for its rock formation, it was thought worthy of some designation: we therefore called it "Thunder Island," in memory of a storm of hail and rain, accompanied by lightning and thunder of more than ordinary violence, which made us very uncomfortable for the rest of the day and during the ensuing night. It was the last of twenty thunder storms which we had encountered since entering the prairies on the 14th of June, and was only second to one in violence and sublimity.

Anxious to get on, we pulled at the sweeps until after dusk, having reached an island about four miles from Thunder Island. We found a sheltered cove, and all slept in the boat, there being no spot on the boulder-beach, or barrier, on which we could discover six feet of level ground.

Sept. 29th.—When morning dawned, which it did in a drenching, cold rain, we found the boat attached to one of the stony barriers which protect certain aspects of the islands, or main shore. The ever-present marsh lay between us and the timber we needed for fuel; but the wind now rising to a gale, compelled us to remain satisfied with an exploration of our boulder barrier to its utmost limits. It was about one hundred yards broad, two to three miles long, and consisted of waterworn masses of limestone and gneiss, with limestone gravel between them. The marsh which separated it from the island was full of weeds, and harboured wild fowl, some of which we succeeded in killing.

The steersman found great difficulty in discovering the mouth of Partridge Crop River, or St. Martin River as it is also called. A maze of rushes extending inland, as far as the eye can see, hides it from view. Half a mile up the stream we saw the houses of the Mission, established but afterwards abandoned, by the Rev. Mr. Cowley. All the houses were in ruins, and tenantless. The whole country is very low, and liable to be flooded in the autumn and spring. There are not more than a few hundred acres of land, elevated four or five feet above the river, and fit for agricultural purposes. The spot was one, however, of great resort among the Indians of this part of the country, and hence the probable reason why a selection of this site was made for the establishment of a Mission. On landing, we found one Indian family who are determined to continue the cultivation of the little fields which have been cleared and enclosed. They had accumulated three small stacks of hay, were possessed of a yoke of oxen, and were living in one of the least dilapidated houses.

We took to our boat at the beginning of Partridge Crop River, having secured a guide from the fleet of canoes in the rear, to take us through a narrow passage between beds of rushes which cover many square miles, and constitute the "Crop," so called by the Indians on account of the resemblance which the outline of this reedy expanse bears to the "crop" of a partridge. We threaded our way through the mazes of a marsh supporting rushes so tall that, without climbing the mast of the boat, it was impossible to see beyond the masses which enclosed us. The rushes measured from ten to twelve feet in length, and grew so thickly together that they formed a compact green wall, past which the current flowed as if they were formed of solid, stable materials. Through little openings, which were now and then disclosed, we saw tranquil ponds, with a scarcely perceptible stream. Here reveled hosts of ducks of many species.

We arrived at Fairford at 3 P. M., having occupied about two hours in passing through the Crop.

Fairford is very prettily situated on the banks of Partridge Crop River (a continuation of the Little Saskatchewan), about two miles from Lake Manitobah. The banks are here about twenty feet high, and show alluvial clay with boulders, but a short distance in the rear of the river the limestone approaches the surface, and is covered with eight to ten inches of vegetable mould. Although the appearance of the country is attractive, the shallowness of the soil will not permit of extensive agricultural operations. The dip of the rock is towards the south-west, but at so small an angle as to be almost imperceptible, except when a surface of several square yards is exposed. Fossils are few in number, and obscure :

Fairford, or Partridge Crop.

the limestone breaks up into thin slabs, being very compact and hard.

We attended evening prayers in an excellent schoolhouse, which serves the purpose of a chapel. There were forty persons present, consisting of Indians and Halfbreeds. The service comprised a hymn and a chapter from the New Testament, respectively sung and read in the Ojibway language, an exposition of the chapter by means of an interpreter, and a concluding prayer. The Lord's Prayer was repeated aloud in Ojibway by the whole congregation.

There are one hundred and twenty Christians, adults and children, at this Mission. The houses, fifteen in number, are neat, comfortable, and in excellent order, and several new dwellings are in process of erection. The appearance of this Mission is very promising, and in every way most creditable to the unceasing labours of the zealous missionary, the Rev. Mr. Stagg. Miss Harriet Thompson, a young lady from my native place, Nottingham, England, is residing at the Mission, and devotes herself with exemplary industry, in connection with Mrs. Stagg, to the education and care of Indian and half-breed children. It was a very unexpected and pleasant incident in these remote wilds to meet a young lady so recently from England, acquainted with my relatives and friends, and engaged in a work involving so much strength of character, self-denial, and true Christian sympathy. Miss Harriet Thompson has devoted herself to her difficult task in the right way; although only a few months at the Mission, she has made considerable progress in the language spoken by the Indians who visit Fairford, and I was much surprised and impressed when I heard her repeat aloud at evening service the Lord's prayer in the Ojibway tongue, leading a score of dusky worshipers, who, in the absence of that commiseration which made this incident possible, would in all probability have still been worshipers of Manitou, and slaves to a degrading superstition, in place of possessing a growing faith in "Our Father, which art in heaven." Miss Thompson gave me a few numbers of the "Nottingham Journal," which subsequently served to while away most pleasantly several hours round the camp fires on Dauphin Lake and the summit of the Riding Mountain. The farm at the Mission is in capital order, and although the area adapted for cultivation is not likely to induce the establishment of a large settlement, yet Fairford will become an important centre. We were supplied with potatoes, onions, turnips, fresh bread, and butter, and otherwise most hospitably entertained by Mr. and Mrs. Stagg.

The Hon. Hudson's Bay Company have a post at this mission, but it is matter of deep regret that the heathen Indians who come to barter their furs here should be permitted to have access to rum. The little fleet of canoes before spoken of, arrived during the evening, and at nightfall the sounds of drunken revelry told how terribly the debasing influence of this traffic must operate against the Christian and humanizing influence of the missionary. The post had been but recently established, and the distribution of intoxicating liquors to the Indians appeared to be a subject of deep anxiety and trouble to the Rev. Mr. Stagg.

We reached the mouth of the river at noon on the last day of September, and entered Lake Manitobah with a head wind, which soon compelled a retreat to a low sheltered beach. The exposed aspens were now quite yellow, but a tint of green still remained on groves at some distance from the lake shore. Large boulders are piled up high upon the beach, and behind them is the unfailing marsh. In shallow bays limestone gravel forms a sloping beach to the water's edge, but in the rear is a marsh. It is only at the headlands that rock in position, or firm soil, has been seen as yet.

In the afternoon we set sail, and arrived at Flat Rock Bay, where limestone of Devonian age is seen on the south side. Some of the layers are highly fossiliferous, and hold numbers of *Atrypa reticularis* and *A. aspera.* The stems of crinoids are common, but the species are very few. The rock is nearly horizontal, and the general dip south-west, at a very small angle but many slight undulations occur, giving an inclination of equal extent in an opposite direction. The exposure in the bay is ten feet high, and is worn into caves. The colour is a pale-buff with some reddish-brown layers. Fucoids are abundant, and become, when weathered, yellowish-buff. Small oaks are scattered near the spot where we camped, interspersed with aspen. In the rear tamarac and spruce swamps prevented an examination of the country for more than a few hundred yards from the shore. Where rock in position does not limit the lake, the marginal barrier of boulders is generally found with a beach, a marsh, or a swamp in the rear.

October 1*st.*—Collected fossils, breakfasted, and pulled to Steep Rock Point. Here the limestone (Devonian) is twenty feet high, quite abrupt, with six feet of water at the base of the cliff. The layers are more massive and compact than before noticed; they occur from one to three feet in thickness, are very hard, and hold many organic forms replaced by crystalline carbonate of lime. Three and a half fathoms of water were found within one hundred yards of Steep Rock Point. A number of swans were seen sailing in a little bay to the south of this landmark in Lake Manitobah, which, by the way, the Indians who hunt in this part of the country do not visit, being persuaded that "little men" live in the caves and holes into which the rock has been worn by the action of the waves. We ran on before the wind, past Cherry Islands and Point Pa-oo-nan, or the Waiting Place, until dark, and then made for the shore, soon finding a small sheltered bay in the inside of a boulder beach in process of formation, about two hundred yards from land. Temperature of the lake, 53 deg.; greatest depth of water recorded, twenty-two feet.

Unfortunately the beach consisted of water-worn pebbles strewed over a bank of sand: former experience had established the hopelessness of attempting to procure a night's rest in a bed made on the sand or on small boulders. Sleep may be obtained, but purchased at the price of weary and bruised limbs, and a painful feeling on rising that the night has been utterly thrown away. In preference to sleeping on the sand, we stretched out blankets on the bottom of the boat, and although rather crowded, having made no preparations for this alternative, we nevertheless slept soundly and enjoyed rest.

A fair wind on the 2nd started us at dawn. We steered for the mouth of the Water-hen River, leaving on our left Crane River and Bay, where salt springs exist, and then passed through a narrow channel in a reef of boulders, which stretched from east to west, as far as we could see. The wind being fair, we pressed on, notwithstanding a heavy rain, and landed, rather late in the day for breakfast, on an island near the mouth of Water-hen River, which connects Lake Manitobah with Water-hen and Winnipego-sis Lake. Here we found a pair of white-headed eagles engaged in fishing; and as we came suddenly upon them after rounding a point, one of them dropped a fine white fish he had just caught, which was immediately seized and appropriated by our men for their own breakfast.

We entered one of the many mouths of the river at 2 P.M., and pulled up a broad channel through a vast marsh, whose limits are well defined by a belt of aspens on either hand. Having reached an attractive camping-place, where the woods came down to the edge of the river, we landed with a view to make a short traverse into the country. The river is swift, very broad, and prettily varied with well-wooded islands. At our camp the trees consisted of white spruce, 1 ft. 6 in. in diameter;

poplar, aspen, birch, and tamarac. The land on the river is low, not ten feet above the water ; in the rear we found a tamarac swamp, with belts of white spruce. The channel through which our course lay was about three hundred feet broad and three feet deep, with a flat limestone bottom. The water was clear and brilliant, fish very numerous, and water-fowl abundant.

October 3rd. — *En route* at 9 A.M., — the early part of the morning being employed in drying clothes after the rain of yesterday. We commenced pulling up Water-hen River, which here appears to contain many low islands, and its aggregate breadth must be several hundred yards near our camp. Signs of the approach of cold weather began to thicken around us ; a large flock of pelicans, wheeling in circles far above, suddenly formed into an arrow-headed figure, and struck straight to the south ; and yellow leaves drifted through the air before a cold north wind, promised us, as the half-breeds said, by the beautiful aurora of last night. Islands, low and reedy, continued to appear until we arrived at the Great Bend, where a band of Indians have their winter quarters. The Indians are Roman Catholics, originally from Oxford House. I persuaded one of them to act as guide up Moss River to Dauphin Lake, after we had visited the salt works. Their tents were dirty and excessively odorous. In general, the Indians of Lake Winnipeg and Manitobah, in point of cleanliness, cannot bear comparison with the Prairie Indians.

We met here, also, a freighter's boat, in charge of a French half-breed, who, with his family, was returning from the Salt Springs to Oak Point, with about twelve bushels of salt. We exchanged a little tea and tobacco for ducks and fish ; and on the following morning started by the Middle Branch of Water-hen River for Winnipego-sis * Lake, leaving Water-hen Lake to the north. The river is very broad, shallow, and reedy; low belts of aspens, on either side, indicate the only land visible.

A fair wind drove us swiftly on, and at noon we stopped at Ermine Point, on Winnipego-sis Lake. This is a low sandy beach, with a marsh behind, and remarkable for some fine old elms, crooked and gnarled, yet flourishing near to a salt spring. At 4 we reached Snake Island, where we camped early, for the purpose of examining an exposure of rock, and to collect the fossils which a glance showed it contained in abundance. The Duck Mountain loomed a grand object in the north-west.

The rock exposures on Snake Island are very interesting, not only on account of the fossils they contain, but in consequence of the evidence they afford of a slight upheaval, so rare in the present disposition of the rocks of this region.

The exposure at its highest point does not exceed twenty feet, but it is the centre of a low, narrow anticlinal, running north and south nearly. The dip on the east side is S. 75, E. ∠ 18°; and on the west, W. 20, S. ∠ 5°. The limestone is highly fossiliferous, beautifully stratified, very hard, and bituminous. It holds abundance of *Atrypa reticularis*, *Tellina ovata;* with fossils belonging to the genera *Favosites*, *Euomphalus*, *Productus*, *Gomphoceras*, *Orthoceras*, *Lituites;* together with *Trilobites*, *Crinoids*, &c.

On the morning of the 4th of October we set sail from Snake Island, and arrived at the Salt Works and Springs at noon.

* Winnipego-sis, or Little Winnipeg Lake.

CHAP. XXV.

FROM WINNIPEGO-SIS LAKE TO THE SUMMIT OF THE RIDING MOUNTAIN, AND THE SUMMIT OF THE RIDING MOUNTAIN TO MANITOBAH HOUSE.

Character of the Country.—The Duck Mountain.—The Salt Springs.—The Wells.—The Manufacture of Salt.—Salt Springs and Lagoons.—Moss River.—Rapids.—Character of River.—Valley or Dauphin River.—The Riding Mountain.—Lake Ridge.—Hay Ground.—Dauphin Lake.—Pike.—Snow Birds.—Journey to the Summit of the Riding Mountain.—Marshes.—Ridges.—Character of the Country.—Whisky Jack.—Quaking Bog.—Pitching Track.—Rabbits.—Foot of Mountain.—Cretaceous Rocks.—Terraces.—Conical Hills.—White Spruce.—Brown-nosed Bear.—Summit of the Riding Mountain.—Former Character of the Riding Mountain.—Denudation.—Table Land.—Snow Storm.—Source of the Rapid River.—Indian Superstition.—Descent of Riding Mountain.—Character of the Mountain.—Fish.—Sickness.—Cupping.—Ta-wâ-pit.—Great Bones.—Grasshoppers.—Journey from Dauphin Lake to Lake Manitobah.—Character of the Country.—Bogs.—Aspen Ridges.—Ridge Pitching Track.—Ebb and Flow Lake.—Indian Tent—Interior of.—Supper.—Sleep.—Buffalo Runner.—Manitobah House.

The surface of the country where the Salt Springs are found is only a few feet above the level of Winnipego-sis Lake, and apparently nearly horizontal for many miles inland, on a north-west course. The barren area occupied by the springs and wells is about ten acres in extent, but the open country, with points of surrounding forest converging towards the springs, may include several hundred acres. The trees in the vicinity consist of spruce, aspen, willow, birch, and a few stunted oak. The wells are five feet deep, and the water in them was two feet five inches above the level of the lake on the 5th of October, as ascertained instrumentally. The wells are found upon a slight elevation, which gently inclines and blends with a vast marsh connected with Moss River, in a southerly direction. The woods fringing this marsh approach within a mile of the Springs, west and north-west.

The level country extends across the peninsula from Red Deer's Point, about three miles in breadth, to a deep indentation of Lake Winnipego-sis, about five or six miles broad; after which it continues low and marshy, with tamarac, aspen, and white spruce woods to the foot of the Duck Mountain, a distance of sixteen to eighteen miles. From Snake Island, and even from the level of Winnipego-sis Lake, a few miles from shore, the country between the foot of Duck Mountain and the Lake does not present a single eminence to break the level from which the Duck Mountain rises. It resembles, in every important physical feature, the level tract at the base of the Riding Mountain. These observations apply only to that part visible from Snake Island and the lower portion of Winnipego-sis Lake.

The soil at the Salt Springs is a very retentive yellowish-white clay, containing small limestone boulders and pebbles, with boulders of the unfossiliferous rocks. The wells for obtaining a supply of brine, are sunk wherever a small bubbling spring is observed to issue from this retentive clay. The springs are constantly changing their position, and as the wells become exhausted from time to time, a fresh excavation is made where a new spring is observed to issue. No doubt boring, or deeper wells, would prevent these changes, and not only secure a larger flow of brine, but ensure its permanency. The wells at present are twenty-five in number; but some of them appear to have been lately abandoned, and others have long since ceased to yield brine. They are situated four hundred yards from the lake shore, and were first worked forty years since, by James Monkman. This enterprising individual struggled for many years against the importation of English salt, which was sold in the settlements at a cheaper rate than he could afford to manufacture salt on Lake Winnipego-sis. He has made salt at Swan River and Duck River. The manufacture is now carried on with profit for the Hudson's Bay Company at Swan River, and at Winnipego-sis Lake by Monkman's sons.

At the "Works" there are two small log-houses and three evaporating furnaces. The kettles, of English construction, are well-made rectangular vessels of iron, five feet long, two feet broad, and one foot deep. They are laid upon two rough stone walls, about twenty inches apart, which form the furnace. At one extremity is a low chimney. The whole construction is of the rudest description, and at the close of the season the kettles are removed, turned over, and the furnace permitted to go to ruin, to be rebuilt in the following spring.

The process of making salt is as follows: When a spring is found, a well, five feet broad and five feet deep, is excavated, and near to it an evaporating furnace erected. The brine from the wells is ladled into the kettles, and the salt scooped out as it forms, and allowed to remain for a short time to drain, before it is packed in birch bark roggins for transportation to Red River, where it commands twelve shillings sterling a bushel, or one hundred weight of flour, or a corresponding quantity of fish, pemican, or buffalo meat, according to circumstances.

The brine is very strong, — thirty gallons of brine producing one bushel of salt *; and from one kettle two bushels of salt can be made in one day in dry weather. There are nine kettles at the "Works," seven being in constant use during the summer season. The Half-breeds

* See Chapter on the Devonian Series.

engaged in the manufacture complained of the want of fuel—in other words, of the labour and trouble of cutting down the spruce and poplar near at hand, and the difficulty of hauling it to the furnaces,—an objection of no moment, but characteristic of some of the people, who are generally unaccustomed to long-continued manual labour.

It will be seen that the processes employed in the manufacture of salt are of the rudest description, so that without any outlay, beyond a few days' labour, the quantity might be largely increased. I spoke to John Monkman, who now makes salt here, of pumps and solar evaporation. Of a pump he knew absolutely nothing. He had heard that such an apparatus had been contrived, but had never seen one. He readily comprehended the advantage to be derived from pumping the water into shallow troughs, dug in the retentive clay near the springs, and strengthening the brine by solar evaporation. An Indian guide, who accompanied us up the Moss River, assured me that all along the west coast of Winnipego-sis and Manitobah Lakes, there are salt lagoons and springs. The Indians we met on the Dauphin Lake made the same acknowledgment, but declined to give precise information, alleging that the manufacture of salt drove away the game, and spoiled their hunting.

The extent, character, and importance of the Salt Region in Rupert's Land will be discussed at length in another chapter.

Oct. 6th.—Left the Salt Springs, and sailed before a stiff breeze to the mouth of Moss River. We found four feet of water on the bar, and nine feet at the mouth of the river. A low exposure of limestone occurs near the entrance, and another one mile and a half up the stream. The dip is very irregular. The fossils are few in number, and obscure. In its lithological aspect, it resembles the exposure on Snake Island, seven miles distant, in a north-east direction. The rock is curved, and fractured in places, showing in an exposure 120 yards long and nine feet high, inclinations varying from 20° to 40° east, with short horizontal intervals. Some of the layers are extremely hard, others fissile, others crystalline, with crystals of calc spar between the layers and in the fractures.

Near the mouth of Moss River we came suddenly upon a camp of Swampys or Muskaigoe * Indians. Several birch bark tents were placed round a small log hut, which these Indians have constructed for their permanent winter quarters. This was the first attempt which had come under my notice of Indians constructing a log dwelling-house remote from the settlements, the H. B. Co.'s posts, or Missionary stations. They followed us until we stopped to camp, and asked for a smoke.

We arrived at the first rapids on Moss River during the afternoon. They have a fall of two feet, and consist of an accumulation of boulders resting on rock. The second rapids are formed by similar obstructions. The river is here 120 feet broad and very shallow. The bank, ten feet above the water, sustains fine aspens, with a very thick undergrowth; the soil is a clay, and evidently fertile near the river, but in the rear the country passes into muskeg or swamp. In ascending the second rapids the boat had to be lightened, and hauled up by the men walking in the middle of the stream. The temperature for such work was not conducive to comfort or health, and two of the men caught severe colds, with cramps and pain in their limbs.

Oct. 7th.—A sharp frost during the night. Ice formed on the oars in the morning. Temperature of air at 8 A.M. 30°; of the river, 42°. The thermometer, during

* From "muskeg," a swamp; sometimes spelled "muskaig."

the night, fell to 26°. All the leaves are now fallen, and the country presents a very dreary appearance. The whole of the day was spent in rowing or tracking up Moss River. The bank continues from twelve to fifteen feet high, and supports some fine aspens, one foot to fifteen inches in diameter, with a dense growth of young trees springing up in the place of a former fine aspen forest, of which the large trees are the remains. The river continues very shallow, and contains many boulders of the unfossiliferous rocks.

There is a large area of good land on the west bank, near to Dauphin Lake. This secluded sheet of water we entered at 4 P.M., and came at once in sight of the Riding Mountain in front and the Duck Mountain on our right. Both are very imposing ranges from this point of view, yet presenting similar aspects; both rising from a level country a few feet above Dauphin Lake; and, as far as the eye could judge, both maintaining the same elevation, and showing abrupt wooded escarpments towards the east. They are separated by Valley River, and it is apparent that they were united at a remote epoch, and formed part of the great table-land which stretches south-west toward the Grand Coteau de Missouri.

In the evening we passed the mouth of Valley River, or *Te-wa-te-now-seebe*, literally, "the river which divides the hills," and camped on a beautiful beach, beneath some fine oaks growing upon a ridge close to the water's edge.

On the following morning we started at daylight for a part of the coast nearest to the Riding Mountain. At a distance this magnificent range appears to be clothed with forest, and to rise from a level plain to the height of about 800 feet above the level of Dauphin Lake.

As soon as we arrived opposite to what appeared to be the highest part of the range, we landed, and despatched an Indian to explore the country, and report on the nature of the swamps through which we should have to pass in attempting to reach the summit. During his absence we set nets, and leveled across the ridge which separates the lake from an extensive fertile meadow which lies between it and the mountain. The ridge is 8·96 feet above the present level of Dauphin Lake; the meadow 5·70 feet; and so continues for a distance of one mile, with an almost imperceptible rise, until a second low ridge is reached. The meadow is covered with long, luxuriant grass; a few clumps of poplar and thickets of willow vary its uniformity. There are, no doubt, many thousand acres of excellent hay-ground on the banks of Dauphin Lake, but the breadth of the tract did not appear to exceed two miles. The oak on the ridges occurs in patches, and the trees are from twelve to fifteen inches in diameter.

Dauphin Lake is twenty-one miles long, has a greatest breadth of twelve miles, and an area of 170 square miles. Its approximate elevation above the level of the sea is 700 feet, or seventy-two feet above Lake Winnipeg. It is very shallow.

Our nets produced five splendid pike, weighing about fifteen pounds each.

At the close of the day the Indian returned. He had advanced to the first great ridge, about nine miles distant, and reported eighteen inches of water in the swamps, with ice a quarter of an inch thick.

Snow birds were seen for the first time during the afternoon. They came about our camp in large flocks, but they did not appear to have quite assumed their winter dress. The evening and part of the night were spent in making arrangements for an ascent of the Riding Mountain. We took provisions for four days, a blanket for each man, with a good supply of guns and ammunition. Three of the men were left in charge of the boat, with instructions to cure all the fish they could take, as the danger of being arrested by ice in Lake Manitobah was not improbable. That large body of water has been known to freeze as early as the 25th of October.

At sunrise on the morning of the 9th of October we set out for the ascent of the Riding Mountain. Each man had a pack weighing about thirty pounds. My share consisted of a sledge-hammer for geological purposes, which proved to be an extremely inconvenient article to carry across swamps or through bushes. Once, indeed, when crossing a quaking bog with the hammer on my shoulder, I received a severe blow on the back of the head as I broke through the covering of moss over which we were pushing our steps, when endeavouring to fall flat on the springy surface.

For the first mile and a half the country is quite level and dry, with the exception of three narrow and shallow marshes. The soil is excellent and hay abundant, but no doubt in spring this extensive flat must be very wet, and, probably, to a considerable extent under water. In its present condition the pasturage it affords is very abundant and luxuriant. We soon arrived at a low ridge which marks the limit of the good land, not averaging more than two miles from Dauphin Lake. To the ridge succeeded marshes and willow brakes. These were bounded by low gravelly ridges clothed with aspen, which were again succeeded by marshes.

Finding it quite impossible to outflank the marshes, which appeared to stretch from river to river descending from the mountains, and to be co-extensive with the shores of the lake, we determined to push through to the highest peak, which was in reality the nearest point of the mountain to us, its greater altitude being only apparent on account of its proximity, as we afterwards ascertained. In an hour we arrived at a white spruce swamp, in which many fine trees, fully eighteen inches in diameter, were observed. Beyond the white spruce swamp we came to an old lake ridge, about fifteen feet above the general level, beautifully rounded, and composed of limestone gravel, with many boulders of the unfossiliferous rocks on the south or land side.

This ridge resembled the Big Ridge of the Assinniboine in most particulars. Our Indian guide told us that it extended for many days' journey north and south of Dauphin Lake. It forms the Indian pitching track at the foot of the Riding Mountain.

The term "pitching track" is applied to an Indian trail from one part of the country to another. West of Lake Manitobah, Dauphin Lake, and Winnipego-sis Lake, the "pitching track" follows the ridge described above. It is, in fact, the main and only dry road in this region. On the crest of the ridge there is a narrow well-worn path, which, for many generations probably, has been the highway of the Indians passing from Lake Manitobah to the Assinniboine, through the valley of *Te-wa-te-now-seebe*, or "The River that divides the hills." This pitching track is connected with "the Ridge pitching track," between Ebb and Flow Lake and Dauphin Lake.

The Whisky Jack is numerous on the ridge, and in the spruce swamp were several ravens. Formidable marshes succeed the ridge; we waded through them knee deep for a space of half a mile, and then rested a short time on a small island where stunted aspens grew. We continued to pass through marshes, aspen islands, and over low ridges clothed with willow, until a bog of such a quaking character, and of such great breadth, presented itself, that the men demanded a smoke before attempting to cross. Our Ojibway half-breed, Wigwam, insisted upon carrying the sledge hammer in addition to his pack, declaring that he was accustomed to quaking bogs, and we should have enough to do to get across without carrying anything that might impede our progress. The surface of the bog consisted of a thick elastic covering of moss, sufficiently tough to bear our weight when passing quickly over it, but if we stopped for more than half a minute the moss slowly sank and a pool of water collected around us. We marched, or rather trotted in single file, about ten yards apart. The Indian who took the lead passed nimbly over the surface, so also did the half-breeds; Mr. Fleming and I, however, two or three times broke through with one leg, but succeeded in withdrawing the unfortunate member without further damage than immersion in water and mud emitting a very unpleasant odour. Occasionally we rested on a narrow strip where the tangled roots of a few willows afforded a firm footing. The breadth of this bog was about one mile where we crossed; it was succeeded by a belt of tall reeds growing in water one foot deep; plunging and wading through this, we arrived at a gently sloping ridge about eighteen feet in altitude. On the other side of this ridge a narrow deep swamp separated us from the foot of the mountain; wading through it we ascended a hill about forty feet high, and found ourselves upon a dry terrace, on which we determined to camp, having accomplished a distance of about eleven miles. The men soon dispersed to hunt rabbits; a dozen were killed in a few minutes, skinned and placed on sticks before the fire to roast.

In passing through the swamps, we saw many fresh moose tracks, — hunters' signs which aroused the Indian and half-breeds to a high pitch of excitement, and caused

"WIGWAM," AN OJIBWAY HALF-BREED

them to steal swiftly on their wet and treacherous path with a speed which we found it very difficult to maintain, although they indulged no hope of coming within shot of such noble game, in consequence of an unfavourable wind, even if we had been able to preserve the necessary silence in passing through the haunts of this wary animal.

The timber on the hill at the foot of the mountain consists of aspen, with a few small oak. The soil on the plateau is of excellent quality and the underbrush very luxuriant.

The night promising to be very cold, ice forming on the kettles within a few yards of the camp, we built two large fires and slept between them, having previously dried our wet clothes as far as circumstances would permit. At 8 P. M., the sky was quite free from clouds; the comet shone with brilliant lustre, a flashing aurora gradually spread over the northern sky, the stars glittered like diamonds in the south, and the whole heavens assumed that aspect of silent beauty which renders night in the wilderness so impressive and sublime.

October 10*th.*—Soon after breakfast, we arrived at a steep embankment about seventy feet high, which formed the termination of a terrace about a mile broad, covered with small aspens, and threaded with moose paths. The terrace ascends very gradually and is abruptly bounded by a steep bank, from which a broken hilly tract rises towards the escarpment which forms the eastern limit of the Riding Mountain. This broken tract is covered with aspens and spruce of large size, especially in the hollows. We crossed the beds of two or three streams, which flowed through deep gullies to the plain below. Thus far, the soil consisted of drift clay with many large boulders in the beds of the rivulets; but at an altitude of about 400 feet above Dauphin Lake we arrived at a cliff-like exposure of Cretaceous rocks, through which a stream had cut a channel seventy to ninety feet deep. These rocks seemed to form the boundary of a third terrace on which were numerous conical hills consisting of gravel and boulders of the unfossiliferous rocks. The stratification appeared to be nearly horizontal, with a very slight dip to the south-west. Although a careful search was made for organic remains very few were discovered. These were identical with those found on the Little Souris, and in every particular, except the occurrence of bands holding *Inoceramus*, the rocks on the Riding Mountain resembled the exposures on the Little Souris. The layers containing ferruginous concretions were found, as well as a soft thin band from which the Indians make their pipes. The total thickness of the exposure exceeded 100 feet.

We now followed a moose path until we arrived at a high conical hill, which promised a fine view of the surrounding country. Having reached the summit, the relation of the conical hills and terraces became evident. A wide and deep valley separated us from the table-land of the Riding Mountain, about one mile distant in an air line, and perhaps 200 feet above us. Three terraces were distinctly visible below us; a range of conical hills, the result of atmospheric agencies, lay at the foot of the precipitous escarpment of the mountain, and followed its general direction. Limestone and unfossiliferous boulders were strewn on the summits and flanks of the weather-worn hills, while in the hollows between them, small lakes lay half concealed by a fine forest of white spruce and aspens.

From the brow of the hill where we stopped to dine, the Indian shot a large brown-nosed bear, which suddenly appeared on the terrace below us about seventy yards from our camp fire. His skin was in good condition, and remarkably handsome; the animal might weigh 350 lbs., although not yet fat. Leaving three men to cut up and prepare the meat, we commenced the last ascent, and arrived at the summit of the Riding Mountain at three in the afternoon. The last rise was very abrupt it consisted of a steep escarpment of drift clay with boulders, covered with a fine white-spruce, birch, and aspen forest. At the foot of the escarpment were ponds or small lakes, which fed the mountain streams we had crossed.

The view from the summit was superb, enabling the eye to take in the whole of Dauphin Lake and the intervening country, together with part of Winnipego-sis Lake. The outline of the Duck Mountain rose clear and blue in the north-east, and from our point of view the Riding and Duck Mountains appeared continuous, and preserved a uniform, bold, precipitous outline, rising abruptly from a level country lying from 800 to 1,000 feet below them. The swamps through which we had passed, were mapped in narrow strips far below; they showed by their connection with the ridges, and their parallelism to Dauphin Lake, that they had been formed by its retreating waters. The ancient beach before mentioned as extending far to the north and south, could be traced with a glass by the trees it sustained, until lost in distance; it followed the contour of the lake, whose form was again determined by the escarpment of the Riding Mountain. It required no effort of the imagination to recall the time when the whole of the flat country below us, towards the Laurentides on the east side of Lake Winnipeg, was occupied by the continuation of the Riding and Duck Mountains with their associated ranges to the north, and when the Cretaceous series, superimposed in patches by Tertiary rocks, extended to the basin of Lake

Winnipeg as it now is. The whole of this immense denuded tract of country is a splendid instance of the power of water and ice to remove many thousand cubic miles of rock.

It is very probable that before the Boulder Drift period, the chain of mountains beginning with Turtle Mountain near the forty-ninth parallel, and terminating with the Porcupine and Pasquia Ranges, including the Riding, Duck, and Thunder Mountains, were part of a high table land, composed of Cretaceous and Tertiary rocks, which extended from the Grand Coteau de Missouri to the Laurentides. The areas most affected by denudation are now occupied by Lakes Winnipeg, Manitobah, Winnipe-go-sis, and the valleys of their tributaries. The precipitous eastern escarpment of the chain of mountains show the action of oceanic agencies to which they would be directly exposed, if the country were submerged to more than 1000 feet, and from the distribution of boulders, there can be no question that a submergence to a far greater extent has taken place since the Tertiary epoch. The connection of these ranges will be best seen by an inspection of the map.*

The summit of the Riding Mountains is a vast table land declining in steps to the Assinniboine. The forest which covers the upper plateau consists of very fine white spruce, birch, poplar, and aspen ; the circumferences of some of the trees about our camp, measured five feet from the ground, were as follow : — Aspen, 4 ft. 6 in., 4 ft. 6 in., 4 ft. 1 in., 3 ft. 9 in., 5 ft. ; White Spruce, 7 ft. 3 in., 5 ft. 6 in., 6 ft. 6 in., 6 ft. ; Birch, 3 ft. 6 in., 3 ft. ; Poplar, 4 ft. 9 in., 4 ft. 6 in. These trees represent, as far as my observations permitted me to judge, the

* Vide Chapter on the surface geology of the valley of Lake Winnipeg.

general character of the forest on the summit plateau of the Riding Mountain.

Soon after our arrival at the summit, clouds began to gather in from the north-west, and towards evening a snow-storm set in which continued during the greater portion of the night. Beneath the shelter of the overhanging branches of a spruce we made an excellent camp, and having built a roaring fire, were soon engaged in discussing bear steaks as we reclined on a couch of spruce boughs, under a roof constructed of the same excellent material and impenetrable to snow.

October 11*th*.—When morning dawned we found the country covered with a mantle of snow, six inches deep. This did not prevent us from making a traverse in the direction of the lakes from which the Rapid River takes its rise. The course we had taken led us, as was afterwards ascertained, to within a few miles of the spot reached by Mr. Dickinson when he ascended the valley of Rapid River a few weeks before. This was precisely the result I was anxious to attain. An inspection of the map will show that our explorations when combined, passed through a comparatively unknown country, nearly along the 100th degree of longitude west of Greenwich, and stretching from the 52nd to the 49th parallel of latitude, thus embracing part of Winnipego-sis Lake, Moss River, Dauphin Lake, the Riding Mountain, the Little Saskatchewan or Rapid River, and the Little Souris, to the 49th parallel.

Our progress to the south was soon arrested by a lake, and the lateness of the season made it advisable not to linger too long in this region, lest we should be arrested by ice forming in the great lakes below. Anxious to kill a moose, I endeavoured to persuade the Indian to follow a fresh track, but he declared that the mountain was full of devils, and that the grizzly bear was not unfrequently met with, so that no persuasion could induce him to follow the track unless a half-breed accompanied him. The sky and air became quite free from clouds and mist before 10 A. M., permitting us to take bearings of different prominent points. After measuring a number of trees in the neighbourhood of our camp, we commenced to retrace our steps at noon. Our course was a little to the left of our track on the preceding day; and the following estimate of the ascents, descents, and distances were noted as carefully as circumstances would permit.

The first descent from the summit is about 250 feet deep, and very precipitous; where the snow had not lodged, boulders were seen reposing on unstratified clay and gravel. A narrow gully is then crossed, and an ascent of forty feet made to a terrace sloping towards the east; on this terrace are the conical hills before alluded to. The descent continues for a further depth of 150 feet in a distance of half a mile; this brought us to the edge of a ravine seventy feet deep. At the bottom flows a small stream over gravel and boulders. A rise of thirty feet led us to the top of the opposite bank, along which we traveled until we came to its termination at the beginning of a second terrace about eighty feet below us. This narrow table land is consequently about 480 feet below the summit, and on it we found the second range of conical hills. A gradual descent for a quarter of a mile lowered us thirty feet; we then ascended a bank about twenty feet high, and found ourselves on the edge of a precipitous descent, 150 feet deep, which brought us to the third terrace, and to the edge of a ravine seventy feet deep. On the sides of the ravine, and far above it, exposures of Cretaceous rocks were seen; the highest spot where the rock was observed in position, is probably between 400 and 500 feet below the summit, or about 500 above Dauphin Lake. A sudden descent of 120 feet then occurring, brought us to a fourth terrace, bounded by a steep bank, to which succeeded a narrow step, and then the low ridge where we had formed our camp on the ninth. We arrived there thoroughly wet and uncomfortable; the temperature, however, was much higher than on the mountain, and the snow of the previous night had entirely disappeared as we descended, but a drenching rain instead promised a very disagreeable night, as we could find no friendly spruce near at hand to afford protection.

October 12*th*.—The greater part of this day was spent in retracing our steps to Dauphin Lake. The walk through the marshes and bogs was found to be more fatiguing than during our ascent, on account of the rain and the cold water in the swamps. Two of the men complained of rheumatic pains, and were incapable of doing any work upon their arrival at the camp in the afternoon. During our absence the men left at Dauphin Lake had set nets, and caught some fine pike. The precipitation which had occurred on the Riding Mountain in the form of snow, was here a drizzling rain, which again commenced soon after our arrival, and continued throughout the night. On the following morning one of our best half-breeds was seriously ill; he complained of excruciating pains in the head and limbs, and experienced great relief from cupping, which the Indian performed with a flint and bowl of a tobacco pipe. At noon we started in the boat for an Indian encampment at the west end of the lake, about six miles distant. Here we found Tà-wâ-pit, an old Ojibway, with two sons, and their wives and children. Having made arrangements with Tà-wâ-pit for the hire of two horses and a guide to cross the country to Manitobah House,

Mr. Fleming took charge of the boat, to return by Moss River, while I remained with one half-breed with the intention of making a land journey round the south side of Dauphin Lake to the Company's post on Lake Manitobah, which was to be our rendezvous.

October 14*th.*—Tà-wâ-pit stayed during the greater part of the night by our camp-fire, talking with the half-breed, smoking and drinking tea. He pointed out the spot near to us, where he was accustomed to take salt from the edges of a spring during the summer months. He described also at length the appearance and virtues of some gigantic bones exposed in the bank of Valley River near where it cuts through the old Lake Ridge. Tà-wâ-pit calls these bones a great medicine, and, contrary to the usual custom of the Indians, he now and then takes small fragments, bruises them to powder, and uses them as a medicinal preparation. From his description I infer that the bones are those of a mammoth; his rough drawing of the ribs and teeth in the sand corresponded, in point of dimensions, with those of that gigantic animal.

Tà-wâ-pit and family live a very retired life on the shores of Dauphin Lake. The old man is evidently of a misanthropic turn of mind; he does not associate with other Indians who hunt and live on Moss River, and the northern part of the lake. His potatoes, of which he planted a small patch in the spring, were completely destroyed by grasshoppers; affording another proof of the immense range and devastating progress of these insects in Rupert's Land, during the past two or three years. Tà-wâ-pit showed me a knife he had made out of an old file, and some pipes he was making from a soft shale, procured in the Riding Mountain some miles southwest of his tent. The shale was similar in all respects to a band I had noticed on the Little Souris, also a few days before, in the exposure described as occurring on the flank of the mountain, and from which the half-breeds had taken small blocks to make pipes. A couple of pounds of buck shot which I divided among the old man and his sons, delighted them beyond measure, and in return, Tà-wâ-pit presented me with a new pipe, and the moufle of a moose.

The hop grows in great luxuriance and abundance at the south end of Dauphin Lake; there is fine pasturage as far as the old lake ridge, but the narrow strips of marsh and quaking bog almost on a level with the dry portion show that these extensive flats are liable to be submerged in the spring. Starting at dawn, our course followed for a few miles the shore of Dauphin Lake until we came to Turtle River, which having crossed in a small canoe, we soon after took an easterly direction and entered a region of swamp, ridge, and quaking bog of very formidable character. During the whole of the afternoon our course lay through marshes and bogs, separated by low ridges covered with aspen. The horses were quite useless, and frequently stuck fast; when this occurred we were compelled to carry the bedding and provisions to the nearest ridge and help the wretched animals through the deep bogs into which they sank at every step, breaking through the elastic covering of moss which was generally of sufficient strength to support a man running lightly over it, but not tenacious enough to bear the weight of a horse. Just as night closed in we arrived at a dry gravelly ridge where there was a plentiful supply of dead aspen, from which we made a roaring fire and soon dried our wet clothes and blankets. The night was bitterly cold and the exertion of wading for many hours together through ice-cold water caused every limb to ache; the Indian guide thought nothing of it, and immediately after supper he lay down before the fire and was soon sound asleep. Two or three times in the night I rose to replenish the fire, and found the Indian without any covering but the wet skin clothes he had worn during the day, curled up on the bare ground and enjoying profound slumber.

Early on the following morning we arrived at the Ridge Pitching tract, along which we continued for a few miles, and then again descended into a region of swamps and quaking bogs. In no respect does the Ridge Pitching track between Dauphin Lake and Ebb and Flow Lake differ from the Big Ridge of the Assinniboine except in altitude. It is about one hundred yards across, evenly rounded, composed of gravel, and covered to a great extent with the bearberry. On either side are small oaks and aspens, succeeded by marshes. Its altitude above the marsh is about fifteen feet. The guide said it formed an extension of the ridge on White Mud River described in Chapter XX., and if this be the case no better means of communication by land with this part of the country could be found than the Ridge Pitching track.

Soon after leaving this excellent road we stuck fast in a quaking bog about one mile broad. The horses were mired, and it was only by dint of the greatest exertion and much cruel beating that the Indian and half-breed succeeded in getting them on to dry land. In the afternoon we arrived at Crow Creek, and the country becoming drier we were enabled to make better progress. After passing Sucker Creek, which, with the streamlet before named, flows sluggishly in a trench about ten feet deep, we arrived at a fine open prairie surrounded by tall aspen woods and covered with a splendid crop of wild hay. Here we met an Indian who was setting traps, the hunting season having already commenced. He invited us to his tent which was placed on the shores of Ebb and Flow Lake, not more than twelve or fourteen miles from Manitobah House. It turned out that the half-breed with me knew the Indian well by reputation; he is one of the most successful and industrious hunters in this part of the lake region; his tent was well supplied with Indian luxuries, such as tea, tobacco, and coarse clothing. In the small prairies near us were several fine buffalo runners, and if Indian habits and customs would permit of the accumulation of wealth, our host might soon become a rich man.

His birch-bark tent was roomy and clean. Thirteen

Skin Tents, Cree; Birchbark Tents, Ojibway.

persons including children squatted round the fire in the centre. On the floor some excellent matting was laid upon spruce boughs for the strangers; the squaws squatted on the bare ground, the father of the family on an old buffalo robe. Attached to the poles of the tent were a gun, bows and arrows, a spear, and some mink skins. Suspended on cross pieces over the fire were fishing nets and floats, clothes, and a bunch of the bearberry to mix with tobacco for the manufacture of kinni-kinnik.

Soon after we entered the squaw began to prepare supper, this was done by boiling white-fish and potatoes together; when cooked the whole was poured into a large tin dish and handed to me, together with a cup of tea. Helping myself I passed the dish to the Indian, but he laid it at his feet; as soon as I had finished my supper, the Indian helped himself and the half-breed, and then passed the dish to his squaw, who divided the remainder among the other inmates of the tent. These consisted of an old watchful and restless Indian woman, the mother of the mistress of the tent; a newly married couple related to our host; the Indian guide from Dauphin Lake, and five children. After supper I spread my blanket and lay down, quite overcome with the long continued exertion of wading through swamps and quaking bogs, but too tired to sleep. The half-breed and Indians sat talking for many hours before they turned their feet to the fire, rolled themselves up in a blanket and seemingly at once "found sleep."

On the following morning I rose with a few aches and pains, which the succeeding events of the morning rapidly dispelled. After breakfast my Indian host offered me a favourite buffalo runner to ride to Manitobah House. The exertion required to manage this animal soon removed all unpleasant rheumatic symptoms. Her extraordinary sagacity is related in a subsequent chapter, but her mouth was evidently formed of sole leather and not amenable to persuasions administered through an Indian bridle. The country on the shore of Ebb and Flow Lake is low but well fitted for a limited settlement. There is an abundant supply of aspen timber interspersed with small strips and groves of oak and birch. I arrived at Manitobah House soon after noon, and was cordially received and hospitably entertained by Mr. Mackenzie the gentleman in charge.

CHAP. XXVI.

MANITOBAH HOUSE — MANITOBAH ISLAND — OAK POINT — OAK POINT TO THE SETTLEMENTS ON RED RIVER.

Mr. and Mrs. Mackenzie.—Manitobah House.—Messenger.—Missionary Privations.—Want of Supplies.—Snow Storm.—Indian Summer.—Snow Birds.—Manitobah House.—Rock.—John Campbell.—Whitefish.—Importance of.—Aspect of Country.—The Narrows.—Manitobah Island.—Dimensions of.—Oak.—Rock Formation.—Fossils.—Indian Superstitions. Fairies.—Signals.—Arrival of Boat at Manitobah Island.—Coast of Lake Manitobah.—Old Mission Station.—Unfitness of this part of the Coast of the Lake for Settlement.—Indian Liberality.—Monkman's Point.—Cause of the Formation of Marshes.—H. B. Co. Breeding Establishment.—Oak Point.—Dimensions of Lake Manitobah.—Prairie bordering the Lake.—Shoal Lake.—Character of the Country.—Big Ridge.—Little Ridge.—Arrive at the Settlements.

I REMAINED one week at Manitobah House, waiting for Mr. Fleming, who was detained by contrary winds. To Mr. and Mrs. Mackenzie I am indebted for much generous hospitality, and have great pleasure in possessing this opportunity of acknowledging their kindness and the endeavour they made to give me all the assistance and information in their power. Manitobah House is in a very dilapidated condition, but Mr. Mackenzie has erected another dwelling, which was nearly completed during my stay. A Roman Catholic Mission formerly existed here, but having been abandoned, the buildings were sold to the Hudson's Bay Company, and in the year following the transfer, they were accidentally destroyed by fire.

On Saturday, the 16th October, a messenger arrived from Fairford, on his way to Oak Point, whither he was journeying for the supplies of the Mission in charge of the Rev. Mr. Stagg.

This excellent Missionary has suffered much inconvenience from being disappointed in obtaining supplies of clothing and other indispensable articles for the children and adults, now Christian members of his Mission, and the messenger who arrived at Manitobah House was despatched at his expense to bring the necessaries which had been brought from York Factory to Red River, but not forwarded to the Mission at the usual time by the Hudson's Bay Company's brigade.

In the afternoon of this day a snow storm commenced which continued all night, and covered the ground with nine inches of snow. The thermometer was at the freezing point, but Mr. Mackenzie stated his conviction that the "Indian Summer" not having yet occurred, the snow would soon disappear and we might have fine weather for ten days or a fortnight; a prediction borne out by the rapid disappearance of the snow on the following day, and the occurrence of beautiful weather with frosty nights to near the end of October.

On Monday, the 18th, snow birds were flying about the Post in large flocks; ducks wending their way to the south, and all appearances and signs of approaching winter rapidly following one another.

Tuesday and Wednesday were occupied in writing letters and making up my journal. On Thursday, the 21st, the boat not having arrived, I proceeded to examine the surrounding country. The day was warm and fine, with much smoke from the south-west, coming no doubt from the burning prairies.

Manitobah House is very prettily situated near the Narrows of the lake. Immediately before it is a cluster of low islands on which some fine ash-leaved maple and elm grow; they are the favourite camping grounds of the Indians who hunt and fish in the country about Lake Manitobah. The land in the rear of the House is stony, but good, and there is an area of many thousand acres in extent, well adapted for a settlement. The timber consisting almost altogether of aspen on the mainland, is of fair dimensions, trees from twelve inches to fifteen inches in diameter being common. Near the Post, but on the opposite side of the lake, there is a considerable quantity of balsam-spruce and tamarac. There are no rock exposures visible near the Post, but in making an excavation for a cellar under the new house, the workmen came upon limestone rock, four feet below the surface. It was apparently horizontal, but in the fragments procured no organic remains were visible; its lithological aspect was similar to the rock on Manitobah Island, to be hereafter described. When the surface of the exposed rock was cleaned with a bucket or two of water, well preserved ice groves were visible; their direction was N. 10° W.—S. 10° E.

I visited the house of a freeman named John Campbell a few hundred yards south of the Post, and found there two comfortable log shanties, a potato field, two or three haystacks and some cattle. Campbell's son informed me that it was much easier to live here than at the Settlements. He stated as a proof of the abundance of natural pasture in the swamps that he allowed some of his cattle to remain in the woods and swamps all winter, but they became very poor towards spring. White-fish * are abundant and of excellent quality; they are the mainstay of the people attached to the Post, who cure them for winter consumption by the simple process of drying in the sun. The fishing season had already commenced

* *Coregonus albus.*

and Campbell had caught 500 white-fish, but he wanted 4000 for his winter supply. As soon as the fish are caught in the gill nets, and brought to shore, a slit is made above the tail, through which a pointed stick is pushed. Ten fish are placed on each stick, and the sticks are staged in the open air about nine feet from the ground, beyond the reach of dogs. No curing, cleaning, or any preservative process is employed; the dry air and frost preserve them until they are needed. The importance of the white-fish in this region may be imagined when it is known that not only does it form the chief food of the Indians in the lake region for a great portion of the year, but *three white-fish* per diem constitute the sole daily allowance of the half-breeds attached to this Post; absolutely nothing more. Flour, tea, sugar, &c., are luxuries, which, if they wish to indulge in, must be purchased at high prices; nevertheless, they are healthy, happy, and according to their notions comfortable.

The white-fish I saw staged at Campbell's might average three to four pounds each. They are considered to be superior to those caught in Lake Winnipeg. This important source of food in these regions is well named At-ik-um-aig, or the Reindeer of the Waters, by the Ojibways. It forms a principal article of diet during a large portion of the year, not only of the Indians, but also of the settlers at Red River. The price the frozen fish fetch in the settlement is five for a shilling, or 100 for a pound sterling. During our winter journey to Canada, we purchased them at Red Lake, in Minnesota State, at the rate of $6 (£1 9*s.*) per hundred, to feed the dogs; each dog was allowed one white-fish, and a morsel of pemmican, as long as the pemmican lasted.

Indian summer began on the 21st. The weather was warm, "smoky," and very delightful. No boat being yet within sight, I visited the islands opposite Manitobah House, the marshes at the mouth of Ebb and Flow Lake, and the country in the rear of the Post. Its extraordinary flatness is shown by the great expanse of marsh about the islands, and along the coast north of the Hudson's Bay Company's Post. The level of the lake was three feet below high water mark, and about two feet above the lowest point to which it has been known to fall for many years.

The boat not arriving on the evening of the 22nd, I determined to take a small supply of provisions and go with Whiteway the half-breed, who had accompanied me from Dauphin Lake, as far as Manitobah Island, about twelve miles in a direction due north, and there await its arrival. This part of Lake Manitobah is not more than from three to four miles across, studded with low islands, and on the east side the coast is indented with deep bays. The strait is shallow, twenty-one feet of water close to the Narrows being the greatest depth recorded.

Manitobah Island, from which the lake derives its name, is about 600 yards long, and 200 yards broad; on its north side there is a perpendicular limestone cliff fifteen feet high; a few yards from its edge, a well defined ancient lake beach crosses the island, resembling in most particulars the Ridge Pitching track, or the Big Ridge of the Assinniboine. The part that remains in a good state of preservation is not more than 150 yards long, the breadth of the island being here about 220 yards. From this ancient beach, the land slopes gradually in a southerly direction to the present beach, with its fringe of rushes at the south extremity of the island. The timber consists of oak and birch; many of the first named tree have been cut by the people of Fairford and Manitobah House.

The native carpenter employed to build Mr. Mackenzie's new residence accompanied us to the island, and although very anxious to make the traverse across the lake after passing the Narrows, he remained for a few hours to cut a couple of oak logs which he proposed to take with him to Fairford, to mend the old freighter's boat which had formed his pay for six weeks' labour. He embraced this opportunity, on account of the difficulty of procuring oak timber near the Mission. Although oak was seen several times on the shores of Lake Manitobah, north of the Narrows, yet nowhere was it found of such serviceable thickness (fifteen inches) and length as on Manitobah Island. In the rear of the marshes which border the lake it is known to exist in small quantities.

Among the Devonian fossils procured on the island were *Atrypa reticularis*, *Atrypa aspera*, two species of *Chonetes*, a small *Productus*, an *Orthoceras*, and fragments of a large fish.

I remained on this island with Whiteway for three days; we shot a mink, a few duck, and saw a red fox, but although the island was so small, we found it impossible to kill him. Indians appeared occasionally in their canoes on the north-east coast of the lake, but although they heard our guns and fired in return, yet they would not venture near us. They have all a great aversion to caves and overhanging rocks, conceiving that such places are the abode of fairies, or Manitou. The origin of this superstition in relation to Manitobah Island is due to the sounds produced by the waves as they beat upon the beach at the foot of the cliffs near its northern extremity. During the night time, when a gentle breeze is blowing from the north, the various sounds heard on the island are quite sufficient to strike awe into the minds of the superstitious Indians. These sounds frequently resemble the ringing of distant church bells; so close, indeed, is this resemblance, that several times during the night I woke with the impression that I was listening to chimes. When the breeze subsided, and the waves played gently on the beach, a low wailing sound would be heard from our camping place, about 300 yards from the cliffs where the noise was produced. At night it was peculiarly impressive, and as we lay on the moss-covered rock, it was very easy to comprehend the objection which uneducated Indians, naturally of a fanciful and superstitious turn of mind, should have to land or remain on this "fairy" island.

On the night of Monday, the 25th October, we built as usual a large fire on the beach to serve as a beacon light to Mr. Fleming, and at nine lay down to sleep. Whiteway was telling me about the adventure of Sho-Shons, (Long-ears), whose tent was within a few miles of us, and who was tossed by a buffalo bull when hunting in the prairies during the past summer, when at 10 P. M. three shots were heard, apparently about three miles north of the island. We sprang up and replied with three shots, and proceeded at once to supply the beacon fire with dry wood. Whiteway put his ear to the water's edge, and after a short pause declared that he heard oars. After a few minutes we fired three more shots, and waited the result; in half an hour's time the boat loomed through the gloom, and before eleven o'clock Mr. Fleming and the crew were on Manitobah Island.

They had been detained by contrary winds, but were successful in killing prairie hen *, duck and plover in the upper part of the lake, near Crow Bay. At sunset Mr. Fleming had touched a low point a few miles to the north-east of our island, where Indians were encamped;

* The Prairie Hen or Pinnated Grouse, *Tetrao cupido,* is not often found so far north as lat. 52° in the wooded country.

they told him that they had repeatedly heard shots from the Narrows, but did not care to know who had fired them in that quarter. A quiet admission that the terrors of Manitobah Island were sufficient to check the curiosity even of an Ojibway Indian.

It was past noon on the morning of the 26th when we reached Manitobah House; we remained there for an hour to partake of the hospitality of Mr. and Mrs. Mackenzie, and procure a supply of white-fish and potatoes. In the afternoon we pulled towards McKay's Point, passed between Sugar Island and Birch Island, both low and marshy areas, and camped at sunset on Pelican Island, a circular sandy beach enclosing an extensive marsh, in which duck still remained in considerable numbers. The lake near the coast is shallow, the greatest depth recorded being thirteen feet.

The whole of the coast as far as Swan Creek is very low, and bordered by beaches enclosing marshes. Here and there wooded points ten to twelve feet above the lake level separate the marshes from one another; on one of these points we observed some very fine elm, but the prevailing timber consists of aspen. A mission was established some years since at Elm Point, by the Rev. Mr. Cowley, but abandoned soon after. An attempt was made to open a cart track from this Mission to the prairies near Oak Point, but it was thought that the Indians who professed to guide Mr. Cowley through the driest part of the country, took him through the most swampy portion. The Indians now say that dry ridges exist, with intervening marshes, over and through which a cart track could be established without difficulty; but it is evident that the character of the country on this part of Lake Manitobah is not fitted for farming purposes. Isolated areas like Elm Point are doubtless to be found, but not sufficiently extensive to give to this region any value in an agricultural point of view.

We met an Indian in a canoe near Elm Point, and Whiteway, at my request, told him we were starving. I wished to ascertain the truth of the statement so often made respecting the liberality of these Indians in cases of necessity. The answer was a happy one; approaching our boat in his canoe, the Indian said, "Look, if you see anything to eat, take it." In his canoe were sixty fine white-fish and a few pike. I gave him some potatoes, tobacco, and tea, and accepted a dozen white-fish which he pressed us to take.

The shore continues low as far as Sandy Point; it is bounded by beaches fringed with fine aspen forests in the rear of marshes filled with rushes, which occupy part of every sheltered cove and bay open to the lake. We camped at Monkman's Point, where one of the family has a fishing station. They were catching their winter supply of white-fish. Monkman * pointed out a marsh in the rear of our camp, which he said was once dry ground, and afforded splendid pasturage for horses. This probably occurred during a period of low water. The marsh is separated from the lake by a gravelly beach; and a fall in the level of the lake, to the extent of two feet, would not only drain and dry this marsh, but many hundred square miles of marshy areas formed under similar circumstances and at the same period. Mr. Mackenzie, of Manitobah House, told me that former residents at that Post had seen the lake for a long period of time two feet lower than at present. In fact, before the floods of 1852 the lake was at its lowest level, and the swamps and marshes

* The brother of John Monkman, of Oak Point, a celebrated character at Selkirk Settlement. More will be said of this individual in a future chapter.

fringing its low north-eastern coast were then dry areas covered with rank grass. In the course of a few years this will again take place, and for a long period, perhaps, settlers may enjoy fine pasture lands, destined again to revert to an intermittent condition of swamp or marsh. Monkman informed me that many years since the Hudson's Bay Company had a breeding establishment near this Point, and he remembered the time when 120 horses were pastured in the neighbourhood of Swan Creek, about twelve miles from Oak Point.

On the 28th we passed through an immense expanse of reeds called Marshy Point, threading our way through an intricate channel in which large numbers of duck still lingered. About one o'clock we arrived at Oak Point, where we found John Monkman and a number of settlers from Red River catching their winter supply of white-fish in gill nets.

Lake Manitobah is 120 miles long by 24 broad in its widest part, from headland to headland; but if estimated from Oak Point to the mouth of White Mud River, on the west side, the breadth does not fall far short of thirty miles. The area of the lake is about 1900 square miles, and its approximate altitude above the sea 670 feet, or forty-two feet above Lake Winnipeg. It is remarkably shallow, so that in the parts sounded, which were sometimes twelve to fifteen miles broad, the depth never exceeded twenty-three feet; this occurred half way between Cherry Island and Sandy Point in the upper portion of the lake. In the two traverses between Manitobah Island and Cherry Island not more than twenty-one feet was recorded, while within four miles of the coast in the southern or larger portion of the lake, eighteen feet was the greatest depth found.

The effects of winds on the large surfaces of water exposed by the great lakes of the Winnipeg Basin, is very well seen at the Narrows near Manitobah Island, the Dog's Head (Lake Winnipeg), Waterhen River, and the mouths of the Winnipeg and Red Rivers. The currents produced by the pressure of the wind changing the levels of the lakes has probably exercised an important influence in connecting different parts of the same lake basins.

At the Narrows, Lake Manitobah, a northerly wind will cause a strong current to flow through the straits into the lower or southern half of the lake; while a south wind produces a corresponding effect in the northern portion, and perceptibly increases the volume of water in the Little Saskatchewan. At the Dog's Head the current sometimes approaches the force of a rapid when the wind blows from the north; the great depth of Lake Winnipeg at this point, which, I was assured by half-breeds and Indians who fish there during the winter, exceeds one hundred and twenty feet, is doubtless the result.

At first sight it appears strange that the limestone cliffs should not have been gradually broken away, and the communication between the upper and lower portion of Lake Winnipeg enlarged. But running water exercises comparatively little effect in excavating a deep channel through a rocky barrier, or in widening a water-course; ice, beyond all question, is the main instrument in abrading, denuding, and excavating. At the Dog's Head the ice has little force on account of the proximity of islands, either when acting with a thrust or bearing away masses of rock frozen to its substance. By far the greater portion of the ice formed on this part of the coast is so protected by the islands as to melt before it can be moved by winds with its rocky burdens to distant parts of the lake.

At Manitobah House I observed the water rise fully

eighteen inches before a storm. Canoes left in calm weather on a beach high and dry are not unfrequently washed away when a strong south or north wind sets in, and it often happens that, even before the approach of a change in the direction of the wind is indicated by clouds, the water of the lakes show by rising the operation of a distant pressure which has not yet manifested itself at the point of observation. The Indians and half-breeds in the fall of the year, when winds are variable, frequently notice the mouths of streams or rude registers, such as a stone set up by themselves on the beach, to see if any indications are afforded of a change in the wind, not appreciable by any other means.

In 1823 Mr. Keating, in his narrative of Major Long's Expedition to the sources of St. Peter's River, described the effects of winds on the waters of Lake Winnipeg taking place at the mouth of the Winnipeg River, as follows :—" A question which has been much discussed by travellers, is that of the supposed periodical rises in the lakes ; we do not propose to take part in the discussion at present, but we may state that we observed at Fort Alexander an appearance, such as has probably more than once been mistaken for the effect of a tide.

" On our arrival we pitched our tents upon a sort of wharf projecting into the river, and elevated about two feet above the level of the water. In the afternoon a very high wind blew from the lake and accumulated the waters in the bay, so as to cause them to overflow the wharf and oblige us to remove our tents. The next morning the waters had subsided to their former level."

The splendid prairies bordering on the southern shores of Lake Manitobah may be said to begin at Oak Point. Their boundary is an imaginary line extending south-easterly towards the Indian Settlement on Red River on the one hand, and to the old lake Ridge, where it is cut by White Mud River on the other; a distance in an air line of one hundred and ten miles. North of this line the country is in general marshy, full of reticulating lakes and low aspen-covered ridges.

The settlement at Oak Point contains about a dozen houses; their appearance does not give a stranger a favourable impression of the industry and energy of their occupants. No advantage appears to be taken of the splendid country by which they are surrounded; and with the exception of John Monkman, who at times is a marvel of energy injudiciously directed, they do not seem to have made any progress in improving their dwellings, or in enclosing a farm, since they first established themselves at Lake Manitobah. About ten miles in a south-westerly direction from Oak Point a number of French half-breeds have formed a settlement on the shores of the lake. They enjoy the advantage of having a resident Missionary (R. C.) among them.

On the 29th we made preparations for a journey on horseback to the Settlements, striking diagonally across the prairie region just described. The country in the neighbourhood of Oak Point is very attractive; its general level is about ten feet above the lake; it resembles in every respect the region about White Mud River. Our road, for a few miles, lay across a very rich and fertile tract, until an almost imperceptible ascent introduced us to a low gravelly ridge, upon which aspen woods grow in narrow strips; the forest preserves a uniform outline as far as the eye can reach, in a direction corresponding to the present form of Lake Manitobah, indicating, without glancing at the soil, the direction and extent of the subaqueous ridges, afterwards a low coast line, which were formed over the floor of Lake Manitobah at a higher level. Succeeding this low flat ridge is a broad plateau slightly undulating and studded with straggling clumps of young poplar and small oak, with willows in the shallow depressions. The soil becomes rich in vegetable mould again as we approach Shoal Lake, an extensive sheet of water, shallow, reedy, connected with numerous lakes lying to the north, and a favourite haunt of aquatic birds.

The south shore of Shoal Lake is particularly attractive. Ridges supporting heavy oak fringe the shore, and beautiful meadows, bordered with aspen and oak woods, reveal themselves in making a short traverse to the south. Although the shores of the lake are marshy, yet the oak ridges some few hundred yards south of it are high and dry. For a grazing establishment on the largest scale Shoal Lake is admirably fitted. Wild hay in any desirable quantity exists around its marshy shores, and in the beautiful prairies lying south of it timber of excellent quality for building purposes and fuel may be procured in abundance; in the spring and autumn the lake is covered with wild fowl of every variety. Shoal Lake is a favourite sporting ground of the gentlemen of Fort Garry and the half-breeds of the settlement. It is on the main road to Lake Manitobah, and is probably destined to become a place of some note as a grazing station in the course of time.

On the 30th October I set out with Whiteway in advance of the carts, in the hope of being able to reach the settlements before nightfall. We passed through an excellent prairie country studded with aspen groves, and occasionally relieved by a broad shallow ridge, probably of subaqueous origin, like those already described. The Big Ridge of the Assinniboine is not well defined where we descended it, about eight miles west of Stony Mountain. It appears to be divided into two portions, part expanding into an undulating tract of country a few hundred yards broad, the other preserving the outline and character of the Big Ridge, but named in consequence of its diminished altitude the Little Ridge. The level country at the base of either is everywhere beautiful, fertile, and admirably adapted for settlement. We descended the Little Ridge, a step of the Big Ridge, at about four in the afternoon, and in the distance could see the twin steeples of St. Boniface with their tinned roofs glancing brilliantly in the south-east about fifteen miles off. We then passed through the magnificent prairies lying between Stony Mountain and Red River, reaching the edge of the Big Swamp just before sunset, and arrived at our temporary quarters in the settlement half an hour after dark.

The country between Oak Point and Stony Mountain is not much inferior in point of fertility and fitness for settlement to the prairies of Red River and the Assinniboine bounded by the Big Ridge. In many parts no difference in the rank luxuriance of the grass on these prairies and those south of the Big Ridge could be distinguished, but the area of light or gravelly soil covered with short stunted grass is far greater, and thus diminishes the available extent of soil adapted for agriculture. It is doubtful whether this drawback is not counterbalanced by the proximity of the country north of the Big Ridge to the forest-covered tract between the great lakes, and to the haunts of vast numbers of wild fowl which breed on the borders of the small sheets of water so numerous in this region. This tract, south of the probable limit of the forest, may be regarded as well adapted for agriculture, the groves and strips of aspen and oak on the dry gravelly areas only serving to break a vast level expanse into a series of very attractive plains, apparently bounded by forests, which are found as the traveller penetrates them to be but narrow belts separating one beautiful prairie from another.

CHAP. XXVII.

WINTER JOURNEY WITH DOGS FROM FORT GARRY TO CROW WING.

Arrival of Lord Richard Grosvenor, Lord Frederick Cavendish, Mr. Henry Danby Seymour, M.P., and the Hon. Evelyn Ashley, at Fort Garry.—Buffalo Hunting.—Lord Grosvenor's Expedition to Fort Ellice.—Preparations for a Winter's Journey.—John Monkman.—Cline.—Daily Allowance of Dogs.—A Winter Road.—A Cariole.—A Sledge.—The Driver.—Making the Road.—Prospects of a Race to Crow Wing.—The Start.—Fort Pembina.—Mr. Mackenzie.—The Woods and Prairies in the Winter Season.—Temperature at Pembina.—A Camp in the Snow.—Preparations for the Night.—Mocassins.—The Morning Start.—Making a Cache in Pine River.—Dogs watching the Operation.—They return at Night to break open the Cache.—Terrible Fate of Mr. Mackenzie in Dec., 1859, frozen to Death in attempting to reach Pembina from Pine Creek.—Running across a Prairie with the Thermometer at 20° below Zero.—Appearance of the Party after the "Run."—Appearance of a Camp during the Night.—Watchfulness of the Dogs.—Catching and harnessing them in the Morning.—Treatment of Dogs by the Half-breeds.—Overturning a Cariole.—Traveling in a Snow Storm.—Preparing to Camp in a Snow Storm.—Dogs "lying close" after a Fall of Snow during the Night.—Sagacity of these Animals.—Red Lake.—News of Monkman's Party behind us.—The Roman Catholic Missionary frozen to Death two Days previous to our Arrival at Red Lake Mission.—Indians reading the History of the Missionary's Journey from his Tracks on the Ice.—Indians relating the History of his Journey.—Savage Mimicry.—The Rev. Laurenz Lautiger, the Roman Catholic Missionary.—The Height of Land.—Cass Lake.—Arrival of Monkman's Party at Midnight.—Leech Lake.—A Dance.—The last Night in the Woods.—The last Day's Run.—Pine Woods.—Morning.—A twenty-mile Gallop.—Crow Wing.

Upon our arrival at Selkirk Settlement subsequently to the exploration of Lakes Winnipeg and Manitobah, we heard that a party of English noblemen and gentlemen had reached Fort Garry, and were then preparing for a short trip into the prairies in search of buffalo. The aristocratic hunters consisted of Lord Richard Grosvenor, Lord Frederick Cavendish, Henry Danby Seymour, Esq., M.P., and the Honourable Evelyn Ashley. They were successful in meeting with and killing buffalo a short distance south of the boundary line, in the direction of Devil's Lake, and returned from their expedition in the middle of November.

The hardships and privations inseparable from prairie adventure at this late season of the year, with the thermometer sometimes as low as zero, were not sufficient to deter Lord Grosvenor from undertaking a much longer expedition to the west, after his return from hunting buffalo. He started on the 22nd November in company with James Mackay, one of the most distinguished and enterprising natives of Selkirk Settlement, for Fort Ellice, proposing subsequently to visit the Plain Crees on the Qu'appelle.

This adventurous journey on horseback at so late a period of the year, considering the slender outfit of clothing and provisions which Lord Grosvenor took with him, showed no ordinary courage and confidence in the possession of great physical endurance. A journey in dog carioles over the snow is comfort itself compared with riding in the face of a cutting wind when the thermometer is not far removed from zero.

Lord Cavendish, Mr. Seymour, and Mr. Ashley engaged John Monkman of Oak Point, Lake Manitobah, to organize their dog trains and make arrangements for their journey to Crow Wing on the Mississippi, proposing to return to civilized life as soon as sufficient snow fell to make the country passable for dog trains. John Monkman is the most noted runner in Selkirk Settlement; with his magnificent train of dogs, possibly the best in

FORT GARRY

Rupert's Land, he has accomplished the journey between Pembina and Fort Garry, a distance of sixty-eight miles, in seven hours and a half. The men Monkman engaged, six in number, were generally excellent runners, and provided with good dogs.

The preparations for the winter journey of both parties to St. Paul were rather of a formidable character even at Red River, requiring not less than sixteen carioles and sledges drawn by fifty dogs in all.

It being necessary that I should take with me a number of geological specimens, field apparatus, books, &c. &c., my party required nine carioles and sledges, and a corresponding number of men; each sledge or cariole on a long winter's journey requiring one man to manage it, although the same individual, with well-trained dogs, is competent to conduct two sledges on good roads and for short journeys, when time is not an object and food easily accessible.

I engaged a half-breed of the name of Cline, an excellent runner and a very willing attendant, to organize my trains.

The distance between Fort Garry and Crow Wing is about 400 miles by the winter road, and the only places where supplies can be obtained are at Pembina, Red Lake, and Cass Lake. The chief objection to travel with a large number of dogs is the difficulty of supplying them with food, nor can the several stations be always relied on to furnish the requisite quantity for an unexpected intrusion of many of these hungry animals. Each dog requires daily about two pounds of pemmican or three pounds of white-fish, so that the provisions for a train of carioles employing thirty dogs would involve the carriage of 600 lbs. of pemmican or 900 lbs. of white-fish for a ten days' journey. A train of three dogs will draw 300 lbs. forty miles a day for ten or twelve days in succession if well fed, and the road is tolerably good, over a level country. A winter road, it may be here remarked, is nothing more than a cariole or sledge track caused by the passage of this primitive kind of vehicle over the snow, and is liable to be obliterated by every fresh fall. A cariole is constructed of a very thin board, ten feet long and twelve or fourteen inches broad, turned up at one end in the form of half a circle, like the bow of an Ojibway canoe. To this board a high cradle, like the body of a small carriage, is attached, about eighteen inches from the end of the board or floor. The frame-work is covered with buffalo skin parchment, and painted or decorated according to taste. The inside is lined with a blanket or buffalo robe, and when the traveller is seated in his cariole, with outstretched legs, he is only separated from the snow by the thin plank which forms the floor. A sledge is nothing more than a thin board ten or twelve feet long, twelve inches broad, and turned up at one end. The baggage is attached to it by means of buffalo thongs, and two or three dogs are harnessed to this simple vehicle with the same materials. The dogs attached to a cariole are generally decorated with collars, from which beadwork and tassels are suspended together with a string of small bells. When a train is in motion the driver runs behind the cariole or sledge, guiding it by means of a loop fastened to each corner of the floor; when tired or anxious to ride he sits on the small box containing the traveler's baggage, which is fastened to the projecting floor in rear of the cariole, or else he stands on it if no box is attached. A winter road is uniformly of the breadth of the floor of the cariole, rarely exceeding fifteen inches, and of a depth proportioned to the quantity of snow which has fallen. In making a new road where the snow is deep, a half-breed walks on snow shoes some distance in front of the dogs, which follow his track with the utmost precision through all its windings; after four or five trains have passed, the road is generally considered to be sufficiently hard pressed to admit of the easy passage of the succeeding trains; hence, a great point is gained in dog traveling if a new road has recently been made by a party in advance. To make the road is regarded, indeed, as the chief difficulty in journeying with dogs.

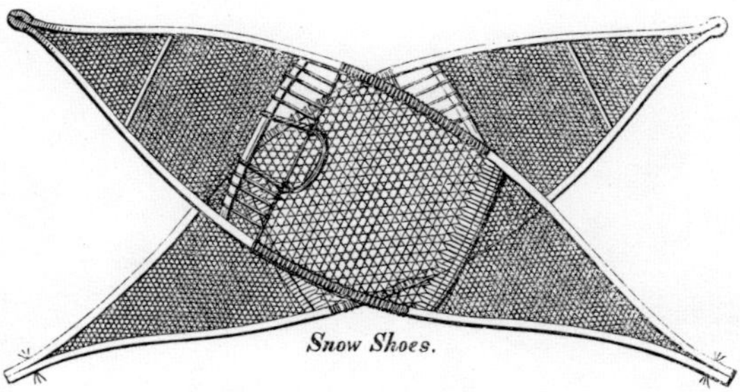

Snow Shoes.

From the hour it was known in Selkirk Settlement that the two parties would probably start nearly at the same time, and great feeling existing among the half-breeds respecting their endurance and the ease and speed with which their dogs could accomplish a long journey, a warm spirit of emulation arose between the men attached to each party, which rapidly communicated itself to their wives and friends. Cline told me he had heard confidentially that Monkman's plan was to give us the start for two days, and then, taking advantage of the road we should make through the untrodden wilderness, pass us triumphantly a day or two before we arrived at Crow Wing. It gradually became evident that the idea of a race from Fort Garry to Crow Wing communicated itself to the gentlemen of both parties, and indeed stimulated more or less all who were to make the journey. Cline and his men appeared extremely anxious that they should not be far behind their competitors, and even indulged the hope that if we got a fair start we might not be overtaken. When the subject of a race was first mooted to me by Cline, I regarded our chances as almost hopeless, considering the baggage with which we were encumbered. Afterwards it

Dog Carioles.

occurred to me that one chance lay in the greater ease with which my party, long inured to exposure and fatigue, would be able to sustain prolonged physical exertion; an hallucination, however, which subsequent experience of the physical capabilities of Lord Cavendish, Mr. Seymour, and Mr. Ashley, during their stay at Red River, served to dispel.

All my trains being ready, we started on Tuesday,

November 30th, at an early hour from Fort Garry, and took the east bank of Red River through the French settlements. Monkman's party expected to follow in the afternoon or early on the following morning. On Wednesday we reached Fort Pembina, and stayed the night with Mr. Mackenzie, the officer in charge of the Post, whose sad fate last December (described further on) is a melancholy proof of the danger attending traveling alone during the winters of this climate. The woods and prairies are then perfect deserts, Indians being at their winter quarters, birds far in the sunny south, and wild animals hybernating, or seeking food and shelter in the thickest parts of the swamps and forests. So complete is this desolation in the interior of many parts of Rupert's Land during the winter, that Mr. Christie, chief factor of the Hudson's Bay Company, when traveling from Athabasca to Red River in December 1859, did not meet with a single Indian throughout a long and dreary journey of 1400 miles.*

The thermometer at Pembina Fort indicated 22° below zero on the morning of December 2nd, when we left the Post. Having procured another train of two dogs at the small village of Pembina, two miles from the Hudson's Bay Post of that name, we struck across the prairie to the "first of the Two Creeks," where we camped. A camp is always made in "woods," if possible, for the sake of fuel and shelter. The first operation is to sweep the snow from the ground, and prepare a place for the fire and blankets. This is easily accomplished with snow shoes; and as soon as an area proportioned to the size of the party is exposed, a fire is made sufficiently long to admit of each man lying for the night with his feet towards it. No tent or

* *Nor'-Wester*, Fort Garry, Feb. 1860.

covering of any description beyond a blanket stretched on poles is admissible, as it would scarcely be possible to fold canvass in the morning, and time does not generally allow of the erection of a hut, nor are the materials always at hand. When pine or spruce is accessible, a very comfortable floor can be made from the boughs, but in the prairie country or on its borders these useful trees are rarely to be seen.

As soon as the fire is made and supper in course of preparation, the dogs are fed. After eating their allow-

Pembina.

ance for the twenty-four hours, the dogs seek for warm spots for themselves as near the fire as possible, or if the night be very cold, and any wind stirring, they partially bury themselves in the snow. As soon as supper is discussed, which with the half-breeds almost uniformly consisted of cold pemmican and tea, mocassins are taken off, dried if damp, and put on again; the fire is replenished, and one by one each man or two together cover themselves completely with their blankets or robes, and go to sleep. Mocassins are necessary in making a winter journey, leather boots or shoes would be too cold and probably become hard, neither do they admit of that freedom of circulation which makes a soft and pliable mocassin of dressed buffalo or moose skin so warm and comfortable. The feet rarely get wet in traveling in these regions; the intense cold preserves the snow perfectly dry, and it is only near a fire that moisture penetrates a mocassin during cold weather. During a thaw, a mocassin is wet through immediately, and the discomfort must be endured from camp to camp, but colds are unknown from this cause if exercise be persevered in. Under the mocassin, the half-breeds wear a square piece of flannel or blanket wrapped round the feet, to serve as a stocking. The gentlemen wore one pair of worsted stockings, a half boot of duffel, and buffalo skin mocassins, and no one complained of cold feet at any time.

The first thing on waking in the morning is to make up the fire and prepare for breakfast. We generally rose at five, and were ready to start by daylight; breakfast, repacking the bedding, catching and harnessing the dogs usually occupying two hours and a half.

On the following day we arrived at Pine River, where Cline made a cache of pemmican in the river, some distance from our track, to be taken up on his return. A hole was cut through the ice, then about fifteen inches thick, and a buffalo hide thong having been tied round the bag, and fastened to a stick, it was let down into the water, just below the ice, the stick being stretched across the orifice. Lumps of ice were then piled on the bag of pemmican, and water poured on them. The temperature of the air being at the time considerably below zero, the water froze the instant it touched the ice, and bound the masses together in one block. Fresh ice being added, and water poured on it, the hole soon became filled with a solid mass; the operation was continued until a mound one foot above the frozen surface of the river was accumulated over the cache. As I was returning with Cline to the place where we had left the carioles and dogs, we observed two of these sagacious animals who had been unharnessed by mistake, the driver supposing we were going to camp, quietly watching our proceedings from the bank of the river. Cline, with an exclamation of anger and surprise, remarked, "Now, if we don't take care those *sacrés chiens* will try to get at the pemmican to-night, and the rascals will tell the others; I know them of old, they served me that trick before; we'll tie them to night." The conscious dogs, with tails depressed, started at a gallop back to the carioles, when they saw Cline's threatening mien.

In order to avoid the breaking open of the cache by the dogs, which might be the source of great inconvenience, and perhaps suffering to the men on their return, we pushed on for several miles before we camped. Cline counted the dogs after supper, but neglected to tie the spies, having found that none had strayed, he thought they might be trusted, and, wrapping himself in his blanket, he soon went to sleep. Rising long before daylight, according to our custom, several dogs were soon ascertained to be missing, and not being found after a rapid search in the neighbourhood, Cline instructed two of the men to make a circle round the camp, and examine all tracks by torch-light. In a very short time one of them came back, stating that fresh dog tracks pointed in the direction of Pine River, where we had cached the pemmican. Three or four men instantly started back and found the missing dogs busily engaged in scratching at the cache. It was so thoroughly frozen and compactly made, that they had produced little impression on the small mound of ice, but no doubt time and perseverance would have enabled them to reach the supplies beneath.

Pine River crossing is the spot from which Mr. Mackenzie, who had so hospitably treated us at Fort Pembina, started on the morning of the 29th December 1859, on his ill-fated journey in search of assistance. He and some companions were escorting an engineer from George Town to Fort Garry, who was traveling thither to make alterations and repairs in the steamer Anson Northrup, then laid up for the winter near the Indian settlement. The party fell short of provisions, and Mr. Mackenzie pushed on in the hope of being able to send supplies from Pembina. After leaving his companions, he appears to have followed the trail for some distance, and at the approach of night to have lost his way. His beaten track showed that in order to keep himself from freezing, he had spent the night in running round in a circle. At the break of day he started again across the trackless waste, but in a direction considerably to the eastward of his proper course. A second day of fruitless wandering was followed by a night more dreary than the first. The third day's journey brought him near to Roseau Lake, far to the east of his destination; here his strength appears to have failed him, for having hung some shreds of his coat on a tree, to mark his last resting-place, he lay down beneath it, where his frozen body was found, with one hand on his heart and the other grasping a compass.*

On the day succeeding our camp near Pine River, we crossed a very bleak and desolate prairie about eight miles

* An account of this melancholy journey is given in the *Red River Nor'-Wester*, for January 14, 1860.

broad. Fortunately the wind was blowing at our backs, otherwise it would have been insupportable, the temperature of the air having fallen to 20° below zero. As it was almost impossible to endure the inaction involved by being tightly packed in a small cariole, notwithstanding a liberal supply of blankets and robes, we preferred to run after the dogs across that inhospitable prairie which, however beautiful it may be in summer, is an exposed and desolate wild in December. Mutual astonishment was expressed at the appearance presented by each individual after this long run. Our eyebrows, beards, moustaches, hair, and eyelashes were uniformly frosted; the moisture from the breath had formed icicles down our beards, which were firmly frozen to the hoods of our blanket coats; patches of frost-bites on our cheeks, nose, or ears demanded instant friction with snow; and the outside of the coat of each man, from the nape of the neck to the loins, was covered with a narrow sheet of hoar-frost, formed by the solidification of the insensible perspiration the moment it reached the outer air.

The appearance of a winter camp during the night, when men and dogs are buried in profound slumber, is very wild and savage. Throwing a few dry sticks into the fire to light up the scene, the silent, slumbering forms of the travelers are seen stretched in two parallel rows with their feet to the fire; between the men one, two, and sometimes three huge dogs have crept; some are lying on the legs of the half-breeds for the sake of warmth, others have found a snug berth close to the fire but in imminent danger of burning their fur, a few lie coiled outside of the circle half buried in the snow. The cold is so intense that their faces are white with frozen breath, and scarcely distinguishable. The fire, even when in full glow, has not power to melt the snow more than a few inches from it, without it is exposed to direct and prolonged radiation. Now and then a watchful dog raises his head, probably disturbed by some slight movement of the sleepers; he looks once round and buries his face again. Sometimes a dog will utter a low warning growl, when three or four other dogs, probably old stagers, will rouse themselves for an instant, listen and growl, generally all looking in one direction and snuffing the air. A half-breed sits up, looks at the dogs, observes their mien and actions, and after a moment's pause, uttering the word "wolves," he quickly coils himself under his blanket again.

The most disagreeable part of the daily routine of a long winter's journey is the catching and harnessing of the dogs. Some of these animals at the beginning of winter, when fresh at their work for the season, are exceedingly restive under coercion of any description, and not unfrequently snap at their masters, who invariably arm themselves with very strong mittens of buffalo or deer hide when harnessing a savage and powerful animal. They require long-continued and most severe punishment to make them obedient to the word of command. The treatment to which many of the poor beasts are subjected would give them a fair claim to the protection of a law against cruelty to animals. The faces of some of our dogs were dreadfully disfigured by the blows which their unfeeling and thoughtless masters inflicted on them. An Esquimaux whip is the instrument which every driver should be compelled to use, but the half-breeds trust to sticks and stones, or any object within reach on the road, which is picked up as they pass and thrown at the dogs. It is painful to witness the sudden start of terror with which each animal, looking over his shoulder as he trots along, watches the mien and motions of the driver as he poises the stick, which he knows how to throw with certain dexterity at the terrified animals. All the dogs give a simultaneous jump on one side as the missile flies past them, when directed at the leader of the train; and not unfrequently would the cariole be overturned if it were not for the strength and skill of the driver in holding the loop with which he steers it. When this occurrence takes place and the dogs are at full speed, the only plan left for the helpless traveller is to draw his arms close to his sides, and wait until the cariole is righted by the driver; but any attempt to right the cariole by putting out an arm is a dangerous operation, which might occasion a broken limb. In descending steep hills, it is always advisable to walk or run, which all would prefer for the sake of exercise, except when the road is very good, and the trains can proceed for many miles at a gallop without fatigue.

A heavy snow storm is a serious matter in the prairie. It is then absolutely necessary for all the trains to keep close together; the drifting snow soon obliterates the road; and, although the dogs, by means of their sensitive noses, will follow the tracks of the leading cariole even when completely hidden from view by a light fall, yet when drifts accumulate they are at fault.

Preparing to camp in a snow storm is not an agreeable operation, or suggestive of that comfort and safety which a camp almost always presents. When the fire is well lighted, supper cooked and eaten, and the party "turned in," then it does not matter much how heavily it snows, the trouble being reserved for the following day. After a heavy fall during the night, men, dogs, carioles and sledges are all covered with a thick mantle of pure white; a sudden shout from the guide enlivens many of the apparently lifeless forms, recognized only by their outline; but some of the sagacious dogs take advantage of the concealment afforded by the snow, and, quite neglectful of the whistles and shouts of their masters, "lie close." We were detained for more than three hours on one occasion after a heavy snow storm, by some of the dogs preserving perfect silence and a motionless position under their covering of snow, within thirty yards of our camp fire. They were found by dint of walking systematically round the camp fire in a continually enlarging circle, the half-breeds being quite aware of the advantage which these cunning animals would take of their accidental concealment. A loud shout every now and then announced that a searcher had stumbled over a truant, whose depressed mien and conscious look showed how well he knew that he had been the cause of anxiety and trouble.

On the 8th of December we arrived at Red Lake, and had a splendid gallop of twelve miles across the ice from the mouth of Red Lake River to the Ojibway village and Missionary station. An Indian from Pembina caught us just as we were about to commence the traverse, and brought the information that Monkman's party had left Fort Garry on the day following our departure, and were steadily pursuing the road we had made. The Indian thought they would catch us in a day or two. Soon after our arrival at the Red Lake Mission we learned that the Roman Catholic Missionary had been frozen to death two days previously, in an attempt to cross the ice during a snow storm, from a promontory about two miles away from the Mission. He had been visiting a camp of Ojibways, who warned him of the perils of a return across the ice during the storm, and invited him to pass the night in their wigwams; but the missionary thought that he would not incur any danger of freezing during so short a traverse, although the thermometer indicated a temperature of 25° below zero, at the opposite station.

He was frozen within two hundred yards of the Mission House, near to which were a number of log houses, tenanted at the time by half-breeds and Indians, When the body was found on the following morning, a number of Indians set themselves to trace his steps from the Ojibway camp across the ice, a difficult undertaking, in consequence of the high wind which was blowing at the time having, to an inexperienced eye obliterated all traces of his steps. With astonishing accuracy these wild men read the brief history of his journey, and related the incidents to me as we stood on the banks of Red Lake, with the Ojibway village and the course of the unfortunate missionary in view. "There," said my dusky informant, pointing to the ice not more than half a mile from the houses, "there he first turned his back to the wind, and there he knelt to pray," the Indian suiting the action to the word, and kneeling in the attitude which the track showed the missionary had assumed. Now he faced the wind and ran against the blinding snow and pitiless storm; here he turned his back again; there his tracks showed how he slipped and fell, and once again where he knelt to pray. The marks of his fingers were seen on the crust of snow lying in frozen patches on the ice. Once more he fell, rose again, knelt for a while, and made a last effort to push against the storm. They came at length to where he had fallen for the last time, and subsequently knelt with his hands on the ice, his head touching the snow. He was found with hands clasped in the attitude of prayer, his head bent upon his breast. The barking dogs at the Mission must have been aware that he was approaching, notwithstanding the gloom of evening and the drifting snow, for they bayed fiercely in the direction he was coming about the time he was supposed to have fallen. The half-breeds heard the dogs and looked out in expectation of seeing the missionary approach, but as the dogs soon ceased to bark they thought it was a false alarm, and did not go to meet and assist him.

It was painfully interesting to watch the Indians relate the narrative of this short but terrible journey from the information they had gathered on the almost trackless ice and snow. The imitation of the actions and motions of the poor missionary, his attitude of prayer, his drooping head touching the cold ice, his backward wanderings, were all so faithfully represented, so true to nature, that the reality seemed to be occurring before me, rather than the solemn mimicry of a savage.

After the Indian, who was most active in impersonating the missionary, had finished his mournful tale, he quietly took a lighted pipe from one of his companions standing by, and drawing his blanket over his head seated himself upon the prostrate trunk of a tree, and without any expression of feeling covertly glanced in my face to see the effect of his narrative; and when I asked him through the half-breed interpreter where the body was lying, he coldly pointed with one finger to a log-hut close by, without withdrawing the pipe from his mouth or showing any further interest in the matter.

The name of the missionary was Laurenz Lautiger, from Krainburg in Carniola; he had been placed by the half-breeds in a rough coffin made from half-a-dozen pine boards, and, as he lay robed in his priestly vestments, calm, and without any trace of suffering, it seemed almost impossible to realize that he had just met with death in such a terrible form.

When we arrived at Crow Wing, a few days after this sad occurrence, I went to see his brother missionary stationed there, and related to him all the known circumstances of his death as I had received them from the half-breeds and Indians of Red Lake, and then learned that the Rev. Laurenz Lautiger was an Austrian, a man well skilled in many languages, pious, devoted and most enthusiastic in his endeavours to christianize the Ojibways of Red Lake. He had not been long in the country or he would have had the experience necessary to guide him on such a night as that in which he met with his untimely fate. At the hour of his death we must have been encamped on Red Lake River, about fifty miles from the station, having made an early camp in consequence of the piercing cold. My thermometer in the woods showed twenty-two degrees below zero just before I rolled myself in my buffalo robe for the night; but as we had selected close woods for our night's resting-place, it is not improbable that the temperature on the exposed ice of Red Lake was as low as thirty below zero.

Our course from Red Lake lay through the woods bordering Red Lake River. We traversed three very picturesque sheets of frozen water, and then arrived at Opashkwa Lake, which lies at the foot of the dividing ridge constituting the height of land, the waters on one side flowing into Hudson's Bay, on the other into the Gulf of Mexico. After passing this natural boundary we crossed Turtle Lake, and numerous dilatations of Turtle River abounding in wild rice, and camped on the evening of the 10th December close to the borders of Cass Lake, and near a cluster of Indian wigwams. Shortly after midnight our dogs began to bark furiously, and the Indian camp seemed suddenly to receive an unusual influx of visitors. Cline came to me and whispered, "Monkman's come." At daybreak on the following morning I recognized Monkman's dogs fraternizing with Cline's, and a short stroll to the Indian camp satisfied me that his party had all arrived in good condition, and were distributed among the different wigwams. I found Lord Cavendish and Mr. Ashley in one tent, and Mr. Seymour in another, as comfortably "fixed" as it was possible to expect or wish for under the circumstances incident to a midnight journey from the summit of the Mississippi slope, with the thermometer at zero.

A mission was established at Cass Lake in 1844, and the condition of the Indians is reported to have been much ameliorated by the exertions of the missionaries. The Indian village is situated on both sides of Turtle River, and contains several log cabins erected by the natives. No spirituous liquor is permitted to be introduced into this part of the Indian Reserves in Minnesota. The prosperity of the Indians, and their gradual approach to civilization, is probably largely due to this prohibition.

Starting several hours in advance of Monkman, we arrived early in the afternoon at the Indian agency and village on the south shore of Leech Lake, and were very cordially received by the agent. The other travelers came up with us before nightfall, and the half-breeds of the united parties decided upon having a dance. Fiddles were soon procured, a capacious store was cleared of goods and packages, and the female population of the agency and village, which included some very passable half-breed girls and "wives," having promptly assembled, a series of lively dances commenced, which were kept up until midnight. In the absence of whisky, that curse of savage and half-civilized life, strict propriety of demeanour was maintained throughout, although there was no lack of merriment, joke, and song. The scene was rendered more striking and characteristic of the wild life of these remote woods, by the presence of some Indians who were attracted from a neighbouring camp by the unusual sounds of music and dancing.

Monkman started a few minutes before us on the following morning, and we had the advantage of the road he made, which, however, was not much gain, as a good track had already been formed between Leech Lake and Crow Wing. After a fair day's journey through magnificent pine forests, both parties camped close together, about forty-four miles from our destination.

Cline began to pluck up his courage, which had received a check when he found Monkman pushing on until midnight in order to overtake us at Cass Lake. If we had not been compelled to wait for the heavily laden sledges we should have made much longer daily journeys, and no doubt Monkman would have traveled more rapidly. Fifty miles a day with light carioles and good dogs, would not have been out of the way. On the night of the 12th we slept for the last time in the woods, and both travelers and half-breeds looked forward with some anxiety to the race on the morrow. Monkman and Cline had agreed that both parties should start together in the order in which they camped, and our fire was made about one hundred yards in advance of Monkman's.

We rose at three and started at five; the thermometer showed a few degrees below zero, but the air was still, and the almost illimitable pine forest completely hushed, not the slightest breath disturbing the tall and light tops of the trees. We ran on in silence for half an hour, the hungry dogs themselves appeared to be sensible that their long journey was drawing to a close, for scarcely a note of encouragement or reproval was uttered by the half-breeds. My cariole led the way, the others of my party following close in the rear. Some forty minutes after we had started, the dogs being thoroughly warm at their work, we heard a yelp far behind us. Cline whispered to me, "Monkman's passing them;" and during each succeeding five minutes yelp after yelp announced that the other trains were being passed, until in little more than an hour from our start I heard Monkman's well-known voice close behind me. Without speaking a word he and the cariole he was driving passed mine. A thrill of excitement ran through me when I felt the warm breath of his powerful dogs beat upon my face, as the noble creatures swiftly trotted past. Neither dogs nor men, nor the muffled-up inmates of the carioles made any sign. One after the other Mr. Seymour, Lord Cavendish, and Mr. Ashley, flitted noiselessly by. Cline whispered again, "They'll make a good road, my dogs will not be far behind."

It is impossible to describe the feelings which this rapid, silent, gliding through those vast pine forests inspired. Morning dawned slowly, but the gloom of the forest seemed to grow more intense as I occasionally caught glimpses of the brightening sky above. The sun rose without a cloud, gilding the tops of tall trees with an indescribable lustre, beautifully reflected by the snow wherever the golden light penetrated. After a run of twenty-six miles Cline came up with Monkman a few minutes after he had stopped for dinner. Mr. Dickinson followed close behind me, and in another hour both parties were together again. The next run was to bring us to Crow Wing, between nineteen and twenty miles distant. Starting in the order in which we arrived at the camping ground, we hurried at a rapid gallop down the Mississippi slope; and here the race began in earnest. The road was in excellent condition, the day bright and cold, the dogs eager and hungry, and the men and travelers in good condition and in excellent spirits. After a splendid gallop of twenty miles we entered Crow Wing in the following order and close together :—Lord Cavendish first, Mr. Seymour second, Mr. Hind third, Hon. Mr. Ashley fourth, Mr. Dickinson fifth, Mr. Fleming sixth, the rest nowhere.

CHAP. XXVIII.

INDIAN WEALTH. — THE BUFFALO. — THE HORSE AND THE DOG.

The Bison or Buffalo.—Its Value.—Two kinds of Buffalo reported to exist by Half-breeds.—The Plain Buffalo and the Wood Buffalo.—The Lithuanian Bison.—Characters of.—Former range of the Buffalo.—Modern Range of.—The Red River Bands.—The Saskatchewan Bands.—Wintering Quarters of the north-western Bands of Buffalo.—Summer Ranges.—Systematic Migration of.—Buffalo Hunt.—Census of Red River Half-breed Hunt.—Blind Buffalo.—Crossing of Buffalo with domesticated Cattle.—Character of mixed Breeds.—The Horse.—Training of Horses.—Docility of.—Illustrations.—Attachment of Indians to their Horses.—Hopplings.—Smokes.—The Dog.—Its Uses.—The Midnight Howl.—Dog Feasts.—Dogs at the Hudson's Bay Posts.—Voracity of.—Cross with the Wolf.—Sacrifice of Dogs.

The flesh of the bison or buffalo either in the form of "fresh meat," "dried meat," or "pemmican" has been the main stay of the fur trade in Rupert's Land for more than three quarters of a century. Without the buffalo it would have been impossible to have supported the large establishments on the Saskatchewan, or to have made the long voyages from the Rocky Mountains to Hudson's Bay, much less to have carried on that vast system of internal communication which employs upwards of two hundred boats, averaging four tons each, and manned by more than one thousand two hundred voyageurs.

It is, however, to the wandering and savage tribes of the great prairies that the buffalo is all-important. This huge and unwieldy animal, together with the horse and the dog, occupy the same relation to Prairie Indians that domesticated animals and the productions of the farm and the forest bear to civilized races. During the greater part the year the Prairie Indians follow the buffalo, and not only subsist upon its flesh, but from its skin and sinews they make their tents, clothing, saddles, bow-strings and dog harness. The hide cut into strips serves them for cordage, the sinews split into threads for twine. The dried dung is often their only fuel for weeks together on the treeless plains between the Assinniboine and the Grand Coteau, and on the South Branch of the Saskatchewan. Dried meat, pemmican, marrow, soft fat, sinews, dressed skins and robes, all from the buffalo, form their articles of commerce, in exchange for which they demand tea, which is now becoming a most coveted luxury, tobacco, powder and shot, and if possible, rum. It may truly be said that they exist on the buffalo, and their knowledge of the habits of this animal is consequently essential to their preservation.

The existence of two kinds of buffalo is firmly believed by many hunters at Red River; they are stated to be the prairie buffalo and the buffalo of the woods. Many old hunters with whom I have conversed on this subject, aver that the so-called wood buffalo is a distinct species, and although they are not able to offer scientific proofs, yet the difference in size, colour, hair, and horns, are enumerated as the evidence upon which they base their statement. Men from their youth familiar with these animals in the Great Plains, and the varieties which are frequently met with in large herds still cling to this opinion. The buffalo of the plains are not always of the dark and rich bright brown which forms their characteristic colour. They are sometimes seen from white to almost black, and a grey buffalo is not at all uncommon.

Buffalo emasculated by wolves, are often found in the prairies, where they grow to an immense size; the skin

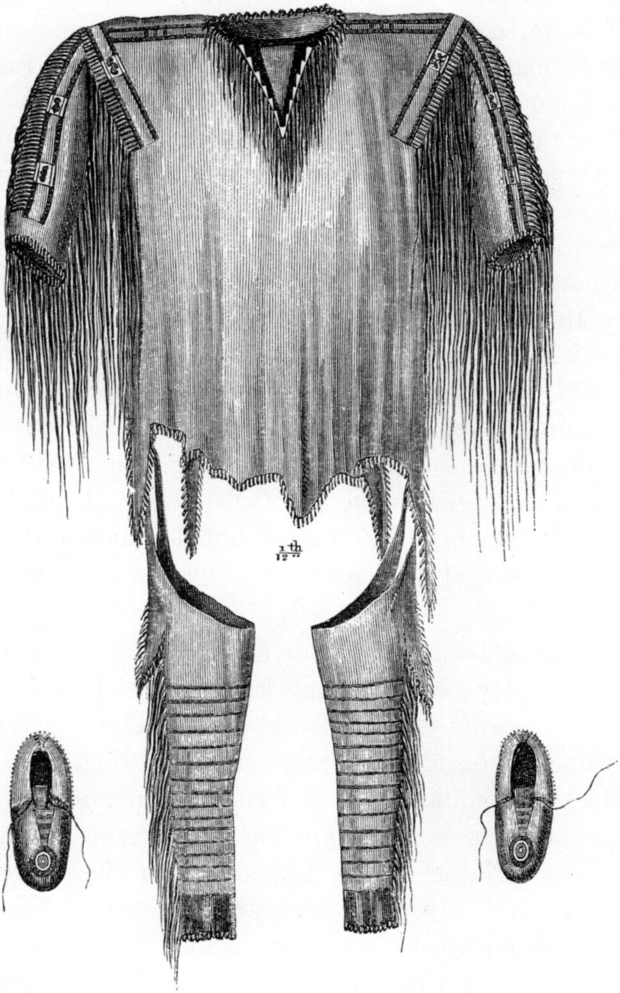

Sioux Dress and Mocassins..

of the buffalo ox is recognized by the shortness of the wool and by its large dimensions. The skin of the so-called wood buffalo, is much larger than that of the common animal, the hair is very short, mane or hair about the neck, short and soft, and altogether destitute of curl, which is the common feature in the hair or wool of the prairie animal. Two skins of the so-called wood buffalo, which I saw at Selkirk Settlement, bore a very close resemblance to the skin of the Lithuanian bison, judging from the specimens of that species which I have since had an opportunity of seeing in the British Museum.

The wood buffalo is stated to be very scarce, and only found north of the Saskatchewan, and on the flanks of the Rocky Mountains. It never ventures into the open plains; the prairie buffalo, on the contrary, generally avoids the woods in summer and keeps to the open country; but in winter they are frequently found in the woods of the Little Souris, the Saskatchewan, the Touchwood Hills, and the aspen groves on the Qu'appelle. There is no doubt that formerly the prairie buffalo ranged through open woods almost as much as he now does through the prairies.

Great Slave Lake is the northern limit of the buffalo, and the country between that large body of water and the Saskatchewan is partially wooded. The buffalo are now found in considerable numbers on the east flank of the Rocky Mountains. The former limits of the wanderings of these animals are carefully recorded in the narrative of Major Long's Expedition to the sources of the St. Peter's River. It appears that the buffalo was formerly found throughout the whole territory of the United States, with the exception of that part which lies east of Hudson's River and Lake Champlain, and of narrow strips of coast on the Atlantic and the Gulf of Mexico. These were swampy, and had probably low thick woods. That it did not exist on the Atlantic coast is rendered probable from the circumstance that all the early writers whom Mr. Colhoun the astronomer to Major Long's Expedition consulted on the subject, do not mention the buffalo as existing there, but further back. There can be no doubt that the animal approached the Gulf of Mexico, near the Bay of St. Bernard, for Alvar Nunez, about the year 1535, saw them not far from the coast, and Jontel, one hundred and fifty years afterwards, saw them at the Bay of St. Bernard. It is probable that this bay is the lowest point of latitude at which this animal has been found east of the Rocky Mountains.

Its existence on the west side of the Rocky Mountains is well ascertained, although it was not seen by the early explorers of that region.

At present it is scarcely seen east of the Mississippi, and south of the St. Lawrence. Governor Cass's party found, in 1819, buffaloes on the east side of the Mississippi, above the falls of St. Anthony. Every year this animal's rovings are restricted. In 1822, the limit of its wanderings down the St. Peter was Great Swan Lake.*

The ranges of the buffalo in the north-western prairies are still maintained with great exactness, and old hunters, if the plains have not been burnt, can generally tell the direction in which herds will be found at certain seasons of the year. If the plains have been extensively burnt in the autumn, the search for the main herds during the following spring must depend on the course the fires have taken.

Red River hunters recognize two grand divisions of buffalo, those of the Grand Coteau and Red River, and those of the Saskatchewan. Other ranges of immense herds exist beyond the Missouri towards the south, as far

* Major Long's Expedition to the Sources of the St. Peter's River.

as Texas and Mexico. The north-western buffalo ranges are as follow. The bands belonging to the Red River Range winter on the Little Souris, and south-easterly towards and beyond Devil's Lake, and thence on to Red River and the Shayenne. Here too, they are found in the spring. Their course then lies west towards the Grand Coteau de Missouri, until the month of June, when they turn north, and revisit the Little Souris from the west, winding round the west flank of Turtle Mountain to Devil's Lake, and by the main river (Red River), to the Shayenne again. In the memory of many Red River hunters, the buffalo were accustomed to visit the prairies of the Assinniboine as far north as Lake Manitobah, where in fact their skulls and bones are now to be seen; their skulls are also seen on the east side of the Red River of the north, in Minnesota, but the living animal is very rarely to be met with. A few years ago they were accustomed to pass on the east side of Turtle Mountain through the Blue Hills of the Souris, but of late years their wanderings in this direction have ceased; experience teaching them that their enemies, the half-breeds, have approached too near their haunts in that direction.

The country about the west side of Turtle Mountain in June 1858 was scored with their tracks at one of the crossing places on the Little Souris, as if deep parallel ruts had been artificially cut down the hill-sides. These ruts, often one foot deep and sixteen inches broad, would converge from the prairie for many miles to a favourite crossing or drinking place; and they are often seen in regions in which the buffalo is no longer a visitor.

The great western herds winter between the south and the north branches of the Saskatchewan, south of the Touchwood Hills, and beyond the north Saskatchewan in the valley of the Athabaska; they cross the South Branch in June and July, visit the prairies on the south side of the Touchwood Hill range, and cross the Qu'appelle valley anywhere between the Elbow of the South Branch and a few miles west of Fort Ellice on the Assinniboine. They then strike for the Grand Coteau de Missouri, and their eastern flank often approaches the Red River herds coming north from the Grand Coteau. They then proceed across the Missouri up the Yellow Stone, and return to the Saskatchewan and Athabaska as winter approaches, by the flanks of the Rocky Mountains. We saw many small herds, belonging to the western bands, cross the Qu'appelle valley, and proceed in single file towards the Grand Coteau in July 1858. The eastern bands, which we had expected to find on the Little Souris, were on the main river (Red River is so termed by the half-breeds hunting in this quarter). They had proceeded early thither, far to the south of their usual track, in consequence of the devastating fires which swept the plains from the Rocky Mountains to Red River in the autumn of 1857. We met bulls all moving south, when approaching Fort Ellice; they had come from their winter quarters near the Touchwood Hill range. As a general rule the Saskatchewan bands of buffalo go north during the autumn, and south during the summer. The Little Souris and main river bands, go north-west in summer and south-east in autumn. It is almost needless to remark again that fires interfere with this systematic migration, but there are no other impediments which will divert the buffalo from their course. The half-breeds state that no slaughter by large parties of hunters or Indians can turn large herds from the general direction they have taken when on the march; want of food is alone able to make them deviate from the course they have taken. The approach of numerous herds can be recognised by a low rumbling sound they occasion, if the weather be calm, fully twenty miles before they arrive, this warning is best perceived by applying the ear to a badger hole. During the rutting season they can be heard bellowing for a great distance on a still night. When we arrived at the Sandy Hills on the South Branch, the Crees, on being asked if the buffalo were numerous near at hand, answered, "listen to-night and you will hear them."

The summer and fall buffalo hunts are the grand events of the year to the Red River settlers, in fact the chief dependence for a livelihood of the greater part of the population. The start is usually made from the settlements about the 15th of June for the summer hunt, the hunters remaining in the prairie until the 20th August or 1st of September. One division (the White Horse Plain) goes by the Assinniboine River to the "rapids crossing place," and then proceed in a south-westerly direction. The other, or Red River division, pass on to Pembina, and then take a southerly direction. The two divisions sometimes meet, but not intentionally. Mr. Flett in 1849 took a census of the White Horse Plain division near the Chiefs' Mountain, not far from the Shayenne River, Dacotah Territory, and enumerated 603 carts, 700 half-breeds, 200 Indians, 600 horses, 200 oxen, 400 dogs and one cat.

Mr. Ross * gives the following census of the number of carts assembled in camp for the buffalo hunt at five different periods :—

In 1820.	Number of carts	assembled for	the first trip	540
In 1825	,,	,,	,,	680
In 1830	,,	,,	,,	820
In 1835	,,	,,	,,	970
In 1840	,,	,,	,,	1210

* The Red River Settlement, its rise, progress, and present state. London. 1856.

After the start from the settlement has been well made, and all stragglers or tardy hunters have arrived, a great council is held, and a president elected. A number of captains are nominated by the president and people jointly. The captains then proceed to appoint their own policemen, the number assigned to each not exceeding ten. Their duty is to see that the laws of the hunt are strictly carried out. In 1849, if a man ran a buffalo without permission before the general hunt began, his saddle and bridle were cut to pieces, for the first offence; for the second offence of the same description his clothes were cut off his back. At the present day these punishments are changed to a fine of twenty shillings for the first offence. No gun is permitted to be fired when in the buffalo country before the "race" begins. A priest sometimes goes with the hunt, and mass is then celebrated in the open prairies. At night the carts are placed in the form of a circle with the horses and cattle inside the ring, and it is the duty of the captains and their policemen to see that this is rightly done. All laws are proclaimed in camp, and relate to the hunt alone. All camping orders are given by signal, a flag being carried by the guides, who are appointed by election. Each guide has his turn of one day, and no man can pass a guide on duty without subjecting himself to a fine of five shillings. No hunter can leave the camp to return home without permission, and no one is permitted to stir until any animal or property of value, supposed to be lost, is recovered. The policemen, at the order of the captains, can seize any cart at night-fall and place it where they choose for the public safety, but on the following morning they are compelled to bring it back to the spot from which they moved it the evening previous. This power is very necessary in order that the horses may not be stampeded by night attacks of the Sioux or other Indian tribes at war with the half-breeds. A heavy fine is imposed in case of neglect in extinguishing fires when the camp is broken up in the morning. In sight of buffalo, all the hunters are drawn up in line, the president, captains, and police being a few yards in advance, restraining the impatient hunters. Not yet, not yet, is the subdued whisper of the president; the approach to the herd is cautiously made. Now! the president exclaims, and as the word leaves his lips the charge is made, and in a few minutes the excited half-breeds are among the bewildered buffalo.

Blind buffalo are frequently found accompanying herds, and sometimes they are met with alone. Their eyes have been destroyed by prairie fires; but their quickened sense of hearing and smell, and their increased alertness enable them to guard against danger, and makes it more difficult to approach them in quiet weather than those possessing sight. The hunters think that blind buffalo frequently give the alarm when they are stealthily approaching a herd in an undulating country. When galloping over stony ground blind buffalo frequently fall, but when quietly feeding they avoid the stones and boulders with wonderful skill. The domestication of the buffalo is a subject of much interest to the future population of Red River, and the following information on that subject may be implicitly relied on.

Humboldt* says that Albert Gallatin, who, before he appeared in Europe as a distinguished diplomatist, had obtained by personal inspection great knowledge of the uncultivated part of the United States, assures us that "the mixed breed was quite common fifty years ago in some of the north-western counties of Virginia; and the

* Aspects of Nature.

cows, the issue of that mixture, propagated like all others." "The favourite food of the buffalo is *Tripsacum dactyloides* (buffalo grass) and an undescribed species of clover nearly allied to *Trifolium repens*, and designated by Barton as *Trifolium bisonicum*. According to the statement of Gomara, there was still living in the north-west of Mexico, in latitude 40°, an Indian tribe whose principal riches consisted in herds of tame bisons or buffalo. But notwithstanding the possibility of taming the bison, notwithstanding the quantity of milk it yields, and notwithstanding the herds of lamas in the Cordilleras of Peru, no pastoral life or pastoral people were found when America was discovered, and there is no historical evidence of this intermediate stage in the life of nations ever having existed there."

In a description of domesticated herds of buffalo, and the results of crossing with the common cow, from the Patent Office Reports, it is stated that the mixed breeds are of various colours; striped with black on a grey ground, like the zebra; some others brindled red; some pure red, with white faces: and others red, without any markings of white. The mixed bloods have not only produced from the tame and buffalo bull, but it is known that the half-bloods reproduce, viz. those that were the product of the common cow and wild buffalo bull. At the first settlement of the country, cows that were considered the best for milking were the half-blood down to the quarter, and even eighth, of the buffalo blood. But the writer's experiments have not satisfied him that the half buffalo bull will produce again. That the half-breed heifer will be productive from either race, he has tested beyond the possibility of doubt.

"The domesticated buffalo retains the same haughty bearing that distinguishes him in his natural state. He will, however, feed or fatten on whatever suits the tame cow, and requires about the same amount of food. I have never milked either the full-blood or mixed breed, but have no doubt they might be made good milkers, although their bags or udders are less than those of the common cow; yet, from the strength of the calf, the dam must yield as much, or even more milk, than the common cow."*

Next to the buffalo the horse is the mainstay of the prairie Indians. Good horses are not very common among the Crees; they are, however, very intelligent and well trained. A good buffalo runner is invaluable to them, for although it does not require a fast horse to catch a bull, the cows, possessing greater speed, often outstrip them. A good Indian horse possesses some excellent characteristics, the result of training, which it may be interesting to enumerate, for the purpose of exhibiting how admirably this animal serves his rude and savage masters. When galloping after a buffalo, an Indian horse watches the animal as intently as his rider, always swerving when he observes the buffalo's tail begin to vibrate, and breaking into short gallop at his utmost speed when he sees the tail erect, a sure indication of an immediate charge. The rider may with safety entrust himself to his horse if mounted on a trained buffalo runner; he will be carried within three yards of the flanks of the animal, and safely withdrawn when danger is threatened. If the horse stumbles and throws his rider, the sagacious animal stops instantly and waits for him to mount again. A happy instance happened to myself when riding a fiery grey mare an Ojibway Indian lent me to gallop from his tent to Manitobah House, a distance of ten miles. "She is my favourite buffalo

* U. S. Patent Office Report.

runner, said the Indian, and will not need the thong." She ran away with me, however, as soon as we reached a grassy opening about a mile across, and in the midst of her gallop the belly band broke, and the little Indian saddle slipping round, I was thrown at once on the soft turf. The mare stopped immediately, turned round and stood by my side, waiting until I had risen and adjusted the saddle. As soon as I mounted she started off again, as if my sudden and unexpected descent had been intentional. At another time, when driving a small cariole over the frozen waters of Red River, the horse, an Indian one, not being roughshod, slipped and fell, but without an effort to rise remained perfectly quiet until I had loosened the harness, when he scrambled up, gained a rough portion of ice, and quietly waited to be harnessed afresh.

Indian horses are excellent watchers by night; our half-breeds were accustomed to note with care the aspect of the horses before retiring to rest; if they showed the least signs of uneasiness, such as staring about them instead of feeding quietly, or, when feeding with the "bite" in their mouth, stopping to listen, or snuffing the air, or approaching the fires when the flies where not troublesome, they would look for the cause and sometimes set watchers. When during the night, however dark, the horses suddenly approached the carts, the half-breeds would go to them, caress them, and watch the direction in which they fed or looked, knowing that their heads would be turned towards the danger, whether of Indians, or bears, or wolves.

One more instance will suffice to show the docility and training of Indian horses. I was riding a small horse which we had procured from the Crees on the Qu'appelle, in company with a Blackfoot half-breed, some distance before the carts, in the valley of Long Creek.* As we ascended a low hill we saw a bear 250 yards before us. My companion could speak but few words of English, so with signs he motioned me to dismount, and, having satisfied himself that the horses saw the bear, he led them a few yards aside behind a clump of willows, and tying their bridles together he patted them on the neck and pointed to the bear, caressed them again, and afterwards motioned me to follow him. The horses, with pricked ears, followed with their eyes every movement of the bear now slowly moving from us, but occasionally stopping to crop the twigs of willow. We crawled to leeward, and got within seventy yards of the bear, he then perceived us, I fired and sent a ball through his lungs. We waited to see if he would rise again, but finding that he lay struggling on his back, we approached and despatched him. On looking round for the horses they were seen standing in the same place intently watching us. My companion called them, they came slowly up and stopped within forty yards, eyeing the bear all the time. Finding that we approached it and handled it, they began to feed, evidently being satisfied that it was harmless.

Prairie Indians become very much attached to their horses, if they succeed in getting possession of a valuable animal. They often keep him in a tent when in the neighbourhood of an enemy's country or among noted thieves of their own tribe. During the daytime, when the camp is well supplied with meat and the buffalo are near, they tether him in the prairie, and indolently stretching themselves at full length on the grass, patiently watch him feed, removing the stake to a fresh spot as soon as he has cropped the best portion of the area limited by his tether. At night, when it was not thought necessary to tether our horses, we always hoppled them,

* Long Creek flows into the Main Saskatchewan, near Fort à la Corne.

that is tied their fore-feet together with dressed buffalo hide. Iron hopplings are in great request among half-breeds on their hunting expeditions. They can then more safely allow their horses to feed some distance from the camp, but instances have been known of Indians who have succeeded in approaching and catching a horse furnished with iron hopplings, in revenge for their disappointment at not being able to gallop away with their prize, sending an arrow through the animal or otherwise seriously injuring him. During the fly season, smokes are made every night for the horses, and if this precaution is neglected they will remind their masters of their want of care by surrounding the camp fire and standing with their heads in the smoke. It is this habit of crowding round the smoke of a fire to avoid the torment of flies which makes Indian horses so difficult to drive from a prairie on fire. Many are burned every year on account of their being unable to comprehend the danger which threatens them. The buffalo are more wary, the smell of fire is often sufficient to drive them from pastures where they have been quietly feeding.

Next to the horse, the dog is the Prairie Indian's most valuable friend. The dog is the great stand-by of the squaws, who have to attend to all the duties of the camp, the men employing themselves solely in hunting and fighting. The dogs drag on poles the camp furniture, the provisions, the little children, and all the valuables of the family. It is a very amusing sight to witness several hundred dogs solemnly engaged in moving a large camp. They look wistfully at passers-by, and take advantage of the least want of attention on the part of their mistresses to lie down, or snarl and snap at their companions in the work. They nevertheless obey the word of command with alacrity and willingness if not fatigued.

The midnight howl of three or four hundred dogs is an awful and appalling sound. It rises suddenly from a low prolonged whine to a deep melancholy howl, caught up again and again to the distraction of tired travelers anxious to take rest in sleep. When any great event takes place, a dog feast is proclaimed, and it is sufficiently disgusting to see the men handle and feel the unfortunate animals as if they were sheep, with a view to select the fattest, so powerful are early habits and associations in directing our feelings and tastes. Although some of the Indian dogs we saw among the Crees of the Sandy Hills are large and ferocious looking animals, we never found them vicious or inclined to attack us; they were always deterred from approaching by the sight of a stick or a feint at picking up a stone.

Although I made many inquiries, the Indians could give no information respecting the occurrence of hydrophobia among their dogs, and the same observation, as far as I could discover, applies to the dogs so numerous at Red River, and at the different Posts of the Hudson's Bay Company. Large numbers of dogs are kept at the Company's Posts to haul sledges during winter; in summer time they are fed on fish at fishing stations; in the prairie they feed upon the offal of buffalo. Dogs will go for a week without food, and yet get into condition for traveling, if well fed, in a fortnight or eighteen days. At Manitobah House I saw them devour large pike alive, which were thrown to them as they were taken from the nets. Indian dogs are terrible thieves, especially those originating from a cross with the wolf. It was necessary to place out of reach or under cover every article bearing the least resemblance to leather when we were among the Crees. A careless half-breed would wake in the morning and find his harness eaten, or his whip devoured; and it sometimes happened that the long tether of buffalo hide would be found partly consumed by dogs if their appetite had not been lately appeased. The wolves have this trick also when food is scarce, especially when the tether is allowed to trail loosely from the horse's neck without being attached to a stake, thus leaving him at liberty to wander some distance from the camp during the night. The intelligence of dogs when traveling

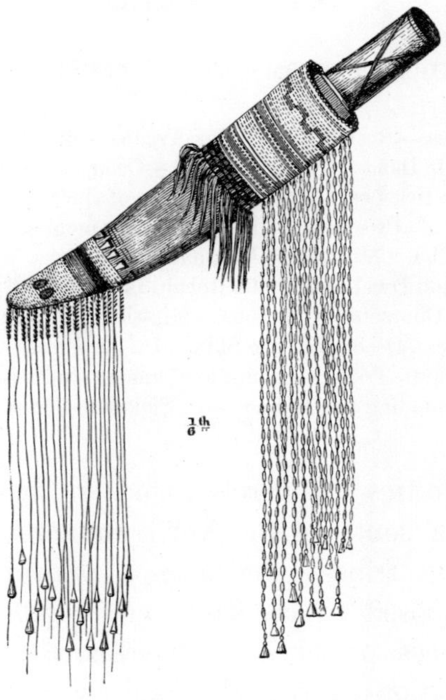

Sioux Knife Sheath.

during the winter is astonishing, several curious instances occurred during our homeward journey, which will be found in the narrative of the winter journey to Crow Wing.

With Crees, Ojibways, Swampys, and Sioux, the dog is supposed to be the most acceptable sacrifice to offended deities, five dogs being the common number for a propitiatory offering.

CHAP. XXIX.

INDIAN CUSTOMS AND SUPERSTITIONS.

Indian Antiquities.—Result of the Fur Trade.—Ojibways Invaders of the Prairies.—Scalp Dance.—Wood Indians.—Occupations of Indians.—Indian Cruelty.—Mis-tick-oos, Chief of the Crees of the Sandy Hills.—The Fox. —Treatment of Prisoners.—Medicine Ceremonies.—Happy Hunting Grounds.—Indian "Medicine" Men and "Medicines."—Influence of Conjurors.—Manitou Dwellings.—Manitobah Island.—The Rev. Mr. Cowley. —Sacrifices.—Character of Indians.—Mis-tick-oos' best Wife.—Mis-tick-oos' Son's Wife.—Decorating the Skin.—Indian Pipes.—Tà-wâ-pit's Pipe. —Pipes peculiar to Tribes.—Salutations among Indians.—Indians in the Prairie.—Impounding the Buffalo.—In Sickness.—Idea of Lightning.

Indian antiquities are rarely found in the valley of the Saskatchewan south of the North Branch. The customs of wandering tribes inhabiting a prairie country are generally opposed to the rude arts which exist among barbarous races preserving a fixed abode. Not even at the fishing stations on the lakes and rivers, where different tribes have congregated at certain seasons of the year, probably for centuries, do we find any lasting memorials of individual handicraft or combined labour.

Antiquities to be ascribed to different races than those which now occupy the country exist here and there. Such are the underground houses on Rainy River, the Mandan houses with their intrenchments on the Little Souris; but with these exceptions no other ancient monuments were seen during the explorations.

The rings of stones marking the site of Cree encampments on the Qu'appelle are of comparatively modern date, and belong doubtless to the ancestors of the present races now in possession of the country.

Rude pottery and arrow heads have been found at Red River settlements, about two feet below the surface of the soil. The fragments resemble those common in many parts of Canada, and from their numbers lead to the inference that at a remote period the banks of this stream

Indian Hunters' Tents in the rear of Fort Garry.

were peopled by races familiar with the art of making vessels from clay.

One result of the active pursuit of the fur trade for upwards of a century in the valley of the Saskatchewan, is seen in the blending of the different tribes by intermarriage. The Crees of the Plains and the Ojibways and Swampys of the Woods, although speaking different languages, are often found hunting the buffalo in company, and not unfrequently form family connections. The Ojibways of Lake Winnipeg may now be discovered, summer and winter, near the Grand Forks of the Saskatchewan, having emigrated 400 miles west of Red River, where they have permanently established themselves. All the Ojibways now found west of the Lake of the Woods, and the east coast of Lake Winnipeg are invaders of the country. The real home of the Ojibway is the region about the south, west, and north of Lake Superior; the habits of life of the emigrants or invaders have been adapted to the character of the country they now occupy, and being no longer dependent upon the forest for food and clothing, many of them, on the banks of the Assinniboine, Red River, Lake Manitobah, and Dauphin Lake, possess horses, and join the half-breeds in their annual spring and fall hunts. Notwithstanding this intercourse and blending of different nations, most of the superstitions and customs peculiar to each are still maintained and practised.

Nearly one hundred years ago (1770), Mr. Hutchins, of the Hudson's Bay Company's service, framed an enumeration of the tribes between Lake Winnipeg, and within one hundred miles of James's Bay, speaking the Ojibway tongue. The tribes enumerated have evidently derived their names in conformity with long established usage, from their hunting and fishing stations.

It is often asked whether the thrilling descriptions of savage life in Cooper's delightful romances are imaginary or real; and if real, whether they exist now among the tribes which have long been familiar with civilized man, such as the Plain Crees, the Sioux, the Swampys, and the Ojibways. It is enough to visit the secluded Ojibway graves on the banks of Red River, and contemplate Sioux scalps decorated with beads, bits of cloth, coloured ribbons, and strips of leather, suspended at the extremity of a long slender stick near the head of the grave, to feel satisfied that one barbarous custom still prevails. But to be an eyewitness of a scalp dance, or a skull dance, is more than sufficient to press home the conviction that the fiendish passions so faithfully described by Cooper, still find expression in violent gesture, loud vociferation, triumphant song and barbarous feasting, with undiminished strength and bitterness, even after a century's intercourse with civilized man.

Some of the incidents narrated in the following pages will show how far old superstitions and customs prevail among the Indians occupying the country between Red River and the South Branch of the Saskatchewan.

Early in the spring of 1858 the warlike bands of Ojibways called the Lac la Pluie Indians, were thrown into a state of savage excitement by the arrival of messengers from their friends on Red River, with tidings that two Sioux had been killed and scalped in the Plains. In testimony of this triumph, they brought with them two fingers severed from the hands of the unfortunate Sioux. The announcement of the intelligence that the scalps would be sent, after their Red River brethren had celebrated war dances over them, was received with wild clamour and shouting. After the scalps had been carried from hand to hand and the victory that won them triumphed over with dancing, singing, and feasting, they would be returned to the warriors who took them, and finally suspended over the graves of relatives or friends mourning the loss of any of their kindred by the hands of the Sioux.

The Crees, Ojibways, and Swampys, belonging to the great body of Wood Indians, assemble in the spring of the year to celebrate their medicine feasts and other ceremonies, which are generally determined by the arrival of migratory birds, or the time when the sturgeon begin to ascend the rivers. The day on which the annual goose dance takes place is regularly entered in the journals of

Fort à la Corne and other Posts of the Hudson's Bay Company. In the woods on the main Saskatchewan behind Fort à la Corne, as well as in several places on the banks of the Little Saskatchewan, Winnipego-sis, and Dauphin Lakes, we found the carved and painted posts which are made to play an important part in these observances. (Vol. I. p. 402.)

During the summer they separate into families or small

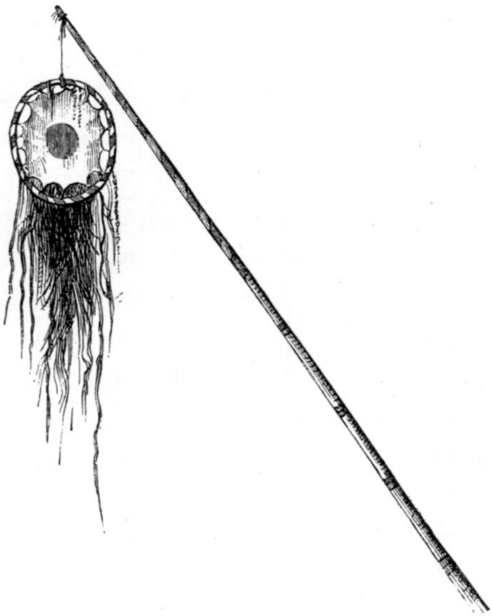

Sioux Scalp from the Graves at Red River.

bands, and hunt, fish, or go to the Plains in search of buffalo. At the approach of winter, they "take debt" or otherwise obtain supplies at the different Posts of the Company, and retire to their winter quarters to trap the fur-bearing animals. The Plain or Prairie Indians follow the buffalo, and vary the monotony of their existence by forming war parties against their enemies, such as the

Plain Crees against the Sioux and the Blackfeet, the Ojibways against the Sioux, and Assinniboines.

In August 1858, when camped near the Elbow of the South Branch of the Saskatchewan, we found the Plain Crees hastening from the west to the east bank of the river, with a strong war party of Blackfeet in pursuit. Scouts were posted on some of the outlying sand hills on the Qu'appelle west of Sand Hill Lake, where a large party of Crees, under the direction of their chief, Mis-tick-oos*, were impounding buffalo. Both by day and night a strict watch was kept in order to guard against a surprise. Mis-tick-oos pointed out some of his band who had penetrated through the Blackfeet country to the Rocky Mountains two years ago, and returned with several scalps, grizzly bear-claw necklaces, pipes, and other trophies of success. He also related with much feeling how twenty-five young warriors had gone on a similar excursion during the summer of 1856, but none had yet returned.

About a month before we arrived at the Sandy Hills on the Qu'appelle, a large body of the Plain Crees met a portion of the Blackfeet tribe at the Eagle Hills, on the North Branch of the Saskatchewan, to arrange terms of peace. All matters went on smoothly, and the representatives of the two nations separated as friends. Some of the Crees, however, incapable of resisting the opportunity, stole some horses from the Blackfeet. They were pursued, and three of them taken. One was killed instantly, the others were led back in triumph to the camp of the Blackfeet; they were stripped, their hands were tied behind their backs, a hole was bored through both wrists and a stick passed through them and so tightly fastened

* Shortstick, or "the little tree."

that it could not be removed without assistance; the captives were then separated and dismissed singly to find their way to their friends. One only reached his tribe, and was lying in a tent not far from the spot where we were encamped.

Mis-tick-oos, when relating these adventures, raised the pipe he held in his hand and exclaimed, "This is what my Blackfoot friend gave me one day, the next he killed my young men; he is now my enemy again." I expressed a wish to purchase the pipe; the chief's reply was "Take it," handing it to me with a gloomy frown, and silently extending his hand for the Wapekan-cuspwägān, or clay pipe, which I was smoking at the time.

The great chief of the Plain Crees is styled "the Fox;" he is held in high estimation by all the Plain Indians with whom he comes in contact, either in peace or war. He is dreaded by the Sioux, the Blackfeet, the Bloodies, the Fall Indians, the Assinniboines, and the tribes who occasionally hunt on the Grand Coteau de Missouri and the South Branch of the Sackatchewan.

The barbarous and cruel treatment of prisoners so often described in narratives of Indian warfare, is common even now in the prairies south of the Qu'appelle and the Assinniboine. On that part of Red River which lies in the State of Minnesota, Indian warfare, with all its horrors, is constantly carried on between the Ojibways and Sioux.

Not a year passes without the loss of several Red River half-breeds by the scalping knife of the Sioux; and, as was the case in the autumn of 1858, quite close to the settlement of St. Joseph, near the boundary line, about thirty miles west of Red River. When a prisoner is taken, the Sioux sometimes adopt a terrible

"THE FOX,"—CHIEF OF THE PLAIN CREES

mode of death during the summer season. They have been known to strip a half-breed, tie him to a stake on the borders of a marsh in the prairie, and leave him exposed to the attacks of millions of mosquitoes, without being able to move any part of his body. When the agony of fever and the torment of thirst come upon him, they leave him to die a dreadful, lingering death with water at his feet, and buzzards hovering and circling around him in greedy expectation.

By way of illustrating the character of the medicine or conjuring ceremonies which may be witnessed during all seasons of the year when several families are encamped together, I shall describe a scene of which I was an eye-witness near the Hudson's Bay Company's Post at the Touchwood Hills. The conversation was carried on in Cree, but, I believe, faithfully interpreted to me by the officer then in charge of the Post, who was present during the incantations. The interpretation was pronounced exact by one of the Cree half-breeds attached to my party.

At the time of my arrival at this Post, a conjuror of some celebrity was endeavouring to cure a sickly woman by the exercise of his cunning. The unfortunate invalid was lying in a buffalo skin tent, while the conjuror, painted and decorated, and wearing his medicine bag, employed himself in beating a drum within a few feet of her, and singing at intervals the following words, first uttered slowly, with a pause between each word, then as in ordinary conversation, lastly, with energy and rapidity :—

"Great—is—the—man—who—walks—
In—the—middle—of—the—earth,—
He—is—the—only—true—Lord."

The word "Lord" is not employed in the sense of supreme master, but is rather intended to convey an idea of independence and individual power, and is better expressed in English, as the half-breeds informed me, by the word "gentleman."

The conjuror occasionally came out of the tent, and whenever the supposed Manitou or Fairy who was the alleged cause of the woman's illness approached, a little bell suspended from the poles supporting the tent tinkled, and gave the alarm; the conjuror immediately seized his

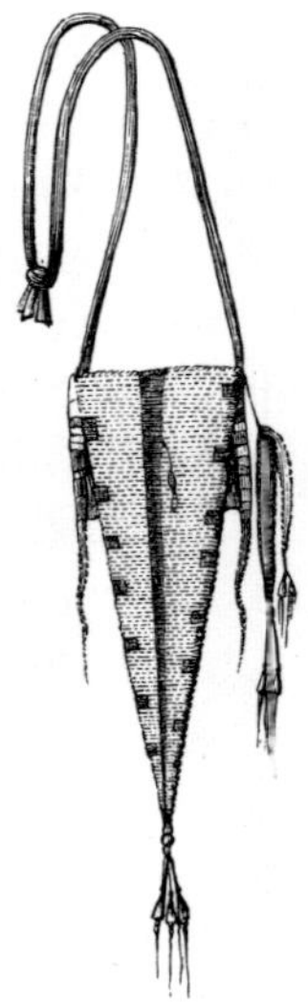

Cree Medicine Bag.

drum, commenced his song, and by his incantations, succeeded in pacifying the Manitou. These proceedings continued for two nights; at the close of the second night, after a prolonged ringing of the little bell, violent shaking of the tent-poles, loud beating of the drum, and chanting of the words before quoted, the conjuror announced that he had discovered the reasons of the Manitou's anger, and the means to appease it.

You had a dream, said the conjuror, and when you rose in the morning you promised to make an offering to the Manitou; you have forgotten your pledge, and you are suffering in consequence of your neglect.

The woman demanded what she had dreamt, and what she promised, avowing her ignorance of both dream and promise. The conjuror told her that when the buffalo were around her tent last winter and no fear of starvation before her eyes, she had dreamed that the buffalo would always surround her, that famine and sorrow were always to be strangers to her, and, in gratitude, had vowed to make a sacrifice of her best robe. The woman, wearied no doubt with the conjuror's unceasing drum and song, probably, too, believing that a false confession was the lesser evil, as it might bring the promised relief, acknowledged that the conjuror was in the right. The penalty she was told to pay consisted of the sacrifice of "throwing away" two robes, or double the amount of the promise she had made; after which her health was to be restored.

Scenes similar to the one just described may be witnessed whenever several families are camping together; but the sacrifices required to be made depend upon the ability of the deluded creatures to satisfy the demands of the conjuror.

"The happy hunting-grounds," the heaven of Indians, so often spoken of by writers of fiction, are an actual reality in the imagination of Crees and Ojibways, as well as of other north-western tribes. A Plain Cree on the Qu'appelle gravely informed one of my men, that he had once been dead and visited the spirit world. His narrative was to the following effect:—"I was sick and fell asleep. I awoke on the bank of a deep river, whose waters were flowing swiftly and black from a great mist on the south to a great mist on the north. Many other

Indians sat on the banks of the river gazing at its waters, and on the gloomy shore which lay wrapped in mist on the other side. Time after time the mist before us would roll away and reveal the mouth of another great river pouring its flood into the one on whose banks I was sitting. The country to the south of this river was bright and glorious, to the north, dark and gloomy. On the one side were the happy hunting-grounds, on the other the hunting-grounds of bad Indians. Time after time my companions tried to cross the swift stream before us, in order to reach the happy hunting-grounds; some arrived in safety, others reached the north bank, and disappeared in the mist which overhung the bad country. I tried at last, but the current was too strong for me, the recollection of bad deeds prevented me from stemming the current, and I was swept on to the north shore of the opposite river. I scrambled up the bank, and spent many moons in hunting in that dreary land; always on the point of starving, or being hurt by enemies, or wet and cold and miserable. At length I came upon a river like the one I had crossed, with mists and the mouth of a great stream opposite to me; breaking clouds soon revealed happy hunting-grounds on one side, and a more gloomy and terrible country on the other. Many Indians were there before me, looking at the river and trying to cross; some succeeded, but a few were swept to the bad country, these were very wicked Indians. I tried to cross. I knew I had been a good Indian in this dreary hunting-ground. I took courage, and swam strong against the stream. I reached the happy hunting-grounds; all my sorrow disappeared as I climbed to the top of the bank and saw before me Indians numerous as grass leaves, buffalo on the distant plains thick as raindrops in summer, a cloudless sky above, and a warm, fresh, scented, happy breeze blowing in my face. I sank to sleep, and woke alone in my tent in these prairies again."

Whatever faith the Indian medicine men possess in the efficacy of their charms, it is certain that they entertain great respect for the white man's medicine. A laughable incident occurred at the Touchwood Hills. The conjuror, of whom mention has just been made, entered a room at this Post where I was sitting with Mr. and Mrs. H., who were temporarily in charge. The Indian and a companion seated themselves upon one of my boxes which contained a small medicine chest. Mrs. H. asked me to give her some sticking plaster. I crossed the room to open the medicine chest, when Mrs. H. (a half-breed) said to her husband, in the Cree language: "Will his medicines do me any harm if I stop here while he opens them?" Mr. H. answered jestingly, "Yes, you had better go into the other room." On motioning the Indians to move, they rose, and I opened the chest. The moment they saw the bottles, they hurried out of the room, hastened to the summit of a neighbouring hill, and, divesting themselves of every article of clothing, shook their garments repeatedly, and after hanging them on bushes in the sun, squatted on their haunches to await the deodorizing influence of the breeze.

In the valley of the Qu'appelle River, we frequently found offerings to Manitou or Fairies suspended on branches of trees; they consisted of fragments of cloth, strings of beads, shreds of painted buffalo hide, bears' teeth and claws, and other trifles. Our half-breeds always regarded them with respect, and never molested or liked to see us molest these offerings to Manitou. This custom prevails everywhere in the valley of Lake Winnipeg, and on the banks of the settled parts of Red

River, where the medicine drum and rattle may be heard more frequently in some parishes than the sound of church bells.

A conjuror celebrated for the potency of his charms will often exercise a very injurious influence over an entire band, consisting of ten or twelve families, in deterring them from frequenting particular hunting or fishing-grounds if they offend him. From numerous instances of this dangerous influence, I select the following, which occurred on the Little Saskatchewan, or as it is sometimes called, Dauphin River. When ascending that stream, we

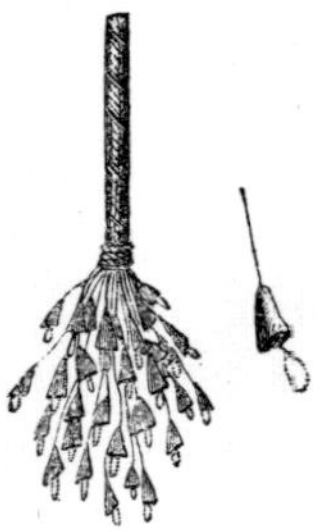

Medicine Rattle.

came upon a large camp of Swampys, who were on their way to the Hudson's Bay Company's post at Fairford. Their usual wintering-place was at the Pike's Head near the mouth of Jack-fish River, an excellent fishing station on Lake Winnipeg, but they had abandoned the intention of wintering there, in consequence of a threat which had been conveyed to them from a noted conjuror styled "the Badger," of the Grand Rapid of the Saskatchewan, to the effect that "if the band ventured to winter at the Pike's Head, he would do something." This ambiguous threat was quite sufficient to deter them from visiting their old haunts, and would probably be instrumental in producing much suffering, if not actual want, to many of the band.

There are many places on Lake Winnipeg and Manitobah which the Indians who hunt and live on the shores of those inland seas dare not visit. There is scarcely a cave or headland which has not some legend attached to it, familiar to all the wanderers on these coasts.

On the west side of Lake Winnipeg, in the long, dark, and gloomy chambers formed by fissures in the limestone, bad spirits are supposed to dwell, according to the belief of the Indians who hunt on the coast, and he would be a powerful charmer who could induce a heathen Indian to approach, much less enter, the abodes of these imaginary Manitous.

Near Limestone Cave Point, on Lake Winnipeg, are several of these supposed fairy dwellings. When an Indian approaches them in his canoe, he either lays an offering on the beach or gives them as wide a berth as possible.

Steep Rock Point, on Lake Manitobah, is also a noted dwelling-place for the "Little Men." Some of the traditions connected with these places are very absurd, and appear to have little meaning to civilized men; nevertheless, among the barbarous tribes of those regions, they are associated with their past history, or with the history of the race that preceded them. Manitobah Lake, a body of water of very imposing dimensions, having an area of 1,900 square miles, derives its name from one of these superstitions. I stayed for three days on Manitobah island, where a Manitou dwells, but although Indians passed and repassed, heard and answered our shots, yet they could not be persuaded to land. The only evidence of fairy presence which I met with, was the "fairy-like music" of the waves of Lake Manitobah, beating upon the hard limestone shingle on the beach, and producing a very beautiful and melancholy resemblance to distant church-bells. All night long this ringing musical sound was heard, and would, no doubt, in the active imagination of Indians, suggest the existence of those Manitou with which they people the air, the water, the forests, and the caves of the earth.

The able and zealous missionary at the Indian Settlement, Red River, the Rev. A. Cowley, has had sad experience of the gross superstitions which darken the intellect of the Ojibways of Lake Manitobah. In 1842 he proceeded to what appeared to be a promising station on the shores of this lake, where he had an opportunity of observing some remarkable instances of heathen faith in dreams and charms. Mr. Cowley writes: "One day I saw something hanging on a tree and went to look at it. It consisted of twenty small rods, peeled, and painted red and black, and fastened together on a plane, with cords of bark. A piece of tobacco was placed between the tenth and eleventh rods, and the whole was suspended perpendicularly from a branch of the tree. It belonged to the old chief, who told me that when he was a young man he lay down to dream, and that in his dream, the moon spoke to him, and told him to make this charm, and to renew it every new moon, that he might have a long life. He had regularly done so ever since, till the preceding summer, when he almost forgot it, and was taken so ill as to be near dying; but he remembered it, his friends did it for him, and he recovered." *

Sacrifices and offerings are of very frequent occurrence among the Indians of the Saskatchewan Valley. The customary offering consists of two, three, and sometimes five dogs. At the mouth of the Qu'appelle River, an

* Quoted by S. Tucker, in "The Rainbow of the North."

Indian, in June 1858, set his net and caught a large fish of a kind different to any with which he was familiar; he immediately pronounced it to be a Manitou, and carefully restoring it to the water again, at once sacrificed five valuable dogs to appease the anger of the supposed fairy. On approaching Long Lake, an arm of the Qu'appelle River Valley, the Crees warned us not to visit the lake by night, as it was full of devils. They told me very extraordinary tales, which are too absurd to be worth relating, of the dimensions and power of these devils, and they appeared to live in awe and terror of them.

Like most heathen and barbarous races, Indians suffer much from their superstitious fears. When the weather is fine, and their tents are well supplied with provisions, they are an independent and joyous people. Full of frolic and fond of relating anecdotes, they laugh immoderately at any trifling joke or absurdity, and seem thoroughly to enjoy existence.

A ridiculous incident occurred in the tent belonging to the Cree chief, Mis-tick-oos, in which I played a more prominent part than I should have selected had any choice been offered me. I heard of this incident again hundreds of miles from the spot where it occurred, as we journeyed homewards from the Grand Forks.

It happened during a visit I paid to Mis-tick-oos (Vol. I. p. 362), after a long and tedious talk, which lasted nearly seven hours, relating to the object we had in view in visiting the country. Three of Mis-tick-oos' wives were visible, and with their children formed altogether a party of eighteen or twenty. I rose from a buffalo robe where I was seated by the side of Mis-tick-oos to examine some arrows which one of his sons was making, and when my curiosity was satisfied, I sat down on what I thought to be a bundle of buffalo robes, close to the young Indian.

I was not a little astonished to feel the robes slowly move and undulate beneath me, and before I could rise and look into the cause, I found myself projected into the middle of the tent among the embers, by means of some violent spasmodic action from beneath the supposed pile of robes. Mis-tick-oos and his three wives with the other inmates, shrieked with laughter, vociferating some words in Cree. Meanwhile, the buffalo robes were slowly thrown on one side, and, to my astonishment, were revealed the huge proportions of the chief's fourth, youngest and best wife. She shook a mass of hair from her head, and joined in the laughter at my discomfiture. Other Indians hearing the noise came in, and Mis-tick-oos, with tears in his eyes, told his friends how "the white stranger had sat upon his best wife, thinking she was a pile of robes, and how she tossed him into the middle of the tent like a buffalo bull pitching a colt."

During our stay with the Crees of the Sandy Hills on the South Branch, when passing the door of the tent belonging to the chief's eldest son, who was my companion at the time, I observed a young squaw leaning upon sticks, evidently in great trouble, and weeping bitterly. The moment she saw us she hobbled into the tent, with a low cry of pain, and closed the entrance. I asked the interpreter what this meant. After some conversation with her husband, he said that the woman was suffering from a beating he had given her for a violation of her faith during his absence in the spring on a war excursion. "I would have killed her," muttered the husband, "but I thought it a pity to kill two at once. She had her choice whether she would have her hair, her nose, or her ear cut off, or whether she would have a beating. She chose what she has got; but I would have killed her had I not known I should regret having killed both." It is needless to add that the woman soon expected to become a mother.

Smearing the skin with different coloured pigments is a universal custom among the wood and prairie Indians. Sometimes the operation is very tastefully performed. Warriors on the "war-path" often paint the figure of the hand over the mouth, as used in sounding the war-whoop; this is a distinctive sign that the Indian so decorated has been recently, or is still engaged in the pursuit of his enemies. Vermilion is the most coveted colour; the Ojibways particularly are very fond of decorating their faces with this brilliant pigment. The Plain Crees are partial to white, green, and blue, and not only paint the face, but also the chest and arms. They cut and gash the skin and flesh on the arms, sides, chest, and legs, as a token of grief for any deceased friend or relation. My friend Mis-tick-oos' body was dreadfully disfigured by scars from wounds made by himself in manifestation of his grief.

Ornamenting the skin of the arms and breast with the figures of birds, quadrupeds, or symbols of different kinds, is common among the Plain Crees. The operation is performed with a needle, a thorn, the point of a knife, or the edge of a flint. The surface of the skin is cut or penetrated with the instrument used, and the colour rubbed in, as in the process often adopted on a small scale by sailors. The effect is to produce a permanent representation of different objects on the skin, but it does not resemble the ridges and furrows produced by the tattooing process of the New Zealanders.

Nothing appears to contribute so largely to the comfort and enjoyment of Indians, whether of the woods or the prairies, as the pwahgăn of the Ojibway, the uspwägān of the Cree, or the pipe of the English.

When inhaling the fumes of tobacco, the bear-berry or the inner bark of the red willow, the Indian relinquishes himself to the narcotic influences of the "weed," a term by the way applied to the bear-berry, and to the dry and gravelly ridges where that pretty little creeping plant flourishes; the local names, "weed-ridge," "weed-hill," being not uncommon in Rupert's Land. It has been well said that "the tobacco pipe constitutes the peculiar and most characteristic symbol of America, intimately interwoven with the rites and superstitions and with the relics of ancient customs and historical traditions of the aborigines of the New World. If Europe borrowed from it the first knowledge of its prized narcotic, the gift was received unaccompanied by any of the sacred or peculiar virtues which the Red Indian still attaches to it as the symbol of hospitality and amicable intercourse, and Longfellow, accordingly, with no less poetic vigour than fitness, opens his "Song of Hiawatha," with the institution of the "Peace-pipe" by the Great Spirit, the Master of Life.*

Pipe No. 1 was presented to me by Tà-wâ-pit, an old Indian of Dauphin Lake. He had another in his pouch nearly completed, made from the soft shale which crops out on the Riding Mountain. I asked Tà-wâ-pit "what he would do for a smoke" until he had finished the new pipe? After the half-breed with me had made him understand my question, he rose to his feet, and walking to the edge of a swamp close by, cut three or four reeds, and joining some pieces together, after he had made a hole through the joints, he gently pushed one extremity in a slanting direction into the earth, which he had previously made firm by pressure with his foot; he then cut out a small hole in the clay above the extremity of the reed,

* "Narcotic Usages and Superstitions of the Old and New World," by Daniel Wilson, LL.D. Canadian Journal, new series, vol. ii.

and moulding it with his fingers, laughingly said, "Now give me tobacco, and I will show you how to smoke it." He filled the hole with a mixture of tobacco and the bear-berry, placed a live coal on the top, and stretching

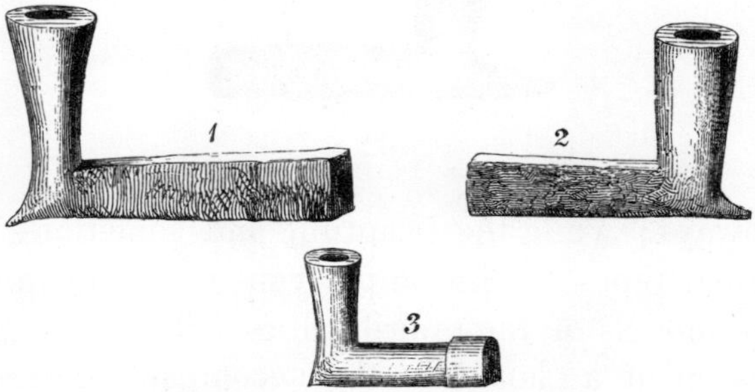

Tobacco Pipes of the Swampys of Lake Winnipeg.

himself at full length on the ground, with his chin supported by both hands, he took the reed between his lips and enjoyed a long smoke.

Different tribes of Indians affect peculiar shapes and ornaments in the manufacture of their pipes. The Ojibways

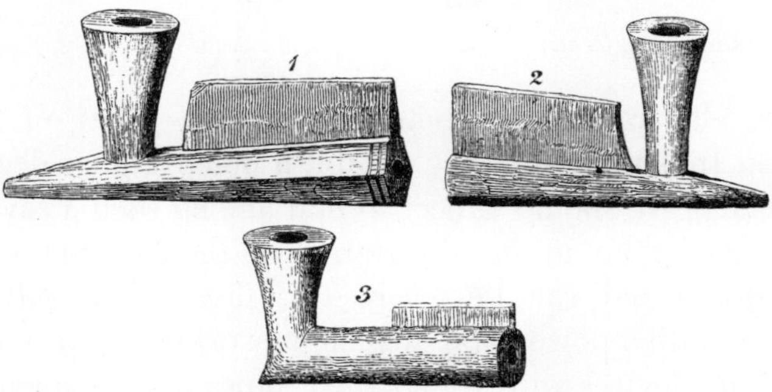

Tobacco Pipes of the Ojibways of Rainy Lake, &c.

of Rainy Lake and those of the same nation living on Lake Winnipeg, have different patterns, but preserve a certain resemblance, which appears to be characteristic of this people.

The pipes manufactured by the Plain Crees and the Blackfeet are formed after the same model; those of the Sioux approach more to the form almost universally adopted by

Sioux Pipe.

the Ojibways; while the beautiful and sometimes highly ornamented pipes of the Chipewyans resemble more the favourite models of the prairie tribes. These distinctions will be seen at a glance in the accompanying drawings, which are all reduced from pipes in my possession.

Chipewyan (1 and 2), Plain Cree (3), and Blackfoot (4) Pipes.

The elaborate and sometimes beautiful pipes of the Babeen Indians, while they exhibit a much higher degree of art than we should expect to find among such a savage race, are by no means illustrative of their superstitions or customs, and can be received as illustrations only of their imitative power and ingenious workmanship. The grotesque devices with which their pipes are ornamented can generally be traced to objects which they have seen since they became familiar with the traders belonging to the Hudson's Bay Company on the north-west coast.

The customary salutation among the Ojibways who have been brought in contact with the French Canadian voyageurs of the old North-West Company is, "boujou! boujou!" from bon jour. Among the Plain Crees, with

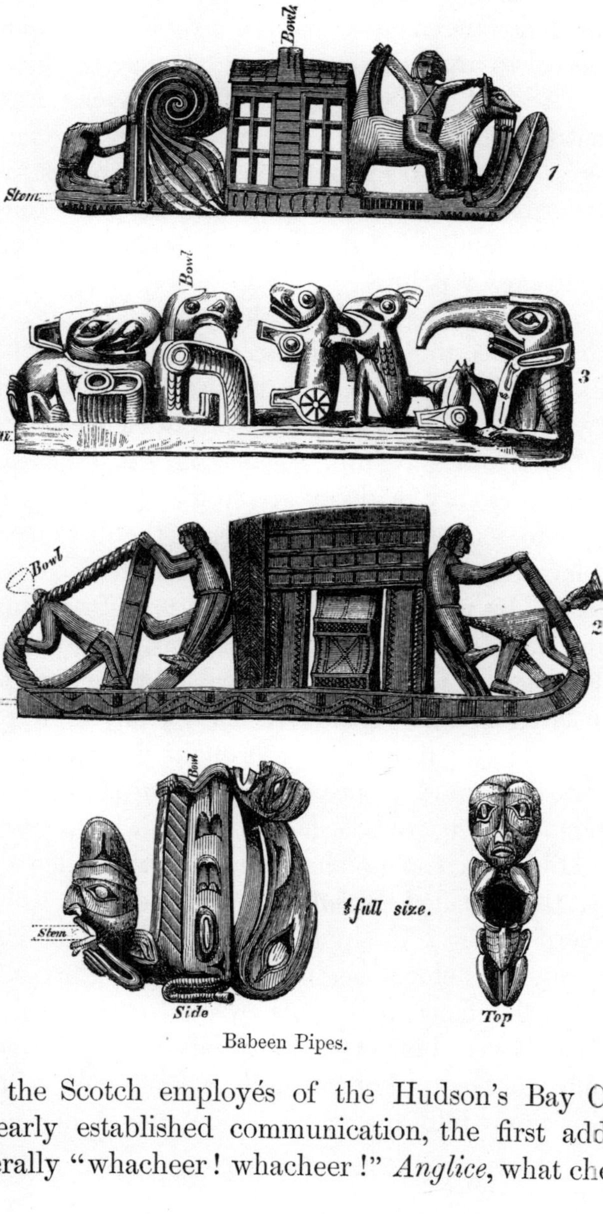

Babeen Pipes.

whom the Scotch employés of the Hudson's Bay Company early established communication, the first address is generally "whacheer! whacheer!" *Anglice*, what cheer?

Shaking hands is customary both on meeting and parting. The usual preliminary to a council or a series of questions, is a smoke; and nothing smoothes the way to an affable arrangement in case of a difficulty so quickly as a proposal to argue the point and arrange matters over a pipe. A great point is gained if the traveler is able to present the chief of the party with a plug or canister of tobacco to replenish his pipe, and when he offers to return it, a courteous intimation that he may keep the remainder or hand it round to his young men, is often a very advantageous stroke of policy.

In order to understand the character and nature of wild prairie Indians, they must be seen in their tents when well supplied with provisions, and disposed to be cheerful and merry. In the prairies when on horseback, they are often quiet and watchful, always on the look out, and when even twenty or thirty are in a band, they generally manage to see a suspicious object in the distance at the same moment, so that a simultaneous note of exclamation is uttered by most or all of the party. In hunting the buffalo they are wild with excitement, but no scene or incident seems to have such a maddening effect upon them as when the buffalo are successfully driven into a pound.* Until the herd is brought in by the skilled hunters, the utmost silence is preserved around the fence of the pound: men, women, and children, with pent-up feelings, hold their robes so as to close every orifice through which the terrified animals might endeavour to escape. The herd once in the pound, a scene of diabolical butchery and excitement begins; men, women, and children climb on the fence, and shoot arrows or thrust spears at the bewildered buffalo, with shouts, screams, and yells horrible to hear. But when the young men, and even women jump into the arena amidst the dying and the

* The half-breeds call these enclosures "ponds."

dead, smear themselves with blood, thrust their arms up to the shoulders into the reeking bodies of their victims, the savage barbarity of the wild prairie Indian shows itself in its true colours. Not even a scalp dance over many fallen foes, affords such a terrible picture of degraded humanity as a large band of prairie Indians, some hundreds in number, during and after the slaughter of buffalo in the pound.

The condition of the Indians of the Saskatchewan Valley at the present day is very different to what it

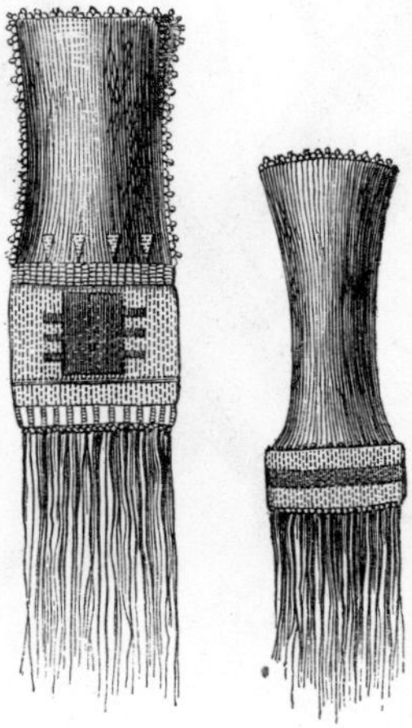

Cree Fire-bags.

was even half a century since. Not only have imported diseases greatly diminished their numbers, but game of different kinds has become so scarce that during some seasons starvation is no fiction. In the northern parts of Rupert's Land a great mortality took place in 1816, 1817, and 1818, from small-pox and measles. Vaccine inoculation was then introduced by the Hudson' Bay Company, and small-pox has been unknown in the country since.*

* Sir George Simpson. Blue Book on the affairs of the Hudson's Bay Company.

The Sioux south of the boundary are said to be all vaccinated.

In sickness prairie Indians are much depressed, and often seek consolation in the monotonous drum of the medicine man and his heathenish incantations ; an infliction which the grossest and most debased superstition alone

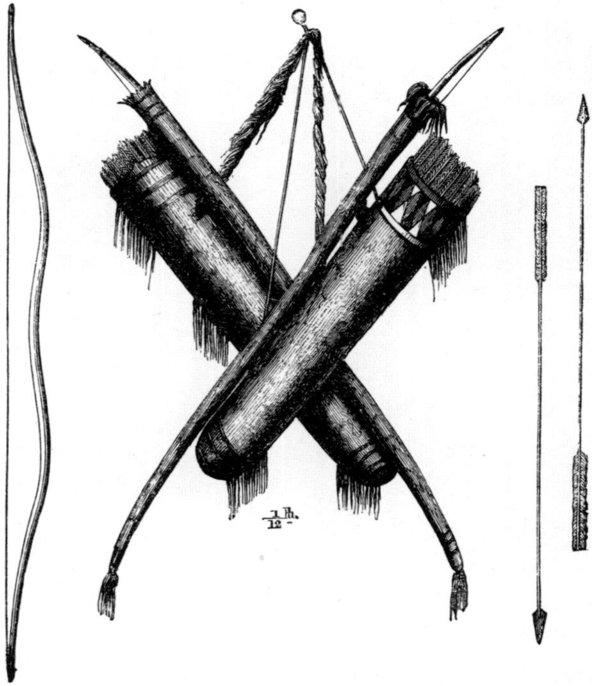

Sioux Quiver, Bow, aud Arrows.

would tolerate; it is submitted to with confidence and hope, however, by men who are anxious and timid during the roll of thunder, invoking the Great Bird by whose flapping wings they suppose it to be produced, or crouching from the blink of his all-penetrating eye, which they allege is the lightning's flash.

CHAP. XXX.

INDIAN POPULATION OF BRITISH AMERICA.

Origin of Indian Races.—Kindred and Relationship.—Iroquois Customs.—Iroquois Institutions.—Iroquois League.—Indian Population of Rupert's Land.—Probably over-estimated.—Number of Indians frequenting particular Posts.—Prairie Indians.—Colonel Lefroy's Estimate.—The Sioux or Dakotah Indians.—Principal Bands.—Conjurors.—Months.—Language.—The Blackfeet.—Country occupied by the Blackfeet.—Blackfeet Tribes.—Indians near the Boundary Line.—Indians of British America.—Indians of the United States.—Early History of the Indians.—Mutability of Indian Nations.—The Hurons and Iroquois.—The Prairie Tribes.—The Remnant.

THE origin of the aborigines on this continent still remains enveloped in thick darkness. Many of their manners, superstitions, and customs correspond to those of Orientals, and it is not improbable that modern ethnologists may be on the right track in their efforts to solve this deeply interesting question.

The ties of kindred and relationship are of a very complex character among the Ojibways; in more than one instance a singular exemplification of cross-relationship occurred during our voyage in 1858 on Lakes Winnipeg and Manitobah, which may serve to show the permanency of ancient customs and traditions among families now dwelling nearly a thousand miles west of the hunting-grounds of their ancestors.

Near the mouth of the Little Saskatchewan, we met an Indian family journeying in a small canoe towards the mouth of Red River. The family consisted of a young Indian, his wife, and two little children. The father was born on the shores of Lake Winnipeg, and had never traveled east of that lake. After a few words had passed between him and a half-breed Ojibway from Lake Superior (Wigwam), they shook hands and proclaimed themselves related to one another. Each belonged, as I was informed, to the tribe whose "*totem*" or insignia was the "Bear," and having by some means, which Wigwam could not or would not explain, ascertained this fact, they spoke to one another as brothers. A similar relationship was established between Wigwam and another Ojibway on Moss River, solely, as he assured me, because he and his newly found friend belonged to a tribe whose "*totem*" was the "Bear." The Cree half-breeds told me that in their communication with the Ojibways of Lake Winnipeg, and farther to the west, this recognition of relationship not unfrequently took place between individuals who met for the first time, and who were born and lived in districts far apart. In connection with this singular kind of consanguinity and the bearing it may possibly have upon the origin of the Indian races, I append the following extract from an ethnological paper, read at the Montreal Meeting of the American Association for the Advancement of Science*:—

"It has occurred to me, after a careful examination of the system of consanguinity and descent of the Iroquois, that we may yet be able, by means of it, to solve the question whether our Indian races are of Asiatic origin. Language changes its vocabulary not only, but also modifies its grammatical structure in the progress of ages; thus eluding the inquiries which philologists have pressed it to answer; but a system of consanguinity once matured and brought into working operation is, in the nature of things, more unchangeable than language;—not in the names employed as a vocabulary of relationship, but in the ideas which underlie the system itself.

---

* By Lewis H. Morgan, Esq., of Rochester, N. Y.

The Indo-European nations have one system, identical in its principal features, with an antiquity of thirty-five centuries, as a fact of actual record. That of the Iroquois is original, clearly defined, and the reverse of the former. It is, at least, to be presumed that it has an antiquity coeval with the race. That of the Chippewa is the same as the Iroquois, with slight modifications; thus establishing the fact of its existence in two of the principal generic stocks. Besides this, there are traces of the same system among the Aztecs, Mohaves, Creeks, Dacotahs, Delawares, Winnebagoes, and other races, all tending to show that the system has been, and now is, universal upon this continent. Should this last fact be established, the antiquity of the system as coeval with the Indian race upon the continent, will also become established. Upon the basis of these two facts, and assuming that these races are of Asiatic origin, we may predict the existence of the same system in Asia, at the present moment, among the descendants of their common ancestors, if any remain.

"A brief explanation of the principal features of the system of the Iroquois is annexed, which will assist in working out every other, particularly if they are founded upon the same ideas.

"The institutions of the Iroquois were founded upon the family relationships; in fact, their celebrated league was but an elaboration of these relationships into a complex system of civil polity. At the base of this were their laws of descent. They were unlike both the civil and the canon laws, but yet were original and well defined. The chief differences were two: first descent among the Iroquois followed the female line, or passed through the mother; while in each of the former systems it follows the male, or passes through the father. In the second place the collateral lines, with the Iroquois, were finally brought into or merged in the lineal; while, in the other cases, every remove from the common ancestor separated the collateral lines from the lineal, until after a few generations actual relationship ceased among collaterals.

"To bring out distinctly this code of descent, it will be necessary to give a brief explanation of the division of the Iroquois into tribes, the union of the several tribes into one nation, and of the several nations into one league. Without a reference to their civil organization, it would be impossible to present it in an understandable form.

"In each of the five nations who composed the original league there were eight tribes, named: Wolf, Bear, Beaver, and Turtle; Deer, Snipe, Heron, and Hawk. The Onondaga nation, therefore, was a counterpart of the Cayuga, each having the same number of tribes, and of the same name; so also, interchangeably, of the Oneida, the Mohawk, and the Seneca nations. In effect, the Wolf tribe was divided into five parts, and one-fifth part of it placed in each of the five nations. The remaining tribes were subjected to the same division and distribution. Between the individual members of the Wolf or other tribe thus divided, or, in other words, between the separated parts of each tribe, there existed the tie of consanguinity. The

Mohawk of the Turtle tribe recognized the Seneca of the Turtle tribe as a relative, and between them existed the bond of kindred blood. In like manner the Oneida of the Hawk tribe received the Onondaga or the Cayuga of the same tribe as a relative, not in an ideal or conventional sense, but as actually connected with him by the ties of consanguinity. Herein we discover an element of union between the five nations, of remarkable vitality and power. A cross-relationship existed between the several tribes of each nation and the tribes of corresponding name in each of the other nations, which bound them together in the league with indissoluble bonds. If either of the nations had wished to cast off the alliance, it would have broken this eight-fold bond of consanguinity. Had the nations fallen into collision with each other, it would have brought Hawk tribe against Hawk tribe — in a word, brother against brother. The history of the Iroquois exhibits the wisdom of these organic provisions; for, during the long period through which the league subsisted, they never fell into anarchy, nor even approximated to a dissolution from internal disorders.

"At no time in the history of the Iroquois could a man marry a woman o his own tribe, even in another nation. All the members of a tribe were within the prohibited degrees of consanguinity; and to this day, among the descendants of the Iroquois, this law is religiously observed. Husband and wife, therefore, were in every case of different tribes. The children were of the tribe of the mother. Here, then, we discover, one of the central ideas of their laws of descent: to place the father and mother in different tribes, and to assign the children to the tribe of the mother. Several important results followed, of which the most remarkable was, the perpetual disinheritance of the male line. As all titles, as well as property, descended in the female line, and were hereditary in the tribe, the son could never succeed to his father's title of sachem, nor inherit even his tomahawk.

"A tribe of the Iroquois, it thus appears, was not, like the Grecian and Roman tribes, a circle or group of families, for two tribes were necessarily represented in every family; neither, like the Jewish, was it constituted of the lineal descendants of a common father; on the contrary, it involved the idea of descent from a common mother; nor has it any resemblance to the Scottish clan, or to the canton of the Switzer. It approaches, however, nearer to the Jewish. Denying geographical boundaries, a tribe of the Iroquois was composed of a part of a multitude of families, as wide spread as the territories of the race, but yet united together by a common tribal bond. The mother, her children, and the descendants of her daughters, in the female line, would, in perpetuity, be linked with the fortunes of her own tribe; while the father, his brothers and sisters, and the descendants in the female line of his sisters would be united to another tribe, and held by its affinities. No circumstances could work a translation from one tribe to another, or even suspend the nationality of the individual. If a Cayuga woman of the Hawk tribe married a Seneca, her children were of the Hawk tribe and Cayugas, and her descendants in the female line, to the latest posterity, continued to be Cayugas and of the Hawk tribe, although they resided with the Senecas, and by successive intermarriage with them had lost nearly every particle of Cayuga blood. Neither could intermarriage with one of a foreign nation confer the Iroquois nationality upon the wife or children of the marriage, and the same *vice versâ*. If a Mohawk married a Delaware woman, she and her children were not only Delaware still, but ever continued aliens, unless naturalized as Mohawks, with the forms and ceremonies prescribed in case of adoption."

The difficulty of obtaining reliable information respecting the Indian population has been acknowledged by all who have given attention to this subject. I am convinced that the number of Indians inhabiting Rupert's Land has been considerably overrated. The estimates published in the Appendix to the Report from the Select Committee on the Hudson's Bay Company furnish the following result:—

Thickwood Indians on the east side of the Rocky Mountains	35,000
The Plain Tribes (Blackfeet, &c.)	25,000
	60,000

The Indian population of Rupert's Land is estimated at 42,870. Over the plain or prairie tribes the Hudson's Bay Company profess to have no control, and they are returned as numbering 25,000 souls. Excellent authorities, noticed in the following pages, do not assign more than half that number to the most numerous tribes of Prairie Indians who hunt on the Saskatchewan and Missouri and their tributaries.

The Plain Crees and Thickwood Indians are under the control of the Company, but I think that their numbers are also over estimated, and the grounds on which this opinion is advanced are stated in the following paragraphs.

The basis of the census for the Thickwood Indians and the Plain Crees is the number frequenting the establishments of the Hudson's Bay Company in 1856, and the following enumeration at certain Posts chiefly visited by the Plain Crees is given:—

Post.	No. of Indians frequenting it.
Fort Ellice	500
Qu'appelle Lakes	250
Touchwood Hills	300
Fort à la Corne	300
	1,350 Indians.

Upon perusal of the foregoing table the reader would infer that 1,350 Indians visited the Posts named. It happens, however, that many Indians trade with two or more Posts, although every effort is made to limit them to one particular station. Their names appear on the books at different establishments, and in the enumeration of the Indians inhabiting certain districts, some of them are counted twice and even three times. I ascertained beyond doubt, that this practice existed to an extent which would affect the census in a marked degree. The custom of giving credit to Indians encourages this system, while a natural desire to attach additional hunters to a Post on the part of the traders, induces less caution than would otherwise be exercised. As the result of very careful inquiries wherever opportunities offered, of obtaining exact information, I am inclined to think that the estimate of 42,870 is about one-fourth too high.

The estimated number of Indians frequenting particular establishments of the Hudson's Bay Company referred to in these volumes, during 1856, is given in the following table:—

Locality.	Number.
Fort William	350
Pigeon River	50
Fort Frances	1,500
Rat Portage	500
Lac de Bois Blanc	200

Locality.	Number.
Shoal Lake	200
White Dog	100
Fort Alexander	300
Lac de Bonnet	50
Fort à la Corne	300
Cumberland House	250
The Pas	300
Fort Pelly	800
Fort Ellice	500
Qu'appelle Lakes	250
Shoal River	150
Touchwood Hills	300
Egg Lake	200
Manitobah House	200

On the North Branch of the Saskatchewan, where the Prairie Indians assemble, the following enumeration is given in the Blue Book:—

Locality.	No. of Indians.
Edmonton	7,500
Carlton	5,000
Fort Pitt	7,000
Rocky Mountain House	6,000

This census may approximate to the actual number of Indians visiting a particular Post, yet there is strong reason to suppose that the same individuals are to a large extent enumerated twice if not thrice.

The Plain or Prairie Indians belong to the following principal tribes:—

Blackfeet,
Bloodies,
Piegans,
Fall Indians, or Gros Ventres,
Crees,
Assinniboines,
Sioux.

The Wood Indians of the Saskatchewan Valley belong to the great family of Crees and Ojibways. The Sioux, Blackfeet, Bloodies, and Piegans are Dakotahs.

Mr. Harriet, a chief-factor of the Hudson's Bay Company, who had passed his life among the Blackfeet, estimated the six or seven tribes going by that general name as mustering 1,600 to 1,700 tents, at eight per tent, 13,000.*

Mr. Rowand, one of the oldest resident traders, estimates the Blackfeet tribes as follows: —

Blackfeet proper	300
Piegans	400
Bloods	250
Gros Ventres, or Fall Indians	400
Circes	45
Cotones, Small Robes } Mountain Tribes	250
At 8 persons per tent, 13,100.	1,645 tents.

The Assinniboines are divided into Strongwood and Plain Assinniboines, or Stonys.

Mr. Harriet, in 1842, estimated the

Strongwood Assinniboines	at 80 tents =	640
Mr. Rowand, the Plain Assinniboines	„ 300 „ =	2,400
	380 tents =	3,020

The Strongwood Crees about Edmonton Mr. Rowand estimated at . 400 tents, at 10 per tent =	4,000
Crees of the Plains . 200 „ „ „ =	2,000
	6,000

Colonel Lefroy† states that the aggregate of the tribes inhabiting the plains on British territory was estimated in 1843 at not more than 23,400. Since that period they have diminished in numbers, and some of the Blackfeet bands have stationed themselves permanently on the Missouri. In succeeding pages, recent estimates of the Blackfeet tribes, and the limits of their hunting-grounds are given.

* Colonel Lefroy.

† "On the probable number of the native Indian population of British America," by Captain (now Colonel) J. H. Lefroy, R.A. Canadian Journal, vol. i. Old Series.

The Sioux and the Blackfeet being the most warlike tribes of the north-west, and retaining their ancient customs to the fullest extent, the following brief notices of these formidable native races are introduced. The Plain and Wood Crees and the Ojibways are almost altogether amenable to the influence of the Hudson's Bay Company, and are in fact the hunters upon whom they rely for a considerable proportion of their furs, robes, skins, and provisions.

THE SIOUX OR DAKOTAH INDIANS.

The nation of the Sioux Indians or Dakotahs * is composed of seven principal bands. Their aggregate number probably does not exceed twenty-five thousand. Their hunting-grounds extend from the Mississippi River to the Black Hills in Nebraska, and from the mouth of the Big Sioux River to Devil's Lake. Although the Sioux have no dealings with the half-breeds of Red River, or with the Hudson's Bay Company, yet they often cross the 49th parallel in pursuit of the buffalo, and more frequently in search of a scalp from their hereditary enemies, the Ojibways and Crees. As the most dreaded invaders of the prairies north of the boundary line, this powerful nation deserves a special notice.

The name Dakotah signifies the "Allied," and they speak of themselves as the "Oceti sakowin" or "Seven Council Fires." The following enumeration of the principal bands which compose the nation, by the members of the American Dakotah Mission, will be found at length in the Grammar and Dictionary prepared with so much care, labour, and zeal, under the editorial management of

* See Introduction to a Grammar and Dictionary of the Dakotah Language, published by the Smithsonian Institution.

the Rev. S. R. Riggs, A. M., Missionary of the American Board of Commissioners for Foreign Missions.

1. The Mdewakantonwans, *Village of the Spirit Lake.* The name is derived from Mdewakan (Spirit or Sacred Lake), Mille Lacs (Minnesota), in the country now claimed by the Ojibways. This band numbers about two thousand.

2. The Wahpekutes, *Leaf Shooters*, five hundred.

3. The Wahpetonwans, *Village in the Leaves*, twelve hundred.

4. The Sisitonwans, *Village of the Marsh*, two thousand five hundred. Their hunting-ground is about the Coteau des Prairies, and they subsist on the buffalo.

5. The Ihanktonwanna, *the End Village Band*, four thousand. Their country is on the north-east of the Missouri, as far as Devil's Lake. These are the great enemies of the Red River half-breeds.

6. The Ihanktonwans, *the Village at the End*, two thousand four hundred. Their country is west of the Missouri. They are frequently termed Yanctons.

7. The Titonwans, *the Village of the Prairie*, twelve thousand five hundred. Their hunting-ground is west of the Missouri. They are divided into seven bands: the Sicaugu, *Burnt-Thighs;* the Itazipco, *Bow-pith;* the Sihasapa, *Blackfeet;* the Minikanye wozupi, *Those who plant by the water;* the Oohenoupa, *Two-boilings;* and the Oglala and Hunkpapa.

The conjurors believe that their dreams are revelations from the Spirit World, and they aver that their prophetic visions are the mental revival of occurrences in a former state of existence. Years with them are enumerated by winters; a distance is estimated by the number of nights a man will sleep on the way. The Ojibways have the same method of expressing time and distance. They divide the year into moons, but weeks are unknown to them. The Dakotahs of the valley of the Minnesota have the following months in the year *: —

1. Wi-tehi, January; the hard moon.
2. Wicata-wi, February; the racoon moon.
3. Istawicayazan-wi, March; the sore (eye) moon.
4. Magaokada-wi, April; the moon in which the geese lay eggs.
5. Wozupi-wi, May; the planting moon.
6. Wazustecasa-wi, June; the moon when the strawberries are red.
7. Canpasapa-wi, July; the moon when the choke cherries are ripe.
8. Wasutou-wi, August; the harvest moon.
9. Psinhnaketu-wi, September; the moon when rice is laid up to dry.
10. Wi-wazupi, October; the drying rice moon.
11. Takiyuha-wi, November; the deer rutting moon.
12. Tahecapsun-wi, December; the moon when the deer shed their horns.

The Dakotahs have a common and a sacred language. The conjuror, the war prophet, and the dreamer employ a language in which words are borrowed from other Indian tongues and dialects; they make much use of descriptive expressions, and use words apart from the ordinary signification. The Ojibways abbreviate their sentences and employ many elliptical forms of expression, so much so that half-breeds, quite familiar with the colloquial language, fail to comprehend a medicine man when in the full flow of excited oratory.

The American missionaries, in their admirable written Dakotah language, employ five vowels, and twenty-four consonants, among which are two *c*'s, two *g*'s, two *h*'s, two *k*'s, two *n*'s, two *s*'s, two *t*'s, and two *z*'s. The repetition of the same letter is used to denote a guttural, an aspirate, an emphatic, or a nasal sound. Thus, *c* is both an aspirate and an emphatic letter; *g* like the English *g* and guttural; *h* like the English *h* and guttural; *k* as in English and emphatic; *n* as in English and nasal; *p* as in English and emphatic; *s* as in English and aspirate;

* See Grammar and Dictionary before referred to

t as in English and emphatic; *z* as in English and aspirate.

All syllables are enunciated plainly and fully, but accentuation often determines the meaning of a word. There are three numbers: singular, dual, and plural; the dual including the person speaking and the person spoken to. The proper names of the Dakotahs are words, simple and compounded, which are in common use in the language. The son of a chief when he succeeds his father usually takes the name of his father or grandfather. As with the Ojibways and Swampys, their proper names consist of a single noun or a noun and adjective. The Ojibways have, however, distinct family or "*totem*" names which they employ when speaking of their ancestors; as I am of the family of the Bear, the Eagle, the Thunder-cloud, &c. The Dakotahs have no surnames, the children of a family have particular names which belong to them in the order of their birth up to the fifth child. In counting they use their fingers, bending them as they enumerate until they reach ten. They then bend down a little finger to record one ten and begin again; when the second ten is counted they put down a second finger, and so on.

Dakotah verbs have only two forms of tense, the indefinite and the future; the other tenses are expressed by the help of adverbs, and the context. Words in a sentence are thus placed, first the noun, second the adjective, third the verb, thus:—

Ateunyanpi mahpiya ekta nanke chin
Father-we-have heaven in thou-art the;
Nichaze kin wakandapi kte;
Thy-name the holy-regarded shall;
Nitokichonze kin u kte;
Thy-kingdom the come shall.*

* See a Grammar and Dictionary of the Dakotah Language, published by the Smithsonian Institution.

THE BLACKFEET.

Mr. James Doty, who resided for many years in the country of the Blackfeet, and who is acquainted with a large portion of that nation, gave the following boundaries of their country and estimate of the numbers of the people to Governor Stevens in 1853.* The country in which they reside and hunt is bounded as follows: "By a line beginning on the north, where the 50th parallel crosses the Rocky Mountains, thence east on said parallel to the 106th meridian, thence south to the headwaters of the Milk River, down said river to the Missouri, up the Missouri to the mouth of the Judith, thence up the Judith to its source in the Rocky Mountains, and north along their base to the place of beginning."

The country between the Missouri and the headwaters of the Yellowstone is unoccupied. It is the war road of the Blackfeet to and from the Crows, Flatheads, and Snakes. It may also be considered as a transient hunting-ground of the Flatheads, who hunt buffalo there for a time in the fall.

The Blackfeet nation is divided into four distinct tribes or bands, whose names, numbers, and localities † are as follows: —

	Lodges.	Population.	Warriors.
The Blackfeet . . .	250	1750	625
The Blood ‡ . . .	350	2450	875
The Piegans . . .	350	2450	875
The Gros Ventres . .	360	2520	900
Total . . .	1310	9170	3275

* Explorations and Surveys for a Railroad Route from the Mississippi to the Pacific, p. 443.

† The country occupied by these tribes is evidently more extensive than supposed by Mr. Doty; their permanent lodges are found far beyond the limits given in the text.

‡ Called by the half-breeds "Bloodies."

The Bloods and Blackfeet occupy the country between Milk and Marias Rivers, to the 50th parallel of latitude.

The Piegans occupy the country between the Milk and Marias Rivers, and between the Teton and the Missouri.

The Gros Ventres occupy the country bordering upon Milk River from its mouth to the territory of the Piegans. The Bloods, Piegans, and Blackfeet speak the same language, the Gros Ventres the Arapahoe language; they were adopted by the Blackfeet about thirty years since, having seceded from their own nation. On the Upper Missouri, near the great bend, the Gros Ventres have a large village of mud houses. Some of the lodges are capable of supporting 100 persons; one part is appropriated to their horses, dogs, cattle, and chickens, another to their sleeping apartments; the lodges are built entirely by women. The Gros Ventres formerly hunted on the Assinniboine. Mr. J. M. Stanley, the artist of Governor Stevens' Exploration, states that the Blackfeet proper are divided into three distinct bands: the Blood band, 400 lodges; the Piegan band, 430 lodges; and the Blackfeet band, 500 lodges, averaging ten to a lodge, and amounting in all to 13,300 souls. The Piegans and Bloods hunt, trade, and winter on American soil, while the Blackfeet extend their hunt as far north as the Saskatchewan, and trade as frequently with the British as with the American Posts.*

The following census of the Indian tribes of the United States, inhabiting the states and territories adjoining the 49th parallel, is abstracted from the statistics of the tribes as reported to the Bureau of Indian Affairs †:—

* Explorations and Surveys for a Railroad to the Pacific, p. 449.

† See the History, Condition, and Prospects of the Indian tribes of the United States, by H. R. Schoolcraft, LL.D.

Name of tribe.	Numbers.	
Assinniboine	. 8900	Extending from the Missouri into Rupert's Land.
Blackfeet	. 9530	Nebraska.
Bloods	. 1612	Upper Missouri.
Crees	. 800	Upper Missouri.
Sioux (Ihanktonwanna) .	. 4000	Dakotah territory.
Gros Ventres	. 2500	Between the Missouri and the Saskatchewan.

The subjoined enumeration of Indian tribes inhabiting British North America, is from a paper "On the probable number of the Native Indian Population of British America," by Colonel Lefroy, referred to on page 152 :—

Chipewyan tribes—namely: Chipewyans proper, Dogribs, Hare or Slave Indians, Yellow Knives, Beaver Indians, Da-ha-dumies and Carriers	7,575
Northern Indians of the *Kutchin* stock	6,082
Indians of the Plains (Blackfeet, Assinniboines, &c.) .	23,400
Chipeways (Ojibways) and Crees, exclusive of the above .	8,675
Indians of the Seaboard and Islands of the Pacific . .	63,840
Indians of New Caledonia—Interior	2,000
Indians of Canada	13,000
	124,518

The number of Indians frequenting the establishments of the Hudson's Bay Company, in 1856, are thus classified in the Blue Book * : —

Thickwood Indians on the east side of the Rocky Mountains .	35,000
The Plain Tribes (Blackfeet, &c.)	25,000
The Esquimaux	4,000
Indians settled in Canada	3,000
Indians in British Oregon and on the north-west coast . .	80,000
Total Indians	147,000

The census of the Indian tribes of the United States with whom intercourse was kept up by agents in the year

* Report from the Select Committee on the Hudson's Bay Company.

1855 is stated in the Annual Report of the Indian Bureau of that year to include 314,622 souls.

An approximation to the total Indian population of the United States and British America will be as follows : —

Indians of the United States	314,622
Indians of British America (Colonel Lefroy) . . .	124,518
Total	439,140

Or,

Indians of the United States	314,622
Indians of British America, according to the census of the Hudson's Bay Company	147,000
Total	461,622

The records of the early history of the Indians who formerly occupied Canada and the northern States of the Union prove that their numbers, during the first half of the seventeenth century, must have at least quadrupled the entire aboriginal population now occupying the vast territories under the control of the Hudson's Bay Company.

The extraordinary mutability of nations in the savage state, and the rapidity with which one race supplants another over large areas, is thus noticed by a recent writer on the early discoveries of the French in North America * : —" When Cartier arrived in the St. Lawrence, he described large and permanent Indian villages at Stadacona and Hochelaga ; but little more than half a century afterwards, when Champlain visited the same localities, he apparently found few Indians about Quebec, and none permanently settled at Montreal. There may have been some exaggeration in Cartier's account, but the main fact remains, and it may probably be accounted for by

* " On the Early Discoveries of the French in North America," by John Langton, M.A., Auditor of Public Accounts, Canada. A paper read before the Canadian Institute, and published in the " Canadian Journal," New Series, No. 12.

the increasing power of the Iroquois, which made those places dangerous abodes, and compelled the tribes which formerly occupied them, to retreat into the interior. Again, the country north of Lake Ontario is described by Champlain as affording signs of having been formerly extensively cultivated and thickly inhabited, but in his day it was entirely deserted, and only used as a hunting ground by the neighbouring tribes. But the country of the Ottawa, and across to the northern shore of Lake Huron, as also the Western Peninsula, is described as full of Hurons, and of Algonquin, Ottawa, Nipissing, and other allied tribes. Amongst the Hurons alone, in the limited area between Matchedash Bay and Lake Simcoe, he reckons eighteen walled villages, numbering 2,000 fighting men, and Sagard puts the whole population down at 30,000 or 40,000 souls.* Yet, within thirty years from that time this region was also a desert, and the remnants of the former inhabitants had retreated to the Northern Lakes, and as far west as the Sioux. The Hurons indeed were almost exterminated, and the paltry remnant which had not been either destroyed or incorporated with other tribes, were collected and brought down to Quebec, where their descendants still occupy the village of Lorette. All the tribes of the Western Peninsula, and the Eries on the south shore of that lake, seem also to have been utterly exterminated, as well as the greater part of the Illinois,

* It would not appear that this estimate can have been very greatly exaggerated, from the account given of the missionary establishments. They numbered in their most flourishing period, about the year 1645, forty-two missionaries besides their attendants. Of these two or three only remained at the principal station of Ste. Marie, at the mouth of the Wye, five other villages were called residences, where one or two missionaries remained permanently, and the rest moved from village to village often having as many as ten under their charge. As several of these villages are mentioned as containing from 100 to 200 cabins, and four to five families residing in each, the whole population cannot have fallen far short of 30,000.

and other western tribes; and the Iroquois were dominant over all Upper Canada, and all the northern part of New York and Ohio. All this occurred without the intervention of the white man; and there has been no disappearance of a savage race since, from the diseases and vices which civilization brings in its train, which has surpassed, even if it has equalled in completeness and rapidity, the desolation which the conquering Iroquois spread around them. They, too, have now nearly vanished from the scene of their former power under other influences, and may soon, like the Eries and Hurons, be remembered only by a name; but when we find such extraordinary vicissitudes occurring during the brief space, of which we have any certain record, we cease to be so much surprised at the total disappearance of the mound builders and other prehistoric races."

The Mandans and Assinniboines who hunt on both American and British soil, and who are essentially prairie Indians, were estimated in 1783 to be capable of sending into the field 25,000 and 40,000 fighting men respectively. In 1786 the small-pox, coming north from the Mexican provinces, almost depopulated the country. In 1838 the same disease swept off at least one half of the prairie tribes. Five-and-twenty years ago, before this epidemic, aided by constant wars, had reduced the Plain Crees to one-sixth or eighth of their former numbers, Fort Ellice was often the scene of exciting Indian display. The officer in charge in July 1858, remembers the time when the entire tribe who now hunt on the Qu'appelle and South Branch would approach the Fort to receive their supplies, preceded by 800 mounted warriors, singing their war songs. Twenty-five years ago the tribe numbered 4000, in 500 tents; at the present day they do not exceed 120 tents, which represent a population of 960 or 1000 souls.

As stated in a preceding chapter, small-pox and measles produced a great mortality among the Wood Indians north of the Saskatchewan in 1816, 1817, and 1818. The ravages of this scourge began to tell in a fearful manner upon the native races in the valley of the St. Lawrence as early as the latter part of the seventeenth century. Charlevoix relates in 1670 that there were rarely less than 1200 Indians to be seen encamped at Tadousac, the entrepôt of the fur trade at that period during the trading season. The small-pox put an end to the trade by almost annihilating the Indians. Some tribes were quite exterminated, others amalgamated with surviving tribes, or carried their furs to the English Fort on Hudson's Bay.

When the Iroquois, who formerly occupied permanent villages on the south shore of Lake Ontario and the south bank of the St. Lawrence, were first known to Europeans, they alone were estimated by La Hontan at 70,000 souls. The numbers of the Indian population of British North America, on the east side of the Rocky Mountains, amounted to 67,000 in 1856, according to the Hudson's Bay Company. During that year the small-pox again visited the plain tribes, coming up the Missouri River, and destroying fully 3000 from among those who hunt on the Upper Missouri and between both branches of the Saskatchewan. In 1857 and 1858 it still lingered near the foot of the Rocky Mountains, and despair was stamped on the faces of the unhappy Mandans who were visited by Lieut. Warren soon after the scourge appeared amongst them. Wars, disease, and starvation have reduced to at least one-twentieth part of its former numbers an aboriginal population which two centuries ago occupied this vast area. How long will the remnant be preserved to minister to the cupidity of the white race who now rule and sustain them only so far as they are subservient to the objects of the fur trade?

Before the dispersion of the Hurons, about the year 1650, the customs observed by that people relative to the dead were as follows. Under the impression that the spirit, notwithstanding its separation from the body, did not immediately take its departure, the women were accustomed to frequent the grave of the deceased with tears,

Indian Graves covered with Split Sticks.

moans, and other outward signs of grief. The corpse was placed in a burying-ground called by the Hurons Oi-go-sa-yé. If the death had been natural, each corpse was encased separately in birch bark and elevated on four poles. They remained there until the celebration of the "Feast of the Dead," which took place every eight or ten years. At this period the inhabitants of a village taking down each bier in their Oi-go-sa-yé, carefully removed the dried flesh from the bones and wrapped the skeleton in furs and skins. The bones of the dead having been thus gathered from a wide extent of country * were placed with much solemnity in a large excavation richly decorated with furs. Valuables of different kinds were deposited with the remains, under the belief that they would be required by those to whom they belonged in another world or state of existence.

When the death has been violent or unnatural, the corpse was burned or buried immediately; and should a Huron have been frozen to death, the corpse would be carefully dissected, and the skeleton buried, never to be exhumed. The Hurons believed that the spirits of those who died in war, or suffered a violent death in any other way, enjoyed no communion in a future life with those who died in the ordinary course of nature.†

An ancient ossuary of the Hurons was opened near Penetanguishene in 1846. Its appearance externally was that of a mound about twenty-eight feet in diameter, covered with large trees which had grown upon it after its construction. An immense shroud of beaver skin enveloped the sacred deposit. Twenty-six copper kettles, hatchets, marine shells, bracelets or belts of wampum, &c., were placed near to the bones.

The Jesuit Missionary, P. de Brebeaf, who assisted at one of the " Feasts of the Dead " at the village of Ossossane before the dispersion of the Hurons, relates that the ceremony took place in the presence of 2000 Indians, who offered 1200 presents at the common tomb in testimony of their grief. The people belonging to five large villages deposited the bones of their dead in a gigantic shroud, composed of forty-eight robes, each robe being made of ten beaver skins. After being carefully wrapped in this rich shroud, they were placed between moss and

* Compare the description of ossuaries in Western Canada, Vol. I. p. 90.

† Relations Abrégées.

bark. A wall of stones was built round this vast ossuary to preserve it from profanation. Before covering the bones with earth, a few grains of Indian corn were thrown by the women upon the sacred relics.

According to the superstitious belief of the Hurons the souls of the dead remain near the bodies until the "Feast of the Dead;" after which ceremony they become free, and can at once depart for the Land of Spirits, which they believe to be situated in the regions of the setting sun.*

* Relations Abrégées.

INDIAN GRAVES

CHAP. XXXI.

INDIAN TITLE TO RED RIVER.

Indian Title in Canada.—Importance of the Question in Rupert's Land.—Cost of Indian Wars to the United States' Government.—Advance of Settlements towards the West.—Probability of a War with the Sioux.—Indian Races occupying the Country available for present Settlement in Rupert's Land.—Restlessness of these People.—The Right Hon. E. Ellice, M.P., on Indian Title in Canada.—Proclamation of 1763.—Opinion of the Canadian Commissioners on Indian Affairs with Respect to Indian Title in Canada.—Title to Red River.—Grant to Lord Selkirk.—Treaty between Lord Selkirk and the Crees and Saulteaux of Red River.—Peguis.—His Letter to the Aborigines' Protection Society.—His Address in 1859 to the "Great House."—M. MacDermott's Statement.—Meeting of the Half-Breeds of Red River.—Opinion respecting Indian Title.—Importance of the Question.—Treaty of the Americans with the Saulteaux for the northern part of Minnesota on Red River.

The question of Indian title is one of very great interest and importance in regard to the future peace of the colony, and as much misapprehension appears to exist respecting the territorial rights of different tribes of Indians, and their title to the land they now claim, the present condition of the question may be noticed here, as far as the slender and unconnected evidence at command admits. In Canada much trouble, great expense, and endless inquiry have been created by Indian claims, which even now remain in part unsettled, and are a source of many incidental expenses to the Government, which might have been avoided if proper arrangements had been made at the right season. In Rupert's Land, where disaffected Indians can influence the savage prairie tribes and arouse them to hostility, the subject is one of great magnitude; open war with Sioux, Assinniboines, Plain Crees or Blackfeet, might render a vast area of prairie country unapproachable for many years, and expose the settlers to constant alarms and depredations. The Indian wars undertaken by the United States Government during the last half century, have cost infinitely more than the most liberal annuities or comprehensive efforts for the amelioration of the condition of the aborigines would have done; and in relation to the northern prairie tribes, war is always to be expected at a day's notice.

The encroachments of western settlers upon Indian lands are constant and increasing in the United States, and there is no reason to suppose that these encroachments will diminish for many years to come. Already the Red River south of the boundary line, as well as its south-western tributaries, are invaded from the valley of the Mississippi, and as the territory of Dakotah has not yet been ceded to the United States Government, the prospect of a war with the Sioux, whose hunting grounds embrace it, becomes daily more imminent. Lieutenant Warren, who has conducted several United States' exploring expeditions in Dakotah and Nebraska territories, remarks: "The advance of the settlements is universally acknowledged to be a necessity of our national development, and is justifiable in displacing the native races on that ground alone. But the government, instead of being so constituted as to prepare the way for settlement by wise and just treaties of purchase from the present owners, and proper protection and support for the indigent race so dispossessed, is sometimes behind its obligations in these respects; and in some instances Congress refuses or delays to ratify the treaties made by the duly authorized agents of the government. The result is, that the settler and pioneer are precipitated into the Indian's country, without the Indian having received the first consideration promised him; and he often, in a manner that enlists the sympathies of all mankind, takes up the tomahawk in defence of his right and perishes in the attempt." * The same officer states that there are so many inevitable causes at work to produce a war with the Dakotahs (Sioux) before many years, that he regards the greatest fruit of his explorations to be the knowledge of the proper routes by which to invade their country and conquer them, but at the same time he thinks that many of the causes of war with them might be removed by timely action in relation to the treaties made with them.

The country of the Dakotahs borders on British territory, some of the tribes (the Ihanktonwanna, par. 5, p. 154, Vol. II.) are the confirmed enemies of the half-breeds and Ojibways of Red River; peace has often been made, but as often broken again upon trivial and even accidental grounds.

The frontier tribes can muster at least two thousand warriors by uniting with several of their more southern allies. Being the most warlike and numerous Indians in the United States territories, and their hunting grounds interlocking with those of the Crees in British America, they will probably yet play an important and active part in the future of the colony and the new adjoining territory of Chippewa.†

* Preliminary Report of Explorations in Nebraska and Dakotah in the years 1855-6 and 1857, by Lieut. G. K. Warren, Top. Eng. U.S. army.

† The new territory of Dakotah is to be composed of a part of the present territory of that name and a portion of Nebraska, and bounded as follows:— The forty-sixth meridian of north latitude on the north, Minnesota and

Thickwood Crees, Swampy Crees, Plain Crees, and Ojibways are the Indian nations who now occupy that part of Rupert's Land where settlements would first be made. These nations are friendly to one another and hostile to the Sioux. They are, in fact, the hunters of the Hudson's Bay Company, and consequently friendly with that body, who have never sought to extend the settlements of the white race in Rupert's Land; but of late years since the questions relating to title to lands, annuities, and compensation have been raised, they are becoming dissatisfied, suspicious, and untrustworthy.

The Right Honourable Edward Ellice, M.P., in reply to a question put by Mr. Christie during his examination before the Select Committee on the Hudson's Bay, respecting the extinction of the Indian title in Rupert's Land, stated that "the English Government never extinguished the Indian title in Canada when they took possession; the Americans, while they have been extending their possessions, have extinguished the Indian title, but in Canada there has never been any treaty with the Indians to extinguish the title, the Crown, retaining certain reserves for the Indians, has always insisted upon the right to

Iowa to the mouth of the Big Sioux on the east, on the south following the Missouri river from the mouth of the Big Sioux to the mouth of the Nebraska, and along the Niobarah to the one hundred and second meridian of west longitude, along the hundred and second meridian to the forty-third parallel north latitude, thence along the forty-third parallel to the crest of the Rocky Mountains, and on the west by Washington Territory. This territory will consist of about one hundred and thirty-five thousand square miles. This does not include or interfere with any of the settled portions of Nebraska.

Chippewa is an entirely new territory, and is composed of the northern part of Dacotah and Nebraska, bounded as follows:—The British Possessions on the north, Minnesota on the east, the forty-sixth parallel of north latitude on the south, and Washington on the west. This will make an area of about one hundred and thirty thousand square miles.

occupy the lands, and to grant the lands."* This reply to a question now about to assume an importance in regard to the Red River Settlement of the greatest magnitude, is liable to produce a very erroneous impression respecting the Indian title in Canada, and the respect which has been paid to it during the history of that country. It is desirable that this point should be clearly stated, and with this object the proclamation of 1763 as far as it relates to the Indians in Canada is appended.

"And we do further declare it to be our royal will and pleasure, for the present as aforesaid, to reserve under our sovereignty, protection, and dominion, for the use of the said Indians, all the lands and territories not included within the limits granted to the Hudson's Bay Company; as also the lands and territories lying to the westward of the sources of the rivers which fall into the sea, from the west and north-west as aforesaid. And we do hereby strictly forbid, on pain of our displeasure, all our loving subjects from making any purchases or settlements whatever, or taking possession of the lands above reserved without our special leave and license for that purpose."

The report of the commissioners appointed to investigate the Indian affairs in Canada in 1847, thus state their views on the question of title to lands :—

"Although the Crown claims the territorial estate and eminent dominion in Canada, as in other of the older colonies; it has, ever since its possession of the province, conceded to the Indians the right of occupancy upon their *old hunting grounds*, and their claim to compensation for its surrender, reserving to itself the exclusive privilege of treating with them for the surrender or purchase of any portions of the land. This is distinctly laid down in

* Question 6001, Blue Book.

the proclamation of 1763, and the principle has since been generally acknowledged and rarely infringed upon by the Government. The same rule has been followed by the Government of the United States, who pay annuities for the surrender of Indian lands to the extent of about 140,000*l.* a-year."*

Great and apparently reasonable doubt exists respecting the Indian title to that part of the valley of Red River and the Assinniboine now occupied by the settlements. The royal charter for incorporating the Hudson's Bay Company, granted by Charles II., A.D. 1670, transferred to the Company the trade, lands, mines, minerals, fisheries, &c., of Rupert's Land. The territory to be reckoned one of his Majesty's plantations or colonies in America, and the Governor and Company to be the Lords Proprietors of the same for ever.†

On the 12th June 1811 the Hudson's Bay Company made a grant of lands to Lord Selkirk included within the following boundaries:—"All that tract of land or territory bounded by an imaginary line running as follows, that is to say, beginning on the western shores of the Lake Winnipeg at a point in 52° 30′ north latitude, and thence running due west to the Lake Winnepego-sis, then in a southerly direction through the said lake so as to strike its western shore in latitude 52°, then due west to the place where the 52° intersects the western branch of Red River, the Assinniboine River, then due south from that point of intersection to the height of land, which separates the waters running into Hudson's Bay from those of the Missouri and Mississippi, then in an easterly direction

* "Report on the Affairs of the Indians in Canada." Legislative Council, Sessional Papers, Appendix T, 1847.

† See the Royal Charter of Incorporation, page 409 of the Report from the Select Committee on the Hudson's Bay Company.

along the said height of land to the source of the Winnipeg River (meaning by such last-named river the principal branch of the waters which unite in Lake Seiganagah), thence along the main stream of these waters, and the middle of the several lakes through which they flow, to the mouth of the Winnipeg River, and thence in a northerly direction through the middle of Lake Winnipeg to the place of beginning."*

Ross, in his "Red River Settlement, its Rise, Progress, and Present State," introduces a treaty made between Lord Selkirk and certain Indian chiefs, Crees and Saulteaux (or Ojibways), on the 18th July, 1817, in which the chiefs agree to give unto the king, for the use of the Earl of Selkirk, a considerable tract of land on the Assinniboine and Red Rivers for the quit-rent of 100 lbs. of tobacco, to be paid annually to the chiefs and warriors of the Cree and Saulteaux tribes then occupying the country.†

In 1857 Peguis, an immigrant from Pigeon River, Lake Superior, at Red River, sent a letter to the Aborigines' Protection Society, London, complaining of the non-fulfilment of this treaty. The following extract from the letter sent by Peguis is published in the Blue Book‡:—

"Many winters ago, in 1812, the lands along the Red River, in the Assinniboine country on which I and the

* Part of this deed is published in the narrative of Major Long's Expedition to the Source of St. Peter's River, 1824. The above extract is from the Blue Book. Major Long's quotation varies in the spelling of a few names of places, and concludes with the words, "which territory is called Ossiniboia."

† This treaty is signed by Lord Selkirk and the following chiefs:—

Moche W. Keocab (Le sonent).	Ouekidvat (Grande oreilles).
Mechudewikonaie (La robe noir).	Kayajiekebinoa (L'homme noir).
Pegowis.	

‡ Page 445, Report from the Select Committee on the Hudson's Bay Company.

tribe of Indians of whom I am chief then lived, were taken possession of, without permission of myself or my tribe, by a body of white settlers. For the sake of peace, I, as the representative of my tribe, allowed them to remain on our lands on their promising that we should be well paid for them by a great chief, who was to follow them. This great chief, whom we call the Silver Chief (the Earl of Selkirk), arrived in the spring after the war between the North-west and Hudson's Bay Companies (1817). He told us he wanted land for some of his countrymen, who were very poor in their own country; and I consented, on the condition that he paid well for my tribe's land, he could have from the confluence of the Assinniboine to near Maple Sugar Point on the Red River (a distance of twenty to twenty-four miles), following the course of the river, and as far back on each side of the river as a horse could be seen under (easily distinguished). The Silver Chief told us he had little with which to pay us for our lands when he made this arrangement, in consequence of the troubles of the North-west Company. He, however, asked us what we most required for the present, and we told him we would be content till the following year, when he promised again to return, to take only ammunition and tobacco. The Silver Chief never returned, and either his son or the Hudson's Bay Company have ever since paid us annually for our lands only the small quantity of ammunition and tobacco which in the first instance we took as a preliminary to a final bargain about our lands." . . . .

In March, 1859, Peguis dictated another letter on the subject of the title of his tribe to a portion of the lands on Red River. This singular communication, as published in the "Aborigines' Friend and Colonial Intelligencer," is as follows:—

"I Peguis, × (his mark), Salteaux Chief of the Indian Settlement at Red River, wish to make my statement to the Great House across the great waters.

"I and my people have our minds much disturbed by the Hudson's Bay Company, because the said Company have never arranged with me for our lands. We never sold our lands to the said Company, nor to the Earl of Selkirk; and yet the said Company mark out and sell our lands without our permission. Is this right? I and my people do not take their property from them, without giving them great value for it, as furs and other things, and is it right that the said Company should take our landed property from us without our permission, and without our receiving payment for the same? I have asked the said Company for payment, through their agents, and I asked Mr. Mactavish for the same thing, last spring, but I got nothing for my lands.

"If I were nearer the Great House, I would speak much and loud. I and my people are disturbed, and will the Great House approve of another Fur Company being chartered from Canada? Will there be another Company for the North, and another for the South? Will the Great House sanction more hostilities as before, when there were two Fur Companies trading in our country? And will another Company take in land for five miles on each side of the great road to be made between this place and Canada, without consulting me and my brother chiefs? I speak loud: listen! We have had enough of all Fur Companies. Please send us out rather mechanics and implements to help our families in forming settlements, and to secure as reserves, &c.

"I, Peguis, × (his mark), moreover, hereby agree with the letters which my brother chiefs, Makasis, Kes-kisimakurs and Wa-was-ka-sis, sent across the great waters to Mr. Isbister, and to the Aborigines' Protection Society last spring about our lands, and pray the great Mother to take us all under her own protection, and to rule the country for us herself.

"PEGUIS, × (his mark).

"Given under my hand this 21st day of March, 1859.

"Signed by Peguis, Saulteaux Chief of the Indian Settlement, in the presence of the undersigned,

"JOSEPH MONKMAN,
"JOHN HOPE."

In reply to this "statement to the Great House," Mr. Andrew McDermott, a well-known, influential, and wealthy Red River trader, who has been in the country since 1812, refers to the treaty with Lord Selkirk, stating that since the date of the treaty, the Indians, or their descendants named therein, have received an annual payment of 8*l.*

sterling from the Hudson's Bay Company*, and calls attention to the notice conspicuously posted by order of the chiefs in various parts of the settlement in 1858, warning the settlers that if they cut hay beyond the two-mile boundary referred to in their treaty with Lord Selkirk, their stacks would be destroyed. The letter called forth a strong expression of opinion from a large meeting of half-breeds convened for the purpose on the 7th March, 1860, when the following resolutions were adopted:—

"1st. That the Cree chief, Senna, who has the best claim to this country, never disposed of it to the Earl of Selkirk or the Hudson's Bay Company.

"2nd. That the Hudson's Bay Company do not, as is alleged, pay 8*l.* sterling per annum to each of the five chiefs mentioned in Mr. McDermott's letter.

"3rd. That the paltry presents given to some or all of these chiefs for many years after 1816, were not given in the way of payment for lands; but merely to keep them friendly towards the Company. The friendship of these chiefs was important, not only because their hostility might have been dangerous, but because they could, by using their influence with their people, bring a large quantity of furs to the Company.

"4th. That presents similar to those given after the year 1816, were given for thirty or forty years before that date, for the purpose of "keeping in" with the Indians; and given not only to the chiefs of this district, but to every influential Indian throughout the country.

"5th. That as no proper arrangement has been made with the native tribes regarding their lands, the 'half-breeds' who are now on the soil, and who, besides being natives, are the immediate representatives of these tribes, ought to use every legitimate means to urge their claims to consideration in any arrangement which the Imperial Government may see fit to make.

"With these points agreed upon, it was resolved to adjourn the meeting until the month of May or June, when the various chiefs referred to would be in the settlement, and when certain persons who were now out wintering, would also be here to corroborate the above facts."

It would be altogether premature to discuss the claims presented by different nations and chiefs to the Red River

* This statement is not substantiated by the evidence of Sir George Simpson, Blue Book, Question 109.

country; but it is clearly evident that the subject will require close investigation and prompt action in order to avoid troublesome disputes. It is also apparent that the calls of humanity, the interests of the new colony, and the claims of the Indians, imperatively demand that the natives should be paid for their lands in such a manner that the future to them may not possess the sad and hopeless aspect which has too long met the gaze of the Indian race in Canada, whose hunting-grounds have been purchased with much apparent commiseration for their condition, but with tardy and inadequate attempts to arrest the fate which, under such neglect, inevitably awaits them.

The Americans have secured a tract of country thirty miles deep on either side of Red River, extending from the boundary line to Buffalo River on the east bank, and Goose River on the west bank. The area of this tract is between 4,000,000 and 5,000,000 acres, and the sum paid for it was $30,000 cash, and $10,000 per annum for twenty years, $2000 of this annuity being reserved by the President, and applied to the improvement of the Indians. The date of this treaty is 1851.

Ross says* that the Assinniboines, Plain Crees, and the Ojibways (Saulteaux) all laid claim to the land, but the title of Ojibways has always been most disputed, they being invaders of the country; yet, being found by the Americans on the spot, they were recognized as the lords paramount of the soil. Their principal chiefs, however, absented themselves, being distrustful, lukewarm, and unwilling to sell their lands; and the treaty was ratified with those of secondary rank who attended the meeting, summoned by the Governor of Minnesota.

* Appendix to the Red River Settlement, &c.

CHAP. XXXII.

MISSIONARY LABOUR AND ITS RESULTS.

Indians in Canada.—Distinction between Indian Nations and Tribes.—The Ojibways and Mistassins.—Families, Nations, Tribes, and Bands.—Indian Families of Rupert's Land and Canada.—The Algonquins and Iroquois.—The Hurons or Wyandots.—Dispersion of the Hurons.—The Iroquois Confederation.—Statistics of Indians in Canada.—Canadian Special Commission.—The Indian Department.—Efforts to ameliorate the Condition of Indians.—The Manitoulin Islands and the Mission at Manitouaning.—The Roman Catholic Missions.—Their School and Village.—Wikwemikong.—Wesleyan Methodist Missions.—Indian Labour Schools.—Cause of the Failure.—Condition of some Indian Villages in Canada.—The Indians of the Northern Coast of Lakes Huron and Superior.—Treaty with these Indians.—Distribution of Annuities.—Hudson's Bay Company.—Sale of Birthrights.—Suggestion with Reference to a Permanent Fund for the Supervision and Instruction of Indians.—Lands surrendered by Indians in Canada.—Testimony of Missionaries and Agents in Relation to Indians.—Advantage of Settled Homes.—Compact Reservations.—Indian Progress in Michigan.—At Red Lake.—At Red River.—Suggestions with regard to the Amelioration of the Condition of Indians generally.—Missionary Labour in Rupert's Land.—The School-house.—Suggestion for the Establishment of a General Store for Outposts.—Native Language.—The Bishop of Rupert's Land.—His Charge, January, 1860.—The Earl of Southesk.—A Christian Assinniboines' Band.—The Church in the Wilderness.

THE prospective condition of the Indian race in Rupert's Land will be greatly dependent upon the steps which may be taken by the future government of the country, to provide for their instruction in the Christian religion; their assumption of a settled mode of life, and their consequent advancement in civilization.

The experience of a century in Canada cannot be

overlooked in any discussion of the measures which appear to commend themselves for encouragement or adoption, in order to secure the amelioration of the condition of this people, whose claims upon the sympathy and protection of the civilized invaders of their hunting-grounds no humane man can dispute. A partial foreknowledge of their destiny, under providence, may be gleaned from an examination of the condition of Indians living in the midst or on the borders of civilized communities, where experiments for their benefit have long been tried, and where the results are perfectly well but not widely known.

It must be borne in mind that there is as broad a distinction between Indian nations and even the tribal bands of the same nation, as between Europeans or Asiatics of separate origin or nationality. The stately and intelligent Ojibways, who formerly occupied the country about Lake Superior and the north shore of Lake Huron, and are now scattered from Lake Ontario to the Grand Forks of the Saskatchewan, ought not to be brought into comparison and classed with the barbarous Mistassins, who hunt and fish on the lower St. Lawrence, although both nations belong to the same great family, and speak dialects of a common language.

Even among different tribes of the same nation, great distinctions are found to prevail, which may be generally traced to the physical characters of the country they inhabit. The Swampy Crees of Lake Winnipeg are far inferior to their brethren of the prairies, and the Ojibways of Tecamamionen or Rainy Lake, are superior to other tribes of the same nation who hunt on the north shores of Lakes Huron and Superior. The same remark applies to "Bands" belonging to different tribes. The Berens River Band, and the Bloody River Band on Lake Winnipeg are low in the scale when compared with the Dauphin River Band, and those who hunt on Lake Manitobah. These minor points arise into importance in reviewing the progress made by different bands, when subjected to missionary influence.

Much misapprehension and perplexity have arisen from the different application by writers of the terms "Family," "Nation," "Tribe," and "Band" in describing the peculiarities of the Indian race. A few words on this subject may not be out of place. At the present day five great Indian "Families," as distinguished by languages of a radically distinct character, occupy the northern portion of the North American continent east of the Rocky Mountains.

They are as follows :—

I. Esquimaux.
II. Chipewyans.
III. Algonquins.
IV. Dakotahs.
V. Iroquois.

The Esquimaux occupy the Arctic coast, and are essentially a maritime people.

The Chipewyans, the region south of the country of the Esquimaux, and live in the woods north of the 55th parallel.

The Algonquins are in possession of the valley of the St. Lawrence, and of the country north of a line drawn from the head waters of the Mississippi, to the north branch of the Saskatchewan ; they are part prairie and part wood Indians.

The Dakotahs occupy the valley of the Missouri, and a considerable portion of the upper Mississippi. They also extend their hunting expeditions to the North Branch of the Saskatchewan, confining themselves throughout this vast region to the prairies and plains. The Iroquois must be regarded as a family nearly extinct, they are half civilized, and from intermixture with the whites, fast losing in many of their settlements, all trace of pure Indian blood.

When the Jesuit missionaries penetrated into Canada about the year 1615 *, they found the country in possession of two only of these families, the *Algonquins* and the *Iroquois*.

The great Algonquin family, whose hunting-grounds then extended from the north west side of the valley of the St. Lawrence to Hudson's Bay, is composed of numerous nations speaking a common language, but embracing many dialects. Among these are included the *Ojibways*, the *Crees*, the *Potawatamies*, the *Shawnees*, the *Lenni-Lenape*, the *Delawares*, the *Ottawas*, the *Nippissings*, the *Abenakis*, the *Amalacites*, the *Montagnais*, the *Sokasis*, the *Mistassins*, and the *Mohegans*.

The *Algonquins* † are generally wanderers, without settled place of abode, living in the woods and subsisting upon wild animals, fish, fruits, roots, and herbs.

The Iroquois family embraced the Hurons, Eries, and Mingoes or Iroquois. The *Hurons* or *Wyandots* speak a language not understood by the Algonquin nations, but allied to that of the *Iroquois*. They lived formerly in stationary villages, and cultivated the soil, growing Indian corn, pumpkins, &c. The country of the Hurons is now the

* Le Père Recollet Joseph Le Caron, in 1615, ascended from Quebec to the country of the Hurons.—*Relations de quelques Missions des Pères de la Compagnie de Jésus dans la Nouvelle France.*

† Algonquins, or Algouméquins, a word derived from *Adirondack*, or Leaf-eaters, derisively applied to a tribe by the Iroquois, and corrupted by the early French settlers into *Algonquins.*—*Ibid.*

county of Simcoe in Western Canada, and the once celebrated missions of St. Ignace, St. Louis and Ste. Marie were respectively situated at, or near the extremities of, Sturgeon, Hog, and Gloucester Bays of Lake Huron. The Hurons occupied this limited area with a population of about 30,000 souls, living in eighteen walled villages, until the year 1650, when they were destroyed or dispersed by the murderous Iroquois. When the remnant fled from their enemies, they separated into five divisions; the first band retired to the Manitoulin Islands, but eventually joined another part of the fugitives who had fled to Quebec, where their descendants occupy the village of La Jeune Lorette. They number now 282 souls, but by intermixture with the whites, they have so lost the original purity of their race, that they scarcely deserve the appellation of Indians. Their agricultural progress may be inferred from the fact, that although they possess 1657 acres, very little is cultivated, and the present farming stock of the tribe amounts to fourteen cows, five horses, and nine pigs. They produced in 1857, 310 bushels of grain, and fourteen tons of hay. They possess one harrow and 63 hoes or spades. Such is their progress after two centuries of instruction and guidance under the Jesuits.

The second band applied to be received into the families of their conquerors, and they were incorporated with the Tsounontonans of the five nations of the Iroquois. The third band found a temporary asylum in the island of Michillimakinac at the upper extremity of Lake Huron, but being followed by the Iroquois, they retreated to the Mississippi, where they encountered the Sioux, who at this time were engaged in extending their territory towards the east, like the Iroquois in the contrary direction. The unfortunate Hurons retired to the shores of Lake Superior, but were followed by the Sioux, and ultimately compelled to retrace their steps to Michillimakinac about the year 1670, where the remnant now dwells. The fourth and most ill-fated of the Huron bands sought refuge among the *Eries*, who occupied the country on the north shores of Lake Erie. They spoke the same language as the Hurons and Iroquois, and lived in permanent villages. The presence of the Hurons among them soon excited the jealousy of the Iroquois, so that after a very short period had elapsed, both the Eries and the greater part of their Huron allies were cut off by the savage, envious, and relentless Iroquois.

The fifth band retired to the new French colony in the lower St. Lawrence, where they were joined by the first band as already stated.

According to the narrative of Sagard, the first historian of the Hurons, they occupied a distinguished position among savage nations. The Hurons represented the nobility of the country, some of the Algonquin nations represented the citizens, and the Montagnais, an Algonquin nation, the poor.*

The Iroquois Confederation consisted originally of a union of five nations, named respectively the *Mohawks* †, the *Oneidas*, the *Onondagas*, the *Cayugas* and the *Senecas*.

They invaded the country of the Hurons about 1650, and were in turn driven back by the Algonquins, into whose hunting-grounds north and north-east of Lake Huron, they appear to have penetrated. After their retreat the Ojibways, and particularly the Missassaguas, the "eagle" tribe of the Ojibways, following them, occu-

* Relation Abrégées, &c. &c.

† The Mohawks, Onondagas, and Cayugas are the oldest members of the Confederacy. Their union precedes tradition.

pied permanently the central portion of the peninsula of Western Canada.

In 1712 the Tuscaroras of North Carolina, subsequently to their defeat by the English, united with the Iroquois and formed with them the confederary of the SIX NATIONS.

At the time of the first discovery of Canada, the Iroquois occupied the south-east valley of the St. Lawrence, especially the south shore of Lake Ontario and the region about the small lakes which still bear the names of the different nations. At the close of the war of independence, being firm allies of the British, they migrated into Canada, and in 1784 a part were established on the Grand River on a tract comprising about 700,000 acres, which was confirmed to them in 1739 by letters patent. In 1845, there remained in the hands of these Indians 52,133 acres, the other portion having been surrendered at different times. This part of the remnant of the famous Six Nations, now numbers on the Grand River 2550, of which between five and six hundred are still pagans. The Canadian Iroquois did not exceed 5000 in 1857.

The following table shows the total number of Indians in Canada at different periods from 1827 to 1857 :—

UPPER CANADA.*		LOWER CANADA.†	
1838	6,643	1827	3,649
1844	6,874	1837	3,575
1846	8,756	1844	3,727
1847	8,862	1852	4,058
1857	9,094	1857	4,396

* Not including the Indians of the north shore of Lakes Huron and Superior, which number respectively, from the Sault Ste. Marie to Pigeon River, six bands, containing 1240 Indians, and from the Sault Ste. Marie to French River, seventeen bands, containing 1422 souls.

† The Indians in the Lower St. Lawrence are not included. They are

In 1857 the Indian census, including settled and migratory tribes, and tribes not within reach of the Missionaries, gave the following numbers of this people:—

Settled Indians in Upper Canada	9,094*
" Lower Canada	4,326
Nomadic tribes visiting the north shore of Lake Huron .	1,422
" " " Lake Superior	1,240
Nomadic tribes of the Lower St. Lawrence	2,500
" not within the reach of missionaries or agents (estimated)	1,000
Grand Total	19,652

The Indians within the reach of government in the year 1818 were estimated at 18,000, but since that period immigration and the extension of the influence of the government will probably account for the apparent increase observed since 1818.

In 1856 the Canadian Government appointed a special Commission to "inquire into and report upon the best means of securing the progress and civilization of the Indian tribes in Canada, and on the best mode of so managing the Indian property, as to secure its full benefit to the Indians, without impeding the settlement of the country."†

The Indian Department takes cognizance of everything relating to Indians in Canada, and in order to carry out the business belonging to it, the province is divided into five districts, each under the charge of a local superintendent.

chiefly *Montagnais* tribes. The *Mistassins* and *Naskapees* number about 2500, of whom 1500 are still pagans, sacrificing to the Deity, who they say inhabits the sun and the moon, a portion of everything they kill.

* Compare these numbers, from official returns, with the estimate by the Hudson's Bay Company of Indians settled in Canada (3000), p. 159.

† In the subsequent notice of the Indians in Canada, much of the information is obtained from the Report of the Canadian Commissioners, appointed in 1842 and 1856.

The first district embraces the whole of Eastern Canada, and a small part of the Upper Province. The second stretches from the western limits of the first to the head of Lake Ontario, comprising also the Saugeen Peninsula, and some of the islands in Lake Huron. The affairs of the six nations, or Iroquois, on Grand River, and those of the Mississaguas in the township of Tuscarora, occupy the exclusive attention of a third superintendent. The tribes resident throughout the western peninsula of Canada are under the charge of a fourth, and the bands in the occupation of the Manitoulin Islands, together with the tribes on Lakes Huron and Superior, form the limits under the fifth superintendent.

No opportunity has been lost, and no pains have been spared, since Western Canada became a province, in planning, suggesting, and advocating schemes for the amelioration of the present and the brightening of the prospective state of Indians in Canada. Secretaries of State, Governors of provinces, and special Commissioners have all interested themselves in the condition of the Indians, and suggested measures for their benefit; but as the Canadian Commissioners say: "With the fatality which seems to have attended this unfortunate race, various obstacles appear to have arisen which prevented these benevolent and judicious projects from being carried out." "We are still groping in the dark, the time for experiments is fast passing away, if it has not already expired." The Commissioners see, however, no reason why the Indian race should not in time take their place among the rest of the population of Canada. A study of their history shows that they have all a greater or less appreciation of the blessings of civilization, and desire their children to be educated like the white man.

The theory of the steady decline of their number is *posi-*

tively controverted by experience, when they are established in villages and properly superintended. There is no inherent defect in the organization of the Indians which disqualifies them from being reclaimed from their savage state. Their general amelioration or marked advance towards civilization, must be the result of long and patient labour, and the development of many years. The absence of action in carrying out the several plans suggested has been most prejudicial to the Indians, yet among the experiments which have been tried, the results in some instances have shown how much can be accomplished for this interesting people when zealous and painstaking officers are intrusted with their superintendence and guidance.

Perhaps no better illustration of the results attending the isolation of Indians, and their instruction in the arts of civilization, can be afforded than that offered by the Manitoulin Islands.*

In 1836 Lieutenant-Governor Sir Francis Head, proposed to collect on Manitoulin, not only the wandering bands on the north shore of Lake Huron, but also the tribes settled in all parts of Upper Canada.†

The scheme was a failure, the only Indians availing

* The Great Manitoulin Island at the head of Lake Huron is 135 miles long, and from 20 to 25 miles broad. The shores of the island are hilly, and clothed with cedar, pine, and birch. The soil of the hills is stony and barren. In the interior of the island are about 20 lakes, some fully 15 to 18 miles long, and from 8 to 10 miles wide, and from 3 to 20 fathoms deep. They abound with trout, pike, white fish, &c. The extent of arable land on the island is about one-third of its area. The trees on the arable land are elm, maple, birch, cherry, and a few oak and beech. The climate is remarkably healthy.—*Description of the Great Manitoulin by the Roman Catholic Missionaries.*

† See Report of the Special Commissioners, appointed on the 8th September, 1856, to investigate Indian affairs in Canada. Printed by order of the Legislative Assembly, 1858.

themselves of the offer being some from the United States, and from the shores of Lake Superior and Huron. The village of Manitouaning was built by the Canadian Government, and placed under the management of a resident superintendent, assisted by a clergyman, a surgeon, and a schoolmaster. Artisans were induced to go there and take charge of workshops, in which Indians were to be taught useful mechanical trades. Contributions from private parties to aid in the work were obtained through the exertions of the missionary and others. The population in 1843 was estimated at 90 individuals. The only other village on the island at that time was Wikwemikong, founded previous to 1836 by Ottawa Indians from Michigan, who had long been converted from heathenism by Roman Catholic missionaries and possessed some knowledge of agriculture before their arrival on the island. In 1843, this village contained 73 Indian houses and was estimated to comprise 376 individuals. There were also a church, a school-house and a saw-mill, together with a house for the missionary and one for the schoolmaster.

In 1857 the total Indian population of the island was 1290, being composed of 977 Catholics, 104 Protestants*, and 145 Pagans, occupying 13 stationary villages and 60 birch-bark tents. With respect to the Protestant village of Manitouaning the Commissioners said in 1857: "Many of the buildings are deserted and ruinous—the school-house is dilapidated and untenable, and the workshops from which the mechanics are withdrawn, are destitute of tools, deserted by the Indians who formerly worked there, and in an utter state of decay. The church is in tolerable repair, *but we found no Indian attending the services.*"

* The Protestant missionaries claim 117 converts on the island.

"The school returns show twenty children as receiving instruction; but the greatest number of days during the last quarter, on which any one child attended the school, was fourteen, and ten of the children do not appear to have been present for a single day." The Rev. Mr. Jacobs stated in 1857, that the number of Indians at his station belonging to his congregation was forty-two.

The Commissioners report his settlement to be "in a much more prosperous condition than Manitouaning, both houses and farms being tidy and kept in better order."

The Roman Catholic Mission is next described by the Commissioners.* Wikwemikong contains a population of 580 all belonging to the Roman Catholic Church. "Although not so well situated as Manitouaning, prosperity smiles upon the settlement.

"The Indians appear respectable in their dress, industrious in their habits, healthy and contented; the services of the church are reported to be numerously attended; the schools were crowded with clean, healthy, intelligent children of both sexes, numbering 125 in the school register. We satisfied ourselves by examination, that the average attendance has been of late for the boys 45 days in the quarter and 56 days for the girls.

* It will be a subject of deep regret to members of the Church of England that the Canadian Commissioners felt it their duty to recommend that a year's gratuity should be given to the resident Protestant missionary at Manitouaning, and that he should be placed on the retired list; also, that the Protestant English school should be abandoned as "practically useless," and simultaneously with this negative evidence of inefficiency and failure, they should feel themselves compelled to recommend that a schoolmaster should be appointed in the Roman Catholic Mission at Wikwemikong. No doubt the recommendation was just, and conveyed an appropriate acknowledgment of the untiring zeal, and unfailing energies of the Roman Catholic Missionaries. With respect to the other villages on the Manitoulin Island, the Commissioners say:—"Notwithstanding that Christianity is making slow and painful progress among them, they must still be considered as almost at the bottom of the scale of civilization."

"The village, which has been erected entirely by the Indians themselves, contains a spacious church, substantially built of stone, plainly finished, and decorated inside in Indian style. The missionaries' house, which has been built for them by their congregation, is also of stone."

In 1848 a manual labour school was built at Alderville in the county of Northumberland, Upper Canada, at a cost of $6,328, and a further sum of $515 has since been laid out in repairs. Another school at Muncey Town was finished in 1851, the amount expended on it being $5,500. In 1856 it was found necessary to enlarge this school, involving an additional outlay of $3,660, and in 1857 repairs to the amount of $640 raised the total sum expended to $9,800. These schools are appropriated to Ojibways and Missassaguas. The management was entrusted to the Wesleyan Methodist Society. The Indian Department pays the Society for the board, clothing, and education of each child from a fund derived from the annuities of the tribes benefited by the schools. The average annual cost of each child has been $64 per head. The Wesleyan Methodist Society engaged to supply furniture, books, stationery, stock and farming implements, as well as to pay the superintendent and teachers, and to provide such assistance as would be necessary to efficiently conduct the institution. The average attendance at these schools has been, at Alnwick, twenty-three boys and sixteen girls; at Muncey Town (Mount Elgin), twenty boys and twenty-one girls. The management has been good, but the results have not answered the expectations of the projectors and sustainers.

In 1857 each farm attached to the different schools had about seventy acres under crop, but the amount expended by the Society in addition to the sum paid by the Indian Department was $2,200. The impression produced on the children is very transient. "They do not seem to carry back with them to their homes any desire to spread among their people the instruction which they have received. They are contented as before to live in the same slovenly manner, the girls make no effort to improve the condition of their houses, nor do the boys attempt to assist their parents steadily on the farm."

The Commissioners think that the following obstacles have impeded the success of the experiment:—

First, the children are too old when they are received into the institution.

Secondly, they remain too short a time at these establishments.

Thirdly, the system does not make any provision for the settlement in life of those who complete their education.

Fourthly, neither the funds at the disposal of the Indian Department, nor those furnished by the Society have been sufficient to enable them to extend the system of practical education so as to include any of the mechanical arts. After seven years' experience of the effects of those schools, the Commissioners recommended the withdrawal of the annual portion of the funds furnished by the Indian Department from Indian revenues, which, it was suggested, should revert to their original object at the commencement of the next financial year.

Such is the cheerless conclusion of an experiment as far as the Indian Department was concerned, where the result might have been very different if one establishment had been assisted with the funds applied to both. One hundred and thirty dollars per annum for each child does not appear too large a sum in order that so important an experiment, promising such excellent results if efficiently carried out, should be thoroughly tested.

The present condition of the several villages which have been built by Government for the Indians in different parts of Canada, reveals a deplorable condition of decay into which long continued neglect has brought them. "The Indian is content to inhabit a hovel so dilapidated as to be hardly tenantable, without exerting himself to put it into repair." This lamentable state of affairs appears to occur only in the absence of strict supervision.

The northern coasts of Lakes Huron and Superior remained in the occupancy of nomadic bands of Ojibway Indians until the year 1850, when the whole of this vast extent of country with the exception of certain reserves was surrendered to the Canadian Government for the sum of $16,640 paid down and $4,400 in perpetual annuity, of which $2,400 is payable to the tribes on Lake Huron and $2,000 is divided among those inhabiting the shores of Lake Superior.

The number of Indians inhabiting the northern shores of Lake Huron in 1850 was only 1,422 and of Lake Superior 1,240, making a total Indian population of 2,662 souls over an extent of country exceeding England in area, and not yet approached by civilization except at the Hudson's Bay Company's posts, and at a few mining locations. Of the 1,422 Indians on the north shore of Lake Huron 317 are Christians, divided as follows:—Catholics 294, Wesleyans 20, Church of England *three*. Many of these Indians, it may be with propriety mentioned here, visit Manitouaning, on Manitoulin Island.*

* The Commissioners state that "These tribes live for the most part by hunting, and on the produce of their fisheries, although they do raise a few potatoes, and a little Indian corn; and they find a market for disposing of their peltries and supplying themselves with necessaries at the posts of the Hudson's Bay Company. They are quite nomadic in their habits; seldom living or remaining long in one spot, and contented with the shelter afforded

In commenting upon the treaty which surrendered to the Canadian Government a territory as large as England, rich in minerals, fisheries, and forests, and tenanted by 2662 Indians, part of the remnant of that great Algonquin confederacy, which two hundred years ago drove the Iroquois to the south shore of Lake Ontario, the Commissioners say, "if we considered that it came properly within our province, we should not hesitate to express our decided regret, that a treaty, shackled by such stipulations, whereby a vast extent of country has been wrung from the Indians for a comparatively nominal sum, should have received the sanction of the Government."

The distribution of the annuities to the Lake Superior Indians is made through the agents of the Hudson's Bay Company, who have "voluntarily assumed" and faithfully discharged this task. It will not escape notice that the voluntarily assumed distribution tends to secure the services of the Indians as permanent hunters to the different fur-trading posts, where they assemble to receive their miserable pay for the vast extent of country surrendered. Many will be disposed to ask why no portion of the annuities or of unclaimed arrears should have been reserved for evangelizing, educating, and establishing in permanent villages, these wandering savages, in the hope of reclaiming a few of their number from barbarism. If the poor remains of once numerous and powerful tribes are encouraged to continue their nomadic habits of life, they will perish one by one from the face of the earth, leaving no records behind them but "treaties" to

by a bark wigwam or a hut of reeds. It is only during the spring and autumn, when they come down from the high grounds to the border of the lake, that they are accessible to those who would urge on them the necessity of Christianity and civilization. There is no difficulty, therefore, in accounting for the small apparent results of the labours of the missionaries."

show with what utter absence of reflection and forethought they sacrificed their country and the rights of their children. This almost wanton indifference to the future shows that they possess an irresistible claim to the care and sympathy of a powerful Christian government. When many thousand pounds yearly are solicited and willingly contributed by the great charities of England to the spread of the Gospel among Indians who still claim and possess the soil, it would surely be true charity, as far as Rupert's Land is concerned, to urge the importance of establishing a fund from the proceeds of the sale of lands, after the Indian title is extinguished as a preliminary to settlement, to be devoted exclusively to the instruction and supervision of those who so recklessly part with their birthright. An addition of one penny per acre to the sum paid to Indians in Canada for the lands they have surrendered, would have produced an annual sum sufficient to afford a permanent maintenance to two hundred ministers, teachers, or superintendents.

The following table shows the aggregate quantity of land surrendered at different times by the Indians of Canada, and the price which has been paid to them per acre :—

Name of Tribe.	Number of Acres surrendered.	Average Price paid per Acre.
Ojibways	7,373,000	$2\frac{1}{2}$*d.*
Missassagas	6,737,750	$\frac{7}{8}$ of a penny.
Ottawas, Pottawatamies, Chippewas, and Hurons	2,001,078	$\frac{3}{16}$ "
Delawares		2*s.*
Saugeen Indians.	1,500,000	$3\frac{1}{2}$*d.*
Ojibways of Lake Superior	Not known. 500*l.*	
" " Huron	" " 600*l.*	
Total	16,137,835 acres.	$1\frac{1}{2}$*d.*

The complete isolation of Indians, in order to shield

them from the contaminating influence of the whites, involves the necessity of stringent police regulations prohibiting the sale of spirits, otherwise the system becomes a failure.* It appears to be an established fact that the health of settled tribes is much better than that of others hovering on the borders of civilization. "Epidemics are less fatal, while the diminished exposure checks the ravages of consumption and of febrile attacks consequent on the hardships inseparable from the precariousness of a hunter's life. The more regular supply of wholesome food is another cause of the improved sanitary condition of the settled tribes."†

The scarcity of the larger animals diminishing their supply of suitable food, and the general adoption of manufactured articles of clothing, such as blankets and cloth, in place of the prepared skins of animals, in which they formerly clothed themselves, have been largely instrumental in making the present nomadic tribes more susceptible of disease than their ancestors.

The testimony of missionaries and agents, embodied in their reports to the head of the Indian Department of the United States, is unanimous in deprecating the system of frequent removal, as practised in the western territories.

Instances are not wanting, either in Canada or the United States, to show that compact reservations surrounded by the whites are a state favourable to the civilization and progress of the Indian. In Michigan, the franchise, and all other rights of citizenship, are exercised by the Indians. They there form an integral part of the population of the state, on the same footing as their white neighbours.

* In Canada the whites follow the Indians with spirits into the bush, and obtain, at a nominal rate, the fruits of months of toil. The same occurs in Rupert's Land.

† Report of the Commissioners.

The Mission established in 1843 at Red Lake for the Ojibways who hunt south of the boundary line, may be instanced as offering a successful illustration of the force of example and the benefit of instruction. The Indians cultivate fields of Indian corn and potatoes. The houses of the missionaries are good and comfortable, and their farm is kept in as good order, and is as well cultivated as any farm in the States. It is really what is intended to be, a "model farm," and the happy results of their example are seen all around them in the well-cultivated fields of the Indians, and the excellent cabins of many of them.*

The Indian settlement at Red River, already described (Vol. I. p. 201), is another happy instance of the excellent results attending zealous missionary labour and supervision.

The Canadian Commissioners notice a curious feature connected with the advance of civilization among the aboriginal population, namely, the taste for agricultural improvement by no means keeps pace with their progress in point of mental cultivation. By giving them a direct interest in the land itself, by securing a fixed location to each family, it is predicted that their attention will be turned to the improvement of it. Hitherto each tribe or band settled within the limits of the province has held the reserves apportioned to them in common, and are by law exempt from taxes and assessments.

The conclusions at which the Canadian Commissioners have arrived, after a very patient and protracted investigation of the condition of Indians in Canada, may be thus briefly expressed †:—

1. In all cases in Western Canada where a final location of a band shall be determined upon, the head of each

* Owen's Geological Report of Wisconsin, Iowa, &c., p. 327.

† Report of the Canadian Commissioners, Blue Book.

family shall be allotted a farm not exceeding twenty-five acres in extent, including an allowance of woodland, where they may obtáin fuel; that for such farm he shall receive a licence giving exclusive occupation of the same to him and his heirs for ever, on condition of bringing into cultivation a certain number of acres in a given time.

2. When bands have been located, and houses built for them by Government, a small portion of their annuities derived from the original sale of their lands should be reserved as a fund to meet the necessary expenses attending repairs.

3. The establishment of industrial schools and model farms should be encouraged and promoted. The reports from the United States Commissioners show the most gratifying results to have arisen from this system.

4. The attendance of children at school should be enforced by fine, or a deduction of a portion of the annuities due to the parents.

5. The appointment of local agents, who should be compelled to reside among the Indians.

6. The establishment of Government stores, with a view to supply the Indians with goods at fixed rates, and to purchase their produce; also the establishment of suitable stores under licence.

7. The disuse of Indian dialects, and instruction in the English language (French in Lower Canada).

8. The gradual extinction of tribal organization.

9. The absolute prohibition of intoxicating liquors.*

The impressions produced, upon a mental review of the condition and progress of the missionary stations between Lake Superior and the Grand Forks of the Saskatchewan,

* All experience tends to show that intemperance is the worst enemy of the Indian race.

nearly all of which I have visited*, have assumed the form of convictions, which I briefly introduce here merely as a traveler's record.

While it is evident that the progress of evangelizing the heathen Indians is decidedly encouraging on the whole, much, very much, under Providence, depends upon the personal character and individual exertions of the missionary. Failure, or rather the want of that degree of success which circumstances have appeared to promise, may generally be traced to an injudicious selection of the site for a missionary station, whence arose insuperable difficulties in providing the Indians with fixed abodes, and the ultimate prospect of maintaining themselves by cultivating the soil. The Bishop of Rupert's Land refers pointedly to the problem which Rupert's Land now presents, in his journal of 1859.† "Difficulties at the two stations remain, nor would I conceal them; but they are the universal Indian difficulties—the want of food, and the consequent inability to maintain the families around a central point. This is the Indian problem, the same throughout the land, which has still to be worked out, and requires much patient thought."

Disappointment in receiving the necessary supplies for keeping up the station, and the competitive opposition of the fur-trade, are formidable difficulties of constant occurrence. Most of all, however, is the absence of success due to a natural incapacity or indisposition to fulfil the duties

* The Roman Catholic Mission of the Immaculate Conception, on the Kaministiquia, Lake Superior; the Church of England Mission at Islington, Winnipeg River; the Indian Missionary Village at Red River; Prairie Portage on the Assinniboine; the Qu'appelle Lake Mission; the Nepowewin on the Main Saskatchewan; Fairford on Partridge Crop River; the churches at Selkirk Settlement, St. John's, St. Paul's, St. Andrew's, the Roman Catholic Cathedral, the Pesbyterian Mission, and the Roman Catholic Red Lake Mission.

† The Church Missionary Record, March, 1860.

imposed by the position in which the missionary is placed. These imperatively demand rare qualities, which in common life it is difficult to find united in one man; patient endurance, much long-suffering, deep heartfelt sympathy for the objects of his care, a warm interest in his work, and unflinching readiness to encounter difficulties, whether they be those of language and superstition, or mere physical hardship and privation.

The school-house appears to be beyond all comparison, the main hope of missionary success. In order to christianize the rising generation, they must be induced to relinquish their wandering habits of life, and settle down in permanent villages. Nomadic Christian Indians are subjected to extraordinary trials, which do not readily occur to one not familiar with their character and manner of living. Indians are extremely susceptible of ridicule, and possess a great degree of family pride; they are consequently often unable to resist the jeers and scoffs of the conjurors, or rise superior to the taunts of those who upbraid them for forsaking the gods of their fathers.

It is very desirable that a missionary station should not be situated near a fur-trading post; a settled life is diametrically opposed to the fur trade, whose stability rests upon the hunters and trappers in its employ. It has happened in Rupert's Land, that when a missionary has succeeded, after years of anxiety and toil, in establishing a station, and gathered round him a little band of Indians who have embraced Christianity, a fur-trading post has been established close by, tending to unsettle and demoralize those who would otherwise have remained quiet and stationary Christians.

In such districts as Mackenzie's River, the missionary is absolutely dependent upon the Hudson's Bay Company, and indeed up to the present year, they have been more or less dependent upon the Company for supplies of every description; now, however, that goods can be conveyed very cheaply by the American steamers on Red River from Minnesota, the time appears to have arrived when the stations south of the Saskatchewan should be no longer dependent upon the Company. The establishment of a general store at Red River under the supervision of the Bishop for the exclusive use of non-resident missionaries south of the Saskatchewan, would be of immense assistance. The store should be placed under the charge of an experienced business man, and be limited in its operation to the outposts, or to missions having an exclusive Indian population. An annual sum should also be at the disposal of the Bishop, to enable him to obtain at all times the means of sending supplies to distant stations.

The want of a general familiarity with the native language is severely felt; interpreters are dangerous, nor do they always rightly understand or render the words of the missionary. The present time is eminently favourable for the selection of permanent village sites, surrounded by a sufficient area of arable and pasture land to provide for the future independence of the Indians who may be induced to settle. In the organization of a new colony, the native races will doubtless not be forgotten. The examples furnished by Canada show how much may be done under judicious management if an early beginning is made, and how much may be lost to these thoughtless people if they are not placed under proper supervision and protected by special laws.

In different parts of the narrative I have adverted to the character of missionary labour in the wilderness, as exemplified by the patient, self-denying constancy of the

Père Chonè of the mission of the Immaculate Conception on the Kaministiquia, Lake Superior, by the unceasing attention and devotion at all times and seasons of the Rev. A. Cowley at the Indian settlement, Red River, by the undaunted energy of the Rev. Archdeacon Cochrane of Prairie Portage, the exemplary perseverance of the Rev. Mr. Stagg* of Fairford, and the talented labours of the Rev. Archdeacon Hunter of St. Andrew's, recently returned from the far-off wilds of Mackenzie's River; and I shall now venture to introduce one suggestive feature in the labours of his lordship the Bishop of Rupert's Land.

Before leaving the settlement I went to pay his lordship a parting visit, and not finding him at his residence, I proceeded, as directed, to the school-house, where he might probably be found, although it was after school hours. On entering the room I found the Bishop seated between two young Cree half-breeds, *teaching them quadratic equations.* His lordship told me that the two lads showed a remarkable talent for mathematics, and for the sake of encouragement, he made a point of giving them instruction in algebra after the daily routine of the school was over, so that this extra tuition should in no way interfere with the more necessary requirements

* "It is always a pleasure to mark Mr. Stagg's earnest and warm-hearted piety: he has always his one great object before him—to win souls to the Saviour, and gain some orphans for his Home. His warmth and affectionateness of manner are very contagious, and he is a great favourite wherever he goes. He wanted to obtain twenty hundred weights of flour in the Red River, of which I gave him but little hope; but, through his persuasion and popularity, he gained his desire, and succeeded in getting the full supply when most others would have failed. He had with him his faithful catechist, Luke Caldwell, and a crew from his own people at Fairford. He is happy with them, and they with him; and I was glad to meet them all, and spend with them a portion of a day."—Journal Letter from the Bishop of Rupert's Land, *Church Missionary Record*, Feb. 1860.

for their future station in life. This little incident forcibly impressed me with a sense of the daily duties and responsibilities which a true missionary may create for himself in remote and barbarous Rupert's Land, even though he occupy the distinguished position of a Bishop in the Anglican Church. In a charge delivered to the clergy of the diocese of Rupert's Land at St. John's, Red River, on January 6th, 1860, his lordship thus sums up the present condition of his diocese:—

"With this measure of apparent outward strength there are some causes and hindrances which, I think you will agree with me, tend to cripple and retard our work.

"There is the very migratory character of our most settled population. This may, in the good providence of God, carry onward the tide of population and scatter it over the wilderness. It may thus ultimately answer a good purpose; but its tendency at the time is felt by most of us very painfully. It weakens parishes and very materially checks education, rendering it more expensive and difficult to be extended to all. It keeps the mass in a state of greater poverty, and prevents their growth and rise. It lessens the amount of public spirit and local attachment, and perpetuates many of the habits of Indian life. It parts and separates, where, if united, all would be combination and strength.

"There is, too, the want of a deeper religious life, even amongst the more advanced Christians. Here there is stagnation instead of movement. The Word is heard with joy and received with readiness; but it is the development of the rich fruit which the minister looks for, and looks too often in vain. Measuring themselves rather by that from which God hath saved them—the condition of the heathen who know not God—than by the standard of by-gone generations and of other countries, they are satisfied with smaller attainments; they rest contented with a lower level, and do not press forward to the measure of the stature of a perfect man. Their condition is a matter of rejoicing to the minister of God at first, as they are eager to hear. It is in their after-course that he suffers disappointment. The building stops before he is prepared; the growth terminates suddenly, after advancing for a time with rapidity, and there is not the higher experience of the divine life.

"There is, morever, an additional check in the Indian work. It is a transition period; change is anticipated. An excitement has seized the Indian mind, and he is little inclined to give a calm and patient attention to the claims of the Gospel. A wider competition is afloat, and baits are held out by the unscrupulous which the poor Indian is too weak to resist. A greater difficulty has thus been found in selecting and planting new stations, while at the old-established missions the steadfastness of the convert has been very sorely tried, if not, in some cases, too successfully shaken. Direct conversions have in consequence been less numerous during the last two years, and I much fear that the next two or three may continue to tell the same tale. At all events, the Indian is less hopeful, and more difficult to act upon, than he was found to be five years ago.

"With these and other causes impeding the progress of our work, and materially affecting its character, the testimony of all of us would, if I mistake not, be somewhat similar to-day; our common acknowledgment would be, that the interval since we last met has not been marked with such distinct success as previous periods; that some of our more sanguine expectations have only been faintly realized. Now, if such be your feelings, brethren, is there no deeper agency to which we may trace this?

Is the condition peculiar to ourselves, or may we throw it under a wider classification, and identify it with what we notice elsewhere on a wider scale? The answer to my own mind is sufficiently clear; the explanation which alone appears to me to account for it, is a greater measure of power put forth by Satan in the days in which we live, not only here but over the whole earth. Can we, then, substantiate this in the world, so as to prove it more than an idle dream?"

The *Fort Garry Nor'-Wester* of January 14th, 1860, in a report of a meeting held at St. John's school-house, Red River, states that the Earl of Southesk, who had just returned from a hunting expedition to the Rocky Mountains, informed the meeting that when in the immediate neighbourhood of that distant mountain region, he fell in with about twelve families of Assinniboine Indians who professed Christianity, and, so far as he could judge, were acting up to their profession. These families were remote from any missionary station, and had not even seen a missionary for many years. Still they showed considerable acquaintance with the Scriptures, and were regular in their morning and evening devotions. A little bell was always rung as their signal for assembling to worship, and the singing of hymns formed part of their religious exercises. At their earnest request his lordship wrote out for them several passages of Scripture. Their knowledge of religion is supposed to have been imparted to them by Rev. Mr. Randall, a Wesleyan Missionary, who went to Fort Edmonton in 1839, and left the country in 1847, on account of ill-health. They have, however, a regular teacher in one of themselves, who has been set apart by them for that purpose.

This is truly the "Church in the wilderness," not unfrequently represented, however humbly, in the prairies of the far West or the illimitable wastes of Rupert's Land, where Christian Indian hunters, when they meet together, often celebrate divine service after their own fashion, meekly repeating the prayers and hymns they have learned from the missionaries. A Plain Cree on the Qu'appelle once astonished me by producing a short notched stick, and after regarding it for a while, he turned to one of my half-breeds and asked if the day was not Sunday. The seed which has been sown, often starts into life after lying dormant for years, and produces a great variety of fruit. It is " bread cast upon the waters, which will be found after many days."

CHAP. XXXIII.

THE HUDSON'S BAY COMPANY.

Incorporation of the Hudson's Bay Company. — Profits of the Company. — The North-West Company of Montreal. — Union of the two Companies. — Profits of the Hudson's Bay Company after the Union. — Proprietors and Stock of the Company. — Administration of their Affairs. — Sir George Simpson. — The Council. — Departments, Districts, and Posts. — Extent of the Administration of the Company for the prosecution of the Fur Trade.

THE Hudson's Bay Company was incorporated in the year 1670, under a royal charter of Charles the Second, which granted them certain territories in North America, together with exclusive privileges of trade and other rights and advantages. During the first twenty years of their existence the profits of the Company were so great * that, notwithstanding considerable losses sustained by the capture of some of their establishments by the French, amounting in value to 118,014*l.*, they were enabled to make a payment to the proprietors in 1684 of fifty per cent., another payment in 1688 of fifty per cent., and a farther payment in 1689 of twenty-five per cent.

In 1690 the stock was trebled without any call being made, besides affording a payment to the proprietors of twenty-five per cent. on the increased or newly created stock; from 1692 to 1697 the Company incurred loss and damage to the amount of 97,500*l.* sterling from the

* See Letter from the Governor of the Hudson's Bay Company to the Lords of the Committee of Privy Council for Trade, February 7th, 1838.

French. In 1720 their circumstances were so far improved that they again trebled their capital stock, with only a call of ten per cent. from the proprietors, on which they paid dividends averaging nine per cent. for many years, showing profits on the originally subscribed capital stock actually paid up of between sixty and seventy per cent. per annum from the year 1690 to 1800, or during a period of 110 years.

Up to this time the Hudson's Bay Company enjoyed a monopoly of the fur trade, and reaped a rich harvest of wealth and influence.

In 1783 the North-West Company was formed, having its head-quarters at Montreal. The North-West Company soon rose to the position of a formidable rival to the Hudson's Bay Company, and the territory the two companies traded in became the scene of animosities, feuds, and bloodshed, involving the destruction of property, the demoralization of the Indians, and the ruin of the fur trade. Owing to this opposition, the interest of the Hudson's Bay Company suffered to such an extent, that between 1800 and 1821, a period of twenty-two years, their dividends were, for the first eight years, reduced to four per cent., during the next six years they could pay no dividend at all, and for the remaining eight years they could pay only four per cent.

In the year 1821 a union between the North-West and Hudson's Bay Companies took place, under the title of the last named. The proprietary were called upon to pay 100*l.* per cent. upon their capital, which, with the stock in trade of both parties in the country, formed a capital stock of 400,000*l.*, on which four per cent. dividend was paid in the years 1821 to 1824, and from that time half yearly dividends of five per cent. to 1828, from 1828 to 1832 a dividend of five per cent., with a bonus of ten per cent. was paid, and from 1832 to 1837 a dividend of five per cent., with an average bonus of six per cent. The distribution of profits to the shareholders for the years 1847 to 1856, both inclusive, was as follows:—

1847—1849, ten per cent. per annum; 1850, twenty per cent. per annum, of which ten per cent. was added to stock; 1851, ten per cent.; 1852, fifteen per cent., of which five per cent. was added to stock; 1853, 18*l.* 4*s.* 6*d.*, of which 8*l.* 4*s.* 6*d.* was added to stock; 1854 to 1856, ten per cent. per annum dividend.* Of 268 proprietors in July 1856, 196 have purchased their stock at from 220 to 240 per cent.†

* Letter from R. G. Smith, Esq., Secretary to the Hudson's Bay Company, to H. Merivale, Esq.—*Appendix to Report from the Select Committee on the Hudson's Bay Company.*

† The capital employed by the Hudson's Bay Company is as follows:—

June 1st 1856.

	£	s.	d.
Amount of assets	1,468,301	16	3
Amount of liabilities	203,233	16	11
Capital	1,265,067	19	4

Consisting of,

	£	s.	d.
Stock, standing in the names of the proprietors .	500,000	0	0
Valuation of the Company's lands and buildings, exclusive of Vancouver's Island and Oregon .	318,884	12	8
Amount expended up to 16th September 1856, in sending miners and labourers to Vancouver's Island, in the coal mines, and other objects of colonization, exclusive of the trading establishments of the Company, and which amount will be repayable by Government if possession of the island is resumed . .	87,071	8	3
Amount invested in Fort Victoria and other establishments and posts in Vancouver's Island, estimated at	75,000	0	0
Amount paid to the Earl of Selkirk for Red River Settlement	84,111	18	5
Carried forward . . .	1,065,067	19	4

The affairs of the Hudson's Bay Company are managed by a Governor-in-chief, sixteen chief factors, twenty-nine chief traders, five surgeons, eighty-seven clerks, sixty-seven postmasters, twelve hundred permanent servants, and five hundred voyageurs, besides temporary employés of different ranks, chiefly consisting of voyageurs and servants. The total number of persons in the employ of the Hudson's Bay Company is about 3000.

Sir George Simpson has been Governor of the Hudson's Bay Company for forty years. He exercises a general supervision over the Company's affairs, presides at their councils in the country, and has the principal direction of the whole interior management in North America. The Governor is assisted by a council for each of the two departments into which the territory is divided.

The seat of council for the northern department is at Norway House, on Lake Winnipeg; for the southern department at Michipicoten, Lake Superior, or Moose Factory, on James' Bay.

The council consists of the chief officers of the Company, the chief factors being ex-officio members of council. Their deliberations are conducted in private. The sixteen chief factors are in charge of different districts in the territory, and a certain number of them assemble every year at Norway House, for the northern department, generally about the middle of June, to meet the Governor and transact business. Seven chief factors, with the

	£	s.	d.
Brought forward	1,065,067	19	4
Property and investments in the territory of Oregon, ceded to the United States by the treaty of 1846, and which are secured to the Company as possessory rights under that treaty—$1,000,000 sterling	200,000	0	0
Total	£1,265,067	19	4

Governor, form a quorum, but if a sufficient number of the higher rank of officers are not present, a quorum is established by the admission of chief traders.

The Hudson's Bay Company's operations extend not only over that part of North America called Rupert's Land and the Indian territory, but also over part of Canada, Newfoundland, Oregon, Russian America, and the Sandwich Isles. The following table exhibits the number of departments and district posts into which this immense territory is divided for the prosecution of the fur trade :—

Country.	Departments.	Districts.	No. of Posts.
Part of Indian territory and part of Rupert's Land.	Northern.	Athabaska . .	4
		McKenzie River .	11
		English River. .	5
		Saskatchewan .	9
		Cumberland . .	3
		Swan River . .	6
		Red River . .	6
		Lac la Pluie . .	7
		Norway House .	3
		York . . .	5
Part of Rupert's Land, and Canada.	Southern.	Albany . .	4
		Kmogumisse .	2
		Lake Superior .	9
		Lake Huron . .	5
		Sault St. Marie .	1
		Moose . . .	4
		East Main . .	3
		Rupert's River .	8
		Temiscamingue .	6
Newfoundland and part of Rupert's Land.	Montreal.	Fort Coulonge .	3
		Lac des Sables .	2
		Lacluire . .	1
		St. Maurice . .	3
		King's Posts .	6
		Mingan . .	3
		Esquimaux Bay .	4
Part of Indian territory, Washington territory, U.S. and Oregon, U.S.	Oregon.	Columbia . .	8
		Colville . .	5
		Snake Country .	3
Vancouver's Island, part of Indian territory and Russian America.	Western	Vancouver's Island	3
		North-West Coast	1
		Thompson's River	1
		New Caledonia .	8
3 Independent Countries.	5 Depmts.	33 Districts.	152 Posts.

From the foregoing table it appears that the operations of the Hudson's Bay Company extend over territories whose inhabitants owe allegiance to three different and independent governments, British, Russian, and the United States. These immense territories, exceeding 4,500,000 square miles in area, are divided, for the exclusive purposes of the fur trade, into four departments and thirty-three districts, in which are included one hundred and fifty-two posts, commanding the services of three thousand agents, traders, voyageurs, and servants, besides giving occasional or constant employment to about one hundred thousand savage Indian hunters. Armed vessels, both sailing and steam, are employed on the North-West Coast to carry on the fur trade with the warlike natives of that distant region. More than twenty years ago the trade of the North-West Coast gave employment to about one thousand men, occupying twenty-one permanent establishments, or engaged in navigating five armed sailing vessels, and one armed steamer, varying from one hundred to three hundred tons in burden. History does not furnish another example of an association of private individuals exerting a powerful influence over so large an extent of the earth's surface, and administering their affairs with such consummate skill and unwavering devotion to the original objects of their incorporation.

CHAP. XXXIV.

THE COMMUNICATION BETWEEN CANADA AND RED RIVER.

THE WINTER COMMUNICATION.—Character of the Country on the North Shores of Lakes Huron and Superior.—Probabilities of a Road being made. Country north of Lake Huron. — Mr. Salter's Survey.—Mr. Murray's Survey.——THE SUMMER COMMUNICATION. — Route proposed by Mr. Dawson.—The Pigeon River Route.—The Old North-West Company's Route.—Cost of improving and opening the Fort William and Arrow Lake Route.—— COMMUNICATION VIÂ THE UNITED STATES. — Arrangement of the Hudson's Bay Company. — Arrangement of Messrs. Burbank and Company.—Captain Palliser's Opinion respecting the Canadian and American Routes. — Objections to his View. — Advantages of the Pigeon River Route.

WINTER COMMUNICATION.

A ROAD between Canada and Red River through British territory is at present impracticable for commercial purposes during the winter season, that is, from November to April, both inclusive. A rocky wilderness, almost destitute of civilized inhabitants, and throughout a considerable part of its area not susceptible of cultivation, extends from the 79th to the 96th meridian. The shortest line of road from the limits of settlement in western Canada *viâ* the north shores of Lakes Huron and Superior, to the north corner of the Lake of the Woods exceeds 1000 miles in length, and would traverse a region where the mean winter temperature does not rise higher than fifteen degrees above zero. When the mineral wealth of the northern shores of Lakes Huron and Superior forms the basis of speculative enterprize on an enlarged scale, a winter communication with those regions will become a necessity, and may ultimately extend westward to Red River. It is not improbable that circumstances now dimly foreseen may expedite the opening of this communication, and make it a matter not only of colonial but also of imperial interest.

The first step towards bridging this wilderness was proposed in 1856, when a bill passed the Canadian House of Assembly incorporating the "North Shore and Quebec Railway Company, and granting 4,000,000 acres of land to the company along the line of route. In May 1860 an act was passed incorporating the "Central Canada Railway Company," which included the North Shore and Quebec Railway within its provisions, no steps having been taken towards commencing that work.

Since the time when the north shore of Lake Huron attracted public attention on account of its mineral wealth, it has been the general custom to draw conclusions respecting the features of the country in the interior from the aspect of the coast, and to predicate a condition of soil and climate, wholly at variance with the facts which have been recently established. Not only did Mr. Salter, who was employed to survey the interior north and north-east of St. Mary's River, find very entensive areas of excellent land covered with a fine forest growth of hard wood trees, but Mr. Murray, of the Canadian Geological Survey, has drawn special attention to the geographical characteristics of a large area in the region indicated. Mr. Murray says*, "It has been remarked in former reports that the north coast of Lake Huron, in many parts picturesque, appears too rocky near the margin to be suited for agricultural settlement, though likely in time to become of importance to the province by the development of the metalliferous ores which the

* Geological Survey of Canada, Report of Progress for the Year 1858.

geological formation of the region is known to contain. But while this description is applicable to the coast line and the margin of some of the rivers and larger lakes of the interior, it is by no means so to the country in general. On the contrary, there are in many parts, especially in the valleys of the Thessalon and its tributaries, extensive tracts of the finest lands, covered with a luxuriant growth of hard wood interspersed with stately pine trees, probably equal in average size to any of the same species known in the province."

The interior of the country north of Lake Superior is now being surveyed for the purpose of laying off mining locations; its extraordinary mineral wealth cannot fail to draw a large mining population, which will cause the agricultural lands in the valleys of the rivers to be sought after; the fisheries are also of great importance. Under these favourable circumstances there is every probability that the country north of Lakes Huron and Superior will speedily grow in importance, and the day is not far distant when uninterrupted communication with the north-western Railroad termini in Canada will be established during the winter months.

THE SUMMER COMMUNICATION.

The opening of a summer communication between Lake Superior and Red River was one of the objects which the Canadian Government had in view in exploring the route. The duty of continuing the examination of the country in 1858, and reporting on the best means of traversing it, was confided to Mr. Simon Dawson.

The following extract from his report will explain the manner in which he proposes to open the communication by carts, boats, and steamers:—

	Land carriage miles.	Navigable miles.
"From Thunder Bay to Dog Lake	28	
Through Dog Lake and River to the Prairie Portage		35
Land road past Prairie and Savanne Portages to Savanne River	5	
Through Savanne River, Lac des Mille Lacs, and the River Seine to the Little Falls . . .		65
Broken navigation on River Seine		59½
Land carriage past the twelve portages on River Seine	7	
From the Seine to the western extremity of Lac Plat, navigable with only one break at Fort Frances .		208
Thence to Fort Garry by land	91½	
Total	131½	367½

"The foregoing does not represent the distance in a direct line, but the extent to be traveled, making allowance for the tortuosities of the route.

"In regard to the means of transport which could be most economically and advantageously used: —

"Waggons or carts would be required on the road between Thunder Bay and Dog Lake.

"On Dog Lake and River boats, such as already referred to as being used by the Hudson's Bay Company, or even a steamer might be employed.

"At the Prairie Portage, carts or waggons would be necessary.

"On the Savanne River, Lac des Mille Lacs, and the River Seine, as far as the Little Falls, after the dam was constructed at the last-mentioned place, there would be an unbroken reach of 65 miles, and on this section it would be advantageous to have a small steamer.

"On the 59½ miles of broken navigation, on the River Seine, between the Little Falls and the twelve portages, boats, as already explained, should be used, while at the land road past the twelve portages, carts or waggons, as on the other sections of road, would be necessary.

"From the River Seine to Fort Frances, a steamer would have a clear run of 50 miles.

"From Fort Frances to Lac Plat, steamers would have an uninterrupted run of 158 miles.

"From the latter place to Fort Garry no provision would have to be made, as the means of transport are to be had in abundance at the Red River Settlement.

"If the communication were opened in the manner above set forth, the journey from Lake Superior to Red River might be performed in about three days, that is, allowing that steamers could be maintained at an average speed of ten miles an hour on the navigable reaches, that the land roads might be passed over at the rate of five miles an hour, and that the interrupted navigation of the Seine could be accomplished at the rate of four miles an hour."*

* * * * * * * *

"According to a rough estimate which I have made, the total cost of opening the communication in the manner which I propose, would not exceed 50,000*l.*, but until the surveys now in progress are completed, and the necessary measurements taken, it is impossible for me to submit a very precise estimate in detail."†

The "Pigeon River route," commonly called the old North-West Company's route, along the Boundary Line, appears to possess many advantages over the Kaministiquia route, or that suggested by Mr. Dawson, *viâ* the river Seine. The objection to which it is supposed to be open in consequence of its lying along the course of the Boundary Line, applies with equal force to the other routes from Rainy Lake to the Lake of the Woods. As yet no route has been surveyed, or is known to exist within British territory, lying altogether north of the Boundary Line. A road fulfilling that condition would have to pass north of the deep indents of Rainy Lake, and cross the Winnipeg at Rat Portage, traversing a region of country through which there is no known water communication, and crossing the rivers, west of the most easterly northern prolongation of Rainy Lake, at right angles, or nearly so, to their directions, and consequently to the ranges of granitic or gneissoid hills which determine the river courses.

* Mr. Dawson's estimate of the time required to make the journey from Lake Superior to Red River (three days) is evidently much too low; since it would be impossible for steamers to travel 10 miles an hour at night on unlighted and often intricate waters. Three miles an hour on such roads as could be constructed with a limited outlay in that broken country would be fully as much as could be accomplished. Mr. Dawson makes no allowance for steamers taking in a supply of fuel, or for the transhipment of cargo; and no time is allowed for men or horses to take meals, or rest on the roads. The uninterrupted speed assumed involves numerous relays, and a system of organization which could not be maintained without enormous outlay.

† Report on the exploration of the country between Lake Superior and the Red River Settlement, and between the latter place and the Assinniboine and Saskatchewan, by S. J. Dawson, Esq., C.E. Toronto, 1859.

The North-West Company cut out a cart road between Point des Meurons, nine miles from Fort William, and Whitefish Lake, lying to the north-east of Arrow Lake, from which it is separated by a portage. This road and all old portage paths require to be cleared, and tramways laid down, over which loaded boats may be hauled by appropriate tackle. The portages on this route are low and short, and might be rendered passable in the manner suggested for a small outlay. The height of land, although more elevated than the Savanne Portage, is nearly one degree further south, and the country through which the route lies is well timbered. Whatever route may be ultimately adopted between Lake Superior and Rainy Lake, steamers must be placed on Rainy River, and a road constructed through the swamps west of the Lake of the Woods to Red River. When these works are completed, the summer communication between Lake Superior and Red River, *viâ* Arrow Lake, will be accomplished within six days, which will bring Fort Garry within twenty-two days of Liverpool.

The cost of improving and opening what may be termed the Fort William and Arrow Lake route to fulfil these conditions, does not exceed 12,000*l.*; the details of this estimate are given in the Appendix. The proposed route involves the following modes of traveling:—

1. Fort William to Arrow Lake by waggon road.
2. Arrow Lake to Rainy Lake by boat of five or six tons or more, crossing the portages on tramways without unloading.
3. Rainy Lake to the Lake of the Woods by steamer with one transhipment at Fort Frances, until the construction of two locks of ten feet lift, and one guard lock, are made to overcome the Chaudière Falls.
4. The North-West corner of the Lake of the Woods to Fort Garry by road.

The Canadian Government have placed the sum of $20,000 per annum, for five years, at the disposal of the

North-West Transit Company, on certain conditions, for the purpose of opening the communication between Lake Superior and Red River. It is not yet determined whether the Company will adopt the route by Mille Lacs and the Seine or the Fort William and Arrow Lake route.

COMMUNICATION WITH THE UNITED STATES.

The summer and winter routes between Fort Garry and Crow Wing are described in preceding chapters. As the importance of this communication progresses the journey will be made with horses and stage coach throughout the winter months by the summer road. During the period of navigation the distance between St. Paul, the present head of navigation on the Mississippi, and Fort Garry, is accomplished in nine days; six days being required to travel from St. Paul to Georgetown, on Red River, by stage coaches, and three days by steamer from Georgetown to Fort Garry. Mr. Burbank's complement of 100 waggons, in brigades of twenty-five each, are now running from St. Paul to Georgetown, conveying the freight of the Hon. Hudson's Bay Company, for which service Mr. Burbank has a contract for five years, whereby he is bound to carry 500 tons annually for that period. The Hudson's Bay Company have relinquished their northern route *viâ* Hudson's Bay, York Factory, and Lake Winnipeg, and adopted the route through the United States, *viâ* St. Paul and Georgetown on Red River to Fort Garry.*

* "On the first day of June next (1860) a new era will open for St. Paul and the north-west. On that day will commence the regular trips of Burbank and Co.'s express, coach, waggon, and steamboat lines between St. Paul and Fort Garry on the Red River of the north, in the Hudson's Bay territory, for the year 1860.

"The first experiment in far north-western passenger and freight transportation was made by Burbank and Co. last year, the result of which is, that

Messrs. Burbank and Co. now undertake the transportation of goods from England or Canada, in bond, collecting all charges on delivery of the goods at Fort Garry; and persons desiring to order goods from England immediately, can have the same forwarded from Liverpool by Allan Brothers and Co., care of J. C. and H. C. Burbank (in bond), St. Paul.

In July 1858, Capt. Palliser was requested by the Under Secretary for the Colonies to state his opinion on four points connected with the country he was engaged in exploring. These were —

"1st. Whether Red River Settlement possesses qualifications which would adapt it for an English colony.

"2nd. What should be the dimensions and the boundary line of such colony, and whether it would be advisable to include the Saskatchewan district in it, so as to establish one great border line from the new colony of British Columbia up to the Red River Settlement, under

---

they have this year on the line from here to Selkirk settlement, appointments sufficient to meet the demands of the Hudson's Bay territory and intermediate country.

"On that portion of the line, from St. Cloud, seventy-five miles above this city, on the Mississippi, to Georgetown, three hundred miles further northwest, on Red River, they will have on and after the first of June, fifteen four horse Concord coaches, making, regularly, tri-weekly trips, and one hundred transportation waggons. The waggons are divided into four brigades of twenty-five each. The whole line will be organized and conducted with military precision. One brigade, commanded by a captain, will leave St. Cloud every week, and another returning, will leave Georgetown. From that point the steamer will connect with Fort Garry, in the Selkirk settlement of the Hudson's Bay territory."—*St. Paul Pioneer*, May 11th.

"The Anson Northup returned to Fort Garry on Monday morning, having made the round trip between this and Georgetown in a little over seven days — the fastest time on record! When she left Georgetown, on Saturday morning last, the heavy teams had not yet come up, the badness of the roads, owing to continuous rains, having kept them back beyond their usual time. Several teams with freight were, however, within a day's journey of Georgetown; but the boat having to make good the time lost on the previous trip, at once returned. Her cargo consisted of 100 kegs of gunpowder, for the Hudson's Bay Company; 21 bales of clothing, for the Royal Canadian Rifles, &c."—*Nor'-Wester*, Fort Garry, June 20th, 1860.

a sway and jurisdiction distinct from the Hudson's Bay Company's authority.

"3rd. What means of access exist for British emigrants to reach this settlement.

"4th. Whether, judging from the explorations you have already made, the country presents such facilities for the construction of a railway as would at some period, though possibly a remote one, encourage her Majesty's Government in the belief that such an undertaking between the Atlantic and the Pacific Ocean could ever be accomplished."

In answer to the third query, viz. "What means of access exist for British immigrants to reach this settlement," Captain Palliser says, "I think there are no means of access to be recommended save those *viâ* the United States."

"The direct route from England *viâ* York Factory, and also that from Canada *viâ* Lake Superior, are too tedious, difficult, and expensive for the generality of settlers. The manner in which natural obstacles have isolated the country from all other British possessions in the East is a matter of considerable weight; indeed, it is *the* obstacle of the country, and one, I fear, almost beyond the remedies of art. The egress and ingress to the settlement from the East is obviously by the Red River Valley and through the States."

Captain Palliser has not had the opportunity of examining and reporting on the Fort William and Arrow Lake route from Lake Superior to Red River. Practical experience of the facilities offered by that route would, I venture to say, have greatly modified the sweeping condemnation of existing and possible means of communication comprehended in the foregoing brief reply to the important question proposed by the Colonial Office. It is a question of great moment to the future colony in the Basin of Lake Winnipeg, of vast importance to Canada, and in view of the extraordinary gold-wealth of British

Columbia, of paramount interest to the Imperial Government, if British laws, institutions, and civilization are to stretch across the American continent, and connect the eastern with the western hemisphere. As an emigrant route, the outlay of a few thousand pounds can make it available for summer communication. An emigrant could then start from Liverpool and proceed to Quebec by steam (eleven days), from Quebec to Collingwood, Lake Huron, by rail (two days), from Collingwood to Fort William by steamer (three days), and from Fort William to Fort Garry *viâ* Arrow Lake and the Boundary Line (six days), or twenty-two days in all from Liverpool to Selkirk Settlement. The route through the United States, *viâ* St. Paul, cannot offer greater advantages as a summer emigrant route than those presented by the Chain of Lakes* along the Boundary Line to the north-west corner of the Lake of the Woods, if improved and served in accordance with the suggestions contained in the Appendix to this volume.† Cattle and live stock generally, will necessarily pass to and from the United States over the prairies of Red River, where food is abundant, and can be obtained without cost, but there is no reason why heavy goods should not in process of time be shipped at Liverpool and proceed direct to Fort William on Lake Superior without transhipment, passing through the magnificent chain of Canadian canals, enumerated on page 16 of the first volume, and thence to Red River, through British territory *viâ* Arrow Lake, Rainy River, and the north-west corner of the Lake of the Woods.

* See profile of the Chain of Lakes on the Old North-West Company's, or the Pigeon River route, according to Mr. Dawson's survey; also plan of the route published in the Blue Book, August, 1860.

† Remarks on the Pigeon River route, with an estimate of the cost of opening the communication from Fort William to Fort Garry. Appendix, Vol. II. No. 12.

CHAP. XXXV.

THE COMMUNICATION BETWEEN RED RIVER AND THE PACIFIC. —THE IMPORTANCE OF THE BASIN OF LAKE WINNIPEG.

The Result of Captain Palliser's Expedition. — Valley of the Saskatchewan. —Character of the Valley. — The Rocky Mountain Passes.—Route across the Continent.—Lieutenant Palmer's Explorations in British Columbia.—Roads in British Columbia.—Lieutenant Richard Mayne's Explorations in British Columbia.—Thompson River.—Present position of British Columbia. — The Customs and Imports of the Colony. — The Gold Fields. —Their extent. — Captain Palliser's suggested Boundary of the New Colony in the Basin of Lake Winnipeg. — Objections to this Boundary. —Importance of the New Colony being conterminous with Canada on the East. — The Natural Boundary of the New Colony. — Importance of the East Coast of Lake Winnipeg.—Necessity for a Natural Northern Boundary.—Suggested Boundary. — THE NATURAL ADVANTAGES OF THE BASIN OF LAKE WINNIPEG.—The Great American Desert.—The FERTILE BELT in the Basin of Lake Winnipeg. — Its Importance. — Its Natural Resources.— Conclusion.

FORT GARRY, RED RIVER TO THE ROCKY MOUNTAINS, THENCE TO THE PACIFIC.

IN the Blue Book recently published *, Capt. Palliser enumerates the result of the explorations of the expedition under his command across the continent, and states that his endeavours, in conjunction with Mr. Sullivan, "to find a route practicable for horses from Edmonton westward across the Rocky Mountains, as far as the longitude of Fort Colvile, and entirely within British territory, have been perfectly successful." In addition to this, Capt. Palliser pursued his way for more than fifty miles to the

* Papers relative to the Exploration of British North America, August 18th, 1860.

westward, still north of the 49th parallel, until he arrived at the camp of the United States Boundary Commissioners in longitude 119° 30′ west.

Captain Palliser describes the region drained by the Saskatchewan in the following words*:—

"The extent of surface drained by the Saskatchewan, and other tributaries to Lake Winnipeg, which we had an opportunity of examining, amounts in round numbers to 150,000 square miles. This region is bounded to the north by what is known as the 'strong woods,' or the southern limit of the great circum-arctic zone of forest, which occupies these latitudes in the northern hemisphere. This line, which is indicated in the map, sweeps to the north-west from the shore of Lake Winnipeg, and reaches its most northerly limit about 54° 30′ N., and longitude 119° W., from where it again passes to south-west, meeting the Rocky Mountains in latitude 51° N., longitude 115° W. Between this line of the 'strong woods' and the northern limit of the true prairie country there is a belt of land varying in width, which at one period must have been covered by an extension of the northern forests, but which has been gradually cleared by successive fires.

"It is now a partially wooded country, abounding in lakes and rich natural pasturage, in some parts rivaling the finest park scenery of our own country. Throughout this region of country the climate seems to preserve the same character, although it passes through very different latitudes, its form being doubtless determined by the curves of the isothermal line. Its superficial extent embraces about 65,000 square miles, of which more than one-third may be considered as at once available for the purposes of the agriculturist. Its elevation increases

* Papers relative to the Exploration of British North America, Aug. 18th, 1860, page 21.

from 700 to 4000 feet as we approach the Rocky Mountains, consequently it is not equally adapted throughout to the cultivation of any one crop, nevertheless at Fort Edmonton, which has an altitude of 3000 feet, even wheat is sometimes cultivated with success.

"The least valuable portion of the prairie country has an extent of about 80,000 square miles, and is that lying along the southern branch of the Saskatchewan, and southward from thence to the boundary line, while its northern limit is known in the Indian languages as 'the edge of the woods,' the original line of the woods before invaded by fire.

"On the western side of the Rocky Mountains, in the country which we examined, there were but few spots at all fitted for the agriculturist, and these form isolated patches in valleys separated by mountain ranges.

"As the next result of our explorations, I shall briefly mention the different passes through the Rocky Mountains which we explored, alluding to the chief advantages and disadvantages of each.

"The Kananaskis Pass and the British Kootanie Pass were examined by myself. Of these I consider the Kananaskis Pass the preferable one, both on account of its direct course through the mountains and its easier ascent.

"The ascent to the height of land from the east is through a wide gently sloping valley, and the immediate watershed is formed by a narrow ridge, which, if pierced by a short tunnel, would reduce the summit level to about 4600 feet above the sea. The descent to the west, into which Kananaskis Pass opens, is comparatively easy.

"The British Kootanie Pass also opens out into the Kootanie River valley, but the altitude here to be overcome is much greater, amounting to 6000 feet. There are likewise two ridges to be passed, which fact would form a very strong objection to this pass.

"The Vermilion Pass, which was traversed by Dr. Hector, presents on a whole the greatest natural facilities for crossing the mountains without the aid of engineering work, as the rise to the height of land is gradual from both sides, a feature which seems to be peculiar to this pass. It would thus be impossible to diminish its summit level (which is less than 5000 feet), as is proposed in the case of Kananaskis Pass, but on the other hand it would be the most suitable for the construction of an *easy waggon road*.

"This, like the other two passes I have mentioned, also strikes the Kootanie River close to its source; but last summer Dr. Hector crossed the mountains by another pass from the head of the north branch of the Saskatchewan, directly to the Columbia River, in the vicinity of the boat encampment.

* * * * * *

"Although I consider this fact established, viz., that a line for a route has been discovered from Red River Settlement to the west coast of the continent, and that line moreover entirely within British territory, yet I wish distinctly to be understood that I think it far from being the best that could be discovered. Time did not admit of a series of attempts in a more northerly direction.

"Dr. Hector's explorations, when within sixty miles of his exit on Thompson's River, were prematurely closed by the advance of winter and the absence of provisions, while forcing his way through timber so thick that he could not penetrate faster than from three to four miles a day, and for a more detailed account of which I refer you to his report accompanying this letter." *

Lieutenant Palmer, R.E., explored the country between Fort Hope on the Fraser and Fort Colville on the Columbia River.† His report is not favourable for a road, although the physical impediments are not such as to be insurmountable.

Lieutenant Palmer states that "westward from the Similkameen valley‡, and nearly to the coast, extends a mountain region, so rugged and bleak, and so inaccessible and devoid of roads of communication, that this valley is the first point east of the Sumas to which attention need be directed.

"It and the Okanagan valley are the main thoroughfares to British Columbia from Washington territory, and indeed east of them, as far as Fort Shepherd, the country affords no known practicable means of ingress.

The main route from Washington territory passes up the valley of the Okanagan River from its junction with the Columbia and forks at the mouth of the Similkameen whence branch the two routes already described.

* * * * *

"A great question now presents itself as to the means of communication between this district and the interior portions of British Columbia.

"Even if it be practicable at an enormous expense to construct a tolerable waggon road across Manson Mountain, or even should a better route be found by following round the valley of the Coquahalla, there still remains the

* Papers relative to the Exploration of British North America, Aug. 1860, p. 17.

† Papers relating to British Columbia, August, 1860.

‡ The Similkameen rises a few miles north of the boundary-line, in long. 120° 40′; it flows into the Okanagan in long. 119° 30′, lat. 48° 56.

dividing ridge of the Cascades, and the route would any way be impracticable for at least seven months in the year. Through Lytton, therefore, or the Kayoosh district, by the valleys of the Thompson and Bonaparte River to Nicolas Lake and Fort Kamloops, and thence to the border, all supplies must eventually pass, and hence the necessity of establishing good means of communication with these points from either the Fraser River or the sea."

"All the practicable means of access to British Columbia, except from the sea, strike the Fraser north of Fort Hope."

"In connexion with this subject it remains but for me to remark, that, from my own observation, and from information afforded me by Mr. M'Donald, I know a frontier road north of the 49th parallel to be practicable from the Similkameen, eastward to the N. Saa-app Lake in the N. Whoyalpitkwu Valley; and further that Captain Palliser's explorations have determined the possibility of extending such a road from that point to Fort Shepherd, though his report on the intervening tract cannot be pronounced as favourable as might be wished."

The communication through British territory between the Atlantic and Pacific has not been lost sight of by Governor Douglas.

His Excellency says, in the Blue Book before referred to (No. 26, Part. III.), "The great object of opening roads from the sea coast into the interior of the country, and from New Westminster to Burrard's Inlet and Pitt River, continues to claim a large share of my attention. The labour involved by these works is enormous; but so essential are they as a means of settling and developing the resources of the country, that their importance can hardly be overrated; and I therefore feel it incumbent on me to strain every nerve in forwarding the progress of undertakings so manifestly conducive to the prosperity of the colony, and which at the same time cannot fail ere long to produce a large increase in the public revenue."

"We hope to complete the last section of a pack-road leading by the left bank of the Fraser, from Derby (Fort Langley) to Lytton, a distance of 170 miles, on or before the 1st day of February next. From Lytton a natural road now exists leading to Red River settlement, by the Coutannais Pass, through the Rocky Mountains, and from thence following the valley of the Saskatchewan, chiefly over an open prairie country of great beauty, and replete with objects of interest to the tourist and the sportsman; a settler may then take his departure from Red River in spring with his cattle and stock, and reach British Columbia by that road in course of the autumn following. This is no mere theory, the experiment having been repeatedly made by parties of Red River people traveling to Colville, from whence there is a good road to Lytton; so much so, indeed, that one of those persons assured me that the whole distance from Lytton to Red River, with the exception of the Coutannais Pass, which is thickly wooded, may be safely travelled with carts. If the Canadian Government would undertake to open a road from Red River to the borders of Lake Superior, which really presents no very formidable difficulties, the connexion between British Columbia and Canada would be complete, and the whole distance might I think, be traveled on British soil."

The town of Lytton and the valley of Thompson River appear to be points of the greatest importance in British Columbia.

Lieutenant Richard Mayne conducted an exploration of the districts bordering on the Thompson, Fraser, and Harrison Rivers.* Lytton is at the Forks of the Thompson and Fraser Rivers, and contains eight or ten stores and a Government House. The Thompson River is about 150 yards wide at its mouth and there is a horse ferry established across it. Fort Kamloops is an important Post situated on the Forks of the Thompson and North Rivers; the Thompson is here 300 yards wide and the North River 320 yards. The head waters of the Thompson are about twenty-two miles east of these Forks and Dr. Hector advanced in 1859 from the east side of the Rocky Mountains within sixty miles of the source of the Thompson. He says that it was his wish "to follow the Columbia River down to its great bend at the boat encampment, and thence following up the valley of Canoe River endeavour to pass to the head waters of the Thompson River and so reach British Columbia." Want of provisions, the approach of winter, and the extraordinary thickness of the forest prevented Dr. Hector from accomplishing this very interesting link between the trails east of the Rocky Mountains and the head waters of the Thompson River; from which point a pack-road already exists to the Pacific considerably to the north of the Boundary Line.

The more or less speedy opening of a line of communication from the Atlantic to the Pacific through British territory, will be very largely dependent upon the progress of British Columbia. Enough is known respecting the prospects of that distant colony, to warrant the assumption that it will soon become a very wealthy and important British dependency. In the official documents recently published relative to the affairs of British Columbia†, a considerable amount of valuable information

* Papers relative to British Columbia, Part III., Aug. 1860, p. 33.

† Further papers relative to the affairs of British Columbia, Part III. Presented to both Houses of Parliament by command of Her Majesty, August, 1860.

is contained, bearing upon the present prosperous state of the colony, the distribution of gold, and the existence of large areas of fertile soil fitted for cultivation.

From the return of the value of imports and customs receipts at New Westminster for the twelve months ending 31st December, 1859, it appears that the imports amounted to 177,219*l.* 7*s.* 5*d.*, and the customs to 18,464*l.* The entire white population of British Columbia does not exceed 5000 men (Oct. 1859) with very few women and children. The value of the present export of gold (Oct. 1859) is estimated at 14,000*l.* a month or 168,000*l.* per annum.

The existence of gold in great abundance and over very wide areas in British Columbia, from the Rocky Mountains to the Pacific, and from the head waters of Fraser River to the Boundary Line, is well established; and "means of communication" embrace all that is required to make this distant colony one of the most important appendages of the British crown.

The second query submitted to Capt. Palliser by the Colonial Office elicited a reply which appears to be open to two serious objections.

"What should be the dimensions and the boundary line of such colony, and whether it would be advisable to include the Saskatchewan District in it, so as to establish one great border line from the new colony of British Columbia up to the Red River Settlement, under a sway and jurisdiction distinct from the Hudson's Bay Company's authority?" Capt. Palliser states in relation to the boundaries of the new colony:—

"If then the united territories of Red River, Swan River, and Saskatchewan, were adopted by Her Majesty's Government, I would suggest the following boundaries:—

"The southern boundary of the colony should be the

49th parallel of north latitude, commencing on east shore of the Lake of the Woods, to where it meets the crest of the Rocky Mountains in long. 115° W. The eastern boundary of the colony should be defined by a line commencing at the 49th parallel on the western shore of the Lake of the Woods, and following the western margin of that lake to the watercourse which unites the Lake of the Woods with Lake Winnipeg, from thence extending around the eastern shore of Lake Winnipeg, and following the watercourse of that lake to the 54th parallel of N. lat. in long. 98° W. The northern boundary of the colony might run from the above point of intersection along the parallel of 54° of N. lat. to the point where it intersects the crest of the Rocky Mountains in 118° of W. long. The whole would thus include a territory of 240,000 square miles."

The adoption of the eastern boundary described in the foregoing paragraph, would effectually isolate the colony from Canada, and interpose between the boundary of the two countries, a territory in the occupation of the Hudson's Bay Company, subject to their jurisdiction, and comprehending the narrow, but fertile valley of Rainy River, and the only practicable route known in British territory between Lake Superior and Red River. The boundary of Upper Canada is the height of land west of Lake Superior. The country drained by those rivers which flow into the St. Lawrence valley belongs to Canada; the country drained by those rivers which flow into Hudson's Bay, constitutes Rupert's Land, and is at present the hunting-ground of the Hudson's Bay Company. Captain Palliser's suggested boundary would leave the entire region between the height of land west of Lake Superior, and the western shore of the Lake of the Woods, nearly five degrees of longitude, in the occupation of the Company. The continuation of the eastern boundary from the outlet of Lake of the Woods, is proposed to be carried down the Winnipeg and along the east shore of that lake to the 54th parallel, in long. 98° W. By thus limiting the colony in an easterly direction, the valuable pine, spruce, minerals, and water-power of the east coast of Lake Winnipeg and of the Winnipeg River, would be legally cut off from the new colony, whose chief want will be timber for building purposes, and water-power to manufacture it on the spot. The jurisdiction of the new colony should manifestly extend over those parts of the country which, from their natural resources, will first attract a certain class of settlers. The lumberers on Lake Winnipeg will one day become relatively as important a body as the lumberers now are, on the Ottawa and the tributaries, to the St. Lawrence in Canada.

In order to secure to the new colony a route to the borders of Canada, and the valuable timber of the east coast of Lake Winnipeg and the affluents of the Winnipeg River, the eastern boundary line ought to be conterminous with that of Canada, as far as the north west corner of that province at the head waters of Neepigon and Firesteel Rivers, in long. 90° 25′, lat. 49° 15′ (approx.); and from this point it should follow the height of land between the Winnipeg Basin and Hudson's Bay, as far as the 54th parallel. The natural boundaries of the new colony might be very simply defined as, "the Basin of Lake Winnipeg north of the 49th parallel of latitude." This determination would give the following limits:—on the south, the 49th parallel from British Columbia to the Lake of the Woods, thence to Canada by the boundary line surveyed according to the treaty of Ghent. The west boundary of Canada stretches from the dividing ridge between South Lake and Gun Flint Lake, to the head waters of Neepigon and Fire-steel Rivers. This would form the eastern boundary of the new colony. From the head water of Fire-steel River, the north-east boundary might follow the dividing ridge between the Basin of Lake Winnipeg and Hudson's Bay, as far as Jack River, down which it should run to Play Green Lake. From this point the northern boundary would be very conveniently and naturally described as passing through, by the shortest course, the head waters of all the rivers flowing into Lake Winnipeg, the main Saskatchewan, and the North Branch of the Saskatchewan.

It is of great importance in an Indian and fur-trading country to possess a natural boundary. Indians and voyageurs can always understand the flow of waters and the term "dividing ridge," but of parallels of latitude they know absolutely nothing. The injunction not to hunt or trap beyond the dividing ridge would be understood by all, but an arbitrary boundary such as a parallel of latitude, might give rise to many disputes in the prosecution of the fur trade, which for some years to come will constitute the most important source of commerce and industry north of the basin of Lake Winnipeg.

THE NATURAL ADVANTAGES OF THE BASIN OF LAKE WINNIPEG FOR A ROUTE ACROSS THE CONTINENT.

It is impossible to examine a correct map of the North American continent without being impressed with the remarkable influence which the Great American Desert must exercise upon the future of the United States and British North America. The general character of this desert south of the 49th parallel is described elsewhere (Ch. XLI. p. 353; Appendix Vol. II. No. VII.), and the important fact has been noticed, that any railroad constructed within the limits of the United States must pass, for a distance of twelve hundred miles west of the Mississippi, through uncultivable land, or, in other words, a comparative desert. Along the 32nd parallel the breadth of this desert is least, and the detached areas of fertile soil greatest in quantity, but the aggregate number of square miles of cultivable land amounts only to 2300 in a distance of 1210 miles (Appendix No. VII.). The northern limit of the Great American Desert is an imaginary line drawn from the Touchwood Hills to the Moose Woods on the South Branch, then south of Battle River as far as long. 112°, when turning south it sweeps along the flanks of the Rocky Mountains in long. 115°. North of this limit of the Great American Desert there is a broad strip of fertile country, rich in water, woods, and pasturage, drained by the North Saskatchewan and some of its affluents, and being a continuation of the fertile prairies of Red River, the eastern water shed of the Assinniboine and Red Deer River, with the outlying patches called the Touchwood Hills, File Hill, &c.

IT IS A PHYSICAL REALITY OF THE HIGHEST IMPORTANCE TO THE INTERESTS OF BRITISH NORTH AMERICA THAT THIS CONTINUOUS BELT CAN BE SETTLED AND CULTIVATED FROM A FEW MILES WEST OF THE LAKE OF THE WOODS TO THE PASSES OF THE ROCKY MOUNTAINS, AND ANY LINE OF COMMUNICATION, WHETHER BY WAGGON ROAD OR RAILROAD, PASSING THROUGH IT, WILL EVENTUALLY ENJOY THE GREAT ADVANTAGE OF BEING FED BY AN AGRICULTURAL POPULATION FROM ONE EXTREMITY TO THE OTHER.

No other part of the American Continent possesses an approach even to this singularly favourable disposition of soil and climate, which last feature, notwithstanding its rigour during the winter season, confers, on account of its humidity, inestimable value on British America south of the 54th parallel.

The natural resources lying within the limits of the Fertile Belt, or on its eastern borders, are themselves of great value as local elements of future wealth and prosperity; but in view of a communication across the continent they acquire paramount importance.

Timber available for fuel and building purposes; lignite coal, though not equal to true coal, nevertheless suitable for many of the different objects to which true coal is applied; iron-ore widely distributed, of great purity and in considerable abundance; salt in quantity sufficient for a dense population. All these crude elements of wealth lie within the limits or on the borders of a region of great fertility, and drained by a river of the first class, navigable by steamer during several months of the year for five hundred miles of its course, and by batteaux for nearly double that distance.

The position which the colony occupying the basin of Lake Winnipeg may assume at the close of the next decade, few will be prepared to define. Bounded on the west by British Columbia, whose gold-wealth will ensure her a marvellously rapid progress, and on the east by the powerful, energetic, and loyal colony of Canada, which now, in conjunction with the sister provinces, contains a population exceeding by ONE MILLION that of the thirteen United States during the revolutionary war, is it likely that British enterprise and patriotism will permit the intervening country to remain a wilderness, or pass into the hands of a foreign government?

It is to be earnestly hoped that the attention of far-seeing, thoughtful, and loyal men will be directed to the present relations of the new colony, its possible future, and the opportunity it presents to plant British institutions and civilization across the North American Continent; thus linking the Atlantic and Pacific oceans together with a chain of British dependencies, distinguished alike by a "SPIRIT OF LOYALTY, OF ORDER, AND OF OBEDIENCE TO THE LAW."*

The occupation and government of the basin of Lake Winnipeg has already become a serious question, and even before the rising generation succeed to the responsibilities of those who now rule the destinies of this great empire, they may have occasion to lament a lost opportunity of inestimable worth, or rejoice in the extension of British dominion over loyal populations, extending in an unbroken series from the eastern to the western hemisphere.

* "Her Majesty has observed with deep satisfaction the spirit of loyalty, of order, and obedience to the law, which prevails among her subjects, both in the United Kingdom and in her dominions beyond the sea; and her Majesty has witnessed with heartfelt pleasure the warm and affectionate reception given to his Royal Highness the Prince of Wales by her North American subjects."—*Her Majesty's Speech,* Aug. 28th, 1860.

GEOLOGY

OF THE

BASIN OF LAKE WINNIPEG

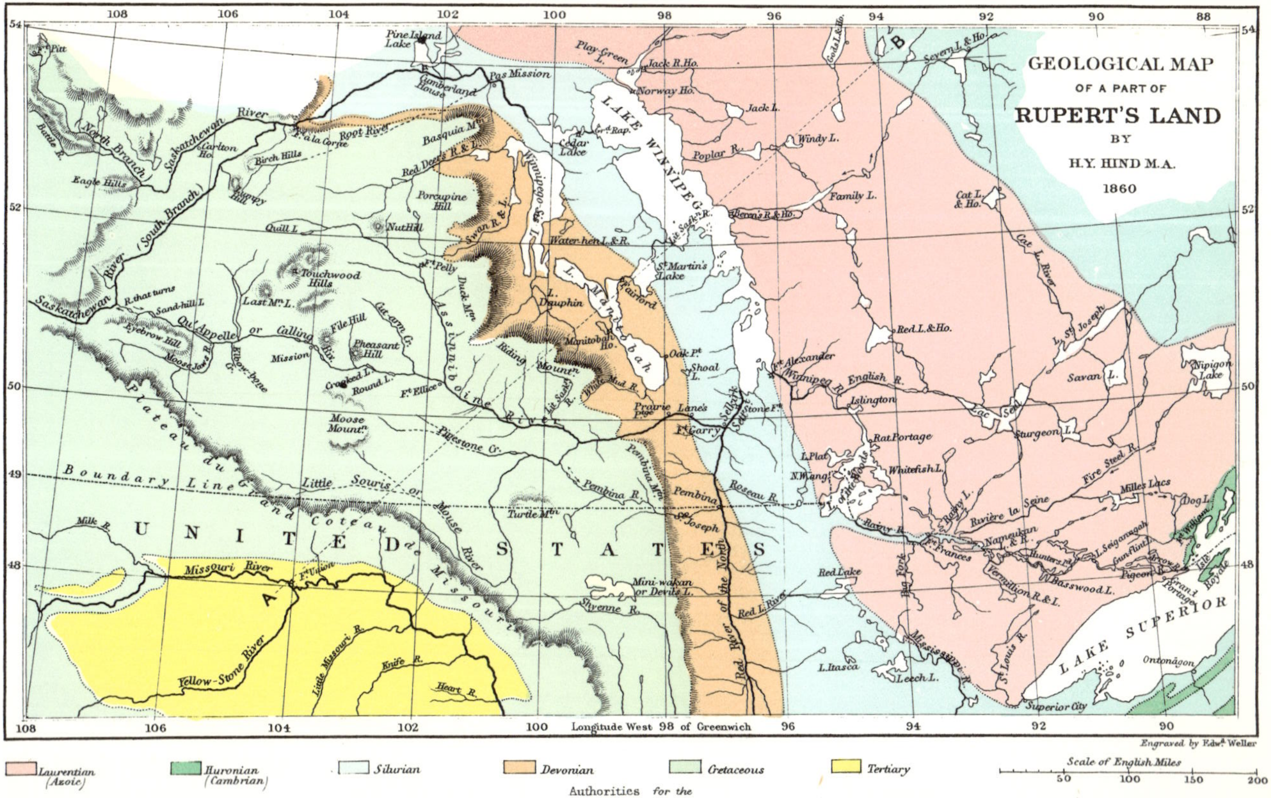

Authorities for the North West Shore of Lake Superior. North of the Boundary Line, Sir W. E. Logan; South Shore, Mess.rs Foster and Whitney; Rainy River, Dr. Bigsby; Tertiary Basin of Missouri Dr. Hayden and Mr. Meek; See also the

CHAP. XXXVI.

SURFACE GEOLOGY.

GEOGRAPHICAL BOUNDARIES. — The Basin of Lake Winnipeg. — Limits. — Elevation of its Boundaries.—Outlet.—Area.—Character of its Rim.—— SURFACE FEATURES.—Area of Low Lake Region.—Terraces of Pembina Mountain.—Escarpment of the Riding, Duck, &c. &c. Mountains.—Prairie Plateau of Rupert's Land. — Plains of the North-West. — Table of Elevations. — Detached Hills. — Country east of Lake Winnipeg.—Lakes. — Direction of Elevating Forces. — Steps to the Height of Land. —— GROOVED, SCRATCHED, AND POLISHED ROCKS. — On Baril Portage. — On Winnipeg River.—On Limestones of Lake Manitobah.——DRIFT.—Over the Country between Lake Superior and Winnipeg.—Lacustrine Deposits. —Drift on the Souris.—On the South Branch.—Forced Arrangement of. —In Blue Clay at Toronto.—Description of this Arrangement.—Sorting of Materials.—Agency of Water and Ice.—Glacial Ice.—Tiers of Boulders on the South Branch.—Layers of Stratified Mud.——ERRATICS.—Distribution of.—Effects of Ice on Lake Winnipeg.—In Lake Manitobah.—On the Red River Prairies. —— BEACHES AND TERRACES. — Great Dog Portage.—Section of.—Beaches on Lake Superior.—On Prairie Portage. —The Big Ridge.—White Mud River Ridge.—Dauphin Lake Ridge.— Prairie Ridges.—Pembina Mountain.—Ridge, character of.—Riding and Duck Mountain Ridges.—Conical Hills. — Bear Hill. — Second Tier of Conical Hills.——SAND HILLS AND DUNES.—Ranges.——CIRCULAR DEPRESSIONS.—Character of.——DENUDATION.—In the Basin of Lake Winnipeg. — Parallelism of bold Limits of Denudation. — Grand Coteau de Missouri.—Niagara Limestone Escarpment of New York and Canada.— Riding and Duck Mountains, &c. — Probable common Origin of. — Dislocations in the Basin of Lake Winnipeg.

GEOGRAPHICAL BOUNDARIES.

THE Basin of Lake Winnipeg, including under this general term the country drained by the rivers flowing into it, extends from the 90th to the 118th meridian.

Its most easterly point being the lake and swamps from which the Savanne River takes its rise, in longitude 90° 14′, latitude 48° 53′, and the most westerly limit from which it draws contributions is probably the Glacier, near Howse Pass, in longitude 117° 35′, latitude 51° 52′.*

The southern extension of its boundary is at Lake Traverse in Dakotah territory, longitude 96° 45′, latitude 45° 58′.† It stretches north as far as Frog Portage, longitude 103° 30′, latitude 55° 26′.‡

This basin consequently extends over twenty-eight degrees of longitude and ten degrees of latitude.

The elevation of its eastern boundary is 1485 feet above the ocean, and the height of land near the sources of the tributary, which rises farthest to the west, is 6347 feet above the same level.§

Its northern boundary is separated from the valley of the Missinnippi by a low portage, over which waters flow during floods‖; while towards the south Lake Traverse, which also sends water into the Mississippi during spring freshets, is only 820 feet above the sea.¶

The outlet of Lake Winnipeg is through the contracted and rocky channel of Nelson River, which flows into Hudson's Bay.

The mean breadth of this great inland basin is about 380 English miles, and its mean length 920 miles, hence its area is approximately 360,000 square miles. Its shape assimilates to that of a truncated section of a circle, lying

* See Captain Palliser's Map in the Blue Book, 1860.

† Military Map of Nebraska and Dakotah, by Lieutenant G. K. Warren, U.S. Top. Eng.

‡ Sir John Richardson, Arctic Searching Expedition.

§ Dr. Hector, Blue Book. Captain Palliser's Explorations.

‖ Sir John Richardson.

¶ Major Long.

in a position the reverse of that produced by the meridians and parallels distinguishing its boundary.

The eastern rim of the basin of Lake Winnipeg is formed by part of the Laurentide range of mountains, which consist of gneiss * interstratified in some localities with bands of crystalline limestone, and much embossed by domes of intrusive granite, syenite, and occasionally with elevations of trap. The highest summit of the Laurentide Mountains, as far as known in this part of their development, does not exceed 1950 feet above the sea, or 1350 feet above Lake Superior.

Striking off in a south-westerly direction from the Savanne Lake, the Height of Land extends to Lake Traverse and is composed of drift hills covering Laurentian, Silurian, and Devonian rocks, with low granitic ranges and exposures of trap. From the Dividing Ridge, at the source of the Pigeon River, south-westerly to Vermilion Lake, the Height of Land Ridge is called by the Indians *Mis-sabe Wa-chu;* in its continuation to the low water-shed between Rainy River and the Mississippi it has the name of *Ish-ko-na-bi Wa-chu.* Continued south-westerly from the Falls of Pokegama, this ridge would pass south of Leech Lake and strike the Red River of the north near the Great Bend, at which point Dr. Owen found Silurian rocks in 1848.†

From this point on Red River the rim of the basin assumes a north-westerly direction as far as the Elbow of the South Branch of the Saskatchewan, along the denuded flanks of the Grand Coteau de Missouri. From the Elbow it turns south-west as far as Chief's Mountain Lake, under the 114th meridian‡, latitude 49°, where it meets

* Under the general term "gneiss," micaceous, hornblendic, and chloritic schists are included.

† Geological Survey of Wisconsin, Iowa, and Minnesota.

‡ Map of the Territory of the United States, by Lieutenant G. K.

with the flanks of the Rocky Mountains, trending first north and then north-westerly. The North Branch and the main Saskatchewan indicate approximately its low northern boundary.

SURFACE FEATURES.

Lake Winnipeg, at an altitude of 628 feet above the sea, occupies the lowest depression of this great central basin, covering, with its associated Lakes Manitobah, Winnepego-sis, Dauphin, and St. Martin, an area slightly exceeding 13,000 square miles, or nearly half as great an extent of the earth's surface as is occupied by Ireland.

The country possessing a mean elevation of one hundred feet above Lake Winnipeg is very closely represented by the outline of Pembina Mountain, forming part of the eastern limit of the Cretaceous Series in the north-west of America, as represented on the geological map.

The area occupied by this low country, which includes a large part of the valley of Red River, the Assinniboine, and the main Saskatchewan, may be estimated at 70,000 square miles, of which nine-tenths are lake, marsh, or surface rock of Silurian or Devonian age, and generally so thinly covered with soil as to be unfit for cultivation, except in small isolated areas.

Succeeding this low region there are the narrow terraces of the Pembina Mountain, which rise in abrupt steps, except in the valleys of the Assinniboine, Valley River, Swan River, and Red Deer's River, to the level of a higher plateau, whose eastern limit is formed by the

Warren, Top. Eng., ordered by the Honourable Jefferson Davis, Secretary of War, to accompany the Reports of the Explorations for a Railroad Route to the Pacific.

precipitous escarpments of the Riding, Duck, and Porcupine Mountains, with the detached outliers, Turtle, Thunder, and Pasquia Mountains. This is the great PRAIRIE PLATEAU of Rupert's Land; it is bounded towards the south-west and west, by the Grand Coteau de Missouri, which forms the north-eastern limit of the PLAINS* of the north-west. The area of the Prairie Plateau, in the basin of Lake Winnipeg, is about 120,000 square miles; it possesses a mean elevation of 1100 feet above the sea, and is approximately represented on the geological map by the Cretaceous Series, coloured green, north-east of the Grand Coteau. The following table shows the relation of these successive levels:—

	Mean Elevation above the Ocean.	Approximate Area in Square Miles.
Low Country (1st Plateau)	700	70,000
Terraces of Pembina Mountain	900—1100	1,500
Great Prairie Plateau (2nd Plateau)	1100	120,000
Great Plain Plateau (3rd Plateau)	1600	———

The Riding and Duck Mountains, with their southern and northern prolongations, probably once formed part of an unbroken level to the Grand Coteau; the intervening depression having been the result of denudation. The isolated ranges of hills, such as the Touchwood Hills, the File Hill, the Pheasant Hill, the Birch Hills, the Lumpy Hill, &c., are parts of this former elevated table-land, and would assume the character of islands in a sea washing the base of the Grand Coteau de Missouri.

On the east side of Lake Winnipeg the primary unfossiliferous rocks along the canoe routes present a broken and extremely irregular surface to the mouth of Rainy River, where the fossiliferous formations are touched. From Rainy Lake to Superior the routes pass through a

* The difference between "Prairies" and "Plains" is stated in Chap. XVI.

water communication, possessing many singular geographical outlines, arising from the geological features of the country.

Both Lake Superior and Lake Winnipeg belong to the same geological basin, and are excavated in its rim, the low Laurentide Mountains separating the great northern basin of fossiliferous rocks from it.

The canoe routes between Lake Superior and Lake Winnipeg form, therefore, two chords on the edge of the gneissoid and granitic boundary dividing these geological basins, and while the general trend of this part of the Laurentide Mountains is geographically north-west, the direction of the undulations, by which it is intersected, is generally north-east and south-west, or at right angles to the former. This disposition is best developed east of Rainy Lake, and on the Pigeon River route, where the agency of the elevating force appears to have acted with greater constancy and uniformity than on the more northern canoe route. The outbursts on the southern route, as well as on the south shore of Lake Superior, are also considerably higher, and the lakes and lake straits more symmetrically arranged.

An inspection of the geographical map will show that the longest axis of nearly all the smaller lakes in the region under review is from N.E. to S.W.

A range of greenstone ridges, forming an anticlinal axis, begins at the Great Bend of the St. Louis River, and runs in a direction N. 30° E. towards the sources of Arrow River and Mountain Lake. The systems of valleys and ridges, resulting from the direction in which the elevating forces have acted in the region about the western extremity of Lake Superior, have been ably discussed by Dr. Norwood.*

* Owen's Report of a Geological Survey of Wisconsin, &c.

GROOVED, SCRATCHED, AND POLISHED ROCKS.

Instances of the action of ice in abrading and polishing extensive surfaces of rock are very numerous on the Laurentides between Lake Superior and Lake Winnipeg. The first wide expanse observed on the west side of the water-shed is at Baril Portage, 143 miles from Lake Superior, and 1500 feet above the sea. Where an arm of Mille Lacs approaches this portage, gneissoid hills and islands about 100 feet high show a well defined stratification dipping north, at an angle of about 15°, and on that side smooth, and sometimes roughly polished; on the south side they are precipitous and abrupt. The same character is observed on the portage itself; the north-eastern exposures of the rocks there are smooth and striated, the southern rugged and often precipitous.

On Sturgeon Lake 208 miles from Lake Superior and 1156 feet above the sea, the north-eastern extremities of hill ranges slope to the water's edge, and when bare are always found to be evenly smoothed and ground down. The aspect of the south and south-western exposures is that of precipitous escarpments.

On a small island about twenty-five miles north of Garden Island, Lake of the Woods, there is a remarkable exposure of greenstone conglomerate, nearly on a level with the water of the lake. The surface of the rock is almost horizontal, beautifully polished, and strongly marked with glacial furrows and scratches. The direction of the furrows is N. 25° E., they are all parallel to one another, some of them half an inch in depth, and nearly double that measure in width. They pursue a uniform direction for many yards until concealed by the bushes which fringe the bare rock some forty or fifty feet from the water's edge.

The conglomerate presents the appearance of an immense table of mosaic work. The pebbles and small boulders enclosed in the matrix have been apparently water worn, some of them, however, showed no lateral abrasion, preserving their angles sharp, and well defined. They were all ground down to one uniform polished surface.

The imbedded boulders and pebbles varied from half an inch to eighteen inches in diameter, and appeared generally to lie with their flat side facing the south-west. The colour of the matrix was a pale green, and of the imbedded pebbles grey, with a tint of green. A vertical section of the rock exhibited the pebbles and boulders, as if resting upon the extremity of the longest axis, with a slight inclination to the east.

From the summit of an abraded granite hill about 250 feet high, on the Winnipeg west of "the Dalles," a very extensive view of the surrounding country is obtained. The broad river, with its numerous deep bays, may be seen stretching far to the north, and all around smooth dome-shaped hills show their bare and scantily wooded summits in every direction. The general surface is either bare and so smooth and polished as to make walking dangerous, or else thickly covered with cariboo moss and tripe de roche.

This general description applies to a vast area drained by the Winnipeg, and on the smoothed and polished rocks glacial grooves may be easily traced for long distances; sometimes also boulders are found resting upon the polished surfaces. On one occasion an attempt to ascend a round dome-shaped mound forming the summit of a granite hill was frustrated by its beautifully polished surface preventing a footing from being obtained. The action of atmospheric agents had only succeeded in dimming its beauty, but had not destroyed its smoothness.

Grooves and scratches occur on the limestones of Lakes Winnipeg and Manitobah, where the surface has been preserved from atmospheric agencies, but whether they are of recent origin or connected with the drift, is not certain.

DRIFT.

The distribution of drift throughout the region lying between Lake Superior and Winnipeg is limited to a few detached areas. The most important of these east of the Height of Land is at the Great Dog Portage, where it occurs in the form of beaches and terraces, which will be described further on. The terraced drift in the valley of the Kaministiquia, below the Kakabeka Falls, is evidently of lacustrine origin. The Height of Land is covered with true boulder drift, having an elevation of about 1500 feet above the present level of the ocean; but on the Height of Land at the sources of the Pigeon River, boulder drift occurs at a considerably higher level.

Small areas of drift may be found at the different portages, and on the islands in Mille Lacs, but in no instance east of Rainy River, were they seen of sufficient extent as to warrant especial notice. In the valley of Rainy River buff-coloured unstratified clay, containing magnesian limestone boulders, and superimposed by stratified clay, occupies both banks for some distance along the river. On the Winnipeg, patches of drift begin at the Islington Mission, and, though far apart from one another, increase in area as the country sinks, until, at the Manitou Rapids, drift clay forms the banks of the river and continues to its mouth.

In the lower part of the Valley of Red River unstratified clay, with boulders from Laurentian rocks and limestone, rises from the water's edge to within four feet of the surface, after which its colour changes, it shows stratification, and is evidently lacustrine. The unstratified clay contains a large per centage of magnesia derived from the underlying rocks.

At Lane's Post on the Assinniboine, the quaternary lacustrine deposit, eight feet below the surface, presents the appearance of irregularly stratified bands of extremely fine material deposited in quiet waters. Upon close examination the bands are found to consist of a series, in which light ash-coloured and dark brown layers alternate, about $\frac{1}{4}$ to $\frac{1}{16}$ of an inch in thickness. In the light-coloured bands minute crystals of sulphate of magnesia and common salt are common. The clay is characterized by its extreme fineness, the marked absence of silicious particles in the form of sand, and the presence of carbonates of lime and magnesia in variable quantities. In certain localities carbonate of lime exists in proportions sufficient to constitute a marl.

The following section is exposed at Lane's Post: —

	Feet.
Vegetable prairie mould	1
Unstratified alluvial clay	5
Stratified dark-coloured clay with light ash-coloured bands	12
Total	18

The subsoil on Red River, near the middle settlement, four feet below the surface, is much darker coloured, not so distinctly stratified, and contains abundance of mica in very fine brilliant plates. The particles of the subsoil on either river are extremely minute, and difficult to separate by washing.

The drift on the Blue Hills of the Souris is chiefly of local origin, and consists almost exclusively of the shales which form the outcrop of the Cretaceous rocks whose limit is defined by the Pembina Mountain. Its age is consequently posterior to that of the true Boulder drift, which is generally distributed over the high prairies to the west, and will be described further on in connection with *beaches* and *conical hills.*

On parts of the Little Souris River the drift is very shallow, indeed it is doubtful whether true Boulder drift is present, the character of the surface material leading to the supposition that it is derived almost exclusively from the subjacent rocks.

The drift on the South Branch of the Saskatchewan, below the Moose Woods, is distinguished by a remarkable peculiarity in the arrangement of the slabs of limestone and boulders of the unfossiliferous rocks distributed through the clay, which do not occupy the position they would assume if dropped from floating ice into soft mud.

Every fact relating to the drift possesses some degree of interest, and may assist in the elucidation of that stupendous phenomenon and its subsequent changes, as well as tend to remove difficulties with which the whole phenomena of the drift are still invested.*

In the blue clay at Toronto, the capital of Western Canada, an arrangement similar to that which exists among the coarser materials of the drift on the South Branch is remarkably well preserved.

During the construction of the esplanade at Toronto, the plan pursued of removing the blue clay was well adapted

* The forced arrangement of blocks of limestone, slabs of shale and unfossiliferous boulders in the blue clay of Toronto, formed the subject of a paper which I read before the Canadian Institute some years ago. As the opportunities for making observations upon this peculiar arrangement were very favourable at that time, I shall here introduce an abstract of the paper, with a view to explain the manner in which slabs and boulders are found arranged in the Drift on the South Branch. (See Report on the Assinniboine and Saskatchewan Exploring Expedition, p. 120.)

to show a perfect sectional view of its components, without the risk of changing in the least degree their relative positions. The clay was cut away until a perpendicular wall was left, varying from ten to twenty feet in height, according to the locality. Wedges were then inserted at the top of the artificial cliff, about two feet from its edge, and driven into the clay until a mass, frequently two feet broad, fifteen or twenty feet long, and twelve or eighteen feet deep, separated and fell. The fresh surface thus exposed was necessarily quite natural in every respect, not having been touched by the tool of the workman or changed by exposure to the weather, and during the years 1855 and 1856, a large area of sectional surface was laid open to view. Two varieties of blue clay exist in the neighbourhood of Toronto, forming deposits quite distinct from one another. The deposit in question overlies rocks of Silurian age, which are exposed in many localities on the lake shore and on the banks of the rivers near the city; it rests upon a blue argillaceous shale, easily recognized as constituting in fragments of different sizes, a large proporton of the substance of the blue clay.

The thickness of this deposit of blue clay varies from ten to twenty-five feet; its upper surface is irregular and undulating, and upon it reposes, in some places, stratified sand and yellow clay, in others, unstratified yellow clay. Resting on the sand or yellow clay, another kind of blue clay occurs, differing, however, essentially from the blue clay which lies at the base of the whole. The lower or inferior blue clay contains quartz sand and small rolled pebbles of granitic rocks, a considerable proportion of blue shale containing fossils belonging to the Silurian shales upon which it rests, and frequently large fragments of the last-named rock, together with more or less rolled or worn masses of granite, gneiss, &c.

The fragments from the Silurian shales often possess sharp and well-defined edges, showing that they have not been water worn or removed far from the rock from which they originated. They are found not only a few inches from the surface of the parent rock, but in numerous instances from fifteen to twenty feet above it, imbedded in a peculiar manner in the blue clay. Some of the larger fragments are scratched and grooved.

A cursory inspection of the artificial cliffs, as they existed during the construction of the esplanade, was sufficient to show that a considerable number of the pebbles and imbedded masses of rock did not occupy the position they would assume if they had not been subjected to some other force besides that of gravity or water in motion. The inclination of the subjacent rock is so slight (thirty feet in the mile) that for all purposes of the present inquiry it may be considered horizontal; and it may be further remarked, that there is no reason to suppose that any material change in position has occurred since or during the accumulation of the blue clay. A large number of the fragments of rock seen in the blue clay are symmetrically inclined at an angle of 60°, 70°, and 80° to the horizon, and frequently lean towards the east and north-east.

What force has thus symmetrically arranged these fragments of shale, &c.? That they now preserve the position into which they were forced by pressure, or that they were brought from a distance and left in that position, is sufficiently evident, as we cannot entertain the opinion that the rock on which the boulder drift rests has materially changed its inclination since or during the Drift epoch.

There can be no doubt that a very large portion of the drift of Canada has been rearranged since it was first deposited. The inferior layer of blue clay is, however, essentially different from the upper layer which is frequently separated from it by a few feet of sand, and in some instances may even directly overlie it and consist of a rearrangement of its materials. The superior blue clay, together with the sand and yellow clay, frequently give evidence of stratification, and thus explain at once the nature of the force to which they have been subjected.

The position of the rock fragments in the inferior blue clay, proves that it cannot have been exposed to the action of water, otherwise they would not preserve the forced arrangement which distinguishes them. Fragments of shale, if submitted to gravity alone, would not have assumed the position in which they were found, had they dropped through water in motion or water at rest, into soft mud. It is well known that shingle, sand, gravel, and clay, either separately or combined, when thrown down an incline, as in the construction of a railway embankment or as in a land slip, will assume a position upon the surface of the embankment, which, if composed of sand, is generally inclined about 45° ; if of harder or coarser materials, at a higher angle. If the embankment or incline be formed under water, like the deltas at the mouths of rivers, this inclination is much less, and is dependent upon the specific gravity of the materials, but under no circumstances is it so high as 45°. If the existence of a current of water sufficiently powerful to move masses of shale and boulders of the unfossiliferous rocks, it cannot be supposed that they would be found deposited upon the slope of a bank at so high an angle as the shale and boulders in the blue clay of Toronto, or the drift on the South Branch of the Saskatchewan, neither is it in the least degree probable that the current which could transport such heavy materials would admit of a *mixture* of clay, sand, shale, and boulders. The materials would be sorted by the current and deposited in the order of their specific gravity. The *sorting* of *materials* is one of the most positive proofs of the action of currents; and where no trace of *sorting* can be discovered, when fine sand, coarse sand, pebbles, and boulders are present, we may reasonably infer that no current assisted in distributing them.

Among the foreign materials entering into the composition of the blue clay, we find granitic masses which have been brought from the outskirts of the fossiliferous rocks in Canada, a distance of at least 100 miles from their present position, and throughout the blue clay we discover also the magnetic oxide of iron.

The materials of local origin exist in great abundance in the form of fragments and masses of shale, limestone, and clay derived from the underlying shales, &c. The nature of the agent which transported the foreign materials from so great a distance is almost universally acknowledged to have been water and floating ice. The finer materials may have been conveyed by water, the coarser drift and erratics would require floating or moving ice. There can be little doubt that both water and floating ice (icebergs and floes) have been instrumental in bearing from northern fossiliferous and unfossiliferous rocks a considerable proportion of the numberless erratics strewn over the surface of a large part of this continent, as well as much of the clayey deposits so plentifully distributed north of the 40th parallel. But the symmetrical arrangement of some of the slabs, pebbles, and boulders in the blue clay at Toronto, in the clay cliffs of the South Branch of the Saskatchewan, and in other localities where the same disposition may be witnessed, points also to the action of *glacial* or *stranded* ice. The phenomena may be explained by coast ice, or the dirt bands of glacial ice, but the entire absence of a sorting of fine and coarse materials, seems to destroy the hypothesis which introduces the agency of currents of water, as the forced but symmetrical arrangement does that of floating ice.

The wide-spread phenomena exhibiting the greater or less action of ice, such as grooved, polished, and embossed rocks, the excavation of the deep lakes of the St. Lawrence basin, the forced arrangement of drift, the ploughing up of large areas, and the extraordinary amount of the denudation at different levels without the evidence of beaches, all point to the action of glacial ice previous to the operations of floating ice in the grand phenomena of the Drift.

The long lines of boulders exposed in two parallel horizontal rows, about twenty feet apart, in the drift of the South Branch, are the records of former shallow lakes or seas in that region. They may represent a coast line, but more probably low ridges formed under water, upon which the boulders were stranded. In the shallow lakes of the Winnipeg basin, the boulders brought year by year by ice from the neighbouring shore accumulate upon long, narrow spits, and ultimately form breakwaters or islands. The same process may have occurred with the boulders on the South Branch. The fine layers of stratified mud, easily split into thin leaves, which lie just above them, show conclusively that they were deposited in quiet water; their horizontality proves that they occupied an ancient coast, floor or ridge below the comparatively tranquil water of a shallow lake. The vast accumulations of sand and clay above them establish the antiquity of the arrangement, and the occurrence of two such layers parallel to one another, and separated by a considerable accumulation of clay and sand, leads to the inference that the conditions which established the existence of one layer also prevailed during the arrangement of the other. These boulders may be distributed over the level floor of a former lake or sea, and cover a vast area; if this be the case it only proves that the agents which brought them operated a second time, after a long intervening period, and with similar results. The fine mud has been submitted to careful microscopical investigation, without establishing the presence of diatomaceæ, or any organic forms from which conclusions might be drawn respecting its origin.

ERRATICS.

The distribution of boulders or erratics in the area explored, may be traced as in Canada to at least two epochs; the Drift and Boulder period, during which by far the larger number were torn from the parent rock and carried by ice to their present resting-places, and the Recent period, including the rearrangement of ancient drift and the distribution of fresh supplies by the action of ice.* The largest boulder was seen in the valley of the Qu'appelle. The circumference of this enormous erratic is seventy-eight feet, and it exposes a

* Where erratics are distributed in large quantities, the locality is indicated on the map accompanying the Reports on the Assinniboine and Saskatchewan Exploring Expedition.

portion above ground at least fourteen feet in altitude. The next largest, one of limestone, was seen on the prairies below the Moose Woods; it is about sixteen feet high, and at least sixty in circumference, is very jagged, and consists of immense slabs, whose edges project two and three feet.* Near it are many others of the same kind, but of smaller dimensions. Near Little Cut Arm Creek, an affluent of the Qu'appelle, large unfossiliferous boulders are very numerous. North of the Assinniboine, near the Big Ridge, boulders are also abundant, and when magnified by refraction look like tents on the level prairies.

The ice on Lake Winnipeg carries off, every spring, fragments of rock belonging to the Laurentian Series which form its eastern shores. Many of these are distributed over the shallows and on the beaches of the western side of the lake; these phenomena resemble in miniature the stupendous operations described by travelers as continually occurring on the shores of the Arctic Ocean.

In Lake Manitobah long lines of boulders are accumulating in shallows and forming extensive reefs; the same operation is going on in all the lakes of this region, and is instrumental in diminishing the area of the lake in one direction, which is probably compensated by a wearing away of the coast in other places. Several of these modern accumulations formed by a re-arrangement of the boulders of the older drift are noticed in preceding chapters. Taken as a whole, and in connection with the destruction of the coasts, they afford a striking illustration

* This erratic was probably one of the series traced by Dr. Hector, from "the Thickwood Hills, in a southerly direction towards the Moose Hills on the South Branch."—*Papers relative to the Exploration of British North America*, 1859.

of the changes now taking place in the relations of land and water throughout the Lake Region.

The boulders and slabs of limestone on the low prairies of Red River and the Assinniboine resting upon lacustrine deposits, were probably brought by ice at a period posterior to the Boulder Drift. They are illustrations of the operations of ice at higher lake levels, similar to those occurring at the present time.

BEACHES AND TERRACES.

The most remarkable beach and terrace between Lake Superior and Lake Winnipeg, showing an ancient coast line, is undoubtedly that which separates Great Dog from Little Dog Lake on the Kaministiquia canoe route.

The Great Dog Portage, fifty-five miles from Lake Superior by the canoe route, rises 490 feet above the level of the Little Dog Lake, and the greatest elevation of the ridge cannot be less than 500 feet above it. The difference between the levels of Little and Great Dog Lakes is 347·81 feet, and the length of the portage between them, one mile and fifty-three chains.

It is stated in Chapter II. that the base of the Great Dog Mountain consists of a gneissoid rock supporting numerous boulders and fragments of the same material. A level plateau of clay then occurs for about a quarter of a mile, at an altitude of 283 feet above Little Dog Lake, from which rises, at a very acute angle, an immense bank or ridge of stratified sand, holding small water-worn pebbles. The bank of sand continues to the summit of the portage, or 185 feet above the clay plateau. The portage path does not pass over the highest part of the sand ridge; east of the path it is probable that its summit is 500 feet above the Little Dog Lake.

Height in Feet.	Distance in Feet.	Margin of Beaches.
ft.	ft.	
163·53	1000	Beginning of 1st plateau.
215·00	1450	Termination of do.
251·74	1650	Beginning of 2nd do.
283·78	2550	End of 2nd plateau, and commencement of sand bank.
468·19	3300	Commencement of 3rd plateau.
472·60	5920	End of 3rd plateau.
490·00	6180	Summit of level and commencement of 4th plateau.
474·00	7400	End of 4th plateau, and commencement of descent to edge of cliff.
395·00	8640	End of descent.
348·00	8712	Bottom of cliff, and level of Great Dog Lake.

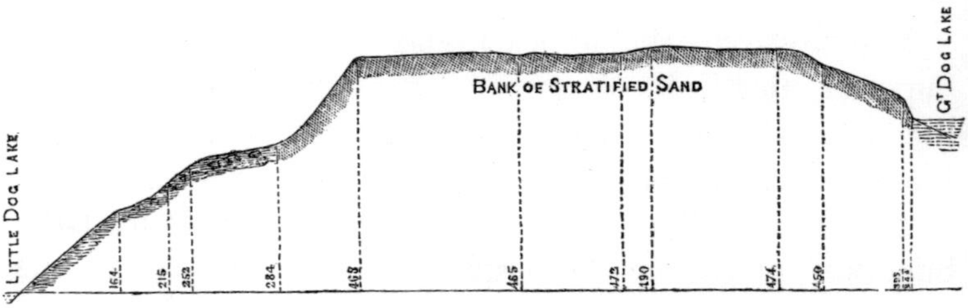

Profile of the Great Dog Mountain.*

Here then we have a terrace 490 feet above Little Dog Lake, or 853 feet above Lake Superior, or 1453 feet above the sea, and furnishing an admirable proof of the value of Dr. Hitchcock's expectation that higher beaches than those measured by Sir William Logan on the shores of Lake Superior would be found in that region. In his Surface Geology, Dr. Hitchcock says, page 63 (Smithsonian Contributions), "I will only add, that if it be admitted that the facts adduced in this paper prove the presence, since the Drift period, of the ocean at a height of 2000 or even 1200 feet, above its present level, then it must have extended over nearly all of our western

* This section was kindly furnished me by Mr. Napier, the engineer to the Red River Expedition, 1857.

country; and unless Professor Agassiz says that he had his eye upon this matter along the shores of Superior, I cannot avoid entertaining the expectation, that what I call beaches will yet be found at a much higher level there, than the 331 feet terrace, measured by Mr. (now Sir William) Logan."

I am inclined to think that another beach and terrace can be recognised at Prairie Portage, one hundred and four miles by the canoe route from Lake Superior; its altitude would exceed that on the Great Dog, being 1485 feet above the ocean. Prairie Portage passes over the height of land, but not the highest land on the route, and its course lies first, south-west up a steep wooded hill, without rock exposure, but composed of drift clays, sand, and numerous boulders; it then enters a narrow valley, which terminates in a small lake, about five acres in area and 20 feet deep, occupying a hollow among the hills on the height of land. The portage path continues on in the same direction until the Height of Land Lake is reached, a small sheet of water, about a square mile in area, and 157 feet above Cold Water Lake. The utmost elevation reached on the Prairie Portage is probably 190 feet above Cold Water Lake or nearly 900 feet above Lake Superior. Portage du Milieu, one hundred and five miles from Lake Superior passes over a low sandy ridge. It is 869 feet above Lake Superior, or 1469 feet above the sea; this ridge may have been contemporaneous with beaches on the summit of the Great Dog.

In the valley of Lake Winnipeg the first prominent ancient beach is the Big Ridge.

Commencing east of Red River, a few miles from Lake Winnipeg, this ridge pursues a south-westerly course until it approaches Red River, within four miles of the Middle Settlement, here it was ascertained by leveling to be

67½ feet above the prairie. On the opposite side of the river, a beach on Stony Mountain corresponds with the Big Ridge, and three or four miles further west it is observed marking the limit of a former extension of the valley of Lake Winnipeg. On the east side of Red River the Big Ridge is traced nearly due south from the Middle Settlement to where it crosses the Roseau, forty-six miles from the mouth of that stream, and on or near the 49th parallel. It is next met with at Pine or Tamarac Creek, in the State of Minnesota, and from this point it may be said to form a continuous and horizontal gravel road, beautifully arched, and about one hundred feet broad, the whole distance to the shores of Lake Winnipeg, or more than 120 miles.

On the west side of Red River, and north of the Assinniboine, I traced the Big Ridge from a point about three miles west of Stony Mountain to near Prairie Portage. Here it appears to have been removed by the agency of the Prairie Portage River and the waters of the Assinniboine, which are said to pass from the valley of that river into Lake Manitobah during very high floods.

Another and higher ridge was observed on White Mud River, about twenty miles west of Lake Manitobah. It resembled in every particular the ridge on the east side of Red River, being about 100 to 120 feet broad, and about twenty-five feet above the level of the prairie. It was again noticed in the rear of Manitobah House, where the same characteristics were preserved. It probably crosses the Assinniboine three or four miles west of Prairie Portage, and is perhaps identical with the lowest ridge or step of the Pembina Mountain.

In the rear of Dauphin Lake, the next ridge in the ascending series occurs, it forms an excellent pitching track for Indians on the east flank of the Riding Mountain. Probably these ridges are found close together, at the foot of the Pembina Mountain, where no less than four distinct steps occur close together near the sources of Scratching River.* The summit of these steps may be the plateau whose altitude was ascertained by Dr. Owen to be 210 feet above the prairie level, and the first steps may be continuous with the Big Ridge, limiting the level prairies of Red River and the Assinniboine.

The prairies enclosed by the Big Ridge are everywhere intersected by small subordinate ridges, which often die out, and are evidently the remains of shoals formed in the shallow bed of Lake Winnipeg when its waters were limited by the Big Ridge. Many opportunities for observing the present formation of similar shoals occurred in Lake Manitobah, St. Martin's Lake, Lake Winnipeg and Dauphin Lake. These, when the lakes become drained, will have the form of ridges in the level country then exposed. Indeed it may be said that the region between Dauphin Mountain and Lake Manitobah in the direction of Ebb and Flow Lake and south of that body of water, is but recently drained, or still in process of draining, being removed from the surface of Ebb and Flow Lake, by a very few feet and covered with water to a large extent in the spring. At present it consists of marsh, bog and ridge, in continued succession. When completely drained, the country will resemble the present prairies of the Assinniboine, with the gentle rich depressions, and the low dry gravelly ridges.

The Pembina Mountain is *par excellence* the ancient beach in the valley of Lake Winnipeg. Dr. Owen thus

* The steps are shown on Sheet 2 of the large map accompanying the Reports on the Assinniboine and Saskatchewan Exploring Expedition.

described it as it presents itself a few miles south of the 49th parallel: "After a hot and fatiguing ride over the plains, we arrived an hour after sunset at the foot of the Pembina Mountain. In the twilight as we stood at our encampment on the plain, it looked as if it might be three hundred feet or more in height; but in the morning, by broad daylight, it seemed less. When I came to measure it, I was somewhat surprised that it did not exceed 210 feet. I observed on this as on many other occasions that a hill rising out of a level plain, appears higher than it really is, especially when, as in this case, the trees on its flank and summit are of small growth. Pembina Mountain is in fact, no mountain at all, nor yet a hill. It is a terrace of table-land, the ancient shore of a great body of water, that once filled the whole of the Red River valley. On its summit it is quite level and extends so, for about five miles westward, to another terrace, the summit of which I was told is level with the great Buffalo Plains, that stretch away towards the Missouri, the hunting grounds of the Sioux and the half-breed population of Red River.

Instead of being composed of ledges of rock, as I was led to suppose, it is a mass of incoherent sand, gravel, and shingle so entirely destitute of cement, that with the hand alone a hole several feet deep may be excavated in a few minutes. The Pembina River has cut through this material a deep, narrow valley, but little elevated above the adjacent plain. Along its banks are precipices of sand, surmounted by gravel and a few boulders. I was told that it was impossible to ascend these banks. So loose is the deposit, that, no sooner is an ascent attempted, than the stones fifty or a hundred feet above, are detached, and come tumbling down at such an alarming rate that the climber is glad to make his escape."*

An inspection of the map will show the contour of the Pembina Mountain as far as ascertained; where Mr. Dickinson ascended it, fifteen miles north of the 49th parallel it occurs in four distinct terraces. It crosses the Assinniboine near the Bad Woods, blends with the Riding and Duck Mountains, and probably appears again on the main Saskatchewan, twenty-two miles from the Grand Forks. The elevation of the entire country east of this long ancient coast line is about 700 feet above the level of the ocean, and it forms the boundary of a distinct tract of lowland, in part surpassingly rich, as over the Red River and Assinniboine prairies, and the region on the main Saskatchewan slightly elevated above the area subjected to annual overflow; part covered with swamp, marsh, or level limestone rock, on which a few inches of soil affords nourishment to small spruce, tamarac, and aspen; and finally, by a shallow water area extending over 13,100 square miles, and embracing lakes which rank with the first class in point of superficies on this continent.

High above the Pembina Mountain the steps and terraces of the Riding and Duck Mountains arise in well-defined succession. On the southern and south-western slopes of these ranges the terraces are distinctly exposed to view; on their north-east and north sides, the Riding and Duck Mountains present a precipitous escarpment which is elevated fully 1000 feet above Lake Winnipeg, or more than 1600 feet above the sea.

Standing on the edge of the escarpment of the Riding Mountain and looking in the direction of Dauphin Lake, a gulf, some two hundred and fifty feet deep, is seen to be succeeded by two ranges of cone-shaped hills covered

* Geological Survey of Iowa, Wisconsin, and Minnesota, p. 179.

with boulders, one lower than the other. The hills are parallel to the general trend of the escarpment, and stand out as bold eminences, showing the extent of the denudation which gave rise to them. These ranges of conical hills correspond with terraces on the west side of the mountain. They are the result of the same denuding forces which have left their impress upon the west flank, and were formed by the unequal wearing away of the more exposed eastern flank, at the time when the terraces on the opposite and sheltered side were in process of arrangement.

I estimated the summit of Bear Hill, one of the most prominent of the conical hills separated from the edge of the escarpment by a deep valley, at 800 feet above Lake Winnipeg; if to this altitude we add 628 feet, the height of Lake Winnipeg above the sea, the elevation of the first terrace below the summit of the mountain, will be about 1,428 feet. This altitude corresponds in a remarkable manner with the sand bank on the Great Dog Portage, 500 miles distant in an air line from Bear Hill. The second tier of conical hills stands upon the second terrace from the summit, and is probably continuous with the Pembina Mountain.

SAND HILLS AND DUNES.

The most extensive of these unstable ranges are described in the narrative.

It is needless to remark that the region they occupy is almost absolutely barren. Many of the hills and dunes are continually exposing fresh surfaces, sometimes beautifully ripple marked. The probability of their being the remains of tertiary deposits, is noticed in a subsequent chapter. The following are the most extensive ranges:—

1. Sand hills and dunes of the Assinniboine, extending

from the Bad Woods to a short distance beyond Pine Creek, forty miles.

2. Sand hills of the Souris.
3. Sand hills and dunes of the Qu'appelle.
4. Sand hills and dunes of the South Branch.
5. Sand and gravel ridges north-west of the Touchwood Hills.

CIRCULAR DEPRESSIONS.

This curious disposition of the drift, probably due to a re-arrangement of its materials, is of not uncommon occurrence south-east of the Touchwood Hills. Circular depressions varying from 100 yards to half a mile in diameter, appear in the prairies, generally surrounded by a ridge of sand or gravel. . Many of them are quite dry, others hold water, which is generally brackish. The deepest and largest depression noticed was about 600 yards across and 40 feet below the general level.

DENUDATION.

An adequate conception of the effects of denudation in the basin of Lake Winnipeg can be best attained if we revert to the period when the Cretaceous shales now forming the summit and flanks of the Turtle, Riding, Duck, Porcupine, and Pasquia Mountains, occupied the basins of Lakes Manitobah and Winnipeg, and found their eastern limits near the present outcrop of the Laurentian Series. In order to complete our view of the extent of this great physical movement, we must conceive the same shales and sandstones, (in part overlaid by Tertiaries filling the depressions or valleys in the Cretaceous rocks, the result of previous denudation) forming an unbroken table-land to the Grand Coteau de Missouri. These relations become more evident upon an inspection of the sections.

LIMITS OF DENUDATION.

A curious and very interesting example of parallelism in bold limits of denudation, can be traced in the basins of Lake Winnipeg and the St. Lawrence. The abrupt escarpment of the Grand Coteau de Missouri preserves a direction throughout the greater portion of its denuded face, parallel to the escarpment of the Niagara limestone which enters Canada from the State of New York at Lewiston, on the Niagara river, and sweeping round the head of Lake Ontario passes up the Indian Peninsula and thence to the Grand Manitoulin Island in Lake Huron. The denuded flanks of the Riding and Duck Mountains with their northern and southern prolongations also preserve the same general direction. Lines drawn on a map of North America to show these well-marked limits of denudation at different periods, will at once suggest the existence of a uniformly acting force operating under similar physical conditions. Like beaches or ridges formed at different sea-levels they have probably a common origin, which the supposition of their being ancient coast-lines is not sufficient to explain, neither do their geographical relations appear to be altogether dependent upon their geological structure. They seem to point to the action of currents, of which the Gulf stream and the compensating Arctic currents are modern illustrations.

DISLOCATIONS.

The basin of Lake Winnipeg from the Laurentides to the Rocky Mountains, has been subjected to a series of dislocations, which have acted very uniformly over areas far apart. The courses of rivers show the general direction of these gradual disturbances, which may have given rise to the numerous "Elbows" which form such a marked feature in the water channels, and consequently in the surface of the rocks of a large part of Rupert's Land. The following table shows the probable effect which has been produced by some of these dislocations :—

Name of Lake and River.	Change of course.
1. St. Martin Lake and the Little Saskatchewan . . .	N. to S.E.
2. Dauphin Lake, Moss River and Waterhen River . . .	N.E. by E. to S.E. by S.
3. South Branch of the Saskatchewan (Elbow) . . .	S.E. to N.E.
4. North Branch of the Saskatchewan	S.E. to N.E.
5. Battle River (Elbow) . .	S.E. to N.E.
6. Bow River (Junction with the South Branch) . . .	S.E. to N.E.

CHAP. XXXVII.

THE LAURENTIAN AND HURONIAN SERIES.

Distribution of Formations. — The Laurentides. —— The Laurentian System.—Description of Laurentian Rocks.—Lime and Soda Felspar.—Titanic Iron-ore. — Crystalline Limestones. — Mineral Species in the Limestones. — Intrusive Rocks. — Economic Materials. — Separation of Laurentian Rocks into two Groups.—Extent of the Limestones in this System.—Area of Laurentian Rocks in the Basin of Lake Winnipeg.—Intrusive Rocks.—Character of the Dividing Ridge.—From Milles Lacs to Rainy Lake.—From Rainy Lake to the Lake of the Woods.—From the Lake of the Woods to the Winnipeg.—The Coast of Lake Winnipeg. — Fundamental Gneiss of Scotland, the Equivalent of the Laurentian Series in Canada.—Adoption of the name Laurentian in British Geology by Sir Roderick Murchison, to represent the Oldest or Fundamental Gneiss of Scotland. — The Huronian Series. — Description of Huronian Rocks.

DISTRIBUTION OF FORMATIONS.

The distribution of series of formations in the order of their occurrence from east to west in the basin of Lake Winnipeg is as follows : —

1. Laurentian Series.
2. Silurian ,,
3. Devonian ,,
4. Carboniferous* ,,
5. Cretaceous ,,
6. Tertiary ,,

* Although this series has not been recognized on the east side of the fossiliferous rocks in the Basin of Lake Winnipeg, yet the occurrence of a *Productus* in a boulder in the bed of Red River affords presumptive evidence that the series is represented there. I have therefore placed it provisionally in the enumeration given above.

THE LAURENTIAN SERIES.

The origin of the name Laurentian and the character of the rock series which compose this system is described by Sir William Logan and Mr. Hunt in the following extract from a "Sketch of the Geology of Canada."

THE LAURENTIDES.*

"The province of Canada is traversed, through its whole length, by a mountainous region dividing it into two basins, which may be distinguished as the northern and the southern basins. These mountains, which have been named the Laurentides, form the north shore of the St. Lawrence, from the gulf as far as Cape Tourmente, near Quebec; from which point they leave the river, and while they follow its general direction become more and more remote, until near Montreal, they are at a distance of ten leagues from the St. Lawrence. Going further westward, this mountainous region follows the line of the Ottawa, and crosses this river near the *Lac des Chats*, fifty leagues from Montreal. Thence taking a southward direction, it reaches the St. Lawrence near the outlet of Lake Ontario, and from this point running north-westward, the southern limit of this formation, reaches the south-eastern extremity of Lake Huron, at Matchedash Bay, and forms the eastern shore of the lake as far as the 47th degree of latitude, where quitting this lake, the formation gains Lake Superior, and extends in a north-west direction to the Arctic Sea.

"To the south of the St. Lawrence, this same region

* A Sketch of the Geology of Canada, serving to explain the Geological Map and Collection of Economic Materials sent to the Universal Exhibition at Paris, 1855, by W. E. Logan, F.R.S., and T. Sterry Hunt, A.M.

covers a considerable space between the Lakes Ontario and Champlain, and constitutes the Adirondack Mountains. With this exception, and perhaps also a small exposure in Arkansas and another near the sources of the Mississippi, this formation is not found to the south of the St. Lawrence, and as it belongs especially to the valley of this river and constitutes the Laurentide Mountains, the Geological Commission of Canada has distinguished it by the name of the *Laurentian system.*"

THE LAURENTIAN SYSTEM.

"The rocks of this system are, almost without exception, ancient sedimentary strata, which have become highly crystalline. They have been very much disturbed and form ranges of hills, having a direction nearly north-east and south-west, rising to the height of 2,000 or 3,000 feet and even higher. The rocks of this formation are the most ancient known on the American continent, and correspond probably to the oldest gneiss of Finland and Scandinavia and to some similar rocks in the North of Scotland.

"The rocks of the Laurentian formation are in great part crystalline schists, for the most part gneissoid or hornblendic. Associated with these schists, are found large stratified masses of a crystalline rock, which is composed almost entirely of a lime and soda felspar. The rock is sometimes fine grained, but more often porphyritic, and contains cleavable masses of felspar, sometimes several inches in diameter; these felspars are triclinic, and have ordinarily the composition of andesine, labradorite, anorthite, or of intermediate varieties. Their colours are various, but the cleavable felspars are generally bluish or reddish, and often give coloured reflections.

Hypersthene is very generally disseminated in these felspathic rocks, but always in small quantity. Titanic iron-ore is also found in them, in a great number of places, sometimes in small grains, but often in considerable masses.

"With schists and felspars are found strata of quartzite, associated with crystalline limestones, which occupy an important place in this formation. These limestones occur in beds of from a few feet to 300 feet in thickness, and often present a succession of thin beds intercalated with beds of gneiss or quartzite; these latter are sometimes quartzite conglomerates, and have in certain cases a base of dolomite. Associated with these limestones, are sometimes found beds composed in great part of wollastonite and of pyroxene, species which evidently owe their origin to the metamorphism of siliceous limestones. Beds of dolomite and limestone more or less magnesian, are often interstratified with the pure limestones of this formation.

"The limestones of this system are rarely compact, and most frequently are coarsely granulated. They are white or reddish, bluish or greyish, and these colours are often arranged in bands which coincide with the stratification. The principal mineral species met with in these limestones, are apatite, fluor, serpentine, phlogopite, scapolite, orthoclase, pyroxene, hornblende, wollastonite, quartz, idocrase, garnet, brown tourmaline, chondrodite, spinel, corundum, zircon, sphene, magnetic and specular iron, and graphite. The chondrodite and graphite are often arranged in bands parallel with the stratification. Beds of a mixture of wollastonite and pyroxene are sometimes met with, which are very rich in zircon, sphene, garnet, and idocrase. The most crystalline varieties of these limestones often exhale a very fetid odour when bruised. The limestones of this formation do not yield everywhere well crystallized minerals; near the Bay of Quinté there are beds met with which still preserve the sedimentary character, and show only the commencement of metamorphism.

"The conditions in which they are sometimes found, indicate that the agents which have rendered these limestones crystalline, have been such as to render the carbonate of lime almost liquid, and that, while in that state, it has undergone great pressure. As evidence of this opinion, we find that the limestone often fills fissures in the adjacent siliceous strata, and envelopes the detached, and often folded fragments of these less fusible beds precisely like an igneous rock.

"The crystalline schists, felspars, quartzites and limestones which we have described, make up the stratified portion of the Laurentian system, but there are besides, intrusive granites, syenites and diorites, which form important masses; the granites are sometimes albitic, and often contain black tourmaline mica in large plates, zircon, and sulphuret of molybdenum.

"Among the economic minerals of this formation, the ores of iron are the most important, and are generally found associated with the limestones."

In 1857 Sir William Logan read a paper before the American Association for the Advancement of Science, in which he referred to a former indication of a "probable separation of the Laurentian rocks of Canada into two great groups: that characterized by the presence of much lime and that without;" but from recent investigations (previous to August, 1857) it has appeared to him "almost certain that the former of these two great groups will be capable of subdivision, and that some of its bands of limestone, with their associate strata, are of a sufficient importance to be represented separately on the map."* At the meeting of the same body in 1859, Sir William Logan exhibited a map on which was delineated in detail the distribution of some of the bands of crystalline limestone interstratified with the gneiss of the Laurentian rocks on the north side of the Ottawa river, about forty miles from Montreal, being a continuation of the work referred to at the Montreal Meeting in 1857. Two additional bands of limestone had been ascertained to underlie the lowest of those previously examined, the whole of the strata associated with these lower three, including the limestones, being supposed to be about 15,000 feet thick.†

* Canadian Journal, January, 1858.

† Until lately the Potsdam sandstone has been supposed to represent the epoch when organic life was first introduced by the Creator on the surface of our globe. Recent discoveries tend to throw back the first peopling of the world into a past so indefinitely remote, that all preconceived ideas of organic history become unsettled and at fault. The following notice is from *The Canadian Naturalist and Geologist,* vol. iv. p. 300 :—

"Although the Laurentian Series has hitherto been considered azoic, a search for fossils in them has not been neglected. Such search is naturally conducted with great difficulties. Any organic remains which may have been entombed in these limestones, would, if they retained their calcareous character be almost certainly obliterated by crystallization, and it would be only through their replacement by a different mineral substance that there would be a chance of some of the forms being preserved. No such instances had been observed on the investigation of the Rouge and its vicinity, but from another locality in the Laurentian formation, Mr. John McMullen, one of the explorers of the Geological Survey, had obtained specimens well worthy of attention. They consisted of parallel or apparently concentric layers resembling those of the coral Stromatocerium, except that they anastomose at various parts; these layers consist of crystalline pyroxene, while the interstices are filled with crystallized carbonate of lime. These specimens had recalled to recollection others which had been obtained from Dr. Wilson of Perth some years ago, and had not then been regarded with sufficient attention. In these similar forms are composed of green serpentine, concretionary, while the interstices are filled with white dolomite. If it be supposed that both are the result of mere unaided mineral arrangement, it would seem strange that identical forms should result from such different

The crystalline limestones in the Laurentian series are daily acquiring increased importance. In the Report for the year 1858, recently issued, Sir William Logan states that in the present state of the investigation there appears to be a sequence of four important bands of crystalline limestone in the Laurentian area examined; but the wrinkled condition of the strata is such that in a space of not more than fifty miles by twenty, one of the bands exhibits an outcrop exceeding two hundred miles in length, which renders it very difficult to determine with precision the volume of rock in which the four calcareous bands are enclosed. The following is Sir William

minerals in places so far apart. If the specimens had been obtained from the altered rocks of the Lower Silurian series there would have been little hesitation in pronouncing them to be fossils. The resemblance of these forms to Stromatocerium from the Bird's-eye limestone, when the coral has been replaced by concretionary silica, is very striking. In the pyroxenic specimens, the pyroxene and the carbonate of lime being both white, the forms, although weathered into strong relief on the surface, are not perceptible in fresh fractures until the fragments are subjected to an acid, the application of which shows the structure running throughout the mass. Several specimens of these supposed fossils were exhibited to the section."

The suppositions embodied in the foregoing extract do not coincide with the views relating to the origin of life expressed by Sir Roderick Murchison in the proceedings of the Geological Society for November, 1859, p. 219. The Cambrian rocks referred to, rest on the fundamental gneiss or Laurentian rocks of the North Highlands of Scotland:—

"The phenomenon relating to these Cambrian sandstones which may well strike the geologist as he passes over the summits of Suilven and Queenaig, is that these very ancient rocks, on which unquestionably the Lower Silurian rocks repose, should be simply sandstones and grits, which have undergone much less change than the sandstone which lies upon them,—the latter having been metamorphosed into quartz-rock. However difficult it may be to account for this fact, it is at all events *most instructive as regards the origin and succession of life in the crust of the earth,* and sustains my view of a beginning. For here (and I have applied the same argument before to the Cambrian sandstones of the Longmynd, which certainly underlie the quartz-rock of the Stiper Stones) the older of the two rocks in Scotland has offered no trace of fossils, whilst the more crystallized structure above exhibits unmistakable signs of former living things."

Logan's approximate estimate of the thickness of the various constituent parts of the mass, arranged in ascending order:—

	Feet.
1. Gneiss	5,000
2. Crystalline Limestone	1,500
3. Gneiss	4,000
4. Crystalline Limestone	2,500
5. Gneiss	3,500
6. Crystalline Limestone	750
7. Gneiss	5,000
8. Crystalline Limestone	500
Total	22,750

The area occupied by the Laurentian series in the Basin of Lake Winnipeg is indicated on the geological map. The country between Lake Superior and the Height of Land has been described by Mr. Alexander Murray, of the Canadian Geological Commission *; the geology of Rainy Lake and of the Lake of the Woods by Dr. Bigsby †; the country on the Pigeon River route by Dr. Norwood ‡; and a general view of the whole country between Lake Superior and Lake Winnipeg by Dr. Hector §, &c.

The intrusive rocks, so numerous throughout the vast extent of this series in the region under review, consist of granites, syenites, greenstones, trap, &c. In Great Dog Lake mica schist rests on each side of syenite outbursts, which have the form of promontories jutting into the lake on the west coast. The valley of Dog River is bounded by low granite ridges which acquire greater

* Report of Progress, 1846.

† "On the Geology of Rainy Lake, South Hudson's Bay," by Dr. J. J. Bigsby, F.G.S. &c.—*Quarterly Journal* of the Geological Society, 1854. Also "On the Geology of the Lake of the Woods," by the same author.—*Quarterly Journal*, 1852.

‡ Geological Survey of Wisconsin, &c.

§ Papers Relative to the Exploration of British North America, by Captain Palliser.—*Blue Book*, 1859.

altitude and breadth at the Height of Land. A similar disposition obtains at Milles Lacs on the western slope, where the elongated dome-shaped intrusions are about 100 feet high and flanked with schists.

On the Pigeon River route the granite and syenite ranges acquire more imposing altitudes, sometimes attaining an elevation of nearly 1300 feet above Lake Superior, but the axis of each range preserves a general N.E. and S.W. direction parallel to the coast of the lake. This uniform disposition is maintained as far as Basswood Lake on the southern route.

The mountain range constituting the Height of Land dies out in a great measure after crossing the St. Louis River in its course south-west; but granitic, metamorphic, and trap rocks are met with at various points along the line of bearing as far as the Mississippi, and reappear again on the Minnesota river.*

At Snake Falls, on the northern route, the river passes over a schist highly inclined to the N.E., and below them, many fine exposures of the same schist occur on islands, frequently projecting like the end of boards of unequal lengths leaning against one another, and varying in thickness from two to five inches. Three miles below Snake Falls, the rock passes into gneiss, and numerous veins and dykes of granite are seen to penetrate it nearly at right angles to the strike; the dip is here N.W. Ten miles below Snake Falls, mica schist again comes into view, intersected with quartz and felspar bands from one to two inches thick. The strike is E. 5° N., and the dip nearly vertical. At the Grand Falls of the Nameaukan, the schists are tilted by steps in the form of the segment of a circle. In Lac Nameaukan, dome-shaped granitic islands parallel to one another, and of oval form, present them-

* Dr. Norwood.

selves not far from the entrance of Lac La Croix. The direction of the longest axis is N. 60° W. A line prolonged through the granitic islands, in a N.W. direction, touches the schists about three hundred yards further on.

Their apparent dip, as seen from the lake, was N.W., at an angle of about 45°. One island, wholly composed of schist, inclined at a high angle, is followed at a distance of about fifty yards by a long flat gneissoid dome. About six hundred yards from the island, the schists dip lightly to the S.E. On the north side, the dip could not be seen; but on the west side they were observed to bend round in a curved form, and from a N.W. dip towards the S.E. On the next island, the gneiss was intersected by numerous joints having a direction N. 70° E., and by quartz and felspathic veins, bearing N. 25° W., or nearly perpendicular to the former. Its surface towards the N.W. by W. was smooth, and inclined at an angle of about 10°. The rock of the Nu Portage is a granite containing mica in plates, and everywhere dotted with numerous beautiful specimens of plumose mica.

Dr. Bigsby thus sums up the geological conditions of Rainy Lake where a change in the direction of the strike is very decided.

"Chloritic and greenstone slates, gneiss and mica slate, in proportional quantities in the order here set down, seem once to have occupied the lake basin, with an E.N.E. strike, and a N.N.W. dip at a high angle usually; but subsequently a very extensive outburst of granite, with some syenite, has taken place to the great disturbance of the stratified rocks, and penetrating them both in intercalations and crosswise; these intrusive rocks occupy a very large portion of the lake."

He divides the region of Rainy Lake for convenience

of description and reference, into six distinct parts, each having its own geological characteristic. The west shore of the lake is mainly occupied by granite, which at the northern portion is finely granular, and porphyritic in equal quantities. On the east coast of its north-westerly extension, are chloritic and greenstone slates; on the eastern arm of Rainy Lake, pale red granite is the prevailing rock, and near the northern extremity of this arm, naked ridges, white as porcelain, and 500 feet high, occur.

In the Lake of the Woods the Laurentian series is separated into two parts by a range of intrusive granite running in an E.S.E. direction from the north-west corner of the lake, as far east as Rainy Lake. The inclined metamorphosed strata on the north side of this granitic anticlinal dip W.N.W.; on the south side to the S.S.W. a direction resulting from the form and position of the intrusive rock.*

The country between the sources of the Winnipeg, and a few miles south of Islington Mission, a distance of nearly thirty miles, appears to be largely occupied by a vast range of intrusive granite and syenite, in the form of dome-shaped hills, varying from 150 to 200 feet high. A view from the summit of one of the highest of these, about fifteen miles due north from Rat Portage presented an unvarying succession of their rounded summits as far as the eye could reach in a westerly direction. The canoe route pursued in 1851 was a short Indian path from Rat Portage to the Great Winnipeg, in a nearly straight north-westerly direction. The country traversed was characterized by great sterility, and an unusual proportion of bare rock. High precipitous mural cliffs, without a trace of stratification observable in them, often

* See Dr. Bigsby, "On the Geology of the Lake of the Woods," before referred to.

formed the boundaries of this branch of the Great Winnipeg.

The series composing the country east of Rainy Lake towards the Height of Land are again reproduced after passing the large area of intrusive granite on the Winnipeg, and continue with some considerable variations of strike and dip occasioned by intrusions, as far as the first falls below the Bonnet Portage, where drift clays conceal the rocks on the banks of the river to its mouth, exposures occurring only at the different falls and portages.

The Laurentian rocks which form the whole of the low east coast of Lake Winnipeg strike off at its north-east corner, and passing to the north of Moose Lake, go on to Beaver Lake.*

The only exposure of Laurentian rocks seen within the area explored west of Lake Winnipeg were observed in St. Martin Lake; they have been described in Chapter XXVI.

In the Quarterly Journal of the Geological Society for August, 1860, Sir Roderick I. Murchison made the following interesting and important announcement respecting the age of the *Fundamental* Gneiss of Scotland, and conferred an appropriate acknowledgment of the services rendered to geology by the distinguished director of the Survey in Canada, by adopting a Canadian geological name for the British rocks of the same age as those which Sir William Logan has named the Laurentian System.

"The changes which are involved in the adoption of my views of the order of succession are, it will be admitted, considerable. In the first place, by showing that mountain masses of sandstone and conglomerate lie unconformably beneath quartzose and calcareous rocks with

* Sir John Richardson's Journal of a Boat Voyage, &c., p. 49. Am. Ed.

true Lower Silurian fossils, we know that the former must be of Cambrian age. We further learn that the old or fundamental gneiss, which lies beneath such Cambrian sandstone, and is entirely unconformable to, and independent of it, is a lower stratified rock than any hitherto recognized in the British Isles. The beginning of the geological alphabet, as applied in the Maps of the Geological Survey to the Cambrian rocks of England, Wales, and Ireland, must therefore be preceded in Scotland by the first letter of some alphabet earlier than the Roman, showing a still lower deep in the north-west of Scotland (as in North America) than exists in England, Wales, or Ireland.

"If this most ancient gneiss required a British name, it might indeed with propriety be termed the 'Lewisian System,' seeing that the large island of the Lewis is essentially composed of it, capped here and there by derivative masses of Cambrian conglomerate; but the term 'Laurentian' having been already applied to rocks of this age in North America by our distinguished associate Sir W. Logan, I adhere to that name, the more so as it is derived from a very extensive region of a great British colony."

THE HURONIAN SERIES.

The Huronian series has not been recognized in the basin of Lake Winnipeg, but as it rests unconformably on the Laurentian rocks for a distance of nearly 500 miles on the shores of Lakes Superior and Huron, occurring both on the north and south shores of the first-named lake, it is not improbable that it will be found to exist in the Winnipeg Basin. In Canada it contains very important metalliferous veins, particularly of native copper. The following description of the Huronian or Cambrian system is from the little work before referred to.*

"The shores of Lakes Huron and Superior offer a series of schists, sandstones, limestones, and conglomerates interstratified with heavy beds of greenstone, and resting unconformably upon the Laurentian formation. As these rocks underlie those of the Silurian system, and have not as yet afforded any fossils, they may probably be referred to the Cambrian system (lower Cambrian of Sedgwick). the schists of this system upon Lake Superior are bluish in colour, and contain beds of cherty silex, marked by calcareous bands, and holding anthracite in its fissures.

"These are covered by a considerable thickness of trap, upon which repose massive beds of red and white sandstone which sometimes becomes conglomerate and contains pebbles of quartz and jasper. Beds of a reddish argillaceous limestone are often interstratified with these sandstones, which are intersected and overlaid by a second eruption of greenstone of great thickness and columnar in its structure. This formation, which, according to the observations of Sir William Logan, has, on Lake Superior a total thickness of about 12,000 feet, is traversed by a vast number of trappean dykes.

"In the corresponding formation on the north shore of Lake Huron, the sandstones are more vitreous and the conglomerates more abundant than on Lake Superior; they are, however, associated with conglomerates and schists similar to those we have just described, and the formation offers great intercalated masses of greenstone. A band of limestone, fifty feet in thickness, forms a part of this series, to which Sir William Logan assigns a thickness

* A Sketch of the Geology of Canada, &c., by Sir William Logan and Mr. Hunt.

of about 10,000 feet. He has shown the occurrence, after the irruption of the interstratified greenstones, of two systems of trap dykes, and a third of granite, intermediate in time between the two irruptions of trap. The formation of the metalliferous veins is still more recent. The principal mineral species of these veins are native copper, quartz, calc-spar, dolomite, fluor, and sulphate of baryta, with several zeolites, of which laumonite is the most common, heulandite, stilbite, thompsonite, apophyllite and analcime are also met with, as well as prehnite and datholite. These veins are only metalliferous where they traverse the beds of greenstones."

CHAP. XXXVIII.

THE SILURIAN AND DEVONIAN SERIES.

Rim of the Silurian Series.—Distribution of Formations.—The Chazy Formation.—Fossils from the Chazy, *Modiolopsis Parviuscula, Orthoceras Simpsoni.*—Bird's-eye, and Trenton Formations.—Hudson River Formation.—The Devonian Series.—Salt Springs.—Salt Springs in Rupert's Land.—Manufacture of Salt.—Salt Trade of the United States.—Fossils from Devonian Rocks in the Winnipeg Basin.

THE SILURIAN SERIES.

THE rim of the great southern fossiliferous basin in North America touches Fond du Lac, the western extremity of Lake Superior.* From this point it makes a bend to the south, and then turning north-westerly comes upon the Mississippi river near its source. The prolongation of the great granitic axis separating the waters flowing into Lake Superior, from those tributary to Lake Winnipeg, penetrates for a considerable distance into the fossiliferous basin, in the northern part of the State of Minnesota, but although the subjacent rock is concealed by drift, the general character of the country, as far as Red Lake, leads to the inference that it is underlaid by the Silurian series. From Red Lake the rim pursues a northerly course to the mouth of Rainy River; it underlies the valley of Rainy River, as far as Rainy Lake, in the form of a narrow and

* Sandstones belonging to the Potsdam and Chazy formations, the lowest members of the fossiliferous rocks in America, are in great force on the south shore of Lake Superior. The basin of the lake is chiefly excavated in the Potsdam sandstone.

very elongated bay.* Sweeping round the south and part of the west coast of the Lake of the Woods, it leaves that body of water and comes on the Winnipeg, above Fort Alexander. The south and west coasts of Lake Winnipeg, and as far as its north-east corner mark the easterly boundary of this basin; it then passes north of Moose Lake, to Beaver Lake, and thence in a north-westerly direction towards and beyond Methy Portage. The Silurian rocks of Great Slave Lake and River present forms chiefly belonging to the upper division of the series. The occurrence of brine springs in that region points also to the proximity of Devonian rocks, which are particularly characterized by brine springs in the basin of Lake Winnipeg. The upper Silurian rocks of the Arctic Archipelago rest everywhere on granitoid rocks with a remarkable red sandstone passing into coarse grit for their base.†

Limestones, sandstones, and shales of Silurian age are exposed from the Grand Rapid of the Saskatchewan to Big Black Island, near the south extremity of Lake Winnipeg. From Big Black Island to the rapids on Red River the formations are concealed by quaternary deposits. On the south-east coast limestone is occasionally seen in position, but its junction with the Laurentian series near the mouth of the Winnipeg is concealed by drift.

The formations which have been recognized on Lake Winnipeg, and in the valley of Red River, are

1. The Chazy Formation.
2. The Bird's-eye „
3. The Trenton „
4. The Hudson River Group.

* See "Geology of the Lake of the Woods."—Dr. Bigsby, Quart. Journ. Geo. Soc. vol. viii. p. 406.

† Rev. Samuel Haughton, F.R.S., Appendix to Capt. M'Clintock's Narrative.

CHAZY FORMATION.

Fine exposures of the Chazy formation occur on Deer Island, Punk Island, and along the west coast of the lake north of Big Grindstone Point as far as the Cat Head. They appear in the form of cliffs, varying from twenty-five to forty-five feet in altitude at nearly all points and promontories. The character of the rock is described in Chapter XXIII. At the Narrows the three limestone promontories, the Bull's Head, Limestone Cave Point, and Whiteway's Point, approach within a few miles of the Laurentian series on the east coast. The strait from Whiteway's Point to the Dog's Head is not more than three miles across. Before this narrow channel was excavated, Lake Winnipeg must have been divided into two parts, like Lakes Manitobah and Winnipego-sis, and it is not improbable that near the Dog's Head a rapid river or falls formerly existed. The relation of the two lakes would then resemble the present relation of Lake Manitobah, the Little Saskatchewan and Lake Winnipeg.

The following enumeration and description (between brackets) of fossils forming part of my collection from the Silurian and Devonian rocks of this region, are from the pen of Mr. E. Billings, F.G.S., Palæontologist to the Canadian Geological Survey.

Plantæ.

[Two species of Fucoids from Punk Island in Lake Winnipeg, resembling forms which occur in the Chazy sandstone.

Zoophyta.

The only coral is a species allied to *Columnaria alveolata.* It is from Grindstone Point, Lake Winnipeg.

Echinodermata.

Columns of a large *Glyptocrinus* allied to *G. ramulosus* occur at Punk Island and Grindstone Point; and besides these, at the latter locality were found several plates of a *Glyptocystites* closely allied to *G. multiporus.*

Brachiopoda.

Two specimens of a plaited *Rhynconella* a little smaller than *R. plena* were found at Punk Island.

Lamellibranchiata.

MODIOLOPSIS PARVIUSCULA. (N. S.)

This species closely resembles *M. modiolaris* (Conrad) but is always much smaller. It is transversely elongate, anterior extremity small, rounded half the width of the posterior; the latter obliquely truncated and somewhat straight from the end of the hinge line for rather more than half the width, then rounded at the lower posterior angle. Hinge line straight or a little arched, full three-fourths the whole length of the shell. The umbones are less than one-fifth the length from the anterior extremity. The valves are moderately convex, obscurely and obliquely carinate from the umbones towards the lower posterior angle. In many specimens the ventral margin is concave near the anterior extremity, as if for the purpose of a byssus. Surface with obscure concentric undulations of growth. Length of large specimen, one and a half inch. In general they are a good deal smaller.

This shell so much resembles *M. modiolaris* that I have long hesitated as to the propriety of giving it a separate name. It is very widely distributed, since we have specimens from Lake Winnipeg at Punk Island, from the Pallideau Islands in Lake Huron, where it occurs in strata which hold fossils of the Chazy, Black River and Trenton limestones, and from near Cornwall and the Island of Montreal in the Chazy.

Besides the above there are several small nearly circular fossils from Punk Island, which appear to be casts of some lamellibranchiate shell.

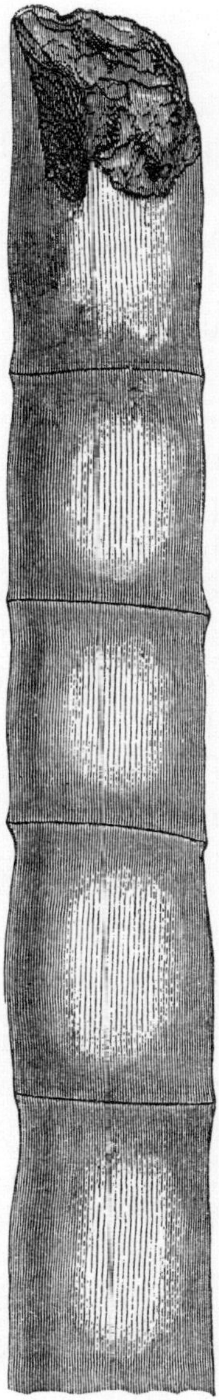

ORTHOCERAS SIMPSONI.
A fragment of the siphuncle of this species.

Gasteropoda.

Trochonema umbilicata (Hall, Sp.) This species occurs at Lake Winnipeg and at the Little Saskatchewan in considerable numbers. A species allied to *Pleurotomaria rotuloides* (Hall) is common at Punk Island, and a *Maclurea* allied to *M. Logani* (Salter), but with more slender whorls was found at Punk Island and the Little Saskatchewan. One of the specimens has the operculum in place, but is destitute of the shell and somewhat distorted. None of the Gasteropoda have the shell preserved.

Cephalopoda.

ORTHOCERAS SIMPSONI. (N. S.)

The specimen is a portion of the siphuncle, nine inches and one-fourth in length, eleven lines in diameter at the larger extremity, and ten at the smaller. It is nearly cylindrical with a broad, shallow constriction above and below each of the narrow annulations which mark the attachment of the septa. There are eight of those septal rings at the following distances from each other, commencing at the smaller extremity :—Between the 1st and 2nd, fourteen lines; 2nd and 3rd, twelve lines; 3rd and 4th, ten and a half lines; 4th and 5th, thirteen and a half lines; 5th and 6th, fifteen lines; 6th and 7th, thirteen and a half lines; 7th and 8th, twelve and a half lines. The annulations are nearly at right angles to the length, and we must infer from this fact either that the septa are scarcely at all concave, or that the siphuncle must be central, or very nearly so. If in an orthoceratite the septa are flat, then no matter whether the siphuncle be central or not, the septal annulations must be at right angles, but if the septa are concave then the annulations will be oblique if the siphuncle be at all removed from the centre. My impression is, that this is a large orthoceratite with distant septa and a nearly central siphuncle since the annulations have a scarcely perceptible obliquity.

It is one of those species in which the siphuncle became gradually filled with a solid calcareous animal secretion, with the exception of a narrow cylindrical channel along the centre. This central canal is clearly indicated in the specimen, and has a diameter of nearly two lines.

Dedicated to Sir George Simpson, Governor of the Hudson's Bay Company.

Locality and formation.—Cat Head, Lake Winnipeg, supposed to be Silurian.

A small serpulite appears to be common at Punk Island; it much resembles the large species of the Chazy limestone.

The occurrence of *M. parviuscula*, *H. umbilicata*, the

Maclurea, and *Glyptocystites* are quite sufficient to show that the localities where they have been collected are Lower Silurian, and most probably about the age of the Black River and Chazy limestones.]

THE BIRD'S-EYE AND TRENTON FORMATIONS.

The whole of the coast on the north-west side of Lake Winnipeg is represented by Sir John Richardson to be occupied by the Bird's-eye Limestone.* Near the first and second Rocky Points the strata contain many gigantic orthoceratites which have been described by Mr. Stokes in the Geological Transactions.†

In Pine Island Lake, there are exposures thirty feet in altitude, containing *Orthocerata* and *Receptaculites Neptunii*.‡ The strike is south-west by west, being at right angles to the general direction of the Laurentides.

THE HUDSON RIVER FORMATION.

This formation appears in cliffs five-and-twenty feet high at the Stone Fort, Red River. It is also exposed near the rapids. Most of the forts and churches in the settlements are constructed of stone from this rock. The

* "As bearing on the resemblance of these (Silurian) quartzose and calcareous rocks to their equivalents in America, I may mention that a recent comparison of several of the American so-called fucoids has enabled Mr. Salter satisfactorily to refer them in many cases to the vertical tubes, or rather filled-up burrows, of large marine worms. He particularly cites the *Phytopsis tubulosus* of the 'Calciferous Sand-rock' of the American geologists; the celebrated 'Birds' Eyes' (*Phytopsis cellulosus*) of the 'Birds'-eye Limestone,' and the *Buthotrephis succulens* of the Trenton group, as being of this nature. The part played by annelides in the older palæozoic epoch was, we thus know, very conspicuous."—*Sir R. I. Murchison, Quar. Jour. Geol. Soc.*, Aug. 1860.

† Journal of a Boat Voyage through Rupert's Land.

‡ Ibid.

colour of its weathered surface is a pale yellowish grey, but of fresh surfaces, a white gray. Dr. Owen, who visited the Red River Settlements in 1848, enumerates the fossils he found near the Stone Fort in his Report published in 1852. He says:—

"About twenty miles below the mouth of the Assinniboine, near lower Fort Garry, solid ledges of limestone are exposed of a light buff colour, sometimes mottled, spotted, or banded with light brown. Immediately opposite the Fort, a considerable amount of rock has been quarried, and used in the construction of the building. In these beds, I succeeded in finding several well-defined and characteristic fossils, sufficient to establish, without the least doubt, the age of the Red River limestones.

They are: *Favosites basaltica; Coscinopora Sulcata;* hemispherical masses of *Syringopora; Chætetes lycoperdon;* a *Conularia;* a small, beautiful undetermined species of *Pleurorhynchus; Ormoceros Brongniarti; Pleurotomaria lenticularis* (?); *Leptæna alternata; Leptæna plano-convexa* (?); *Calymene senaria;* and several specimens of the shield of *Illænus crassicauda.*

Many of these are identically the same fossils which occur in the lower part of F. 3, in Wisconsin and Iowa, in the blue limestones of Indiana, Ohio, Kentucky, and Tennessee, and also in the lower Silurian of Europe.

The *Coscinopora* is precisely the same as the coral, which is particularly characteristic of the lower beds of the upper magnesian limestone of Wisconsin. The specimens of *Favosites basaltica* cannot be distinguished from those which abound in the upper magnesian limestones of Wisconsin and Iowa, and the lower coralline beds of the Falls of the Ohio. It is also worthy of note that these limestones of Red River, like their equivalents in Iowa and Wisconsin, are highly magnesian, containing from seventeen to forty per cent. of the carbonate of that alkaline earth."

* * * * * * *

" On the south shore of that lake (Winnipeg), however, I again had an opportunity of inspecting fossiliferous limestones in *situ*. At the two localities where I succeeded in obtaining a view of them, they were much disturbed, dipping either at a high angle, or standing vertically. On Poplar Point, they are quite thin-bedded, and contain, besides small *Entrochites*, large varieties of *Endoceras*. In a small bay, near Big Swamp Point, the limestone is seen jutting out beneath heavy, loose masses of crystalline rocks, some of which weigh hundreds of tons. The surfaces of many of the limestone slabs at this locality are crowded with well-preserved specimens of the characteristic fossil *Leptæna alternata*."

THE DEVONIAN SERIES.

In consequence of the extreme flatness of the country the junction of the Silurian and Devonian Series has been only approximately determined, chiefly by the occurrence of the saline springs which distinguish the Devonian Series in this region. In all cases where saline springs were seen issuing from rock in position, Devonian formations were recognized by characteristic fossils. The parallelism between the courses of Moss River, Waterhen River, Partridge Crop River, and Dauphin River, suggests at once the line of junction or a dislocation ; if the former, their courses probably indicate the junction of different Devonian or Silurian formations, but in no instance the limits of either series, as was ascertained by the characteristic fossils found at different localities on those rivers.

In 1823, Mr. Keating* noticed the salt springs in Minnesota State and Dakotah Territory, far south of the boundary line. Even at that early period in the history of the settlements on Red River, 500 dollars were earned by one individual during one winter from the sale of the salt he had manufactured from springs near Pembina. The price of salt in the settlement was then six dollars per barrel weighing eighty pounds. At a spring on Saline River, south of the boundary line, Major Long's party found the *Salicornia herbacea* growing very abundantly around it. "Mr. Schweinitz states, on the authority of Mr. Nuttall, that this is the only inland locality of this plant, besides the Onondaga Salt springs in the State of New York." The bottom of shallow lagoons containing brine on Winnipego-sis Lake were thickly covered with a soft and wavy carpet of a species of *Naius.*

In the valley of La Rivière Sale, salt springs are very numerous, and the ground in their vicinity is frequently covered with a thick incrustation. Many years since the half-breeds of the settlement used to collect salt from this valley for domestic purposes. The names Saline Creek and Salt Point on Red River, north of the 49th parallel, were given in consequence of springs strongly impregnated with salt occurring there, but south and west of Stony Mountain no rocks in position have been observed east of Pembina Mountain. The whole country is nearly horizontal, having a mean elevation of about 100 feet above Lake Winnipeg.

Subjoined is a table showing the localities, north of the 49th parallel, where salt springs occur, distinguishing between springs from which salt has been and has not

* Major Long's Expedition to the Sources of St. Peter's River.

been manufactured or collected as a crust on the surface of the ground:

1.	Salt Brook	Red River.
2.	Salt Point	"
3.	La Rivière Sale	Collected from incrustations by the sides of springs. These incrustations are often two inches in thickness.
4.	Salt Point	Winnipego-sis Lake.
5.	Turtle River	Dauphin Lake—collected by Indians.
6.	Crane River	Manitobah Lake — collected by Indians.
7.	Monkman's Salt Works.* . .	Winnipego-sis Lake—manufactured by John Monkman.
8.	Swan River	Manufactured for H. B. Co.
9.	West Coast of Winnipego-sis Lake in many places.	
10.	West Coast of Lake Manitobah in many places.	
12.	The Pas Mountain.	

It has been already stated that the processes employed in the manufacture of salt in Rupert's Land are of the rudest description. By the employment of simple artifices the yield might be greatly increased, and its market value reduced to one-fourth the price it brings at the settlements. In the valley of La Rivière Sale, about twenty-six miles from Fort Garry, springs issue from the sides of the hills in positions very favourable for the employment of solar evaporation in shallow basins, which might be excavated at a lower level than the spring, and salt extracted without the employment of artificial heat; an immense

* The bottom of the evaporating pans at Monkman's Works contained a thin stratum of an extremely hard substance, which subsequent examination showed to consist of silica, a little sulphate of lime, and a very considerable quantity of bromides.

advantage in a country where fuel is scarce and labour dear.

In the State of New York between 500,000 and 600,000 bushels of salt are now made annually by solar evaporation. Wooden vats are employed, with movable roofs, so that the brine may be protected at the approach of unfavourable weather. The average daily supply of brine at these works during six months of the year is 2,000,000 gallons, and the cost per barrel of 300 lbs. is one dollar. Salt made by the boiling process weighs 56 pounds to the bushel, solar (that made by evaporation) salt 75 pounds. By the boiling process at Onondaga the cast iron kettles, holding from 50 to 70 gallons each, are disposed in double rows above suitable furnaces technically called "blocks." Each block contains from 50 to 70 kettles, and manufactures during eight months of the year from 20,000 to 25,000 bushels of salt.

In 1800 the number of bushels of salt made at the Onondaga Salt Works was 50,000; in 1810, 450,000 bushels; in 1830, 1,435,446 bushels; in 1840, 2,622,305 bushels; in 1850, 4,268,919 bushels; in 1857, 4,300,000 bushels; and in 1859, within a fraction of 7,000,000 bushels.

The strength of the brine is measured by a "Salometer," whose zero is distilled water, and maximum, water saturated with common salt, which is represented by 100. The brines of Onondaga vary from 76° to 44°. Wells which do not furnish brine above 50° are not considered worth working.

The sea-water at Nantucket gives a bushel of salt to every 380 gallons; at the salt springs of Zanesville, Ohio, 95 gallons furnish the same quantity of salt, while the old wells of Onondaga yield one bushel from 40 to 45 gallons, and the new wells at Syracuse the same quantity from 30 to 35 gallons of brine.* Mr. Donald Gunn, of Red River Settlement, says that in the dry season 24 gallons of brine from the Winnipego-sis springs will produce one bushel of salt.

The value of the salt trade in the United States may be inferred from the following statistics:—

	Bushels.
In 1840 the quantity of foreign salt imported was . .	8,183,203
In 1850 " " . .	11,224,185
In 1857 " " . .	17,165,704

The value of the foreign salt consumed in 1857 amounted to nearly 2,000,000 dollars, and the value of foreign and domestic salt exported from the States during the same year was 230,000 dollars.

The most eastern exposure of the Devonian Series, recognized by fossils of that age, occurs on Thunder Island, St. Martin Lake; the most westerly exposure is seen on Moss River, and it is between these two points that, as far as known, brine springs are most numerous. Barren areas surrounding brine springs are of frequent occurrence at the foot of the range of hills from the Riding Mountain to the Pas. In a country nearly horizontal, where the attitude of the rocks conforms to the general surface, it will be at all times very difficult to determine the precise line of junction between succeeding series, and fortunately in the present instance the brine springs, which undoubtedly have their source in Devonian rocks, afford an excellent guide in determining the outcrop and extent of the series.

As far as my observations enabled me to judge there is no difference in the general aspect of the country occupied

* The History, Commerce, Sources, Manufacture, and economical Value of Salt consumed in and exported from the United States; by William C. Dennis, of Key West, Florida. Patent Office Report, 1857.

by the Silurian and Devonian Series in this region. The rock of either age almost everywhere approaches the surface and is covered with a few inches of vegetable mould. Where fires have occurred the soil is burned away and the bare surface exposed. Very few areas of drift were seen, the most imposing being some low hills on St. Martin Lake. Denuding forces appear to have cut down the surface of the country to one nearly uniform level from the Riding Mountain ranges to the Laurentides. The upper extremity only of this excavated valley being covered many feet deep with quaternary deposits through which Red River, the Assinniboine and White Mud River have cut their channels.

The uninterrupted continuity of the rim of the Devonian rocks through the United States, Canada, Rupert's Land and the valley of Mackenzie River can scarcely be doubted. Its outcrop from the Atlantic to the Arctic Ocean would extend over sixty degrees of longitude and thirty degrees of latitude. The Devonian rocks in the valley of the Mackenzie River have many characteristics in common with those of Lake Winnipeg.

[The following are the fossils from Snake Island in Lake Winnipego-sis.

Atrypa reticularis (Linne) in abundance, both the common form with moderately coarse ribs and the more finely striated varieties, *Atrypa aspera* (Schlotheim). The specimens very closely resemble those figured by Professor Hall, in his new work, the " Geology of Iowa," plate vi. figs. 3, *a*, *b*, *c*, *d*, but are a little more pointed in front. A fine *Orthis* agrees well with the figures and descriptions of *O. Iowensis*, (Hall,) Geology of Iowa, plate ii. fig. 4, but is a little longer. The proportions are the same, but the length, breadth and depth are each two lines greater than the figures. Besides these there are fragments of several other Brachiopods, among which are two small species of *Productus*.

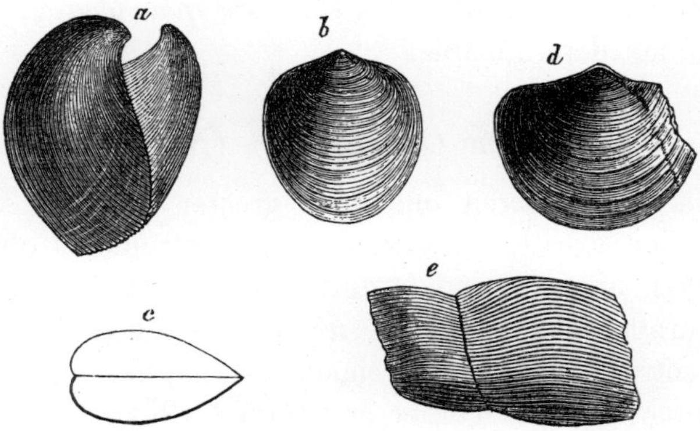

Fossils from Snake Island.

a, *Orthis Iowensis* (Hall) side view.
b, *Lucina occidentalis* (Billings).
c, *do.* *do.* Outlines of same, side view.
d, *Lucina elliptica* (Conrad).
e, *Loxonema nexilis* ?

The Lamellibranchiate shells are *Lucina elliptica* (Conrad) a species of the Corniferous limestone and Hamilton groups of Canada and New York, and a new species of the same genus, which I propose to call *L. occidentalis.*

Of Gasteropoda there are two species of *Euomphalus*, and a fragment of a *Loxonema*, most probably *L. nexilis.*

The Cephalopoda consist of fragments of *Orthoceras*, *Gomphoceras* and a species of *Nautilus* or *Gyroceras.*

Although we have none of the characteristic spirifers, corals, or trilobites to guide us, yet I think that upon the evidence of the above fossils we can safely say that this locality is Devonian, and most probably about the age of the Hamilton group.

The fossils from Manitobah Island are mostly the same as those of Snake Island, with the addition of two species of *Chonetes* and fragments of a large fish. There is also here a large *Stromatopora*, probably *S. concentrica*.

At Thunder Island, St. Martin Lake, the *Stromatopora* occurs, with abundance of a small *Strophomena* and some corals, not determinable.

Lucina Occidentalis. (*N. S.*)

Oval, length about one-ninth greater than the width, hinge line greatly convex, cardinal extremities obtusely rounded, anterior and posterior margins gently convex, sub-parallel ventral margin rounded or a little pointed in the centre; beaks central, small, pointed, incurved, nearly in contact with each other, and turned a little towards the anterior extremity; both valves moderately convex and marked with concentric undulations of growth.

Length of specimen, nine lines, width eight lines, depth of both valves, five lines. The greatest width is at about one-fourth the length below the beaks, from which level the margins converge but little, until within two-fifths of the length of the front, when they become more strongly curved.

Locality and Formation, Snake Island, Lake Winnipegosis. Devonian.]

CHAP. XXXIX.

THE CARBONIFEROUS SERIES. — JURASSIC FORMATION.

Evidence of the Carboniferous Series. — 'Productus.'—Probability of the Carboniferous Series being represented on the Flanks of the Riding Mountain.—Occurrence in Nebraska Territory.—In Kansas Territory.—In the North-West generally.—On the West Edge of the Fossiliferous Basin. — Ammonites from the McKenzie River, probably from Jurassic Rocks. —Ammonites Barnstoni.—Ammonites Billingsi.

THE CARBONIFEROUS SERIES.

The western limits of the Devonian Series are shown on the map to follow the boundary of the Great Cretaceous table-land so well defined by Pembina Mountain, Riding Mountain, Duck Mountain, Porcupine Hill, the Pas Mountain, and the high plateau similar to Pembina Mountain which stretches from the Pas to the Main Saskatchewan, near and below Fort à la Corne. The country at the base of this continuous boundary is uniformly horizontal, and while Devonian rocks in position were seen within thirty miles, and brine springs within ten miles of Cretaceous shales on the precipitous flanks of the Riding Hill Range, yet no evidence of any intermediate formation was visible.

During the ascent of the Riding Mountain, a very careful search was made for traces in the drift of the higher series, in the hope of obtaining evidence of the existence of Carboniferous rocks, but without success. The boulders, so numerous on the ridges and the successive terraces, were carefully examined, but they were found to be derived altogether from the Laurentian Series, the limestones of Lake Winnipeg or the superior Cretaceous shales.

The importance of any evidence of the Carboniferous Series in the Valley of Lake Winnipeg cannot be too highly estimated. A *productus* was given to me by a half-breed at Red River Settlement, who declared he procured it from "solid rock," which is known to occur in two places only in the valley of Red River, at the Stone Fort, and above, at and below the rapids, from which places both Dr. Owen and I obtained Lower Silurian fossils from rock in position. It is probable that the specimen figured below, was procured from a boulder; but boulders are brought

Productus?

SUPPOSED TO BE FROM CARBONIFEROUS LIMESTONE.

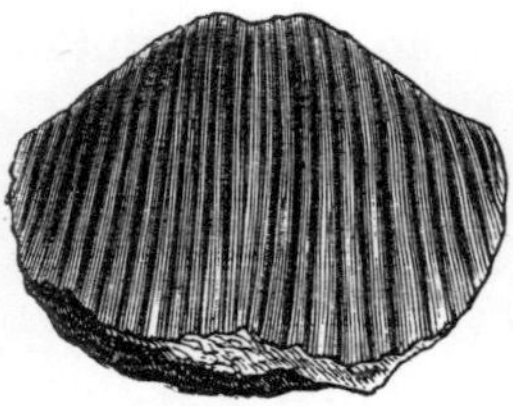

Fig. 1.

Fig. 2.

Fig. 3.

Fig. 1, Productus from Red River.
Fig. 2, ,, Side view.
Fig. 3, ,, Rostral extremity.

north each year from Minnesota by the ice of Red River; it is therefore quite possible that the specimen was brought by ice from the *south*. Under any circumstances, its presence within thirty miles of the mouth of Red River is an important fact, and affords good ground for hope, that if the Carboniferous Series are not represented on the flanks of the Riding, Duck, and Porcupine Mountains, they will be found in the State of Minnesota, or

Dacotah, on the north side of the Height of Land, and in the valley of Red River.

[There is some evidence of the existence of at least a portion of the Carboniferous system in this region. The fossil procured from the half-breed, who said he collected it from "the solid" rock, at some place on the Red River, is a *Productus* of the group *Semireticulati*, all of which appear to be confined to the Carboniferous Series. The specimen is not worn and presents all the appearance of having been freshly broken from the rock.] *

The presence of fragments of any particular rock in the drift of Canada and elsewhere, affords presumptive evidence of the existence of the parent rock in position some distance to the north of the place where the detritus is found, but where the erratics occur in the bed of a river flowing to the north, in which ice of great thickness accumulates, and which moreover is subject to annual overflows, the evidence, under other circumstances, strongly presumptive, must in this case be received with great caution, and regarded rather as an encouragement to continue the search for the Carboniferous Series in both directions from which the boulder might have been carried.

If rocks occupying a position between the Devonian and Cretaceous Series exist on the flanks of the Riding Mountain, it is probable that traces would have been discovered in the drift. The space in which members of the Carboniferous Series or superior formations might occur, is narrowed down to a strip ten miles in breadth between the salt springs south of Dauphin Lake and the outcrop of the Cretaceous shales on the flanks of the mountain.

* The following label was attached to this specimen, "Given to me by a half-breed, who declared he picked it from 'solid rock' in Red River."
(Doubtful.) H. Y. H.

At least seven miles of this distance is so nearly horizontal that it does not rise twenty feet above Dauphin Lake, and the dip of the Devonian strata is uniformly at a very small angle to the south-west, where exposures were seen on Manitobah Lake. (Small local deviations from a uniform dip on Snake Island and Moss River are noticed in Chapters XXIV. and XXV.) The Cretaceous shales were found exposed on the flanks of the mountain, about 400 feet above Dauphin Lake, and the rise from the level country at the foot of the mountain to that altitude is embraced within two and a half or three miles; yet within this narrow limit the drift on the slopes between each terrace, on the terraces themselves, or in the bottom of gullies excavated by mountain streams, gave no evidence of other rocks than those already named. It must be admitted that the time I could devote to an examination of the boulders was short, and a more minute search might give other results.

With this negative evidence in view, it appears probable that the Carboniferous Series is not represented in the only locality where it may be looked for with much chance of success. Nevertheless, between the Devonian and Cretaceous Series in the basin of Lake Winnipeg there is still a vertical section fully 400 feet in altitude, which is concealed by drift on the flanks of the Riding Mountain, covering a horizontal area two and a half to three miles broad. It is possible that within this narrow limit, or further to the north where the area may be much broader, rocks of Carboniferous, Permian, Triassic, or Jurassic age, may yet be found. With a view to show the relation which the Cretaceous and Carboniferous Series have to one another in lower latitudes, the following brief notice of their occurrence in Nebraska and Kansas is introduced.

In Nebraska the Carboniferous Series, or the coal

measures, are exposed at the mouth of the Platte *, and extend up the river about fifty miles, when they dip beneath the water level of the Missouri. They are overlaid by No. 1 of the Nebraska section of the Cretaceous Series in lat. 41°·5, long. 96°. Cretaceous and Tertiary formations then occupy the valley of the Missouri as far as Fort Benton, lat. 47°·54, long. 110°, and extend into British America. Hence it appears that, ten degrees of latitude south of the Riding Mountain, the Cretaceous Series repose on the Carboniferous without the intervention of Permian, Triassic, or Jurassic rocks.

In Kansas territory, on the Kansas and Smoky Hill Rivers, an elaborate section has been made by Messrs. F. B. Meek and F. V. Hayden †, commencing with the Cretaceous sandstones on the summit of the Smoky Hills, lat. 38° 30′ N., long., 98° W., and descending through the various intermediate formations seen along the Smoky Hills and Kansas River to the mouth of the Big Blue River on the Kansas. This section, over 1000 feet vertically, passes from the Cretaceous to the upper coal measures, and includes rocks of Permian age. Messrs. Meek and Hayden remark, in relation to this section, " It will be observed we have in this general section, without attempting to draw lines between the systems or great primary divisions, presented in regular succession the various beds with the fossils found in each, from the Cretaceous sandstone on the summits of the Smoky Hills, down through several hundred feet of intermediate doubtful strata, so as to include the beds containing Permian

* Notes explanatory of a Map and Section illustrating the Geological Structure of the Country bordering on the Missouri River, &c., by F. V. Hayden, M.D.

† Geological Explorations in Kansas Territory, by F. B. Meek and F. V. Hayden, published in the Proceedings of the Academy of Natural Sciences at Philadelphia.

types of fossils, and a considerable thickness of rocks, in which we find great numbers of upper coal measure forms. We have preferred to give the section in this form, because, in the first place, the upper coal measures of this region pass by such imperceptible gradations into the Permian above, that it is very difficult to determine, with our present information, at what particular horizon we should draw the line between them, while on the other hand it is equally difficult to define the limits between the Permian and beds above, in which we found no fossils." *

Jurassic or Triassic formations may occur above the Permian in the section just referred to. Messrs. Meek and Hayden state that "between No. 5 (of the Section) and the Cretaceous above, there is still a rather extensive series of beds in which we found no organic remains; these may be Jurassic or Triassic or both, though as we have elsewhere suggested, we rather incline to the opinion that they may prove to belong to the former."†

Formation No. 1 of the Nebraska Series of the Cretaceous rocks has not yet been recognized in Rupert's Land. This formation reposes on Jurassic rocks in Nebraska territory at the Black Hills.‡ It rests, as before stated, upon the limestones of the coal measures on the Missouri, near the 42nd parallel.

"There is at the base of the Cretaceous system, at distantly separated localities in Nebraska, Kansas, Arkansas, Texas, New Mexico, Alabama, and New Jersey, if not indeed everywhere in North America where that system is well developed, (at any rate east of the Rocky

* Page 19, Geological Explorations in Kansas.

† Page 21, *ibid.*

‡ On the Lower Cretaceous Beds of Kansas and Nebraska, by F. B. Meek and F. V. Hayden.—Proceedings Acad. Nat. Sci. Phil., Dec. 1858, published in Am. Jour. Sci. page 219, 1859.

Mountains,) a series of various coloured clays and sandstones, and beds of sand, often of great thickness. In this series organic remains, excepting leaves of apparently dicotyledonous plants, fossil wood, and obscure casts of shells, are very rarely found, but it everywhere preserves a uniformity of lithological and other characters, pointing unmistakably to a similarity of physical conditions during their deposition, over immense areas.

"Although the weight of evidence thus far favours the conclusion that this lower series is of the age of the Lower Green Sand, or Neocomien, of the old world, we yet want *positive* evidence that portions of it may not be older than any part of the Cretaceous system."

Judging therefore, solely from the relation which the Cretaceous series bears to formations beneath them in their development through Rupert's Land, Nebraska, and Kansas, we might expect to find on the Riding Mountain in the vertical section concealed by drift, beneath formation No. 4, either formation No. 1, 2, and 3 of the Nebraska section, or members of the Jurassic and Permian as well as the Carboniferous series.

In the Report on the Geological Survey of the State of Iowa, published in 1858, the State geologist, James Hall, advances a highly important view of the relation of the Cretaceous formations to the Carboniferous Series. "Towards the Rocky Mountains, the palæozoic rocks are overlaid by the Cretaceous formation, which, in its lower arenaceous members, stretches from the northern limits of the United States territories to the Gulf of Mexico, and throughout a great part of this extent rests unconformably upon the coal measures.

"The line of junction of these two formations is obscured by the denudation of the higher one, and its finely comminuted materials are widely spread over the lower strata, forming with other materials the broad expanse of prairie deposit of the west." *

It is very gratifying to know that on the western side of the great basin between the Laurentides and the Rocky Mountains, and within the limit of the Saskatchewan valley, the Carboniferous series are represented. Sir Roderick I. Murchison, in his address at the Anniversary Meeting of the Royal Geographical Society, in referring to the splendid results of the Palliser Expedition, says, "Thus in addition to the determination of latitude, longitude, and the altitude of the mountains and two of their passes, Dr. Hector presents us with a sketch of the physical and geological structure of the chain, with its axis of slaty sub-crystalline rocks, overlaid by limestones of Devonian and Carboniferous age, and flanked on the eastern face by Carboniferous sandstone, representing, probably, our own coal fields, the whole followed by those Cretaceous and Tertiary deposits which constitute the subsoil of the vast and rich prairies watered by the North and South Saskatchewan, and their affluents." †

Ammonites.

FROM M^cKENZIE'S RIVER, PROBABLY FROM JURASSIC ROCKS.

[The two *Ammonites* from McKenzie's River, are not alone sufficient to determine the age of the rock from which they were obtained; the larger one bears considerable resemblance in form and general appearance to several Jurassic species though they may belong to the Cretaceous epoch. It is very desirable that a good series of specimens should be obtained from this remote

* Remarks on the Tertiary and Cretaceous Formations of Nebraska, &c. &c., by F. B. Meek and F. V. Hayden, M.D.

† Page 318, Proceedings of the Royal Geological Society, vol. iii. No. 4.

northern locality, not only for the purpose of determining the age of the formation, but for the light they might throw upon far more interesting questions respecting the probable climatic conditions in these high northern latitudes during the secondary period.]

After the above paragraph was published in my report on the expedition, page 182, Mr. Meek has written to me to say: "Since writing the paper on these fossils for your report, I have seen some figures of Jurassic species from the Aleutian Islands as high north as about the 58th parallel. Amongst these there is an Ammonite (*A. Wosnesseuski*, Grewingk.—Trans. Mineralogical Society of St. Petersburgh, 1848–9, Pl. IV.) resembling one of those I have described in your report (*A. Barnstoni*) so nearly in some of its characters, as to strengthen my suspicions that the rock from which these fossils were obtained on McKenzie's River may prove to be of Jurassic age. In the peculiar deep conical characters of the umbilicus of *A. Wosnesseuski*, it is very similar to *A. Barnstoni* but it is a more compressed shell, with straighter and more simple costæ; while its septa, if accurately drawn differ from those of *A. Barnstoni*, I am rather inclined to the opinion that these rocks on McKenzie's River will prove Jurassic, but we must wait for additional evidence before any positive opinion can be given."

The Rev. Samuel Haughton, F.R.S., discusses the question of climate in the Arctic regions in relation to the Liassic fossils found in *situ* on Prince Patrick Island and elsewhere north of the 75th parallel, in the geological account of the Arctic Archipelago, published in the Appendix to Captain McClintock's Narrative. Professor Haughton says, "But what are we to say as to the question of temperature? It was certainly necessary for an Ammonite to have a sea free from ice, on which to float and bask in the pale rays of the Arctic sun; and therefore I claim a temperature for those seas, at least similar to that which now prevails in the British Islands; and I may add that the Ammonite, from its habits, was essentially dependent on the temperature of the air as well as on that of the water."

"There is at present a difference of 49° 5′ between the mean annual temperature of Point Wilkie and Dublin; and if this change of temperature be supposed to be caused by a change of the relative positions of land and water, the temperature of Dublin, or of some place on the same parallel of latitude, must be supposed to be raised to 99° 5′, while the temperature of the thermal equator will exceed 124°,—a temperature only a few degrees below that requisite to boil an egg. I reject, without scruple, a theory that requires such a result, which must be considered as a minimum, as it is probable that the Ammonite required a finer climate than that of Britain for the full enjoyment of its existence.

"The theory of central heat, also, appears to me to be open to the same objection, as a mode of explaining this remarkable geological fact; for it will simply add a constant to our present climates, leaving the difference to remain, as at present, to be accounted for by latitude and distribution of land and water.

"The astronomical theory of Herschel, also, which would account for former changes of climate by changes in the radiating power of the sun, would only increase the temperature at each latitude, leaving the differences as at present.

"The only speculation with which I am acquainted which is capable of solving this *opprobrium geologicorum*, is the hypothesis of a change in the axis of rotation of the earth, the admission of which, as a geological possibility, is mathematically demonstrable, and which has recently had some singular evidence in its favour advanced by geologists. In 1851 I brought forward, at the Geological Society of Dublin, a case of augular fragments of granite, occurring in the carboniferous limestone of the county Dublin, and explained the phenomena by the supposition of the transporting power of ice. In 1855 Professor Ramsay laid before the Geological Society of London a full and detailed theory of glaciers and ice as agents concerned in the formation of a remarkable breccia, of Permian age, occurring in the central counties of England; and still more recently the same agent has been employed by the geological surveyors of India to account for the transport of materials at geological periods long antecedent to those in which ice transport is commonly supposed to have commenced. The motion of the earth's axis would reconcile all the facts known, and it must be regarded as a geological desideratum to determine its amount and direction, and to assign the cause of such a movement. The solution of this problem I regard as quite possible." *

This very interesting question has already been solved by the distinguished French mathematician, M. Leverrier. At p. 163 of tome ii. of the *Annales de l'Observatoire Impérial de Paris*, M. Leverrier gives 4° 51′ 42″ AS THE SUPERIOR LIMIT OF THE INCLINATION OF THE EARTH'S ORBIT UPON THE ECLIPTIC OF 1800, subject, however, to small corrections depending on any possible inaccuracies in the values of the planetary masses employed in his researches. Mr. J. R. Hind, Superintendent of the Nautical Almanac, to whom I am indebted for the subjoined table, states that the amount of correction cannot well exceed 8′, and

* Appendix to a Narrative of the Discovery of the Fate of Sir John Franklin and his Companions, by Captain McClintock, R.N., p. 395.

probably is much less. The elements of the earth's orbit for periods of 10,000 years, reckoning backwards from A.D. 1800, are subjoined, from which it will be seen that the axis of rotation of the earth oscillates between certain small limits (4° 51′ 42″), not sufficient to account for the change in the climate of the Arctic regions which the presence of the fossils there appears to involve. Probably the influence of currents, similar to the Gulf stream of the present epoch, may have been instrumental in elevating the temperature of those regions, and producing a climate which would admit of the existence of Ammonites and plants, which can only flourish in a temperate zone.

ELEMENTS OF THE EARTH'S ORBIT, ACCORDING TO M. LEVERRIER.

Years before 1800.	Excentricity.	Longitude of the Perihelion.	Inclination.	Longitude of the Node.
		° ′	° ′ ″	° ′
— 100,000	0·0473	316 18	3 45 31	96 34
— 90,000	0·0452	340 2	2 42 19	76 17
— 80,000	0·0398	4 13	1 18 58	73 47
— 70,000	0·0316	27 22	1 13 58	136 8
— 60,000	0·0218	46 8	2 36 42	136 29
— 50,000	0·0131	50 14	3 40 11	116 9
— 40,000	0·0109	28 36	4 3 1	91 59
— 30,000	0·0151	25 50	3 41 51	66 49
— 20,000	0·0188	44 0	2 44 12	41 34
— 10,000	0·0187	78 28	1 24 35	16 39
0	0·0168	99 30	0 0 0	0 0

*** The longitudes are referred to the Ecliptic and Equinox of 1800.

Mr. James Hall*, the present State Geologist of Iowa, considers that the great development of the lower mem-

* Mr. James Hall, the author of that magnificent work, the Palæontology of the State of New York, &c. &c.

ber of the Cretaceous formations in Western Iowa and the adjoining territories of Nebraska and Kansas, and its extension below the line where any well-marked Cretaceous species have been found, " suggests very strongly that we shall yet find lower geological formations, or those of the age of the Jura, or Oolite of Europe."*

The discovery of Oolitic or Liassic rocks in Exmouth Island by Captain Belcher ; of *Ammonites*, *Spirifer*, *Pecten*, &c. by Captain McClintock in Prince Patrick Island, lat. 76° 30′, long. 117° W. in localities corresponding to the northerly trend of the Carboniferous limestone of the Rocky Mountain region, and the great thickness of the rocks on an outlier of the Llana Estacado below any well marked Cretaceous fossils, indicate, Mr. Hall thinks, " the probable occurrence of lower fossiliferous rocks, or those of Jurassic age, along the whole length of the Rocky Mountains, and probably coextensive with the lower members of the Cretaceous Series."

Ammonites Barnstoni. (N. S.)

[Shell compressed, subglobose, broadly rounded on the dorsum, and prominent or subangular around the umbilicus which is deep, conical, and nearly as broad as the outer whorl. Volutions having their greater diameter at right angles to that of the shell ; each of the inner ones about three-fourths hidden in the profound ventral groove of the succeeding turn. Surface ornamented by distinct regular costæ, which are sharply elevated around the umbilicus, into small elongated subnodose prominences ; and at less than half the distance across the sides of the whorl, their number is increased nearly threefold by division and

* Geological Survey of Iowa, page 144.

implantation; after which they become of uniform size, and arch gently forward in passing over the dorsum.

The septa are deeply divided into five principal lobes and six saddles, which are crowded together, and variously branched and subdivided. The dorsal lobe is a

AMMONITE BARNSTONI. Side view, showing the deep umbilicus.

little longer than wide, and has three branches on each side, the two terminals of which are nearly straight and parallel; the first two lateral branches above these, are nearly of the same size, but more diverging; while the third pair are much smaller—and all sharply digitate, and more or less subdivided. The dorsal saddle is longer than wide, contracted in the middle, and irregularly divided into four unequal branches, the two terminals of which are subdivided into two branchlets each, and all obtusely digitate, and variously sinuous in the margins.

The superior lateral lobe is longer than the dorsal lobe, but very irregularly branched, and, like it, provided with

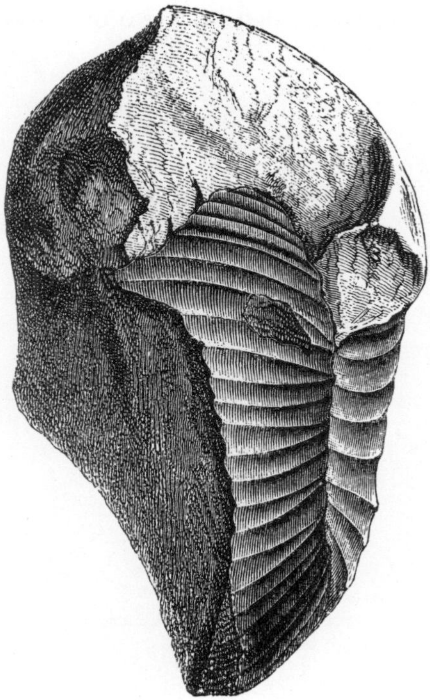

Front view of the same specimen.

numerous sharp digitations on all its divisions; at the extremity it has three very unequal branches, the middle one of which is much longer than the others, and very slender; the other two are small, unequal, opposite, and diverging;—that on the right being subdivided nearly to its base: above these there are several other unequal alternating lateral branches, one of which on the right side is much larger than the others. The lateral saddle is rather smaller than the dorsal, and divided at the extremity into two very unequal branches, of which the one on the left is larger than the other, and again deeply divided into two bifid and deeply sinuous branchlets. The inferior lateral lobe is much smaller than the superior, and very irregularly divided into two or three alternating

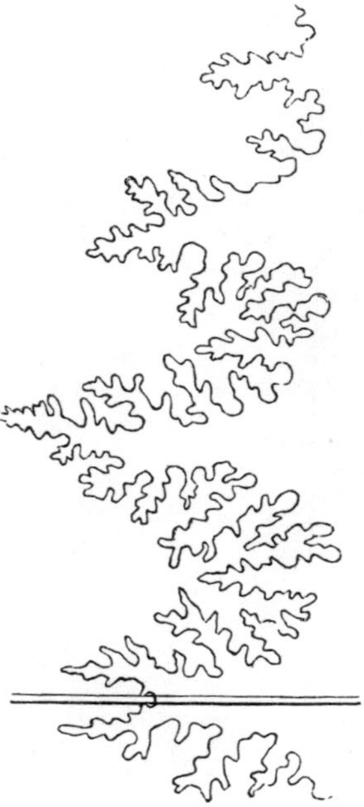

Diagram of one of the septa of AMMONITES BARNSTONI.

unequal lateral branches on each side, and one terminal branch, with numerous sinuosities. The ventral lobe is very small, and simply digitate.

This species bears considerable resemblance in form, and in the size and character of its umbilicus, to the Jurassic species *A. irens*, D'Orbigny (Pal. Franc., tome i. p. 562, pl. 222), but differs in having the costæ pinched up into little subnodose prominences around the umbilicus, and bifurcating on the sides; they are also much more arched in passing over the dorsum. It is quite different from any of the described species from the Nebraska rocks, though I think I have seen some fragments of it in Lieut. Warren's collections from No. 4 of the Nebraska cretaceous subdivisions.

The specific name is given in honour of Mr. George Barnston, chief factor of the Hudson's Bay Company, who discovered it in the valley of Mackenzie's River. It is probably a Cretaceous species, but may be of Jurassic age.]

Ammonites Billingsi.

Fig. 1.

Fig. 2.

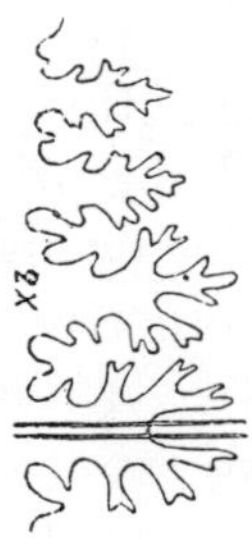

Fig. 3.

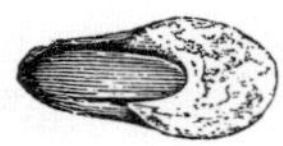

Fig. 1.—Side view. Fig. 2.—One of the septa enlarged.
Fig. 3.—Front view of same specimen.

[Shell moderately compressed, or subdiscoidal; dorsum rounded; umbilicus very small; volutions having their greater breadth at right angles to the shorter diameter of the shell, increasing rather rapidly in size, or more than doubling their diameter each turn; inner ones entirely embraced, and hidden in the ventral groove of the last turn: surface apparently smooth, but showing very faint traces of radiating costæ, which arch a little in crossing the dorsum.

Dorsal lobe longer than wide, provided with three branches on each side, the two terminal of which are much longer than the others, and each subdivided,—the subdivisions being short, and each having two or three small digitations; the first two lateral branches above these are small, opposite, very diverging, and bifid or digitate; and the third pair very small, and apparently simple. The dorsal saddle is as long as the dorsal lobe, but narrower, and has three or four short obtusely rounded branches on each side. The superior lateral lobe is nearly as large as the dorsal saddle, and has three subequal branches at the extremity,—that on the dorsal side being bifurcate, with digitate divisions; and the middle, and other lateral divisions, are provided with three or more small digitations each. The inferior lateral lobe is much smaller than the superior lateral, and has much the same form, excepting that its terminal division is proportionally larger, and the principal lateral division on the dorsal side is not so deeply divided. The ventral lobe is a little smaller, but in other respects very similar to the inferior lateral lobe; between it and the umbilicus there appears to be one or two smaller auxiliary ventral lobes, which seem to show a tendency to branch in the same way as the principal ventral lobe.

The specimen from which the foregoing description was made out, is evidently a young shell; consequently, adult individuals of the same species may be expected to possess much more distinct costæ. The lobes and saddles of the septa, in old shells, will also be found much more deeply divided and more complex, but the mode of branching probably remains the same from the time the principal divisions are formed.

As the specimen described was found in the matrix filling the umbilicus of *A. Barnstoni* (being only 0·67 inch in its greatest diameter), it might be supposed by those who know how widely the *Ammonites* sometimes vary at different ages, that it may be the young of that species. It presents fundamental differences, however, in the mode of branching of the lobes and saddles of its septa, that cannot be due to different stages of development. In addition to this, I found along with it a much smaller specimen, evidently the young of *A. Barnstoni*, which shows that the young of that species did not vary in form materially from the adult, and is quite different from the species now under consideration.

It has much the form of *A. Halli*, Meek and Hayden (Proceed. Acad. Nat. Sci. Phil., vol. viii. p. 70), and there are no differences in the structure of the dorsal lobes of the two, that might not be due to different degrees of development. Their superior lateral lobes and dorsal saddles, however, present radical differences, such as we never see in the same species, however widely they may differ in size or age.

I have named this species in honour of Mr. E. Billings, the accomplished Palæontologist of the Canadian Geological Survey.]

CHAP. XL.

THE CRETACEOUS SERIES.—TERTIARY FORMATIONS.

Distribution of the Cretaceous Series in Western North America.—The Nebraska Section.—Vertical Section in Nebraska and Rupert's Land.—Formation No. 1.—Formation No. 2.—Formation No. 3.—Formation No. 4.—Great Development of Formation No. 4 in Rupert's Land.—Fossils.—Steatitic Minerals.—Analysis of.—Anomia Flemingi.—Inosceramus Canadensis.—Leda Hindi.—Scaphites Nodosus.—Clay Iron-stone.—Bands of this Formation.—Distribution of.—Richness of.—Analysis of.—Formation No. 5 of the Nebraska Section.—Its Occurrence on the Flanks of Rocky Mountains.—Fossils of.—Avicula Linguæformis.—Avicula Nebrascana.—Nautilus Dekayi.——TERTIARY FORMATIONS.—Lignite on the Riding and Duck Mountains.—Sand Dunes probably derived from Tertiary Rocks.—Tertiary Coal.—Lignite Basin of the Missouri.—Lignite on the North Saskatchewan.—On Red Deer River.—On the South Branch.—Lignites of Oregon and British Columbia.

THE Cretaceous Series are more extensively distributed in Rupert's Land, than any other fossiliferous formations. This remark, indeed, is susceptible of extension to the continent of North America, where the extraordinary development of this series may be inferred from the fact, that they have been traced continuously from beyond the North Branch of the Saskatchewan to the Gulf of Mexico, and under the 52nd parallel, nearly across the fossiliferous basin, or over 12° of longitude. Its eastern boundary in Rupert's Land is shown approximately by the Pembina Mountain, and the continuation of that ancient sea coast, to the main Saskatchewan below

Fort à la Corne, thence north-westerly. Cretaceous rocks were recognized by Dr. Hector * at the Snake Portage, in lat. 54° long. 111° 30′; also on Little Red Deer River, at the base of the Rocky Mountains, in long. 114°.

This important series, as it occurs in Nebraska, has been carefully studied by Messrs. Meek and Hayden. In the notes explanatory of a map and section illustrating the geological structure of the country bordering on the Missouri River, Dr. Hayden has described the rocks of Nebraska Territory †, where the Cretaceous Series is best developed; and as this division, styled the NEBRASKA SECTION, forms the standard to which the Cretaceous rocks of the north-west are referred on the American Continent, the following division of the series in Rupert's Land is based upon the Nebraska Section.

The vertical section of the geological formations of Nebraska Territory, with their extension into Rupert's Land, as far as determined, is given on pp. 320, 321.

FORMATION NO. 1, OF VERTICAL SECTION.

The following excellent descriptions, enclosed in brackets, of the formations constituting the Nebraska Section, are from the clear and concise "Notes Explanatory of a Map," &c., by F. B. Meek and F. V. Hayden, M.D. They are introduced here for the purpose of comparison with the development of the same formations in Rupert's Land.

* Papers relative to the Exploration of British North America. Blue Book, 1859.

† Notes explanatory of a Map and Section illustrating the Geological Structure of the Country on the Missouri River from the Mouth of the Platte River to Fort Benton, in lat. 47° 30′ N., long. 110° 30′ W., by F. V. Hayden, M.D. Proc. Acad. Nat. Sci. Phil., May, 1857.

VERTICAL SECTION OF THE GEOLOGICAL FORMATIONS OF NEBRASKA TERRITORY, AS FAR AS DETERMINED, WITH THEIR EXTENSION INTO RUPERT'S LAND.

System	Formation	Subdivisions.	Localities.	Estimated thickness.	Localites in Rupert's Land.
TERTIARY SYSTEM.	MIOCENE.	Light-coloured indurated clays, with occasional beds of sandstone, conglomerate and whitish limestone. Great numbers of *mammalian* and *chelonian* remains with a few fresh-water and land shells.—(Bad Lands of White River.) Beds of clay, sand, sandstone, and lignite, containing great numbers of fresh-water and land *mollusca*, with a few marine or estuary shells; remains of plants, *Saurians*, *Trionyx*, &c.—(Great Lignite Basin.) Sand, sandstone, clays, and very impure lignite, with remains of fresh-water, land, and a few estuary shells, *Saurians*, fishes, *Trionyx*, &c.—(Bad Lands of Judith.)	*Mauvaises Terres* of White River. Great extent of country on both sides of the Missouri between Heart and Milk Rivers; on the Yellow Stone. Bad Land at the mouth of Judith River, &c.	About 900? feet.	Grand Coteau de Missouri.
CRETACEOUS SYSTEM.	No. 5.	Grey and yellowish arenaceous, clays, and sandstones, sometimes weathering to pink colour; containing *Belemnitella bulbosa*, *Nautilus Dekayi*, *Ammonites placenta*, *A. lobatus*, *Scaphites Conradi*, *Baculites ovatus*, and great numbers of other marine mollusca.	Moreau trading post, and under the Tertiary of Sage and Bear Creeks. Fox Hills.	100 to 150 feet.	Yellowish grey concretionary sandstone, underlaid by greenish grey concretionary sandstone. Concretions contain *Avicula Nebrascana* and weather reddish-brown. Selenite in veins, fibrous lignite, impressions of leaves.—Calcareous band on the Qu'appelle at the height of land composed of *Avicula linguæformis*.—South Branch of the Saskatchewan, *Scaphites Conradi*, *Nautilus Dekayi*, *Avicula linguæformis*, *Avicula Nebrascana*.
	No. 4.	Bluish and dark grey plastic clays, containing *Nautilus Dekayi*, *Ammonites placenta*, *Baculites ovatus*, and *B. compressus*, with numerous other marine mollusca,—	Great area about Fort Pierre and along the Missouri below there. Under No. 5, at Sage and Bear Creeks. Great	350 feet.	Lead grey and purple argillaceous shales, characterized by vast numbers of nodules of clay iron-stone.—Little Souris River, containing *Anomia Flemingi*, *Inoceramus Canadensis*, *Leda Hindi*.—Two Creeks, Assinniboine River. *Natica obliquata*.

CRETACEOUS SYSTEM.	No. 3.	20 feet	Dark, very fine unctuous clay, containing much carbonaceous matter, with veins and seams of selenite, sulphuret of iron, fish and scales (local).	Bluffs along the Missouri below the Great Bend. Extends to Big Sioux River, and occurs along the latter stream.	150 feet. Wanting in New Jersey and Alabama.	North Branch of the Saskatchewan at the Coal Falls (?) Sharks' teeth, scales of fish, *Inoceramus*.
		100 feet.	Lead grey calcareous marl, weathering above to a yellowish tint. Scales and other remains of fishes—*Ostrea congesta*—passing downwards into			
		30 feet.	Light grey or yellowish limestone, containing great numbers of *Inoceramus problematicus*, fish scales and *Ostrea congesta*.			
	No. 2.		Dark grey laminated clay; scales and other remains of fishes, small Ammonites, *Inoceramus problematicus*? *Serpula*, small oyster-like *O. congesta*, &c.	Along the Missouri Bluffs, from 10 miles above James River to Big Sioux River.	90 feet. Wanting in New Jersey and Alabama.	Assinniboine, scales of fish. North Branch of the Saskatchewan at the Coal Falls (?)
	No. 1.		Yellowish and reddish friable sandstone, with alternations of dark and whitish clays. Seams and beds of impure lignite, fossil wood, impressions of dicotyledonous leaves; *Solen*, *Pectunculus*, *Cyprina*, &c. This bed is not positively known to belong to the Cretaceous System.	Near the mouth of Big Sioux River, and between there and Council Bluffs. Near Judith River.	90 to 100 feet, or more.	Not recognized in area explored.
CARBONIFEROUS SYSTEM.			Yellow limestone, containing *Fusulina cylindrica*, *Terebratula subtilita*, *Spirifer Meusebachanus*, *Allorisma regularis*, and other fossils of the coal measures.	Forms shoals in the Missouri River at De Soto; 15 to 20 feet exposed at Council Bluffs, at low stages of the river.	Unknown.	Not recognized in area explored.

[In the order of superposition, formation No. 1 rests directly upon the true limestones of the coal measures. Its first exposure seen along the Missouri is at Wood's Bluffs, right bank, about eighty miles above the mouth of the Platte, and it dips beneath the water level of the Missouri, a few miles below the mouth of the Vermilion. Its general character is a coarse grained, friable sandstone, very ferruginous, of a yellow or reddish-yellow colour, with thin beds of impure lignite and various coloured clays. It contains very few fossils, mostly of the genera *Solen*, *Cyprina* and *Pectunculus*, also fossil wood, and numerous impressions of dicotyledonous leaves, similar to the common willow. Its entire thickness is estimated at ninety to one hundred feet, but it may be more.]

This formation has not yet been recognized in Rupert's Land. In Nebraska it reposes upon the upper members of the Carboniferous Series near the mouth of the Platte (lat. 41° 40′), and it overlies Jurassic rocks at the Black Hills.*

FORMATION NO. 2 OF VERTICAL SECTION.

[This formation is first revealed in thin outliers below the mouth of Big Sioux River, and on that stream six miles above its mouth, it caps the bluffs, apparently mingling to some extent with the succeeding bed, and containing at this locality large numbers of *Inoceramus problematicus* and fragments of fishes. Near the mouth of Iowa Creek and above, it shows itself worthy of a

* Descriptions of the Species and Genera of Fossils collected by Dr. F. V. Hayden in Nebraska Territory, under the direction of Lieut. C. K. Warren, U.S. Topographical Engineer, with some Remarks on the Tertiary and Cretaceous Formations of Nebraska, and the parallelism of the latter with those of other portions of the United States and Territories, by F. B. M., and F. V. Hayden, M.D.

separate position in the series. It is composed of a dark leaden grey, laminated, plastic clay, containing few fossils, but great quantities of the sulphate of lime in crystals, assuming a variety of beautiful forms. Its greatest thickness is seen five miles below the mouth of James River. At Dorion's Hills it is seen at low water mark. Entire thickness estimated at ninety feet. Fossils, *Ammonites*, *Inoceramus*, *Cytheria*, *Serpula*, *Ostrea*, and abundant fish remains.]

This formation has been recognized on the Assinniboine.

On the North Branch of the Saskatchewan, a few miles above the Grand Forks, huge masses of a dark coloured, almost black shale, with sharp, well-preserved edges jut out of the banks and are exposed whenever portions of the face of the clay cliffs fall into the river. Their appearance is such as to justify the expectation that rock in position from which they originated is close at hand. Some specimens which I procured and sent to Mr. Meek, contain, according to that gentleman, fish scales, sharks' teeth and *Inoceramus*, which renders it almost certain that the masses were detached from rocks belonging to Formation 2 or 3, of the vertical section. I have therefore assigned in the foregoing table, the locality Coal Falls, North Branch of the Saskatchewan, with a note of interrogation, as the probable outcrop of one or both of these divisions of the vertical section.

FORMATION NO. 3 OF VERTICAL SECTION.

[The geographical distribution of this formation and its influence on the scenery render it one of the most interesting on the Missouri. It is first seen in thin outliers near the mouth of Big Sioux River, and becomes quite conspicuous on the summits of the bluffs, ten miles above the Iowa Creek. At Dorion's Hills, it reaches to the water's edge, and is the prevailing formation from thence to the foot of the Great Bend, where it passes by a gentle dip beneath the water level of the Missouri. At Dorion's Hills there is a fine section of this bed about eighty feet exposed above the water's edge, containing its most abundant and characteristic fossil, *Ostrea congesta.* In many places, as opposite the mouth of Running Water, it assumes the form of a long series of precipitous bluffs, giving a pleasing variety to the general monotony of the scenery. This is one of the principal characteristic external features of this formation.

The upper portion of this rock is a yellowish and grey calcareous marl, very soft and yielding, so that it is easily cut up into numerous ravines by the temporary streams, and thus the bluffs along this part of the Missouri often present the appearance of a series of cones. The lower stratum, however, is more compact, and forms a soft bluish grey limestone.

Though so well developed and covering so wide an area, the middle and upper portions, at least, of this rock can never be made useful for building purposes. Quite soft and friable in places, when detached, it absorbs moisture rapidly and crumbles in pieces. Being a rich calcareous marl, it may be used at some future time as a fertilizer.

The fossils of this formation, though belonging to few species, as far as is yet known, are numerous in individuals. A species of oyster (*Ostrea congesta*) is found in great quantities throughout the bed, and in localities *Inoceramus problematicus* is abundant. Fish remains, though consisting mostly of scales and obscure fragments, are disseminated throughout the deposit, several species of which have already been identified and described by Dr. Leidy. Entire thickness of this bed about 150 feet.

Near First Cedar Island, a very singular bed makes its appearance, superimposed on No. 3, which we shall consider as probably forming a local upper member of that formation. It extends up the Missouri River to a point near the Great Bend, a distance of about eighty miles. Lithologically it is a dull black, unctuous clay, destitute of any grit, and does not effervesce with an acid. It contains some carbonaceous matter and great quantities of selenite in crystals.]

This formation, as stated above, probably occurs on the North Branch of the Saskatchewan, at the Coal Falls.

FORMATION NO. 4 OF VERTICAL SECTION.

[This formation is the most important one in the Cretaceous system of the north-west, not only in regard to its thickness and its geographical distribution, but also in its influence on the agricultural capabilities of the country. It is only second in interest to the succeeding bed in the number, beauty, and variety of its organic remains. Commencing about ten miles above the mouth of James River, where it is seen only in thin outliers capping the distant hills or bluffs, it continues gradually assuming a greater thickness as we ascend the Missouri until reaching the Great Bend, where it monopolizes the whole region, giving to the country underlaid by it a most gloomy and sterile aspect. At the Great Bend it attains a thickness of 200 feet, and continues to occupy the country bordering on the Missouri to the mouth of Grand River, where, in consequence of the dip of the strata, it passes gradually beneath the level of the river.

After dipping beneath the water level between Grand and Cannon Ball Rivers, this formation again rises to the surface about thirty miles below the mouth of Milk River (far up towards the sources of the Missouri), by a reversed dip of the strata, from beneath the northern portion of the Great Lignite Basin, as will be seen by reference to the section on the map. Near the mouth of the Mussel Shell River it occupies the whole country for a distance of about eighty miles, and thins out upon the tops of the hills near the mouth of the Judith River.

In summing up the extent of country underlaid by this great formation, we find that south of the Lignite Basin, it occupies an area of 200 miles in length and 100 in breadth, or 20,000 square miles. North of the Great Lignite Basin, commencing at its first appearance near Milk River, we find it covering an area of 200 miles in length and 60 in breadth, or about 12,000 square miles. I have been thus particular in estimating its approximate limits and extent of surface on account of its influence on the future destiny of that region. Wherever this deposit prevails, it renders the country more completely sterile than any other geological formation I have seen in the north-west. We see from the above estimate that it renders barren over 30,000 square miles of the valley of the Missouri.

The fossils of this formation are too numerous to mention in detail. The upper and lower members appear to be exceedingly fossiliferous, while the intervening portions of considerable thickness contain only a few imperfect specimens of *Cephalopoda* and the bones of *Mosasaurus Missouriensis.* The entire thickness of this formation may be estimated at about 350 feet.]

The formation is probably more extensively developed in Rupert's Land than any other member of the Cretaceous Series.

The most easterly exposure, where it holds characteristic fossils, is on the Little Souris. Fifteen miles from the mouth of that river it consists of a very fissile, dark-blue argillaceous shale, holding numerous concretions containing a large percentage of iron. Some very obscure fossils were found in it, with fragments of *Inoceramus Canadensis.** The shale weathers ash-white; and the exposure on the Little Souris is seventy feet thick in horizontal layers.

Where the river has excavated a passage through the Blue Hills of the Souris, the rock frequently occurs in cliffs, the dip being 3° south. Fragments and perfect forms, but very fragile, of *Inoceramus Canadensis*, are very common. The ferruginous concretions are disposed in regular layers, and constitute a marked feature of the rocks of this valley.

A few miles west of the Blue Hills the dip of a very remarkable exposure of shale, with bands of ferruginous concretions, facing the south, was leveled with the utmost care, and found to be perfectly horizontal. At the base of the exposure, and on a level with the water's edge, a layer occurs full of gigantic *Inoceramus*, probably the same species as those before-mentioned. One specimen measured eight inches and a half in diameter, it was very fragile; but the peculiar prismatic structure of the shell was remarkably well preserved. On attempting to raise it, it separated into thousands of minute prisms.

A search for fossils here was more successful, and resulted in the discovery of several new species, which were examined, named, and described by Mr. Meek of Washington.

On the Two Creeks, an affluent of the Assinniboine, the same formation exists. Among the fossils collected there were *Natica obliquata*, *Avellana concinna*, *Ammonites* (sp. undt.), &c. &c.

* See figure on p. 336.

At the second, or most westerly of the Two Creeks the shales are exposed, and the following section occurs:—

Twenty feet of drift, containing fragments of the underlying shales and water-worn boulders of gneiss, &c.

Six feet of ash-grey shales, containing a stratum five inches thick of a pale-green unctuous steatitic substance.

The steatitic stratum has the following composition *: —

$Si\ O_3$	50·83
Al_2O_3	14·45
CaO	1·23
MgO	2·33
FeO	6·35
KO	0·42
HO	24·36
	99·97

On the Qu'appelle River this rock is again seen below the Big Cut-arm Creek, and near the Scissors Creek, also at White Sand River near Fort Pelly, and on Birdstail Creek east of the Assinniboine. Although no distinctive organic forms were procured, yet the lithological aspect of the rock is the same as on the Little Souris. This remark applies also to the outcrop on the Riding Mountain.

An exposure, a few miles below the mouth of the "River that Turns" on the South Branch of the Saskatchewan, contains at its base a hard calcareous sandstone, containing *Avicula linguæformis;* below it is a soft sandstone destitute of fossils. It is not improbable that the strata above the second concretionary layer pass into Formation 5 of the Nebraska section, and represent the upper Cretaceous series in this region.

Formation No. 4, as developed on the Little Souris

* This analysis was kindly furnished me by H. H. Croft, D.C.L., Professor of Chemistry in University College, Toronto.

River, is distinguished by nodular masses of clay iron-stone of great richness and in extraordinary profusion.

The locality where the clay iron-stone was first remarked occurs fifteen miles from the point where the Little Souris debouches into the Assinniboine. Here the rock consists of a dark-blue argillaceous shale with clay iron-stone layers. The nodular masses vary from eighteen inches to three inches in diameter. They are generally of a compressed elliptical form, and vary from two to nineteen inches in thickness. They form horizontal tiers in the rock, and are often separated by steel-grey and light ash-grey argillaceous bands, which also contain thin seams of clay iron-stone of irregular thickness, and not continuous for many yards.

Where the river has excavated a passage through the Blue Hills of the Souris, the nodules of clay iron-stone are seen disposed in regular layers, jutting out of the cliff and presenting different colours, varying from steel-grey to reddish-brown and light buff-brown, according to the length of time they have been exposed to the action of the atmosphere and the freedom of the ore from mechanical admixtures, such as clay, lime, and magnesia.

A few miles west of the Blue Hills of the Souris, near Plum Creek, the nodular masses of clay iron-stone were everywhere distributed in the river and on the beach. They formed, in fact, the chief constituents of the bottom of the river, and it may be said to flow in this locality over coarse clay iron-stone gravel. The section at the spot referred to, was seventy feet thick, exposed in the form of a nearly perpendicular cliff. The whole of the layers were either tinged with peroxide of iron or of a purple hue, according to the degree of humidity to which they had been exposed by drainage from the prairie or springs.

Where the clay iron-stone bands were first observed

the river was flowing in a S.W. by S. direction; this course it maintained for about thirteen miles, and then suddenly turning at right angles, the valley pursues a N.W. by W. course for twenty-six miles, it then turns to the S.W. again. For thirteen miles along the south-westerly course, and twenty-three miles along the north-westerly, course the iron-stone bands were seen whenever the river was approached, and throughout the greater portion of the distance the opposite bank and the rocks composing it were clearly visible, always preserving the same aspect, and showing numerous clay iron-stone bands. Hence we may infer that over this part of the country drained by the Souris, the clay iron-stone is an integral and important part of Formation No. 4, occurring in the form of a series of bands of nodules or as thin sheets of the ore, sometimes possessing thickness sufficient to give them great economic value, at other times thinning out altogether.

On the flanks of the Riding Mountain, about ninety miles N. from the great bend of the Little Souris, Formation No. 4 was recognized, and although in the valley of Ochre River nodular masses of clay iron-stone were seen, yet the layers were not observed forming a prominent feature in the exposure on the flanks of the precipitous and elevated N.E. escarpment of the Riding Mountain. But no search was made for them during the ascent, and snow covered the ground when we returned. From the circumstance that the nodular masses of ore were seen in the bed of Ochre River, it is not improbable that the iron bearing layers extend as far N. as the Riding Mountain. Associated with the clay iron-stone beds on the Souris is a very persistent band of the steatitic mineral already noticed (p. 328), pale greenish-yellow in colour, very unctuous and soft, and easily moulded into forms by the hand. One bank on the Little Souris is four inches thick. Others above and below it vary from half an inch, and of irregular thickness, to three inches. The soap-stone bands were noticed on a small tributary of the Little Saskatchewan, itself an affluent of the Assinniboine, where Formation No. 4 crops out half way between the great bend of the Souris and the Riding Mountain.

It is not improbable that the clay iron-stone bands extend without interruption from the Little Souris to the Riding Mountain. They would be deeply covered with drift or the superior formations between the summit of the Riding Mountain and the Assinniboine, hence, as sources of iron-ore, not available under the present circumstances of the country.

On the Little Souris the drift is very shallow, sometimes composed almost altogether of the ruins of the formation itself, hence, over wide areas in its valley the ore is easily accessible, and many thousand tons of nodules could be picked from the exposed rock or from the bed of the river, to be annually renewed by the wearing away of the surface of the cliffs. A very rapid survey showed that this annual renewal might be expected to take place, to a certain extent, for a distance of thirty-six miles on each side of the river, over an exposed surface of seventy-two miles in all, without reckoning the vast accumulations which must occupy the bends of the river where spits or beaches are formed.

It has been stated at the beginning of this chapter that the Cretaceous rocks extend from the North Branch of the Saskatchewan to the frontier of Mexico, in part flanking the Rocky Mountains. In Nebraska and Kansas their physical and geological characteristics have been more or less studied, and it would be interesting to know whether the clay iron-stone forms there a constituent of Formation No. 4.

Mr. Meek informs me that clay iron-stone concretions occur at many places in the fourth division of the Cretaceous rock in Nebraska, but he has no knowledge of its occurrence there in quantities sufficient to be considered of much value.

The Quartzose sand, highly ferruginous throughout New Jersey, is one of the representatives of Formations 4 and 5 of the Nebraska Section, probably of No. 4.

The richness of the clay iron-stones of the Little Souris may be inferred from the following comparison:—

The principal ore from which iron is smelted in England and Scotland is the carbonate of iron of the coal measures. It occurs chiefly in the form of nodular masses associated with the strata of coal. This ore yields from 30 to 40 per cent. iron of good quality.

The iron-ore of the Little Souris varies in quality, many nodular masses appearing to consist nearly altogether of carbonate of iron, and possess a semi-crystalline arrangement like the sparry iron of the unfossiliferous rocks. An approximate analysis of several specimens gave 75 to 85 per cent. of proto-carbonate of iron, which would answer to 34 per cent. and 41 per cent. of pure iron respectively.

I am indebted to Professor Croft for an exact analysis of a fragment of a nodule which he took, without making any particular selection, from my specimens.

He found 82·720 per cent. of proto-carbonate of iron, which is equal to 39·930 per cent. of iron, or very nearly 40 per cent.

The richest specimen selected by Dr. Colquhoun, from the sparry iron-ores of the Glasgow coal-field, gave 41·25 per cent. of pure iron. This differs by 1·320 per cent. from the specimen analyzed by Professor Croft, and I am convinced that a little care in selection would establish the fact that the iron-ores of the Souris would take a very high rank in point of richness in metal among the most celebrated English and Scotch clay iron-stone ores.

Subjoined is Professor Croft's analysis: —

Clay iron-stone from the Little Souris.

FeO, CO_2	82·720
CaO, CO_2	6·705
MgO, CO_2	trace
Insoluble residue	8·181
Phosphoric acid and loss . . .	2·394
	100·000

The following specimens are from Formation No. 4 of the vertical section in Rupert's Land.

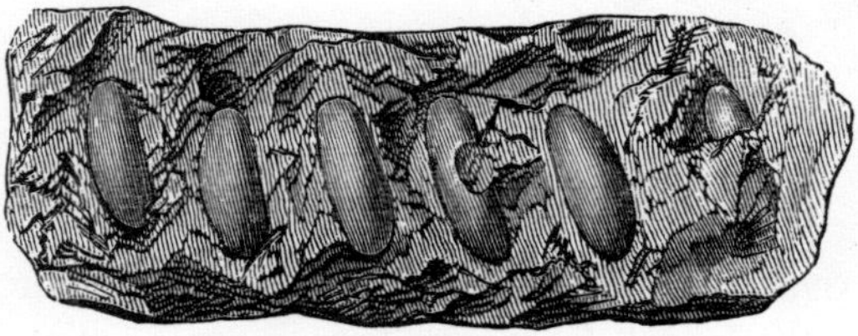

Curious Spiral Fossil from the Two Creeks.

[Several impressions, apparently of the stems of marine plants, occur in the specimens from the locality on the Assinniboine, near the mouth of the Two Creeks.

Along with the above there are also specimens of a very curious spiral body, differing from any fossil I ever before met with. It is a long, slender, slightly flattened, or sub-cylindrical body, measuring in every part of its length about 0·18 inch in its greatest diameter, and very regularly coiled into a spiral form, the turns being widely disconnected, like those of a cork-screw. Each turn measures about 0·58 inch across, and there are five turns in a length of 2·15 inches. It is smooth, and shows no organic structure under a common pocket lens, the organic matter having been replaced by the fine sediment of which the matrix is composed. Unless these are the tendrils or root-like appendages by which some floating plant clung to marine bodies, I can form no conjecture in regard to their nature. (See p. 333.)]

Anomia Flemingi. (*N. S.*)

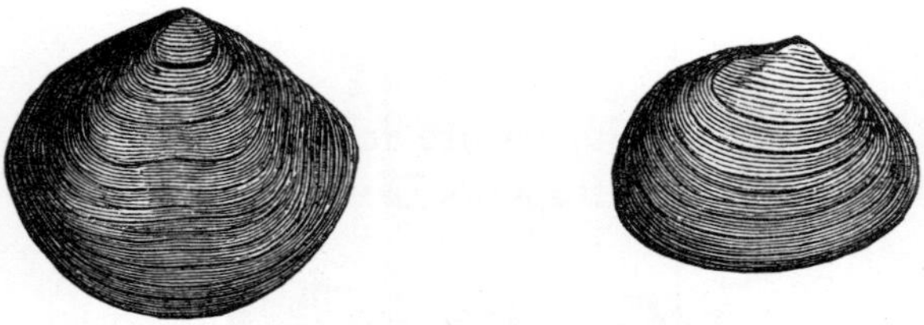

Two different specimens of this species. The fine concentric lines representing shading.

[Shell oval or sub-circular, compressed plano-convex, extremely thin and fragile. Lower valve flat and apparently more nearly circular than the other. Upper valve depressed, convex, rounded in front, and more broadly and less regularly rounded on the ventral side; posterior margin obliquely subtruncate from the dorsal side, rather abruptly rounded, and waved so as to form a broad very obscure fold at its connexion with the ventral margin; beak small, compressed, located near the middle of the cardinal edge, but not projecting beyond it. Surface marked by small obscure lines of growth. Length 1·10 inches; breadth from beak to opposite side, 1 inch.

In Formation No. 4 of the Cretaceous beds in Nebraska there is a species something like this, which Dr. Hayden and I have described (but not yet published) under the name of *A. subtrigonalis*. The species now before me, however, is much more compressed, and more rounded in outline. It differs from *A. tellinoides* of Morton (Synop. Org. Rem., p. 61, pl. v. fig. 11), in being straighter on the cardinal side, and in having the umbo of the upper valve much less prominent and gibbous. Named after Mr. John Fleming, one of the gentlemen connected with the Saskatchewan expedition.

Locality and position.—Little Souris River, in soft lead-grey argillaceous rock, or indurated clay, probably of the age of the fourth division of the Cretaceous Series in Nebraska.]

Inoceramus Canadensis. (*N. S.*)

[Shell broad, oblong-oval, compressed, apparently very nearly equi-valve; anterior side rounded; posterior side longer and more broadly rounded or sub-truncate; base forming a semi-oval curve; hinge straight, of medium length; beaks small, compressed, scarcely rising above the hinge line, located near the anterior side, not very oblique; surface ornamented by small, obscure, irregular concentric undulations, and fine closely arranged rather indistinct lines of growth, which are generally only seen on the outer fibrous layer. Length of largest specimen about 3·35 inches; height near 2·80 inches.

The specimens of this species in the collection are imperfect, but retain enough of its characters to show it is distinct from any of the known species in the Nebraska formations. It resembles somewhat *I. Sagensis*, Owen, (Report, Wisconsin, Iowa, and Minnesota, tab. vii. fig. 3,) but is much more compressed, and longer in proportion to its height.

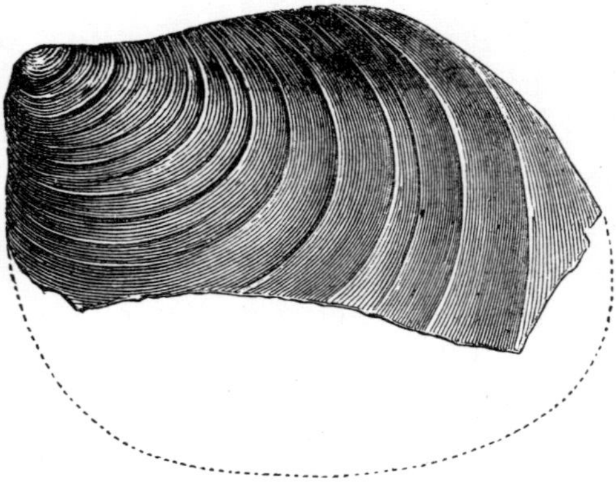

INOCERAMUS CANADENSIS.—Left valve.

It also bears some resemblance to *I. regularis*, D'Orbigny,

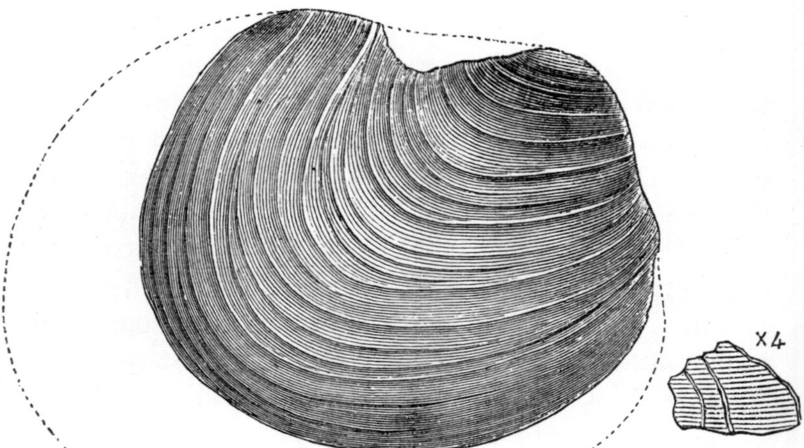

Right valve. The finer concentric lines represent shading. The small figure × 4 shows the fine striæ seen on the outer fibrous layer of the shell, magnified four times.

(Pal. France, T. III. pl. 410,) but is not near so deeply rounded on the ventral border, and is more compressed.

Locality and position.—Same as last.]

Leda Hindi. (N. S.)

Natural size.

The same enlarged.

[Shell small, sub-ovate, compressed; anterior side narrowly rounded; pallial border forming a broad semi-oval or semi-ovate curve, not crenulate within; posterior side a little longer than the other, much compressed, distinctly sinuous below, and provided with a narrow, short, obtusely pointed rostriform extension above; umbones depressed, located a little in advance of the middle; hinge having about twelve teeth in front of the beaks, and probably more behind; surface ornamented by distinct, regularly arranged, rather strong concentric lines. Length 0·35 inch; height 0·18 inch.

This is a very neat little shell, which will be readily distinguished from any of the species yet known in the Nebraska Cretaceous rocks by the distinct sinus in its postero-ventral margin. Even where the border is broken away, the curve of the concentric lines will always show that the sinus did exist in its margin.

The specimen does not show the pallial line, but in form and general appearance the shell is more like *Leda* than *Nucula;* it may, however, possibly belong to the latter genus.

The specific name is given in honour of Professor Henry Y. Hind, of Trinity College, Toronto, in charge of the Assinniboine and Saskatchewan Exploring Expedition, to whose zeal and industry we are indebted for much interesting information respecting the geology and topography of the country explored.

Locality and position.—Little Souris River, from an equivalent to No. 4 of the Nebraska section.]

Scaphites Nodosus [?] *Var.*

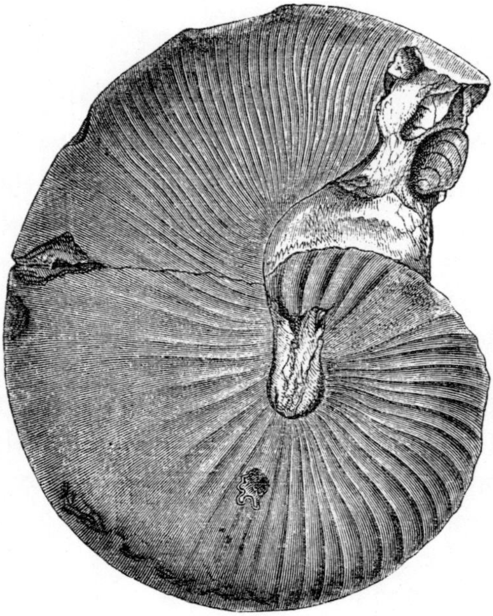

SCAPHITES NODOSUS.—Side view.

[*Scaphites nodosus* [?]—Owen, 1852. Rept. Iowa, Wisconsin, and Minnesota.

Locality.—South Branch of the Saskatchewan, from an equivalent of Formation No. 4 of Nebraska Section.]

Ammonites.—Sp. undt. (fragments).

[*Locality and position.*—Two Creeks, Eq. No. 4 of the Nebraska Cretaceous series.]

Natica Obliquata.

[*Natica obliquata.*—Hall and Meek, Mem. Acad. Arts and Sci., Boston, vol. v. N. S. p. 384, pl. iii. fig. 1.

Locality and position.—Two Creeks, on the Assinniboine; in bed representing Formation No. 4 of the Nebraska Cretaceous series.]

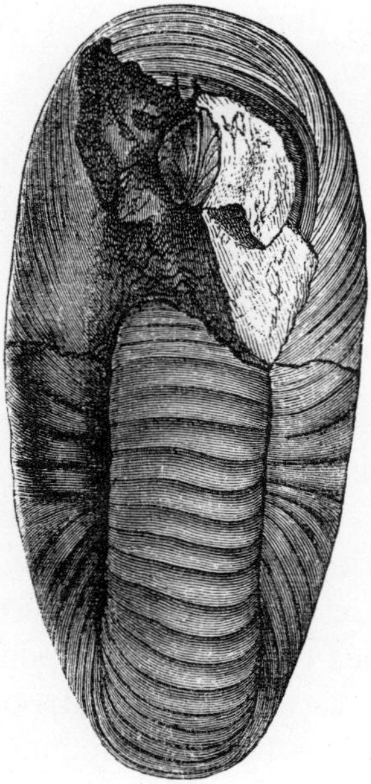

Front view of Scaphites Nodosus. The fine lines represent shading.

The specimen figured shows the remains of two rows of tubercles on the dorsum —one on each side. They are much worn, and have been unfortunately omitted altogether by the artist.

Avellana Concinna.

[*Actæon concinna.*—Hall and Meek, Mem. Am. Acad. Arts and Sci., Boston, vol. v. N. S. p. 388, pl. ii. fig. 6.

The specimen of this species, first figured in the paper above cited, is either a young individual, or the outer lip was broken away; for that now before me, which is evidently the same species, has a strong thickened outer lip; consequently, it cannot be a true *Actæon*, but agrees more nearly with the characters of the genus *Avellana*.

Locality and position.—Same as last.]

Ammonites Placenta.

[*Ammonites placenta.*—Dekay, N. Y. Lyc. Nat. Hist., vol. ii. pl. v, fig. 2; Journ. Acad. Nat. Sci., Phila., vol. vi. p. 88, &c.; Morton, Synop. Org. Rem., p. 36, pl. ii. figs. 1 and 2.

Locality and position.—South Branch of the Saskatchewan, from an equivalent of Formation No. 4 of the Nebraska Cretaceous series.]

FORMATION NO. 5 OF VERTICAL SECTION.

This very interesting bed, though differing lithologically from the preceding one, contains many of the same species of fossils. It is worthy, however, of a distinct position in the series, not only from its extent, thickness, and difference of composition, but also from the more favourable influence that it exerts upon the country underlaid by it. In ascending the Missouri River it first makes its appearance near the mouth of Grand River, about 150 miles above Fort Pierre. Near *Butte aux Grès* it becomes quite conspicuous, acquiring a thickness of eighty or one hundred feet, and containing great quantities of organic remains. Here it forms an extension of what is called Fox Ridge, a series of high hills having a north-west and south-east course, crossing the Missouri River into Minnesota at this point. Its north-eastern limits I have not ascertained. In its south-western extension it continues for a considerable distance nearly parallel with the Missouri, crosses the Moreau River about thirty miles above its mouth, then forms a high dividing ridge between the Moreau and Sheyenne Rivers, at which locality it first took its name. Continuing thence its south-westerly course, it crosses the Sheyenne, and is seen again in its full thickness at the heads of Opening

Creek and Teton River, forming a high ridge from which tributaries of the Sheyenne and Teton take their rise. The little streams flowing into the Sheyenne have a north-westerly course, while those emptying into the Teton take a south-easterly direction. We thus find that this bed underlies an area of about 200 miles in length and fifty miles in breadth, or about 10,000 square miles.

The general character of Formation No. 5 is a yellow arenaceous and argillaceous grit, containing much ferruginous matter, and in localities a profusion of molluscous fossils. It forms a much more fertile soil, more hearty and luxuriant vegetation, sustains a finer growth of timber than Formation No. 4, and abounds in springs of good water.

Like No. 4, this bed yields a great abundance of quite perfect and well-preserved organic remains. Many of the species approximate so closely to Tertiary forms, that did we not find them everywhere associated with Ammonites, Scaphites and other genera which are not known to have existed later than the Cretaceous epoch, we should at once pronounce the formation in which they occur, Tertiary. The whole thickness of this bed is estimated at 100 to 150 feet.]

The rocks at La Roche Percée near the boundary line (long. 103° 59′), with their associated coal beds, described by Dr. Hector, are probably of the age of No. 5, as well as those seen by the same gentleman on the west side of the outer range of hills close to the main chain of the Rocky Mountains.

An exposure of this formation occurs on the Eyebrow Hill stream, where it joins the Qu'appelle Valley. A ferruginous clay in yellow and red layers reposes on a hard greenish-coloured sandstone, seamed with veins of selenite, and containing huge concretions. No fossils were found in the rock.

The upper part of the section on the South Branch, containing concretions full of *Avicula Nebrascana*, is doubtless the representative of No. 5 in this region.

[Several of the specimens obtained near Sand Hill Lake on Qu'appelle River, and the South Branch of the Saskatchewan, are from a green sandstone, which is more indurated, but in other respects more like the green sands of New Jersey, than any I have before seen from north-western localities. In some of these, there are great numbers of *Avicula linguæformis* and *A. Nebrascana*, the first of which occurs in both Nos. 4 and 5 of the Nebraska section, but is more abundant in the latter; and the other is nearly or quite restricted to No. 5, where the two formations are not blended as is sometimes the case. As this rock differs entirely in its lithological characters from Formation No. 4,—while No. 5 is often highly arenaceous, and sometimes assumes a slight greenish tinge, at the higher northern localities in the Upper Missouri country,—the probability is that it represents No. 5, or the most recent member of the Cretaceous Series of the north-west.]

Among the specimens procured from the South Branch were *Scaphites Conradi*, *Nautilus Dekayi*, *Avicula linguæformis*, *Avicula Nebrascana*, *Rostellaria Americana*.

Avicula Linguæformis.

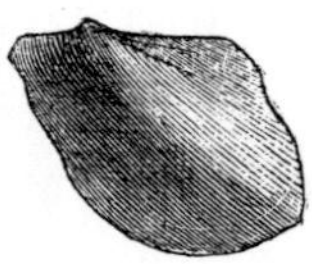

An imperfect specimen of this species. The lines represent shading.

[*Avicula linguæformis*, Evans and Shumard, Proceed. Acad. Nat. Sci. Phila., vol. vii. p. 163.

Locality and position.—Sandy Hills, South Branch of the Saskatchewan. Height of Land in the Qu'appelle Valley, near the Elbow of South Branch of the Saskatchewan.—Upper Cretaceous.]

Avicula Nebrascana.

Two specimens of this species.

[*Avicula Nebrascana*, Evans and Shumard, Trans. Acad. Sci. St. Louis, vol. i. p. 38.

Locality and position.—South Branch of the Saskatchewan.—Upper Cretaceous.]

Nautilus Dekayi.

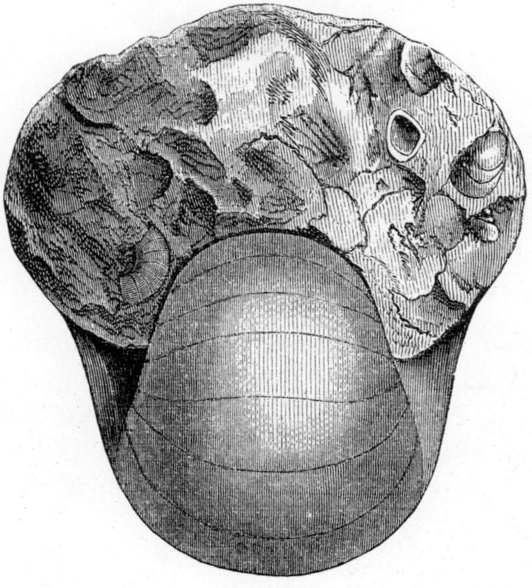

Front view.

Nautilus Dekayi.—Morton, 1834. Synop. Org. Rem., pl. viii. fig. 4, and pl. xiii. fig. 4.

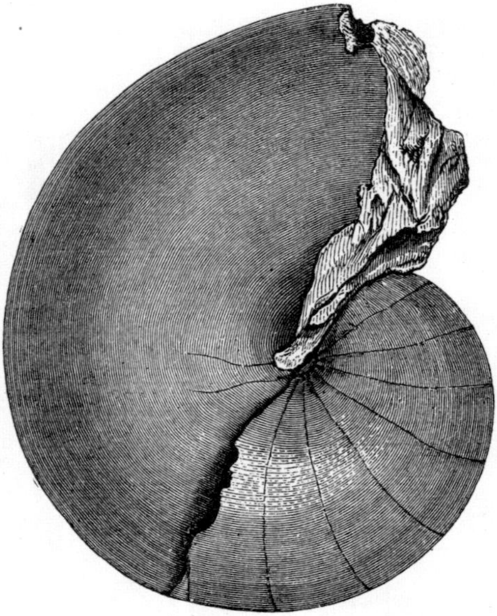

Side view of Nautilus Dekayi.

Locality and position.—South Branch of the Saskatchewan.—Upper Cretaceous.

Rostellaria Americana.

Rostellaria Americana.—Evans and Shumard, Trans. St. Louis Acad. Sci., vol. i. p. 42.

Locality and position.—South Branch of the Saskatchewan.—Upper Cretaceous.

Scaphites Conradi.

Ammonites Conradi.—Morton, 1834. Synop. Org. Rem., p. 39, pl. xix. fig. 4.

Scaphites Conradi.—D'Orbigny, 1850. Prodromus, p. 214.

Ammonites Nebrascensis, &c. — Owen, 1852. Rep. Iowa, &c.

Scaphites Conradi.—Meek and Hayden, 1856. Acad. Nat. Sci. Phila., p. 281.

Locality and position.—South Branch of the Saskatchewan.—No. 5, Nebraska Section, or most recent Cretaceous.

TERTIARY FORMATIONS.

No evidence of Tertiary rocks in position east of the South Branch of the Saskatchewan was obtained during the exploration.

On the crest and abrupt sides of the Riding, Porcupine, and Thunder Mountains, the Indians affirm that beds of Lignite exist, a statement rendered probable by the occurrence of worn fragments in the drift of the valleys of the rivers flowing from those eminences.

But this does not establish the existence of Tertiary rocks on these elevations or their flanks, for Dr. Hector found at La Roche Percée, near the flanks of the Grand Coteau, Lignite bands in a formation which much resembles Formation No. 5 of the Nebraska section, and from the occurrence of the underlying formation on the Riding Mountain, it is not improbable that Formation No. 5 is found in position there with associated beds of Lignite.

The sand dunes, which form so distinguishing a feature near the Elbow of the South Branch, may have been derived from Tertiary sandstones formerly overlying the upper Cretaceous rocks in that vicinity. In a foot-note on page 139 of the Geology and Palæontology of the Mexican Boundary Line, Prof. James Hall says that the drifting sands of the south-west, like those of the north, appear to be derived from the sandstones of the Tertiary period.

In the Upper Saskatchewan, Tertiaries containing Coal or Lignite are well developed; also on the flanks of the Rocky Mountains. They are mentioned in the following notices of the Lignites of the north-west.

TERTIARY COAL, OR LIGNITE.

Although the Lignites are not generally available for economic purposes, yet some seams sufficiently pure for use are known to exist in the great Lignite basin of the Upper Missouri and Saskatchewan. A brief notice of the character of this important material as it occurs in the Tertiary rocks of the North-West, will enable a tolerably accurate judgment to be formed of its probable value as a source of fuel in Rupert's Land.

The great Lignite Basin of the Missouri extends from the 100th to the 108th degree of west longitude, and from the 45th degree of north latitude to an undescribed limit, probably through the Valley of the Saskatchewan to the Valley of the Mackenzie.

Dr. Hayden, who traced the great Missouri formation up that river for a distance of 600 miles, and up the Yellowstone for 300 miles, considers that the fossils obtained from it show conclusively, that it possesses the mixed character of a fresh water and estuary deposit, and that it cannot be older than the Miocene period. It is composed of clays, sands, sandstones, and Lignites. The extent of country known to be occupied by this basin, as it occurs on the Missouri and its tributaries, exceeds 60,000 miles. The beds of Lignite in this extensive formation vary in thickness as well as in purity at different localities. On the Yellowstone they are found seven feet in thickness. At Fort Berthold on the Missouri a two-foot bed is pure enough to be used as fuel.*

Governor Stevens states, in his Report of the Exploration of a route for the Pacific Railway, that Lignite has been traced from the Coulées of the Mouse (Little Souris) River to the head waters of Milk River, a distance of 500 miles, *apparently* underlying the whole of that extensive district of country, with a thickness of bed varying from a few inches to six feet; he regards it as a source of fuel not to be overlooked.†

The Coal or Lignite at Edmonton occurs in three or four beds, the principal of which is from four to six feet thick. Dr. Hector says that it is of very inferior quality, burns with no flame, but rather smoulders away, leaving a plentiful ash. It is used in the forge at the Fort, and found to answer tolerably well. At Rocky Mountain House, Dr. Hector found the beds containing Tertiary coal exposed on the banks of the Saskatchewan, but no bed exceeded two feet in thickness. The coal-bearing strata are exhibited more or less all the way from Rocky Mountain House to Fort Edmonton, 211 miles by the river. Four miles below this place the coal was seen by Dr. Hector for the last time in descending the river on the ice, and eighty miles below the Fort the associated beds disappear, and upper Cretaceous rocks come to the surface conformably.‡

* Page 9. Remarks on the Tertiary and Cretaceous Formations of Nebraska, &c.. by F. B. Meek and F. V. Hayden, M.D.

† Pacific Railway Reports, vol. i. p. 95.

‡ Blue Book.

An extensive deposit of Tertiary coal was found by the same gentleman on Red Deer River, in lat. 52° 12′ N. long. 113° W. associated with the same sands and clays as at Edmonton. One group of three beds measured twenty feet in thickness, of which twelve feet was pure coal, the remainder being carbonaceous clays. This deposit on Red Deer River extends over fourteen miles.

On the South Branch they are said to exist, by Sir Alexander Mackenzie, in long. 116° W.; but as the country between the Elbow and the mouth of Bow River is still in part a *terra incognita*, it is not improbable that important Lignite beds may be found much further east than the longitude specified by that illustrious traveler.*

At Nanino, Vancouver's Island, lignite beds, long conjectured to be of Tertiary age, but now known to be Cretaceous (Dr. Hector), have been worked to some extent for the San Francisco market, and to supply steamers which touch there.†

Some years since Dr. Evans, U. S. Geo., described the coal of Oregon and British Columbia. He says, "These coals do not belong to the true coal measures but to the Tertiary period; they have, however, been altered by volcanic action. The Bellingham Bay coal particularly, in consequence, is of a remarkable crystalline structure, and presents under the magnifier a very singular and beautiful appearance. It will produce excellent coke, and is well-suited to manufacturing and domestic purposes. It burns freely and although rather light for long sea voyages, unless the construction of furnaces should be

* Foot note, page 110, Am. Ed. Sir John Richardson's Arctic Searching Expedition.

† Pacific Railway Report, vol. vi., Geological Report.

changed, lessening the draft, is suitable for river navigation. The coal crops out at various points from the British line to near Port Oxford in Oregon, and is accessible to sail and steam navigation, and almost inexhaustible in quantity. These coals with imperfect machines and facilities for mining can be delivered ready for shipment at from 2$ to 3$ per ton."

Beds of Lignite are found at many places in Texas. A bed four feet thick occurs on the Colorado, near Bastrop, beneath a layer of Eocene fossils. This mineral is also found on the Brazos*, and exists in abundance on the Rio del Norte, the river forming part of the boundary line between the United States and Mexico. Some specimens in the last-named locality are so bituminous as to be of no use in the blacksmith's forge, where it runs together and becomes baked into a solid mass. Seams of Lignite three to four feet thick are exposed on Elm Creek, a tributary of the Del Norte, and have been used and found valuable in a blacksmith's forge. This Lignite occurs in Cretaceous formations.

The occurrence of extensive Lignite beds in the valley of the Mackenzie is well-known, and from the foregoing enumeration of their distribution in Tertiary and Cretaceous formations, it is evident that more or less continuous areas of this important mineral are to be found, along the flanks of the Rocky Mountains, from Mexico to the Arctic Sea.

In Europe it is well-known that Tertiary Lignite deposits possess considerable economic value. They are worked in France, Germany, and Switzerland. In England, the Lignites of Devonshire, associated with beds of clay, are

* Blake, Pacific Railway Reports, vol. ii.

about seventy feet thick. The strata of Lignite coal near the surface vary from eighteen inches to four feet in thickness, separated by beds of brownish-clay of about the same dimensions. The lowermost stratum of Lignite coal is sixteen feet thick.*

* Phillips.

CLIMATE

OF THE

SOUTHERN PART OF RUPERT'S LAND

CHAP. XLI.

Climate of the Laurentides and the Prairies. — Frozen Lakes. — Mean Annual Temperature.—Arid and Humid Region.—Sources of Humidity. —Aridity West of the 98th Meridian.—Mississippi Valley.—Arid Region of the United States.—Humid Region of the Valley of Lake Winnipeg.— Causes of.—Elevation of the Country.—Humid Pacific Winds.—North-easterly Current.—The Arid Region.—Prevailing Winds.—Source of the Humidity.—Rocky Mountain Plateau.—Depression in.—Table of Elevation of Plateau and Passes.—Hail Storms.—Thunder Storms in 1858.— Progress of Dunes.—Summer Surface Wind.—Meteorology of Red River. —Winter Temperatures. — Winter Temperatures at Montreal. — Cold Terms.—Quebec Temperatures.—Climate of the South Branch of the Saskatchewan.—Limit of permanently frozen Soil.—Growth of Forests. —Tail of the Prairies.—Prairies Converted into Forest Land in Missouri. —Character of the Great Plains in the United States.—Major Emory's Statement. — Auroras. — October 2nd. — October 27th. — The Twilight Bow.—Indian Summer.

THE climates of Canada and Rupert's Land under the same parallels of latitude, vary to a considerable extent with the rock formations of the country. Throughout the undulating region of the Laurentides, between Lake Superior and Lake Winnipeg, the proportion of water to dry land is about one to two, not collected into one large water area, but distributed over the surface of the country in the form of countless thousands of lakes, ponds, and marshes. The intense cold of winter is sufficient to solidify the deepest lakes for a depth of several feet, and the thawing of so much ice in spring has the effect of absorbing and rendering latent the heat which would be otherwise expended in warming the soil and advancing vegetation.

Lakes Winnipeg, Manitobah, and Winnipego-sis, together with the smaller lakes belonging to the Winnipeg basin, are deeply frozen every winter; ice often remains in their northern extremities until the beginning of June and greatly retards the progress of vegetation on their immediate shores. The country is clothed with forests giving rise to swamps where water accumulates. The prairies are fully exposed to solar radiation and soon part with their superabundant humidity. Hence one reason that, north of the 47th or 48th parallel, the mildness of the seasons increases rapidly as we advance towards the west, after leaving Red River. The improvement arises not only from greater longitude, but is influenced by the character of the rock formations by which the country is underlaid and surrounded. The soil of the prairies is in general dry and is rapidly warmed by the rays of the sun in spring. The prairies enjoy too, north of the 58th parallel, the genial, warm and comparatively humid winds from the Pacific, which are felt as far north as the latitude of Fort Simpson.*

The mean annual temperature of 40°, as determined by the Smithsonian Institution, passes through Canada and Lake Superior, curves northward and leaves the United States for British America at about the 103rd meridian, crossing the South Branch of the Saskatchewan north of the Elbow.

The country embraced within the limits of our explorations may be divided into two regions in relation to climate; the arid and the humid region. The vast treeless prairie west of the Little Souris lies within that part of the area which receives comparatively a small annual

* Colonel Lefroy; Meteorological observations at Lake Athabasca and Fort Simpson, p. 139.

rain-fall. Its northern limit is roughly shown by the Qu'appelle Valley, or more accurately by an imaginary line drawn from the fishing lakes to the Moose Woods. North and east of this area the precipitation is considerably greater, and supplies the valley of the main Saskatchewan, the Touchwood Hill Range, and the Valley of the Assinniboine with an abundance of moisture, which is protected and treasured by forests.

The valley of Red River east of the Little Souris, or the 101st degree of longitude, receives much humidity from the moist winds coming from the Gulf of Mexico up the Valley of the Mississippi, and over the low height of land which separates the waters of Red River from those of the St. Peter.

The Touchwood Hill Range and the country generally north of the Qu'appelle Valley, and in an easterly direction towards and beyond Lake Winnipeg, are made humid by the south-west Pacific wind, in concurrence with the prevailing east wind of this region. These phenomena are referred to in detail in succeeding paragraphs.

The cause of the aridity and unfitness for settlement of fully one-third of the United States, has been ably discussed by distinguished meteorologists. This remarkable feature, extending over a portion of the American continent within the limits of the United States, of more than 1,000,000 square miles in area, is highly important in relation to the valley of the South Branch of the Saskatchewan, to a large part of which the same peculiarity belongs. The physical geography of the arid region in the United States has been very admirably described by Dr. Joseph Henry.*

"The general character of the soil between the Missis-

* Meteorology in its Connexion with Agriculture, by Professor Joseph Henry, Secretary of the Smithsonian Institution.

sippi River and the Atlantic is that of great fertility, and as a whole, in its natural condition, with some exceptions at the west, is well supplied with timber. The portion also on the western side of the Mississippi, as far as the 98th meridian, including the States of Texas, Louisiana, Arkansas, Missouri, Iowa, and Minnesota, and portions of the territory of Kansas and Nebraska, are fertile, though abounding in prairies and subject occasionally to droughts. But the whole space to the west, between the 98th meridian and the Rocky Mountains, denominated the Great American Plains, is a barren waste, over which the eye may roam to the extent of the visible horizon with scarcely an object to break the monotony.

"From the Rocky Mountains to the Pacific, with the exception of the rich but narrow belt along the ocean, the country may also be considered, in comparison with other portions of the United States, a wilderness unfitted for the uses of the husbandman; although in some of the mountain valleys, as at Salt Lake, by means of irrigation, a precarious supply of food may be obtained sufficient to sustain a considerable population, provided they can be induced to submit to privations from which American citizens generally would shrink. The portions of the mountain system further south are equally inhospitable, though they have been represented to be of a different character. In traversing this region, whole days are frequently passed without meeting a rivulet or spring of water to slake the thirst of the weary traveler.

"We have stated that the entire region west of the 98th degree of west longitude, with the exception of a small portion of western Texas and the narrow border along the Pacific, is a country of comparatively little value to the agriculturist; and, perhaps, it will astonish the reader if we direct his attention to the fact that this line, which passes southward from Lake Winnipeg to the Gulf of Mexico, will divide the whole surface of the United States into two nearly equal parts. This statement, when fully appreciated, will serve to dissipate some of the dreams which have been considered as realities as to the destiny of the western part of the North American continent. Truth, however, transcends even the laudable feelings of pride of country ; and, in order properly to direct the policy of this great confederacy, it is necessary to be well acquainted with the theatre on which its future history is to be enacted and by whose character it will mainly be shaped."

HUMID REGION OF THE VALLEY OF LAKE WINNIPEG.

Prominent among the causes which tend to give humidity, together with an elevated spring and summer temperature, to a part of the valley of Lake Winnipeg, there may be noticed:—First, the comparatively low elevation of the country above the sea level. The prairies of Red River within British territory are not more than 730 feet above the ocean. Those on the South Branch of the Saskatchewan, at the Elbow, do not exceed 1600 feet, and the mean elevation of the country between the South Branch and the Riding Mountain is only 1200 feet above the same level. The influence of the south winds from the Gulf of Mexico on the Red River valley has already been noticed.

Second, the influence of the warm westerly winds from the Pacific Ocean in connexion with the prevailing north-east wind, which is one of the established physical phenomena of this part of British America. It would appear, at first sight, that the snow-capped ridges of the Cascade, Blue, and Rocky Mountains would abstract so much heat from the warm westerly winds coming from the Pacific Ocean, as to neutralize their influence upon the winter and spring temperature of a large part of the country drained by the Saskatchewan. Such, however, is not the case; and happily for the purpose of practically substantiating this apparent anomaly, we have indisputable testimony.

In the magnetical and meteorological observations at Lake Athabasca and Fort Simpson, by Colonel Lefroy, R. A., we find the following important observation, in relation to the phenomena of Pacific winds affecting the climate of the northern regions :—

"A local phenomenon of interest was observed several times at Fort Simpson (lat. 61° 51′ 7″ N., long. 15° 5′ 40″ W.), 460 miles from Sitka (1800 geo. miles from Toronto), in the rapid rise of the temperature of the air, when the wind changed to the south-west from an easterly direction. It appeared as if the warmer air of the Pacific Ocean were transferred across the neighbouring ridges of the Rocky Mountains with little loss of its temperature."

Much of the precipitation in the humid region is due to the Pacific winds, which are not so completely deprived of their moisture in traversing the Rocky Mountain Ranges as in lower latitudes, where the average altitude of these ranges is much higher, and the eastern slope of the mountain of a *far greater mean annual temperature.*

The prevalent winds at Toronto and Lake Athabasca belong, as shown by Colonel Lefroy *, to different and nearly opposite systems. A *north-westerly* current preponderates in the lower latitudes (43° 39′), a *north-easterly* current, inclined at an angle of about 117° prevails

* Magnetical and Meteorological Observations at Lake Athabasca.

in the higher one (58° 45′). Between these latitudes is a region of calm or of variable winds; and there can be no doubt that the north-easterly current materially affects the humidity of the climate of Rupert's Land north of the 50th parallel. The prevalence of north and north-easterly winds during the winter months occasions a great precipitation of snow throughout the humid region. In the Touchwood Hill range, snow not unfrequently accumulates in the woods, where it is undisturbed by winds, to the depth of two feet; on the Riding and Duck Mountain the precipitation is also large, and throughout the humid region very much in excess of the precipitation in lower latitudes.*

Forty-eight inches of rain and thirty-nine inches of snow were registered by Mr. Gunn near the Stone Fort, Red River, between June 1st, 1855, and May 31st, 1856. The precipitation at Toronto during the same period was thirty inches of rain and seventy-two of snow, giving an excess of humidity to the climate of the Red River Settlements, when compared with Toronto, which is represented by fourteen inches, a quantity exceeding the annual precipitation over the greater portion of the eastern flank of the Rocky Mountains south of the great Missouri bend.

The arid region, or Great Plains, west of the 101st degree of longitude, receive a very small amount of precipitation from the humid south winds coming up the Valley of the Mississippi from the Gulf of Mexico. It is too far south to be much affected by north-east winds, or the westerly winds from the Pacific. This vast treeless prairie forms, in fact, the northern limit of the great arid region of the eastern flank of the Rocky Mountains; but

* Compare Lorin Blodget's Rain Charts.

still its humidity is greater than that of the plains south of the Missouri, in consequence of its high northern latitude.

All the thunder-storms we encountered in 1858 in the valley of Lake Winnipeg, came from the west, south-west, or north-west, with one exception. I do not find a single record of thunder-storms with heavy rain coming from the south. This may have been an exceptional year, but the warmth and dryness, often oppressive, of the south wind, west of the 100th degree of longitude, contrasted strongly with the humidity and coolness of winds from the west. This phenomenon is directly opposed to those which prevail in lower latitudes, and may probably be explained as follows: —

Warm air from the Pacific, loaded with moisture, passes at certain periods of the year over the whole range of the Rocky Mountains in British America and in the United States. These Pacific winds occasion but a very small precipitation of rain or snow on the eastern flank of the Rocky Mountains, south of the great Missouri Bend. Similar winds from the Pacific do occasion a considerable precipitation in the northern part of the Saskatchewan Valley. Whence, then, this apparent anomaly? It probably arises from the difference in the temperature of the two regions, the direction of the prevailing winds, and the lowness and comparatively small breadth of the Rocky Mountain ranges in that latitude. In spring and summer, warm westerly winds laden with moisture, in passing over the mountain range south of, say the 46th parallel, are cooled to a certain temperature, and precipitate the greater portion of their moisture in the form of rain or snow upon the mountain ridges. On arriving at the eastern flank of the Rocky Mountains, their temperature rises to that of the region over which they pass, being elevated by the deposition of their moisture and continually increasing density as they descend; but the capacity of air for moisture is well known to be dependent upon its temperature within certain limits; hence the westerly Pacific winds become more warm and more dry as they descend the Eastern Rocky Mountain slope, until they meet the moist winds from the Gulf of Mexico, passing up the Valley of the Mississippi, towards and through the region of the great Canadian Lakes, and over the low height of land separating the waters flowing into Lake Winnipeg from the Mississippi Valley.*

In the latitude of the Valley of the Saskatchewan, however, the moist south-west winds from the Pacific find a depression in the Rocky Mountain range and a low contracted plateau; they lose less humidity than winds passing over the higher and broader ranges to the south, and meet with a prevailing north-easterly wind as they begin to descend their eastern flanks; their temperature is consequently lessened instead of being elevated, and their capacity for moisture diminished; hence, precipitation in the form of rain and hail takes place in descending the slope towards Lake Winnipeg.

There is no doubt that the south-west Pacific winds, passing through the depression in the Rocky Mountains near the 49th parallel, and over the narrow plateau on which they rest, without losing the whole of their moisture, give humidity to the large portion of Rupert's Land they traverse.

The great plateau on which the Rocky Mountain ranges rest, has an average elevation of 4000 feet near the 32nd parallel of latitude, the lowest pass in the most easterly range being there 5717 feet above the ocean. Along the

* Meteorology in its Connexion with Agriculture, by Professor Joseph Henry.

35th parallel the vertical section across the mountain system is of greater width and elevation. The mean height above the ocean is about 5500 feet, and the lowest pass 7750 feet. Between the 38th and 40th parallel the section has an elevation of 7500 feet, and the lowest pass is 10,032 feet above the level of the sea. Beneath the parallel of 47° the base of the plateau is narrow, and has an average altitude of 2500 feet, the lowest pass being 6044 feet above the ocean.* Within British territory north of the 49th parallel the passes in the eastern range are still lower. The recent measurements by Captain Palliser's Expedition show that the height of the Kutanie Pass in latitude 49° 30′ is nearly 6000 feet above the sea level; the Kananaski Pass 5985 feet, and the Vermilion Pass, traversed by Dr. Hector, in latitude 51° 10′, only 4944 feet above the ocean.

The following table exhibits the elevation of the Rocky Mountain plateau, and the height of the lowest Passes above the ocean:—

Table showing the elevation and breadth of the Plateau on which the Rocky Mountain ranges rest, and the height above the Ocean of the lowest Passes, from the 32*nd parallel to the* 51*st parallel, North Latitude.*

	Breadth and Elevation of Plateau between the 32nd and 49th Parallels.						
	3000 and 4000 feet.	4000 and 5000 feet.	5000 and 6000 feet.	6000 and 7000 feet.	7000 and 8000 feet.	8000 and 9000 feet.	9000 and 10,000 feet.
	miles.	*miles.*	*miles.*	*miles.*	*miles.*	*miles.*	*miles.*
32nd parallel . .	170	503	60				
35th ,, . .	185	160	303	235	95		
38th and 39th parallel.	143	725	284	110	155	80	20
41st and 42nd . .	160	580	285	270	107	20	
47th and 49th† . .	130	97	28				

* Dr. J. Henry, "Meteorology in its Connexion with Agriculture."

† Pacific Railroad Explorations.

Summit of the lowest passes above the Ocean from the 32nd to the 51st parallel, North Latitude.

	Feet.		
32nd Parallel	5,717		
35th ,, ...	7,472		
38th and 39th Parallel . .	10,032		
41st and 42nd ,, . .	8,372		
47th and 49th ,, . .	6,044		
Kutanie Pass, lat. 49° 30′ . .	6,000	nearly	Passes discovered by Captain Palliser's Expedition.*
Kananaskis Pass, north of 49th par.	5,985		
Vermilion Pass, lat. 51° 10′ .	4,944		

Not only has the depression in the Rocky Mountain range, north of the 47th parallel of latitude, a remarkable effect upon the climate of the Valley of the North Saskatchewan, but its bearing upon means of communication between the Atlantic and Pacific slopes of the Rocky Mountain ranges, is of the greatest importance.

Hail-storms are not unfrequent during the summer months in Rupert's Land, and the prairies sometimes retain the records of their occurrence for many weeks. On the Grand Coteau de Missouri, hail-storms are so violent that the stones have been known to penetrate the buffalo-skin tents of the Indians who hunt on that elevated plateau. The thunder-storms of 1858 are given in the annexed table.

The progress of dunes affords a very excellent indication of the direction and force of prevailing winds. The Devil's Hills and the sand dunes surrounding that dreary waste on the Assinniboine, in long. 99° 40′ W., showed a bare advancing surface towards the north-east, being pushed in that direction by the prevailing south-west wind. The sand dunes at the Height of Land in the Qu'appelle Valley, in long. 107° W., lat. 51° N , were advancing in an easterly direction; their clean surfaces were facing the east. Had they progressed under a prevailing south-west wind, they would long since have

* Proc. Roy. Geog, Soc. vol. iii. No. 5.

Table showing the number of days on which rain fell, with the character of the Thunder-storms, during the Summer of 1858, *in the Valley of the Assinniboine and Saskatchewan.*

Date.	Time.	Character of Storm.	Locality.
June			
18	2-4 p.m	Heavy rain, thunder . .	Prairie Portage.
19	6 a.m	Slight rain	Do. do.
20	Sunset.	A terrific thunderstorm, heavy rain, high wind . . .	Bad Woods.
21	11 a.m. 1 p.m.	Tremendous thunderstorm, hailstones 1-1½ in. in diameter .	Bear's Head Hill.
21	6 p.m.	Thunderstorm, heavy rain . .	Do. do.
22	3.45 p.m. 6 p.m.	Terrific thunderstorm, continued roar of thunder without intermission for 1½ hours . .	Sandy Hills.
25	8–10 p.m.	Violent thunderstorm, heavy rain	Little Souris.
26–27	Night.	Thunder and rain . . .	Do. do.
29–30	Night.	Thunder and rain . . .	Do. do.
30	6 a.m.	Heavy rain with rolling thunder, without intermission for 1 hour	Do. do.
July			
4	11 a.m.	Rain	Great Prairie.
5	10 p.m.	Lightning in the east, no rain, thermometer in shade 92° at noon	Do. do.
9	9 a.m.	Rain	Assinniboine.
11	3 p.m.	Thunderstorm, hail, and heavy rain	Fort Ellice.
13	7 p.m. 10 p.m.	Thunderstorm of unusual violence and sublimity—*see* narrative .	Qu'appelle Valley.
14	2.30 p.m. to 4.30	Thunder and rain . . .	Do. do.
14–15	Night.	Rain all last night . . .	Do. do.
15	. .	Rain North of Qu'appelle temp. at 6 a.m. 45°	
18	11 p.m.	Heavy rain and thunder . .	Qu'appelle.
19	4 p.m.	Rain North of Qu'appelle Valley	
22	Noon.	Violent thunderstorm with heavy rain and hail	Do.
28	1 p.m.	Rain in torrents . . .	Do.
August			
2	4 p.m.	Heavy thunderstorm with rain .	South Branch.
4	4.30 p.m.	Thunderstorm, heavy rain . .	Do. do.
5	. .	Heavy rain	Do. do.
11	6 p.m.	Violent thunderstorm . .	Long Creek & Main Saskatchewan.
18	6 p.m.	Thunderstorm, rain and high wind	Main Saskatchewan.
25	Noon.	Violent thunderstorm and rain .	Do. do.

invaded and filled up the Valley of the Qu'appelle. These existing records of prevailing winds during the period when the dunes are not frozen, show that while the south-west is the most effective as a summer surface wind in Rupert's Land under the 99th meridian, on the South Branch of the Saskatchewan, eight degrees further west, westerly winds prevail.

THE CLIMATE OF THE VALLEY OF THE RED RIVER.

The climate of the Valley of Red River exhibits the extremes of many characteristics which belong to the interior of continents in corresponding latitudes. High summer temperatures, with winter cold of extraordinary severity appear to prevail in the district called Assinniboia, as in the interior of North-Eastern Europe and Asia. It cannot fail to be noticed, however, that the general absence of late spring and early autumn frosts, with an abundant fall of rain during the agricultural months, are its distinguishing features in relation to husbandry. The melon, growing in the open air and arriving at perfect maturity in August and September, Indian corn succeeding invariably when due precautions are used to ensure ripening before the middle of September, are strong proofs of the almost uniform absence of summer frosts.

A comparison with the climate of Toronto * for corresponding months of the years 1855 and 1856, reveals some very curious and interesting facts, which may possess importance. Limiting our attention at present to the summer months we find from inspection of the following table of comparison, that the summer of Red River during the year of comparison was more than

* This comparison, given on the next page but one, was made by Mr. James Waker, assistant at the Observatory, Toronto.

three degrees warmer than the summer at Toronto, and with this excess of temperature there occurred the unexpected difference of 21·74 inches of rain in favour of Red River during that year.

It must be borne in mind, however, that the results of one year's comparison *are not of much value* in estimating the relative climatic adaptation of regions far apart; nor do they afford sufficient data for a fair estimate of the climate of the locality where the observations were made.

The small difference between the temperature of the spring at Toronto and Red River is an interesting fact. While the summer shows an excess of 3°, the spring gives a deficiency of 2°, so that the mean of the spring and summer months at Red River was nearly one degree higher than the corresponding months at Toronto. No feature in the meteorology of this distant region is likely to excite so much interest as the extraordinary fall of rain during the agricultural months. It is well known that the cause of the sterility of a vast region on the east of the Rocky Mountains, within the limits of the United States, is traced to extreme aridity. The great American desert, which places so vast a barrier between the Mississippi Valley and the west flank of the Rocky Mountains, derives its barrenness from the absence of rain during the summer months. A fall of thirty inches in the Valley of Red River during the summer of 1855, with a corresponding fall of 8·76 at Toronto, shows a remarkable difference in the humidity of the two places, and one which report states is generally maintained in other years.

The following comparisons refer to corresponding months of the same years, and are of course liable to those annual fluctuations to which the climatic elements of all countries are subject. It is very probable that more extended observations will reduce the extremes.

Comparison of the meteorology of Red River Settlement with Toronto, Canada West, with reference to mean temperature, depth of rain and snow, from corresponding observations at both stations, from June, 1855, *to May,* 1856, *inclusive.*

Month.	Mean Temperature.		Rain in inches.		Snow in inches.		Temperature: Rain and Snow at Red River + or — of Toronto.
	Red R.	Toronto.	Red R.	Toronto.	Red R.	Toronto.	
1855.	°	°	°	°	°	°	SUMMER.
June .	69·10	59·93	6·0	4·07	0·0	0·0	Temp. + 3·78
July .	71·16	67·95	12·0	3·24	0·0	0·0	Rain + 21·74 ins.
August .	63·03	64·06	12·5	1·45	0·0	0·0	Snow 0·0.
Summer .	67·76	63·98	30·5	8·76	0·0	0·0	
							AUTUMN.
September	59·26	59·49	5·0	5·59	0·0	0·0	Temp. — 6·94.
October .	42·20	45·39	0·0	2·48	2·0	0·8	Rain — 5·16 ins.
November	21·19	38·58	2·5	4·59	7·0	3·0	Snow + 5·2 ins.
Autumn .	40·88	47·82	7·5	12·66	9·0	3·8	
1856.							WINTER.
December.	–8·31	26·99	0·0	1·85	8·0	29·5	Temp. — 26·42.
January .	–10·55	16·02	0·0	0·00	5·0	13·6	Rain—1·85 ins.
February .	–1·71	15·69	0·0	0·00	6·0	9·7	Snow — 38·8 ins.
Winter .	-6·85	19·57	0·0	1·85	19·0	52·8	
							SPRING.
March .	9·09	23·06	0·0	0·00	6·5	16·2	Temp. — 2·83.
April .	39·83	42·27	6·5	2·78	3·0	0·1	Rain + 3·14 ins.
May .	58·46	50·52	4·0	4·58	2·0	Inap.	Snow – 4·8 ins.
Spring .	35·79	38·62	10·5	7·36	11·5	6·3	
Annual .	34·38	42·50	48·5	30·63	39·5	72·9	

ANNUAL.

Colder mean temperature	8°12
More rain	17·87 inches
Less snow	33·4 „
More moisture and most probably less evaporation .	14·53

At Quebec the difference between the mean temperature of summer and winter is 53°93; at Fort Snelling 56°81; and at Red River Settlement 74°61 in 1855.

The summer temperature of Red River, and the absence of frosts during that season, determine its fitness for agricultural purposes. The following table exhibits a comparison, based upon *one year's observation only*, between the summer temperature of the Settlement and various other well known places in Canada :—

Summer temperature at Red River Settlement . .	67·76
Montreal, Canada	66·62
Quebec	62·91
Toronto	63·98

Assuming that the annual mean of 34·38 deduced from the following tables is within two or three degrees of what would be the results of several years' observation, we find upon inspection of Dove Tables, that there is not one locality within the limits of the United States where so low an annual mean occurs. At Kasan (Russia) lat. 55°48, long. 49°7, the mean of ten years was 35°45, and the difference between the hottest and coldest months 61°33, while at Red River the difference was 82°15 in 1855. The difference between summer and winter at Kasan was 56°00; at Red River 74°61. At Orenburg, lat. 50°46, or in nearly the same latitude as that part of Red River Settlement where these observations were made, and in long. 55°6, the annual mean is 35°6; the difference between the hottest and coldest month 66°38, and the difference between winter and summer 59°66. The following table will exhibit this relation at a glance :—

	Latitude.	Longitude.	Annual Mean.	Difference between Hottest and Coldest Months.	Difference between Summer and Winter.
Red River Settlement (1855)	50°15′	97°W.	34·38	82·15	74·61
Kasan, Russia	55·48	49°7 E.	35·45	61·33	56·00
Orenburg . .	50·46	5 5·6 E.	35·06	66·38	59·66

SEASONS OF THE VALLEY OF LAKE WINNIPEG.

The natural division of the seasons in the Lake Winnipeg Valley is as follows:—

Spring—April and May.

Summer—June, July, August and part of September.

Autumn—Part of September and October.

Winter—November, December, January, February and March.

The natural division of the seasons is strikingly represented by the early and rapid advance of temperature in May in the valley and prairies of the Saskatchewan; and it is also indicated in a very marked degree by the extension northwards to the same valley, between the 95° and 105° of longitude, of numerous plants, whose geographical distribution, east and west of those limits, has a much more southern climatic boundary. The limits of trees rise with the isothermal lines, and these attain a much higher elevation in the interior of British America than on the Atlantic coast.*

In relation to agriculture, the intensity of winter cold is of comparatively little moment. The elevated spring and summer temperature, combined with the humidity of the region in the Valley of Lake Winnipeg, enable Indian corn and the melon to ripen, if ordinary care is taken in selecting soil and in planting seed.

The extraordinary cold of the winter of 1855 and 1856 at Red River, is shown by the tables for December, January, and February, (Mr. Gunn's observations,) which give a mean of $-6°\cdot85$ for the temperature of that season; but if we turn to the records for 1857 and 1858 † we find the mean temperature of the winter to have been

* Gray.

† Meteorological tables recorded by Mr. Dawson's party.

2°·87, showing a difference of eight degrees in favour of the winter of 1857–58. The temperatures recorded were as follows : —

	1855–56.	1857–58.
December	−8·31	−9·11
January	−10·55	−6·2
February	−1·71	−6·68
Winter Mean	−6·85	−2·87

The thermometers supplied to the Red River Expedition in 1857, were made by Negretti and Zambra, and were of the best construction.* They had been compared with a standard at the Provincial Observatory, and their errors recorded, but it does not appear that the proper corrections were made after each observation. Mr. Gunn's thermometer was an ordinary instrument, and, like many of its class, liable to errors at low temperatures. It is, therefore, probable that his winter temperatures are too low, and that the true mean of the winter months at Red River, and consequently the annual mean, may be considerably higher than it appears from his observations.

The cold of February 1858 was exceptional. At Montreal it was the coldest February on record, being 14°·05 below the mean temperature of February, 1857.† But it cannot be denied that the winter cold of Red River is excessive, and the temperature sometimes falls so low as to freeze mercury in a few minutes.

The meteorological register kept at the Stone Fort,

* These thermometers, together with other meteorological apparatus, were furnished to the different members of the Red River Expedition in 1857, by the permission of the Rev. Dr. Ryerson, Chief Superintendent of Schools, from the stock of instruments provided by the Chief Superintendent for observatories attached to the Grammar Schools throughout the Province. The thermometers were compared and their errors determined and tabulated at the Provincial Observatory. A table of errors was attached to each instrument.

† Records of St. Martin's Observatory, Isle Jesus.

Lower Settlement, in 1847, under the superintendence of Captain Moody *, from which extracts were permitted to be made by Dr. Owen, furnish trustworthy evidence respecting the severity of the climate in winter.

The mean temperature for January 1847, was – 12°·5. Observations being taken at 9 A.M., 3 P.M., and 9 P.M. During twenty-two consecutive days of this period, from the 5th to the 26th inclusive, the thermometer never once rose to zero. The lowest temperature reached was – 48°, the highest 30°, giving a range of 78°. On the coldest day, the 20th, when the thermometer showed – 48° at the Stone Fort, and – 47° at Fort Garry, mercury froze in fifteen to twenty minutes when exposed in bullet moulds.

Although there is no record of cold terms in Canada approaching the extreme low temperature and extending over so long a period as those instanced above, yet cold terms of great intensity are not uncommon in Lower Canada. In the excellent observatory of Dr. Smallwood's at Isle Jesus, nine miles west of Montreal, the following records of cold terms have been preserved †:—

Low Temperatures at Isle Jesus, Nine Miles west of Montreal. C.E.

On the 22nd and 23rd Dec., 1854.

22nd.—	8 A.M.,	31·6	below zero.		12 P.M.,	34·8	„
	9 „	27·0	„	23rd.—	6 A.M.,	36·2	„
	10 „	19·1	„		7 „	36·0	„
	11 „	17·8	„		8 „	34·3	„
	12 „	16·6	„		10 „	24·1	„
	1 P.M.,	12·1	„		12 „	13·4	„
	2 „	11·8	„		2 P.M.,	12·6	„
	3 „	8·1	„		4 „	9·1	„
	7 „	23·4	„		6 „	12·6	„
	8 „	26·9	„		10 „	9·1	„
	10 „	28·2	below zero.				

* See Dr. Owen's Geological Survey of Wisconsin, Iowa, and Minnesota, page 181.

† *Canadian Journal* for 1855, and *The Canadian Naturalist* for April, 1859.

On the 9th, 10th and 11th Jan., 1859.

9th.—6 A.M.,	29·9	below zero.
7 ,,	29·0	,,
9 ,,	28·4	,,
12 ,,	23·8	,,
2 P.M.,	21·5	,,
9 ,,	33·9	,,
10 ,,	34·2	,,
12 ,,	36·0	,,
10th.—6 A.M.,	43·6	,,
7 ,,	43·1	,,
9 ,,	41·6	,,
12 ,,	20·1	,,
2 P.M.,	14·3	below zero.
9 ,,	28·8	,,
10 ,,	29·2	,,
12 ,,	31·6	,,
11th.—6 A.M.,	37·1	,,
7 ,,	36·9	,,
12 ,,	24·8	,,
2 P.M.,	19·9	,,
9 ,,	21·0	,,
10 ,,	21·6	,,
12 ,,	18·1	,,

In January, 1859, the thermometer did not rise above zero during a period of 124 hours 30 minutes, or more than five days. Mercury froze in the open air. The mean temperature on the 9th was $-27°{\cdot}8$; 10th, $-29°{\cdot}0$; 11th, $-28°{\cdot}2$. Dr. Smallwood says that this cold term was felt generally throughout Canada and the Eastern States, and seems to have traveled from the west.

The following minimum temperatures were observed at different places :—

Rochester	10·0	below zero.
Brooklyn, (New York,)	9·0	,,
Boston	14·0	,,
Toronto	26·5	,,
Quebec	40·1	,,
Huntingdon	44·0	,,

For the purpose of comparing the monthly mean at Quebec (lat. 46° 49′ 2″, long. 71° 16′) with those of Red River, the table on the opposite page is inserted.*

In the absence of instrumental observations, the progress of vegetation affords the best indication of climate, apart from latitude and elevation above the sea. It has been observed elsewhere that there exists an extraordinary

* See *Canadian Journal*—Old Series.

Months.	Mean.	Maximum.	Minimum.
1854.			
October	46·05	64·7	+31·6
November	31·83	59·8	+10·0
December	13·08	36·6	- 19·2
1855.			
January	16·70	46·0	- 14·0
February	10·55	36·8	- 29·5
March	21·06	47·3	2·4
April	34·14	59·8	5·9
May	49·03	83·0	32·0
June	58·34	88·0	43·2
July	68·86	90·3	51·9
August	61·54	85·0	38·3
September	55·15	81·3	34·7
October	45·43	60·4	28·4
November	28·75	34·3	21·81
December	18·09	40·1	- 19·2
1856.			
January	8·19	27·0	- 16·5
February	11·99	31·9	- 18·0
March	17·60	39·0	-11·0
April	36·90	55·3	6·4

difference between the characteristic fruit tree of the South Branch, the Misaskatomina (*Amelanchier Canadensis*), and the same tree on the North Branch of the Saskatchewan. On the South Branch at the Elbow, and for forty miles down the river, this shrub attains an altitude of twenty feet, with a stem fully three or three and a half inches through; the fruit is large and very juicy; the size of the berry there is equal to the largest black currant, resembling a small grape more than any other fruit.

The period of flowering and fruiting is about three weeks earlier in latitude 51° than between 53rd and 54th parallels west of the 100° of longitude. The prairies of the Assinniboine, of the Qu'appelle, and of the South Branch of the Elbow, are decorated with brilliant spring flowers, and covered with luxuriant herbage, at a time when ice still lingers at the head of Lake Winnipeg, or chills the air and arrests vegetation in Cedar and Cross Lakes on the Main Saskatchewan. Two and a half degrees north of Cumberland, the soil is permanently frozen three feet below the surface. Sir John Richardson relates that in 1851 he did not disengage his canoes from the ice at the upper end of Lake Winnipeg until the 9th of June. At the Touchwood Hills, horses are allowed to remain in the open air all the winter, finding sufficient pasture under the snow to keep them in good condition.

The vegetable in the gardens attached to Fort à la Corne, with a brief notice of the periods of planting and gathering, will show that the climatic adaptation of the North Branch near the Grand Forks is not of a character unfavourable to agricultural operations. As this subject is one of great importance I have introduced some extracts from the journal of the Fort, which are both interesting and valuable. More extended extracts from this journal will be found in the Appendix.

On the 7th August, in the garden attached to Fort à la Corne (about 18 miles below the Grand Forks), potatoes were in flower, and the tubers of early varieties of the size of hens' eggs. Cabbages were well formed; beetroots and carrots quite ready for the kitchen; Indian corn in silk, from seed which was grown in the garden last year, and peas ready for gathering.

No disease has yet been noticed in the potatoes; and the grasshoppers, that scourge of the country south of the Touchwood Hills, have not made their appearance at Fort à la Corne.

In the garden attached to the Nepowewin Mission, under the charge of the Rev. Henry Budd (a zealous missionary of native origin), all the vegetables gave promise of fair and remunerative crops. The potatoes were superb; turnips, both Swedes and white, remarkably fine; Indian corn, from seed grown on the spot last year, in silk; wheat rather too rank in the stalk — it measured five feet three inches in length to the ear, which was well formed but green, and it seemed doubtful whether it would ripen. Mr. Budd speaks very favourably of the soil, climate, and extent of land available for agricultural purposes. Both the mission and the fort are situated within the excavated Valley of the Saskatchewan, and are not so favourably placed for farming purposes as they might be in the Valley of Long Creek. The river, however, is the great highway, and, during the summer season, affords an abundant supply of sturgeon.

Extracts from the Journal at Fort à la Corne, Saskatchewan River.
Lat. 53°·27 ; *long.* 104°·30 W.

1851, October 25th, ice made its appearance in the river. 1852, April 8th, ice solid for the season of the year; 12th, ice started; 13th, ice drifting and lodging on the banks; 21st, ice drifting and disappearing along the banks; 22nd, garden operations commenced. May 14th, first sturgeon caught; 24th, planted potatoes; October 11th, finished taking up potatoes; 25th, fishing season ended; 26th, snow. November 3rd, ice floating in the river. 1854, April 14th, river broke up; on the 15th nearly clear of ice; 28th, garden operations commenced. May 1st, first sturgeon caught; 8th, preparing potato fields; 13th, potato planting. October 2nd, gathered turnips; 3rd, taking up carrots; 10th, commenced taking up potatoes at the mission (190 kegs), turnips, carrots, cabbages—large and good; 11th, cabbages taken up. 1855, May 24th, turnips sown. September 12th, hard frost over night; 27th, took up potatoes—poor crop, much destroyed by grubs; 29th, hard frost; a little ice seen at the gates. October 1st, women digging potatoes; 2nd, ditto; 3rd, taking up turnips; 22nd, ice on the edges of river. 1856, April 2nd, hard frost last night; 4th, water making its appearance on the edges of the river; 7th, froze hard last night; 9th, ice made a start; 17th ice drifting; 23rd, fall of snow during the night: 23rd, nets set, one sturgeon caught; 25th, hard frost. May 2nd, garden operations commenced; 10th, storm of snow; 12th, planted potatoes; 14th, sowed Swedes. September 16th, slight frost last night. October 2nd, commenced taking up potatoes; 22nd, hard frost during the night; 23rd, severe frost during night; 86th, snow in night. November 11th, river full of ice. 1857, April 9th, water appearing on the edges of the river: snow shoes required everywhere; 16th, ice started to-day; 24th, snowed without intermission the whole day. May 3rd, ice drifting all last night; 5th, river full of ice; 12th, planted potatoes and onions; 20th, planting potatoes; three sturgeons caught. June 2nd, hard frost last night; 30th, starvation is staring the people in the face. Have caught no sturgeon for some time back.* 1858, April 21st, ice drifting; large quantities of ice on the banks. May 1st, clearing up of north garden; 7th, preparing potato ground; first sturgeon caught; 12th, planted potatoes; 17th, slight fall of snow; 18th, wind from N. and cold. Think we are going to have a second winter.

The growth of forests is very intimately connected with the climate of a large extent of country. That forests once covered a vast area in Rupert's Land there is no reason to doubt. Not only do the traditions of the natives refer to former forests, but the remains of many still exist as detached groves in secluded valleys, also on the crests of hills, or in the form of blackened prostrated trunks covered with rich grass, and sometimes with vegetable mould or drifted sand. The agent, which has caused the destruction of the forests that once covered many parts of the prairies is undoubtedly fire, and the same swift and effectual destroyer prevents the new growth from acquiring dimensions which would enable it to check their annual progress. Nearly everywhere, with the exception of the treeless, arid prairie west of the Souris, and west of Long Lake on the north side of the Qu'appelle, young willows and aspens were showing themselves in 1858, where fire had not been on the previous year. South of the Assinniboine and Qu'appelle few plains had escaped the conflagration in 1857, and the blackened shoots of willow were visible as bushes, clumps, or wide-spreading thickets where the fire had passed.

The end or tail of the prairies is at Fort Liard, a short distance to the south of Fort Simpson (latitude 61° 51 7″ N.). There is a long high belt of prairie land which

* A common record in the journals at the different posts in Rupert's Land. The cause must be referred to the habits of the people, their occupation, &c., and not to the capabilities of the country.—H. Y. H.

runs as far as the neighbourhood of that locality, at the foot of the Rocky Mountains.

In the State of Missouri, forests have sprung up with wonderful rapidity on the prairies, as the country becomes settled, so as to resist and subdue the encroachment of the annual fires from the west. Missouri lies within the limit of the humid south-west wind coming up the Valley of the Mississippi, and enjoys a greater rainfall than the region west of the 100° of longitude.

GENERAL CHARACTER OF THE REGION WEST OF THE 98TH MERIDIAN IN THE UNITED STATES.

Very great misapprehension has prevailed with regard to the region west of the Mississippi, as well as of the valley drained by the Saskatchewan. Sanguine enthusiasts have laid out new states and territories on the broad map of the Federation, and peopled them in imagination with bustling, industrious, and wealthy communities. Other visionaries have converted the 400,000 square miles drained by the Saskatchewan into a region of unbounded fertility and inexhaustible resources. Whereas a proper appreciation and use of facts will convince the most sanguine, that the larger portion of this area, is, in its present state, unfit for the permanent habitation of man both on account of climate, soil *, and absence of fuel.

The candid opinion of Professor Joseph Henry regarding the adaptation of a large portion of the United States for settlement has been already given; it is confirmed and strengthened by the following excellent summary, from the pen of Major Emory of the United States and Mexican Boundary Commission. It will at once occur to the

* See "Cretaceous Series," Vol. II.

reader that a knowledge of these facts gives great additional value to the truly fertile valleys of Red River, the Assinniboine, part of the Qu'appelle, and portions of the South and North Branch of the Saskatchewan. It determines also the direction in which efforts should be made to people this great wilderness, and guide the progress of settlement in such a manner as will render the country available for that grand desideratum, a route across the continent:—

"In the fanciful and exaggerated description given by many of the character of the western half of the continent, some have no doubt been influenced by a desire to favour particular routes of travel for the emigrants to follow; others by a desire to commend themselves to the political favour of those interested in the settlement and sale of the lands; but much the greater portion by estimating the soil alone, which is generally good, without giving due weight to the infrequency of rains, or the absence of the necessary humidity in the atmosphere, to produce a profitable vegetation. But be the motive what it may, the influence has been equally unfortunate by directing legislation and the military occupation of the country, as if it were susceptible of continuous settlement from the peaks of the Alleghanies to the shores of the Pacific."

"Hypothetical geography has proceeded far enough in the United States. In no country has it been carried to such an extent, or been attended with more disastrous consequences. This pernicious system was commenced under the eminent auspices of Baron Humboldt, who from a few excursions into Mexico, attempted to figure the whole North American continent. It has been followed by individuals to carry out objects of their own. In this way it has come to pass, that, with no other evidence than that furnished by a party of persons traveling on mule back, at the top of their speed, across the continent, the opinion of the country has been held in suspense upon the subject of the proper route for a railway, and even a preference created in the public mind in favour of a route which actual survey has demonstrated to be the most impracticable of all the routes between the 49th and 32nd parallels of latitude. On the same kind of unsubstantial information maps of the whole continent have been produced and engraved in the highest style of art, and sent forth to receive the patronage of Congress, and the applause of geographical societies at home and abroad, while the substantial contributors to accurate geography have seen their works pilfered and distorted, and themselves overlooked and forgotten." * * *

"The plains or basins which I have described as occurring in the mountain system are not the great plains of North America which are referred to so often in the newspaper literature of the day, in the expressions, 'News from the Plains,' 'Indian Depredations on the Plains,' &c.

"The term 'plains,' is applied to the extensive inclined surface reaching from the base of the Rocky Mountains to the shores of the Gulf of Mexico and the valley of the Mississippi, and form a feature in the geography of the western country as notable as any other. Except on the borders of the streams which traverse the plains in their course to the Valley of the Mississippi, scarcely anything exists deserving the name of vegetation. The soil is composed of disintegrated rocks, covered by a loam an inch or two in thickness, which is composed of the exuviæ of animals and decayed vegetable matter.

"The growth on them is principally a short but nutritious grass, called buffalo grass (*Sysleria dyctaloides*). A narrow strip of alluvial soil, supporting a coarse grass and a few cotton-wood trees, marks the line of the water-courses, which are themselves sufficiently few and far between.

"Whatever may be said to the contrary, these plains west of the 100th meridian are wholly unsusceptible of sustaining an agricultural population, until you reach sufficiently far south to encounter the rains from the tropics.

"The precise limits of these rains I am not prepared to give, but think the Red River (of Louisiana) is, perhaps, as far north as they extend. South of that river the plains are covered with grass of larger and more vigourous growth. That which is most widely spread over the face of the country is the grama or mezquite grass, of which there are many varieties. This is incomparably the most nutritious grass known."*

AURORAS.

On the night of October 2nd, when camped on Water-hen River, an Aurora of unusual brilliancy and character, even in these regions, surprised us with the varied magnificence of its display of light and colour. A broad ring of strong auroral light nearly encircled the Pole Star. It possessed an undulatory motion, and continually shot forth, towards and beyond the zenith, vast waves of faint light. They followed one another like huge pulsations—wave after wave—expanding towards the south with undiminished strength, and continuing many minutes at a

* Report on the United States and Mexican Boundary Survey, made under the direction of the Secretary of the Interior, by William H. Emory, Major First Cavalry and United States Commissioner. Washington, 1846, pp. 43-47.

time. Suddenly the waves ceased, the luminous belt or ring increased in brilliancy, lost its regular form, and here and there broke into faint streamers, of a pale yellow colour. The streamers, rapidly increasing, soon reached the zenith, and finally meeting beyond it, shot forth from the luminous arc with swift motion and in rapid succession. Their colour varied from straw to pink. The display of streamers is quite common in this part of the continent. The waves are also not unfrequently seen; but none of the half-breeds or the Indians, whom we saw a few days afterwards, had ever witnessed such a brilliant spectacle as the heavens presented during the early part of the night, when the immense pulsations, 14° to 20° in breadth, and expanding in their apparent ascent from east to west, rolled in tranquil, noiseless beauty, through the heavens overhead.

At 10 P.M., on the 27th of October, when camped on the shores of Lake Manitobah, near Oak Point, a half-breed awoke me to witness a crimson Aurora of surprising magnificence. Unfortunately a few clouds were flitting athwart the sky, which prevented the centre arc from being visible, but perhaps they increased the depth of the colour. The light was generally steady at the edges of the clouds. The appearance of streamers was recognized only in the clear portions of the sky and above the clouds, where the rose or crimson tints were much fainter. It reminded me of the reflection of a vast prairie on fire; the deep rose and crimson tints lasted for half an hour; then gave way to white and straw-coloured streamers, occasionally tinged with pale emerald green.

Coloured Auroras are not unfrequently seen during the summer months, but they rarely possess the extraordinary beauty of those which have just been described. These beautiful "dancing spirits of the dead" impart a solemnity and charm to the still night, which must ever remain one of its most delightful characteristics in these regions.

THE TWILIGHT BOW.

One of the most beautiful celestial phenomena visible after sunset and before sunrise from the north-western prairies is the Twilight Bow. The extraordinary clearness of the nights during summer in this region offers a very favourable opportunity for witnessing the delicate colouring which is communicated to the lower atmosphere by the reflected light from the upper illuminated portions. As the appearance of the Twilight Bow is dependent upon the serenity of the atmosphere to a great degree, its occurrence is not frequently observed or recorded in this country.

The Twilight Bow, and the causes which produce it, are thus described by M. Bravais*:—" Immediately after the setting of the sun, the curve which forms the separation between the atmospheric zone directly illuminated by the sun, and that which is only illuminated secondarily, or by reflection, receives the name of the *Crepuscular curve* or *Twilight Bow*. Some time after sunset, this bow, in traversing the heavens from east to west, passes the zenith; this epoch forms the end of civil twilight, and is the moment when planets and stars of the first magnitude begin to be visible. The eastern half of the heavens being then removed beyond solar illumination, night commences to all persons in apartments whose windows open to the east. Still later the Twilight Bow itself dis-

* Annuaire Metéorologique de la France for 1850; quoted by L. W Meek in the Smithsonian Report for 1856.

appears in the western horizon; it is then the end of the astronomic twilight; it is closed night. We may estimate that civil twilight ends when the sun has declined 6° below the horizon, and that a decline of 16° is necessary to terminate the astronomic twilight."

I often observed the Twilight Bow to be tinged with a delicate rose colour, passing into straw colour, and then into faint emerald green. The line of demarcation between the bow and the illuminated portion of the atmosphere was often very well defined, quite as clearly as in a secondary rainbow. It appeared most brilliant at an altitude of 60° or 70° above the horizon. It descended slowly towards the boundless level, preserving apparently with considerable exactness the form of a parabola. When the Twilight Bow is best developed, the aspect of the prairie is very singular. Towards the east it is cold, cheerless, and gloomy; towards the west it is warm, inspiriting, and suggestive of pleasant thoughts and cheerful anticipations. No wonder the prairie Indians associate delightful dreams of happy hunting-grounds with the setting sun and the beautiful west. They delight to sit silent and thoughtful "in the glory of the sunset," and allow themselves to be transported in imagination

"To the Islands of the Blessed,
To the Kingdom of Ponemah,
To the Land of the hereafter."

INDIAN SUMMER.

Indian summer is a phenomenon of constant yearly occurrence and marked characteristics in the north-west. The following table, kindly furnished from the private memoranda of Mr. James Walker, Assistant at the Provincial Observatory, establishes the fact that the hazy, warm, mellow weather we term Indian summer is a periodical phenomenon in Canada, but the cause does not appear to be quite understood. The characters of Indian summer are more decided in the north-west than in the neighbourhood of Lake Ontario. Sounds are distinctly audible at great distances; objects are difficult to discern unless close at hand; the weather is warm and oppressive, the atmosphere hazy and calm, and every object appears to wear a tranquil and drowsy aspect.

Indian Summer at Toronto.—1840 *to* 1859 *inclusive* [20 *years.*]

Year.	Commencement.	Termination.	No of. Days.	Remarks.
1840 .	1st November .	5th November .	5	
1841 .	29th October .	2nd November .	5	
1842 .	28th October .	4th November .	8	
1843 .	23rd October .	25th October .	3	
1844 .	22nd October .	26th October .	5	And 2nd to 7th Nov. (6 days.)
1845 .	24th October .	29th October .	6	
1846 .	4th November .	7th November .	4	
1847 .	28th October .	31st October .	4	
1848 .	20th November.	23rd November.	4	
1849 .	13th November.	18th November.	6	
1850 .	7th November .	13th November.	7	
1851 .	6th October .	11th October .	6	
1852 .	16th November.	21st November .	6	
1853 .	12th October .	20th October .	9	Well marked.
1854 .	24th October .	28th October .	5	Not well marked.
1855 .	16th October .	26th October .	11	Not well marked.
1856 .	19th October .	22nd October .	4	Very dense fog.
1857 .	5th October .	12th October .	8	And 2nd to 8th Nov. (7 days.)
1858 .	18th October .	28th October .	11	
1859 .	2nd November .	8th November .	7	Well marked.
Mean result.	27th October .	2nd November .	6 days.	J. W.

CHAP. XLII.

THE LOCUSTS AND THE FLOODS.

The Locusts.—General Distribution of the Insect.—Distinctions between Crickets, Grasshoppers and Locusts.—The LOCUST of the North-West.—*Acrydium Femur-rubrum.*—Description of the Insect.—Male and Female.—Accounts of the Appearance of Locusts in the United States and Rupert's Land.—Distribution in 1857 and 1858.—Limits of its Ravages.—Females engaged in laying Eggs.—Vitality of the Eggs.—Power of Flight of this Locust.—Elevation of its Flight above the Sea.—Food of the Insect.—Effect of the periodical Visitations in the Far West and in Rupert's Land.—THE FLOODS.—Flood Years.—Effects of, in 1802.—The Bishop of Rupert's Land Description.—Speculations respecting the Cause of the Floods.—Sudden Melting of an unusual Fall of Snow at the Opening of Spring.

THE LOCUSTS.

THE English name grasshopper is almost universally applied to the insect which forms the subject of the present notice. Its general distribution in the United States, and the dreadful ravages which have been produced at different times by innumerable hosts of these insects in the north-western and north-eastern states of the Union, have led to many accounts of its ravages in the newspapers of the day, in all of which it is described under that name. As the insect is not a grasshopper, but a locust, the following description of the difference between grasshoppers, crickets, and locusts, abbreviated from the excellent and most attractive treatise by Dr. Harris on "Insects injurious to Vegetation," is here introduced, with the object of describing to the general reader the true character of this destroyer.

The abundant and prolific jumping orthopterous insects included by Linnæus in his great genus *Gryllus*, and known by the English names of crickets, grasshoppers, and locusts, may thus be distinguished from one another *:—

"1. Crickets (Achetadæ); wing covers, horizontal; antennæ, long and tapering; two tapering, downy bristles at the end of the body, between which, in most of the females there is a long spear-pointed piercer.

"2. Grasshoppers (Gryllidæ); wing covers, sloping; antennæ, long and tapering; feet, with four joints; end of the body in the females provided with a projecting sword or sabre-shaped piercer.

"3. Locusts (Locustadæ); wing covers, roofed; antennæ, rather short and in general not tapering to the end; feet, with three joints; females, without a projecting piercer.

"The various insects included under the name of locusts nearly all agree in having their wing-covers rather long and narrow, and placed obliquely along the sides of the body, meeting and even overlapping for a short distance at their upper edges, which together form a ridge on the back like a sloping roof. Their antennæ are much shorter than those of most grasshoppers, and do not taper towards the end, but are nearly of equal thickness at both extremities; their feet have only three joints, but as the under side of the first joint is marked by one or two cross lines, the feet, when seen only from below, seem to be four or five jointed. The females have not a long projecting piercer like the crickets and grasshoppers, but the extremity of the body is provided with four short, wedge-like pieces, placed in pairs above and below, and opening and shutting opposite to each other, thus forming an instrument like a pair of nippers, only with four short

* A Treatise on some of the Insects of New England, which are injurious to vegetation, by Thaddeus William Harris, M.D., Boston, 1852.

blades instead of two. When one of these insects is about to lay her eggs, she drives these little wedges into the earth; these being opened and withdrawn, enlarge the orifice, upon which the insect inserts them again, and drives them down deeper than before, and repeats the operation above described, until she has formed a perforation large and deep enough to admit nearly the whole of her abdomen."

The locust of the north-west (*Acrydium femur-rubrum*) has been long noticed in the United States, and particularly in New England. In the State of Maine they appeared in such numbers during the years 1743, 1749, 1754, and 1756 that days of fasting and prayer were appointed on account of the threatened calamity.* In Massachusetts and Vermont they showed themselves in 1797 and 1798. In 1838 they appeared in the vicinity of Baltimore in immense numbers.

The following description is from the pen of Dr. Harris:—

Acrydium femur-rubrum—RED-LEGGED LOCUST.

Grizzled with dirty olive and brown; a black spot extending from the eyes along the sides of the thorax; an oblique yellow line on each side of the body beneath the wings; a row of dusky brown spots along the middle of the wing covers, and the hindmost shanks and feet blood red, with black spines. The wings are transparent, with a very pale greenish-yellow tint, next to the body, and are ruled with brown lines. The hindmost thighs have two large spots on the upper side, and the extremity black, but are red below, and yellow on the inside. The appendages at the tip of the body in the male, are of a long triangular form. Length from $\frac{3}{4}$ to 1 inch, expanded $1\frac{1}{4}$ to $1\frac{3}{4}$ inch.

* Williamson's History of Maine, quoted by Dr. Harris.

The red-legged locust was first described by De Geer, from specimens sent to him from Pennsylvania, and I have retained the scientific name which he gave to it. It is the *Gryllus* (*Locusta*) *erythropus* of Gmelin, and the *Acrydium femorale* of Olivier. It appears to be very generally diffused throughout the United States, and sometimes so greatly abounds in certain places, as to be productive of great injury to vegetation.

The following description was taken down in the prairies west of the Souris, where this insect surrounded us in countless millions, and the air from nine until four was filled with them as with flakes of snow.

Dimensions of the Male Insect.—Length of body, 1 inch; with wings closed 1½ inch; length of wing cases, thirteen lines; of wings, 12½ lines.

Colour.—Of head, bluish-green; of thorax, bluish-green, with two lateral black lines, parallel to the sides of the thorax, and half a line apart; intersected at an acute angle by two short black lines, half a line apart. These marks on the female are more distinct.

Abdomen.—Colour of segments pale bluish-green, with whitish-blue margin; upper portion of each segment dark brown, especially the superior segments; wing cases ash-coloured, with brown spots.

Legs.—Upper surface of posterior legs, brownish-white, with two dark brown spots; outside of the leg red, inside, light red.

Tibia.—Rose-coloured and fringed with two rows of spines. Fore legs yellowish-brown.

Female, differs from the male in the colour of the cheeks, thorax, and upper portion of the abdomen, these parts being of a brighter green. Legs deeper rose-colour; under-side of abdomen yellowish-white. Length of the insect, thirteen lines.

The first authentic account of the appearance of extraordinary swarms of locusts in Rupert's Land that I have met with, assigns the last week of July, 1818, to this event.* Every green herb in the settlements at Red River is stated to have been destroyed by these destructive invaders. In 1819 the young brood hatched from the eggs deposited in the preceding year appeared in the spring and consumed the growing wheat crops. "Every vegetable substance was either eaten up or stripped to the bare stalk; the leaves of bushes and the bark of trees shared the same fate; the grain vanished as fast as it appeared above ground leaving no hope either of 'Seed to the sower or bread to the eater.'" †

Early in 1819 this pestilence disappeared, but in what manner is not stated.

In 1857 the locusts appeared in countless swarms over a large part of North America. They destroyed nearly all the vegetables cultivated at Fort Randall, long. 98° 35′, lat. 43° 4′, and extended their ravages east as far as the state of Iowa.‡ During the same year they devoured the crops in part of Minnesota and advanced as far to the north-east as the Lake of the Woods, where I saw them on Garden Island in August. During the autumn of the same year they appeared on the White Horse plains north of the Assinniboine, where they deposited their eggs. The swarms of this insect must have extended as far west as the South Branch of the Saskatchewan and covered the country in a greater or less degree between the Lake of the Woods and the South Branch, a distance in an air line of 560 miles; the perfect insect in 1857, or the young brood in 1858 having been observed nearly continuously over that wide extent of country.

* Alexander Ross.—"The Red River Settlement," &c., 1856.

† Ibid.

‡ Lieutenant Warren.

In a northerly direction they were seen near the Lumpy Hill within sixty miles of the North Branch, also at Dauphin Lake where they had destroyed the crops of Tàwâpit; they had devoured all the crops with the exception of the potato at Fort Pelly before the 1st of August.

The ascertained limits of this mighty army of insects in 1857 extended from the 94th to the 112th meridian, and from the 41st to the 53rd parallel; from the settlements in Utah territory to near the Valley of the North Saskatchewan, and from the Lake of the Woods to the foot of the Rocky Mountains.

The brood from eggs deposited in 1857 at the Touchwood Hills, rose from the ground and took their flight on the 28th July, after destroying every green leaf in the garden of the Post, and leaving sad traces of their ravages in the prairies lying to the south-east. At the time of my visit not one was to be seen, so general had been their departure.

During the month of September, 1857, I saw the females engaged in laying their eggs. They did not limit themselves to the prairie soil in forming a nest, but riddled the decayed trunks of trees, the thatch of houses and barns, the wood of which they were built, everything, indeed, which they could penetrate with the little blades provided for that purpose. The appearance presented by bare patches of soil, such as the road near the settlements, suggested the idea that a vast number of worms had risen to the surface and then retired again after loosely closing the aperture they had made. When in the act of preparing a nest for her eggs, the female was observed to introduce her abdomen into the soil by repeated thrusts to its full length, and then slowly withdrawing it, eject her eggs to the number of ten or twelve in the form of a half cylinder, loosely covering the orifice after the operation was completed.

In the spring of 1858 the young brood were seen at Prairie Portage, hopping over the newly-fallen snow at the latter end of April. It was thought by the settlers that the cold weather which followed the warm days in the early part of the month when the eggs were hatched, would have destroyed the young brood, but it did not appear to have created any sensible diminution in their numbers.

The extraordinary vitality of the eggs of insects is well-known, but when we reflect that the eggs of the red-legged locust are exposed in Rupert's Land to a temperature lower than that at which mercury freezes or more than 40° below zero, as well as to constant alternations of temperature from the freezing point to below zero in the early spring months, their capacity to resist these influences cannot fail to be regarded as one of the most wonderful features in the life of this insect.

Their power of sustaining long flights is also very remarkable. As stated in the narrative they generally rose from the prairie about nine in the morning and alighted about four in the afternoon. During the intermediate hours I do not recollect one instance in which they were observed to alight, except in anticipation of a thunder-storm, when they would descend perpendicularly from a great altitude. Assuming their speed to have been twenty miles an hour, the distance they would fly in one day probably amounted to one hundred and twenty miles. They have been seen hurrying swiftly to the north at an elevation of 14,500 to 15,000 feet above the sea *, thus manifesting extra-

* Lieutenant Warren.

ordinary capabilities to sustain low temperatures, which would doubtless obtain at that elevation in a region of eternal snows.

Their principal food is the prairie grass and the leaves of shrubs, but they will attack any substances presented to them, even such indigestible articles as leather, traveling bags, woollen garments, saddle girths, and harness. In a few minutes they ate the varnish from the leather case of a telescope I left on the ground in 1858, and so disfigured a valise that the owner who had seen it sound and untouched a few minutes before we stopped to camp, could not recognize it after it had lain ten minutes on the grass. Blankets became instantly covered with them and eaten into holes, the only article of clothing which did not suffer from their voracity was the caoutchouc or gutta percha cloaks and coverings.

The periodical visitations of these locusts have been enumerated among the objectionable features of parts of the Far West, and as some of the obstacles to the settlement of Nebraska.* That they will also exercise an important influence upon the future of the southern part of Rupert's Land, there is but too great reason to fear; already they have twice destroyed the crops in different parts of the Settlements, and in the state of Minnesota in the region about Crow Wing they rendered husbandry hopeless for two years, producing great distress in that newly settled country.

THE FLOODS.

Spring freshets in the valley of Red River sometimes assume the form of wide-spreading devastating floods. The alluvial character of the prairies through which Red

* Explorations in Nebraska and Dakotah, by Lieutenant Warren, U.S. Top. Eng.

River and the Assinniboine flow is quite sufficient to show that extensive inundations have occurred from time to time during many ages. The accumulations of stratified mud containing the buried trunks and branches of trees are the records of overflows similar to those which caused such distress and consternation in 1826 and 1852. There is good ground for belief that floods have occurred in the district of Assinniboia during the following years: —

1776. On the authority of Mr. Nolin.*
1790. Indian Tradition.†
1809. Indians living in the Settlement when Ross wrote, 1856.
1826.‡ Missionary Register, December, 1826.
1852. "Notes on the Flood," by the Bishop of Rupert's Land.

From the level character of the country of the waters' overflow, the deep trench in which Red River glides towards Lake Winnipeg, a shallow wide-spreading lake soon forms. In 1852 the Bishop of Rupert's Land estimated the breadth of the inundated country to be about twelve miles a short distance below Fort Garry. Although the flood of 1852 was not so high as that of 1826, yet its effects were very severely felt in St. John's and St. Paul's parish and about Fort Garry, but in the parish of St. James, St. Andrew, and the Indian Settlement were almost untouched.§ Houses and barns were swept away in the inundated parishes, and the country for miles on either side assumed the appearance of a lake. Some of the settlers took refuge on the Lake Ridge near St. James's Church, which by leveling we ascertained in 1858 to be eleven feet above the gate of Fort Garry;

* Mentioned by Ross, in the Red River Settlement, its Rise, Progress, and Present State.

† Ibid.

‡ Ibid.

§ Notes on the Flood at Red River.

others encamped on Stony Mountain, and others again on Bird's Hill opposite Stony Mountain on the east side of the river. These localities and their altitudes above the level of the prairie have already been described.

The Bishop of Rupert's Land thus describes the scene on the 24th May, when the flood was at its height:—

"My sister and I started for the Stone Fort at 6 A.M. It was necessary to see the Governor and Mrs. Mills on business connected with the continuance of the schools. Breakfasted with Major Caldwell in passing. Changed for a larger canoe at the Upper Fort, and proceeded onwards. Called at my own house, from which we were three canoes in convoy. As it turned out, ours was the only one which escaped damage from the current. The day was very fine, nearly as warm as the preceding. It was like delicious lake scenery at home, but for the want of background. What we here call hills, the Little and Stony Mountains, are but very gentle eminences of a few hundred feet, scarcely visible as heights, though all felt their value on this occasion. Strange sights met our eyes as we proceeded. Some of the bridges we saw four miles below their former locality, and on the opposite side of the river. The railing round some of the graves of the upper church had also been carried down as far. A barn had been tied to a strong tree to secure it, but it eventually floated off. The houses, many of them standing up to the eaves in water, showed less the destructive effects of the water upon them than some weeks after, when the waters had retired. At all the points where the trees had caught the drift wood, it was collected in great quantities, and upon it was a motley assemblage of wheels, hay-carts, tables, doors, chairs, &c.

"The delight was great to reach the first spot where the banks were visible, at Park's Creek, half way between the middle church and the Rapids. Here the current from being confined within narrower limits, became more impetuous, and we had been strongly advised to proceed by land; but not being timid on the water, and having confidence in the skill of our three men, we preferred going on to the Rapids, merely stopping to pay one sad visit to the widow of our poor servant. . . . . At St. Andrew's we called to see Mr. and Mrs. Hunter, and from that drove our own horse down to the Lower Fort. This was the first drive we had had since the carioles were in use over the ice. Here we found a changed scene. . . . . Its chief recommendation (i.e. the Fort) in our eyes, under the circumstances was, that it still stood on a high bank, thirty feet above the river."

Various speculations have been advanced respecting the cause of the periodical floods of Red River. The backing up of the waters during a freshet in consequence of the ice on Lake Winnipeg remaining solid, is sufficiently disproved by the rapidity of current at the Lower or Stone Fort, which the Bishop states was "running at the rate of eight or ten miles an hour," thirty feet below the level of the prairie. The passage of the waters of the Missouri down the Valley of the Little Souris into the Assinniboine, has been a favourite theory among the half-breeds, one however which is not tenable, as the Little Souris does not approach the Missouri nearer than forty miles and no valley or coulée exists between them which would admit of the waters of the Missouri flowing across the Grand Coteau. It is probable that these periodical floods are caused by the sudden melting of an unusual fall of snow in the wooded areas on the east bank of Red River and throughout the large expanse of level open country, drained by Red River and the Assinniboine. The chief source of supply appears to be the valley of

Red River; the Assinniboine does not contribute largely to the flooding of the country; its waters indeed, are said to find their way to Lake Manitobah, down the valley of Rat Rivulet, during freshets, and the prairies at Pembina show, from the trunks of large trees scattered over them, that they were covered with water in 1852. Hence it appears that there is no possibility of guarding against these inundations, and that part of Red River Settlement above the Watermill Creek, subject to them, will always suffer from the disadvantages inseparable from such devastating occurrences.

APPENDIX

I. Table showing the lengths, distances from Lake Superior, heights, elevation above Lake Superior, and the number of the Portages on the Kaministiquia route.

II. A Statistical Account of Red River Colony, taken on the 20th to the 24th May 1856, compared with the Censuses of 1856 and 1843.

III. A Statistical View of the Church Missionary Society's Missions in North-Western America, May 1859.

IV. Stations of Missionaries in the Church Missionary Society's North-West American Missions, May 1859.

V. The Roman Catholic Missions west of Red River.

VI. The Wesleyan Missions.

VII. The Far West in the United States.

VIII. Season at Red River.

IX. Extract from a Journal kept at Fort à la Corne, on the main Saskatchewan, lat. 55·30, long. 104·25.

X. Prices of Furs at Fort Garry in January 1860.

XI. Additional Remarks on the Pigeon River Route, with an estimate of the cost of opening the Communication *viâ* Fort William, Arrow Lake, Gunflint Lake, Rainy Lake, and the North-West Corner of the Lake of the Woods to Fort Garry.

XII. Table of the Portages, Décharges, Rapids, Lakes, Lake Straits, and Navigable Channels on the Pigeon River Route (the old North-West Company's route) from Lake Superior to Rainy Lake, showing their lengths and distances from Lake Superior.

XIII. List of Photographs of Red River; the Churches of Selkirk Settlement; the Houses and Stores of the Settlers; Indian Tents and Graves; the Prairie; the Forts and Stores of the Hon. Hudson's Bay Company; the Native Races, and of the Expedition returning to Crow Wing in Dog Carioles. These Photographs can be procured from Mr. J. Hogarth, No. 5, Haymarket, London.

APPENDIX.

I.

TABLE *showing the lengths, distances from Lake Superior, heights, elevation above Lake Superior, and the number of the Portages on the Kaministiquia route.*

Names.	Lengths.	Distances from Lake Superior.	Heights.	Elevation above Lake Superior.	No. of Portages.	Remarks.
Kaministiquia River.	Mls. Chs.	Mls. Chs.	Feet.	Feet.		
Mouth (Fort William) . .	—	—	—	—	—	
Fort William . . .	0 40	0 40	—	—	—	
Pointe des Meurons . .	9 40	10 0	4·49	4·49	—	
Rapids and Current . .	12 0	22 0	30·00	34·49	—	
Décharges des Paresseux .	0 14	22 14	5·08	39·57	1	
Rapids and Current . .	7 4	29 18	16·63	56·20	—	
Mountain Portage (Kakabeka falls)	0 62	30 0	119·05	175·25	2	
River	0 20	30 20	—	175·25	—	
Rocky Portage (or Ecarté Portage)	0 37	30 57	62·65	237·90	3	
River	2 60	33 27	0·50	238·40	—	
Nicolet Portage . . .	0 6	33 43	6·50	244·90	4	
Rapids and Current . .	1 37	35 0	5·75	250·65	—	
Portage	0 3	35 3	12·62	263·27	5	
River	0 37	35 40	—	263·27	—	
Pot Holes Portage . .	0 13	35 53	6·90	270·17	6	
River	0 22	35 75	—	270·17	—	
Couteau Portage . .	0 5	36 0	19·25	289·42	7	
Trois Décharges . . .	0 35	36 35	10·00	299·42	8	
River	1 0	37 35	0·20	299·62	—	
Poplar Décharge . .	0 5	37 40	3·00	302·62	9	
River	0 40	39 0	0·50	303·12	—	
Décharge	0 9	39 9	3·00	306·12	10	
Rapids and Current . .	9 51	48 60	35·00	341·12	—	
Portage des Maitres . .	0 1	48 61	3·00	344·12	11	
River	0 60	49 41	1·00	345·12	—	
Little Dog Portage . .	0 3	49 44	14·94	360·06	12	
Rapids and Current . .	2 60	52 24	3·00	363·06	—	
Little Dog Lake . . .	1 20	53 44	—	363·06	—	
Great Dog Portage . .	1 52	55 16	347·81	710·87	13	
Great Dog Lake. . .	10 60	65 76	—	710·87	—	
Dog River.						
Mouth	—	65 76	—	710·87	—	
River	30 0	95 76	3·00	713·87	—	
Barrière Portage . .	0 1	95 77	3·50	717·37	14	

Table showing the lengths and distances from Lake Superior, &c. (*continued.*)

Names.	Lengths.	Distances from Lake Superior.	Heights.	Elevation above Lake Superior.	No. of Portages.	Remarks.
Dog River—(continued).	Mls. Chs.	Mls. Chs.	Feet.	Feet.		
River	2 20	98 17	0·20	717·57	—	
Jourdain Portage	0 6	98 23	8·60	726·17	15	
River	0 40	98 63	—	726·17	—	
Prairie River.						
Mouth	—	98 63	—	725·17	—	
River	3 0	101 63	6·50	727·67	—	
Cold Water Portage	0 6	101 69	0·76	728·43	16	
Cold Water Lake	0 14	102 3	—	728·43	—	
Prairie Portage	2 50	104 53	157·12	885·55	17	
Height of Land Lake	0 18	104 71	—	885·55	—	
De Milieu Portage	0 38	105 29	16·39	869·16	18	
Savanne Lake	1 40	106 69	—	869·16	—	
Great Savanne Portage	1 41	108 30	31·69	837·47	19	
Savanne River.						
Mouth	—	108 30	—	837·47	—	
River	13 20	121 50	4·79	832·68	—	
Thousand Lakes	21 60	143 30	—	832·68	—	
Baril Portage	0 17	143 47	1·86	834·54	20	
Baril Lake	8 0	151 47	—	834·54	—	
Brulé Portage	0 21	151 68	47·02	787·52	21	
Upper Brulé Lake (or Cannibals Lake)	8 0	159 68	1·50	786·02	—	
Semi-Décharge	0 3	159 70	2·50	783·52	22	
Lower Brulé Lake	4 20	164 10	1·25	782·27	—	
Great French Portage	1 60	165 70	99·71	682·56	23	
French Portage Lake	1 40	167 30	—	682·56	—	
Pickerel River	2 40	169 70	1·25	681·31	—	
Pickerel Lake	13 0	182 70	—	681·31	—	
Pickerel Portage	0 26	183 16	6·90	674·41	24	
Doré Lake	1 60	184 76	—	674·41	—	
Deux Rivières Portage	0 32	185 28	117·22	557·19	25	
Sturgeon Lake	23 20	208 48	1·00	556·19	—	
Sturgeon River.						
Mouth	—	208 48	—	556·19	—	
Semi-Décharge, 1st Sturgeon Rapids	0 11	208 59	4·51	551·68	26	
River	0 20	208 79	0·25	551·43	—	
Portage, 2nd Sturgeon Rapids	0 3	209 2	6·21	545·22	27	
Rapids and Current	7 8	216 10	10·00	535·22	—	
Semi-Décharge, Minnis Rapids	0 5	216 15	4·50	530·72	28	
Current	5 0	221 15	1·25	529·47	—	
Island Portage	0 3	221 18	10·06	519·41	29	
River	4 0	225 18	2·00	517·41	—	
Nequawquon Lake (or Lac la Croix)	8 0	233 18	—	517·41	—	
Nameukan River.						
Mouth	—	233 18	—	517·41	—	
Current	2 0	235 18	5·00	512·41	—	

Table showing the lengths and distances from Lake Superior, &c. (continued).

Names.	Lengths.	Distances from Lake Superior.	Heights.	Elevation above Lake Superior.	No. of Portages.	Remarks.
Nameaukan River—(continued).	Mls. Chs.	Mls. Chs.	Feet.	Feet.		
Rattlesnake Portage	0 5	235 23	12·14	500·27	30	
Current	3 27	238 50	1·75	498·52	—	
Crow Portage	0 8	238 58	9·88	488·64	31	
Rapids and Current	6 40	245 18	7·00	481·64	—	
Grand Falls Portage	0 6	245 24	16·08	465·56	32	
Current	0 3	248 24	3·00	462·56	—	
Grand Rapids	0 40	248 64	16·00	446·56	—	
River	2 40	251 24	2·00	444·56	—	
Lake Nameaukan	6 40	257 64	—	444·56	—	
Nu Portage	0 6	257 70	8·55	436·01	33	
Lakelet	0 20	258 16	—	436·01	—	
Portage	0 11	258 27	0·21	435·80	34	
River	5 0	263 27	0·50	435·30	—	
Rainy Lake	38 0	301 27	—	435·30	—	
Rainy River.						
Mouth	—	301 27	—	435·30	—	
Rapids	0 40	301 67	3·00	432·30	—	
Currents	1 40	303 27	0·50	431·80	—	
Fort Francis Portage	0 8	303 35	22·88	408·92	35	
River	32 60	336 15	10·00	398·92	—	
Manitou Rapids	0 15	336 30	2·50	396·42	—	
River	6 40	342 70	3·50	392·92	—	
Long Rapids	0 20	343 10	3·00	389·92	—	
River	38 0	381 10	12·10	377·82	—	
Lake of the Woods	72 0	453 10	—	377·82	—	
Winnipeg River.						
Rat Portage	0 13	453 23	15·98	361·84	36	
River	8 7	461 30	1·00	360·84	—	
Les Dalles Rapids	0 20	461 50	3·00	357·84	—	
River	25 0	486 50	2·00	355·84	—	
Grand Décharge	0 20	486 70	6·00	349·84	37	
River	2 0	488 70	2·25	347·59	—	
Terre Jaune Portage	0 5	488 75	22·02	325·57	38	
River	0 55	489 50	0·75	324·82	—	
Charette Décharge	0 2	489 52	3·50	321·32	39	
River	0 78	490 50	1·00	320·32	—	
Terre Blanche Portage	0 10	490 60	8·24	312·08	40	
River	0 28	491 8	0·15	311·93	—	
Cave Rapids	0 2	491 10	2·50	309·43	—	
River	19 0	510 10	4·50	304·93	—	
De l'Isle Portage	0 2	510 12	3·40	301·53	41	
River	22 78	533 10	4·00	297·53	—	
Chute à Jaques Portage	0 3	533 13	12·97	284·56	42	
River	9 57	542 70	1·60	282·96	—	
Point des Bois Portage	0 13	543 3	10·50	272·46	43	
River	0 7	543 10	0·25	272·21	—	
Point aux Chênes Portage	0 5	543 15	19·92	252·29	44	
River	0 75	544 10	1·00	251·29	—	
Roche Brulé Portage	0 3	544 13	7·80	243·49	45	

Table showing the lengths and distances from Lake Superior, &c. (continued).

Names.	Lengths.		Distances from Lake Superior.		Heights.	Elevation above Lake Superior.	No. of Portages.	Remarks.
Winnipeg River—(continued).	Mls.	Chs.	Mls.	Chs.	Feet.	Feet.		
River	4	37	548	50	1·75	241·74	—	
Slave Falls Portage	0	30	549	0	19·80	221·94	46	
River	6	10	555	10	2·25	219·69	—	
Barrier Falls Portage	0	2	555	12	4·97	214·72	47	
River	4	78	560	10	2·00	212·72	—	
Otter Falls	0	1	560	11	3·00	209·72	—	
Current	5	79	566	10	8·00	201·72	—	
The Seven Portages. 1st	0	4	566	14	10·23	191·49	48	
Current	0	16	566	30	1·00	190·49	—	
2nd	0	3	566	33	8·47	182·02	49	
Current	0	37	566	70	2·00	180·02	—	
3rd	0	5	566	75	5·60	174·42	50	
Current	1	15	568	10	2·25	172·17	—	
4th	0	3	568	13	7·68	164·49	51	
Current	0	37	568	50	1·25	163·24	—	
5th	0	2	568	52	2·90	160·34	52	
Current	0	38	569	10	2·00	158·34	—	
6th	0	3	569	13	8·13	150·21	53	
Current	0	7	569	20	1·25	148·96	—	
7th	0	3	569	23	4·75	144·21	54	
River	11	37	580	60	3·00	144·21	—	
Bonnet Lake	4	40	585	20	—	144·21	—	
Bonnet Portage	0	1	585	21	7·31	133·90	55	
River	0	59	586	0	2·00	131·90	—	
Cap de Bonnet Portage	0	4	586	4	5·00	126·90	56	
River	3	16	589	20	3·25	123·65	—	
Big Bonnet Portage	0	50	589	70	34·23	89·42	57	
River	0	30	590	20	1·00	88·42	—	
Petit Roche Portage	0	13	590	33	8·25	80·17	58	
River	3	27	593	60	3·50	76·67	—	
White Mud Portage	0	15	593	75	13·05	63·62	59	
River	3	45	597	40	1·80	61·82	—	
Silver Falls Portage 1st	0	7	597	47	6·06	55·76	60	
River	0	3	597	50	0·25	55·51	—	
2nd	0	13	597	63	15·59	39·95	61	
River	5	47	603	30	1·40	38·55	—	
Pine Portage	0	12	603	42	8·03	30·20	62	
River	11	0	614	42	2·00	28·20	—	
Fort Alexander	0	0	614	42	—	28·20	—	
Mouth of River	1	60	616	22	—	28·20	—	
Winnipeg Lake	41	0	657	22	—	28·20	—	

II.

A Statistical Account of Red River Colony, taken on the 20th to the 24th May 1856, *compared with the Census of* 1856 *and* 1843.

Year.	Number of Families.	Religion.			Country.							Total.		
1856.	Average $6\frac{31}{1082}$ per family.	Episco-palian.	Presby-terian.	Catholic.	England.	Ireland.	Scotland.	Canada.	Norway.	Rupert's Land.	Switzer-land.	Males.	Females.	Total.
1856 . .	1082	488	60	534	40	13	116	92	1	816	2	3225	3298	6523
1849 . .	1052	539	—	513	46	27	129	161	3	684	2	2714	2577	5291
Increase .	30	—	60	21	—	—	—	—	—	132	—	511	721	1232
Decrease	—	51	51	—	6	14	13	69	2	—	—	—	—	—
*1843 . .	870	2345		2798	22	5	110	152	—	571	2	—	—	5143

* The census for this year, given above, is abstracted from the Journal of the Bishop of Montreal, published in 1845, Mr. Smith, the registrar, not having the census for the year 1843 in his possession.—H. Y. H.

A Statistical Account of Red River, &c.—(*continued*).

Year.	Dwellings.			Live Stock.								Implements.					Land.	Machinery.						Public Buildings.			
	Houses.	Stables.	Barns.	Horses.	Mares.	Oxen.	Bulls.	Cows.	Calves.	Pigs.	Sheep.	Ploughs.	Harrows.	Carts.	Canoes.	Boats.	Cultivated land at two bushels wheat per acre.	Wind Mills.	Water Mills.	Thrashing Mills.	Reaping Machines.	Winnowing Machines.	Carding Mill.	Churches.	Schools.	Shops, merchants, &c.	Gaol.
																	Acres.										
1856.	922	1232	399	1503	1296	2726	290	3593	2644	4674	2429	585	730	2045	522	55	8371	16	9	8	2	6	1	9	17	56	1
1849.	745	1066	335	1095	990	2097	155	2147	1615	1565	3096	492	576	1918	528	40	6392½	18	1	—	—	—	—	7	12	—	1
Increase	177	166	64	408	306	629	135	1446	1029	3109	—	93	154	1027	94	15	1978½	—	8	8	2	6	1	2	5	56	—
Decrease	—	—	—	—	—	—	—	—	—	—	667	—	—	—	—	—	—	2	—	—	—	—	—	—	—	—	—
1843.	730	1291 (Stables and Barns)		821	749	—	107	2207	1580	1976	3569	—	—	—	—	—	—	18	1	—	—	—	—	—	—	—	—

COURTS.

Quarterly General Courts, 1855 and 1856.				Total.	Petty Local Courts, 1855 and 1856.																	Total amount of all the 38 cases of Petty Courts for one year.
					Petty Offences.						Debt.											
August.	November,	February.	May.	Number of Cases.	Trespass.	Cases of Damage or Misdemeanor.	Hay Ground Privilege.	Assault and Battery.	Defamation of Character.	Total number of cases.	From 1s. to 5s.	From 5s. to 10s.	From 10s to 20s.	From 20s. to 30s.	From 30s to 40s.	From 40s. to 50s.	From 50s. to 60s.	From 60s. to 70s.	From 80s to 90s.	From 90s. to 100s.	Total number of cases.	£ s. d.
No Cases.	No Cases.	One Case.	No Cases.	1	2	6	1	1	1	11	4	8	5	4	2	1	—	2	—	1	27	46 13 6

III.

Statistical View of the Church Missionary Society's Missions in North-west America, May 1859.

MISSIONS.	NAMES OF DISTRICTS OR PRINCIPAL STATIONS.	FIRST COMMENCEMENT OF OPERATIONS.	NO. OF STATIONS.	CLERICAL AND LAY TEACHERS AND ASSISTANTS. CLERGYMEN. European. English.	CLERGYMEN. European. Lutheran.	CLERGYMEN. Native and Country-born.	LAY TEACHERS AND OTHERS. European. Male.	LAY TEACHERS AND OTHERS. European. Female.	LAY TEACHERS AND OTHERS. E. Indian & Country born. Male.	LAY TEACHERS AND OTHERS. E. Indian & Country born. Female.	LAY TEACHERS AND OTHERS. Native. Male.	LAY TEACHERS AND OTHERS. Native. Female.	TOTAL.	NATIVE COMMUNICANTS.	BAPTISMS DURING THE YEAR. Adults.	BAPTISMS DURING THE YEAR. Children.	BAPTISMS DURING THE YEAR. Adults or Children, but not specified.	BAPTISMS DURING THE YEAR. TOTAL.	SEMINARIES AND SCHOOLS.	NATIVE SEMINARISTS AND SCHOLARS. Boys.	Girls.	Sexes not specified.	Youths and Adults.	TOTAL.
NORTH-WEST AMERICA.	Red River .	1822	4	5	—	1	2	—	1	—	10	—	19	413	No	Re	turns.		14	188	252	—	—	440
	Manitoba .	1842	3	1	—	1	—	—	—	—	3	—	5	41	—	—	13	13	1	33	17	—	—	50
	Cumberland	1840	2	1	—	1	—	—	1	—	4	—	7	138	—	22	—	22	3	25	21	44	—	90
	English River	1852	1	1	—	—	—	—	—	—	—	—	1	65	17	47	14	78	3	—	—	20	52	72
	York Fort .	1854	1	1	—	—	—	—	—	—	1	—	2	35	—	—	—	—	1	—	40	42	—	82
	James' Bay .	1851	1	2	—	—	—	—	1	—	2	—	5	84	—	—	39	39	1	—	—	25	—	25
	Fort Simpson	1857	1	—	—	—	1	—	—	—	—	—	1	—	—	—	—	—	1	—	—	200	50	250
	At Home.	—	—	1	—	—	—	—	—	—	—	—	1	—	—	—	—	—	—	—	—	—	—	—
	Total .		13	12	—	3	3	—	3	—	20	—	41	776	17	69	66	152	24	246	330	331	102	1009

IV.

STATIONS OF MISSIONARIES IN THE CHURCH MISSION SOCIETY.

NORTH-WEST AMERICA MISSION.—May 1859.

RED RIVER.

St. Andrew's . . Ven. Archd. James Hunter, M.A., 1844.
Rev. William West Kirkby, 1852.
Rev. Edwin A. Watkins, 1852.
Mr. Caleb Mayhew, 1856.
1 European Schoolmaster.

La Prairie . . Ven. Archd. W. Cockran, 1825.

Indian Settlement Rev. Abraham Cowley, 1841.

Islington . . . Rev. Robert M'Donald, *Country-born.*

MANITOBA.

Fairford . . . Rev. William Stagg, 1853.
Rev. James Settee, *Native.*

Fort Pelly . . 1 Native Teacher.

CUMBERLAND.

Cumberland . . Rev. Henry George, 1854.

Moose Lake . . 1 Native Catechist.

Nepowewin . . Rev. Henry Budd, *Native.*

ENGLISH RIVER.

English River . Rev. Robert Hunt, 1849.

YORK FACTORY.

York Factory . . Rev. Joseph Phelps Gardiner, 1857.

JAMES' BAY.

Moose Fort . . Rev. John Horden, 1851.
Rev. Thomas H. Fleming, 1857.

PACIFIC.

Fort Simpson . Mr. William Duncan, 1856.

21 *Country-born and Native Teachers in addition to the above-mentioned.*

At Home . . . Rev. William Mason, 1840.

V.

THE ROMAN CATHOLIC MISSIONS WEST OF RED RIVER.

In 1830, Bishop Provencher went to Canada, with a view to induce his church to commence missionary operations in Rupert's Land, west of Red River, and returned the following year with his first missionary, Mr. G. Belcourt. This gentleman at once set himself to acquire the Salteaux dialect, as he was to begin with that tribe, being nearest at hand. He mastered the language thoroughly, and commenced operations above Lane's Fort, at a point which still bears his name. He baptized a number, and the mission got a good start under his management. Others followed him and laboured in hope, but after years of anxious care and toil, the results were not satisfactory. The death of Mr. Darveau, who succeeded Mr. Belcourt, and the better disposition of the Indians elsewhere, induced Bishop Provencher reluctantly to abandon that mission. At present there are no Catholic missions in, or immediately around the settlement: they are all in the interior. It is the universal experience of the missionaries, that the Indians in the immediate neighbourhood of white settlements are much more degraded and less disposed to spiritual matters, than those far from the whites. In 1842 the Roman Catholic missions fairly began. In that year M. Thibeault was sent forth, and was followed in 1844 by Messrs. Lafleche and Bourassa. And so they have increased until now they number sixteen or seventeen, and have five missionary stations.

Lake St. Anne Mission.—This place is known among *voyageurs* by the name of "Devil's Lake," which is a translation of the Indian name usually given in maps, Manitou Lake. It is nearly fifty miles west of Fort Edmonton, the head quarters of the Saskatchewan district. It was first visited by M. Thibeault in 1843, but was formally taken possession of only in the following year by Messrs. Lafleche and Bourassa. There are about forty houses, occupied partly by half-breeds, partly by Crees, Assinniboines, and Blackfeet. The present missionaries are—Rev. Fathers Lacombe, Remas, and Train. Making the St. Anne a sort of rendezvous, they have been through the whole Saskatchewan Valley, and have gone to the source of the Athabasca, to the Rocky Mountains, Red Deer Lake, and Ile-à-la-Crosse. The mission is provided with a chapel (now too small), a good house for the missionaries, a school, and a convent; which last, sisters of Charity will occupy this summer (1860).

Mission of Notre-Dame des Victoires.—The Red Deer Lake, on the border of which is this mission, is in the Saskatchewan district. There are some half-breeds at this mission, besides Crees and Chipewyans from Athabasca and Ile-à-la-Crosse. M. Thibeault visited it in 1845 and subsequently. Rev. Father Remas settled there in 1853. Its present missionaries are Rev. Fathers Maisonneuve and Vissot. The soil and climate permit the cultivation of wheat at this station. The missionaries go at times to Fort Pitt, but find little encouragement, as drunkenness prevails to a shameful extent, both among the Company's servants and the Indians.

St. John the Baptist Mission.—This is at Ile-à-la-Crosse, in the English River district. M. Thibeault first visited the spot in 1845, then in 1846; and in the fall of this year it was formally occupied by Rev. Father Tache and M. Lafleche. Monseigneur Grandin, coadjutor-elect of St. Boniface, and Rev. Father Vegreville, labour there at present among Crees and Chipewyans and the employés of the Company. They annually visit Lac

Vert and Portage à la Loche. There is a handsome church, a substantial house for the missionaries, and one ready for Sisters of Charity.

Mission of the Nativity.—The Rev. Father Alex. Tache was the first missionary that visited Lake Athabasca, at the end of which this mission is situated. He went first in 1847, then in 1848. In the following year Father Faraud was permanently settled there. He understands the Cree and Chipewyan dialects perfectly, and is now assisted by Rev. Father Clut.

Mission of St. Joseph.— It is on the Great Slave Lake. Rev. Father Faraud visited the place first in 1852, and subsequently in 1856; but it was permanently occupied only in 1858, when Rev. Fathers Grollier and Eynard took a position alongside of Fort Resolution. This is the furthest north of all the Roman Catholic missions. In 1858 Rev. Father Grollier went as far as Fort Simpson, the head quarters of the Mackenzie's River district.

The number of baptisms by the missionaries from the year 1842 to 1856 amounted to 5137. Since 1856 there have been no statistics received; they are furnished only at stated intervals. The revenue of the diocese of St. Boniface is not much over 2000*l.*—*Abbreviated from the Nor-Wester. May 14th*, 1860.

VI.

WESLEYAN METHODIST MISSIONS.*

NORWAY HOUSE.

Rev. Robert Brooking, of Rossville.

Number of Members, 164 in 1859.

"Our population is still rapidly on the increase. Since my last report was written there have been thirty baptisms and only five deaths, so that there has been an increase, by births over deaths, of twenty-five; and there has also been an addition by families coming in from other places. During the two years of my resi-

* Thirty-fourth Annual Report of the Missionary Society of the Wesleyan Church in Canada, in connexion with the English Conference, Toronto, 1859.

dence here the population has increased from all sources fully 100. Five marriages have been solemnized during the past year.

"Several new houses were finished last fall, and there are six more now in frame, and the material got out for three or four others, which are to be raised when the boating is over. A larger quantity of ground is fenced in and under cultivation than has ever been before; so that if their prosperity does not keep pace with our wishes, yet we have every reason to thank God and take courage. In fact, there seems a general desire among our people to improve their circumstances."

OXFORD HOUSE.

Rev. Charles Stringfellow, of Jackson's Bay.

"As to the numbers in our Church there is a decrease of sixteen, being last year ninety-nine, the present year they number eighty-three. There are eighteen persons on trial for membership in addition to the eighty-three members. Of this decrease ten have removed to hunting-grounds contiguous to Rossville Mission, and therefore will augment by ten the numbers there in society. Five other persons, four male and one female, have died during the year."

EDMONTON AND ROCKY MOUNTAINS.

Rev. Thos. Wolsey.

VII.

THE FAR WEST IN THE UNITED STATES.

From its mouth to the Great Bend, the Missouri admits of almost continuous settlement on its *immediate* banks; thence, to Fort Union, only about one-fourth could be cultivated; and above Fort Union many extensive but detached bottoms show their adaptation for small independent agricultural areas.

The general westward progress of settlement a few miles west of the Upper Missouri River, is rendered impossible by the conditions of climate and soil which prevail there. The progress of settlement must necessarily be up the Valley of the Mississippi, on the immediate banks of the Missouri, and through the Valley of the Red River of the north, to the cultivable areas in the basin of Lake Winnipeg. The exploration for the Pacific Railroad and the meteorological investigations carried on under the direction of the Surgeon-General of the U. S. army, show conclusively that no settlement of any importance can be established over a vast extent of country, many hundred miles broad, on the eastern flank of the Rocky Mountains, and south of the Great Bend of the Missouri. Owing to the absence of rain, the apparently great rivers, the Platte, the Canadian, the Arkansas, &c., are often converted into long detached reaches or ponds during the summer months, and forbid extensive settlements even on their immediate banks. This great and important physical fact is contrary to popular opinion, which is mainly based upon an inspection of a map, and guided by the glowing but utterly erroneous descriptions which are periodically circulated respecting the wonderful fertility of the Far West, and its capability of sustaining a dense population.

The arid districts of the Upper Missouri are barren tracts, wholly uncultivable, from various causes.* The arid plains between the Platte and Canadian rivers are in great part sand-deserts. The "Sage-plains," or dry districts, with little vegetable growth except varieties of Artemisia, begin on the western border of the plains of the eastern Rocky Mountain slope, and cover much the larger portion of the whole country westward.† The sterile region on the eastern slope of the Rocky Mountains begins about 500 or 600 miles west of the Mississippi, and its breadth varies from 200 to 400 miles; and it is then succeeded by the Rocky Mountain range, which, rising from an altitude of 5200 in feet lat. 32°, reaches 10,000 feet in lat. 38°, and declines to 7490 feet in lat. 42° 24′, and

* Page 684, Army Meteorological Register, U. S.

† Ibid.

about 6000 in lat. 47°. Along this range isolated peaks and ridges rise into the limits of perpetual snow, in some instances attaining an elevation of 17,000 feet. The breadth of the Rocky Mountain range varies from 500 to 900 miles. The soil of the greater part of the sterile region is necessarily so from its composition, and where well constituted for fertility, from the absence of rain at certain seasons. The general character of extreme sterility likewise belongs to the country embraced in the mountain region.* The table subjoined is capable of conveying a very good idea of the great barrier to the westward progress of settlement, which lies between the Mississippi Valley and the Pacific slope of the Rocky Mountains. It is extracted from a table showing the lengths, sums of ascents and descents, &c. &c., of the several routes surveyed for a railroad from the Mississippi to the Pacific, and published in the "Exploration and Surveys," before quoted.

	Length of Railway.	No. of miles of route through arable land.	No. of miles of route through lands generally uncultivable, arable soil being found in small areas.	No. of sq. miles of sums of areas of largest bodies of arable land in uncultivable regions.
	Miles.			
Route near 47th & 49th parallel .	1864	374	1490	1000
" " 41st & 42nd " .	2032	632	1400	1100
" " 38th & 39th " .	2080	620	1460	1100
" " 35th " .	1892	416	1476	2300
" " 32nd " .	1618	408	1210	2300

This table shows that the least distance of uncultivable land through which a railway from the Mississippi to the Pacific must pass, in the United States territory, exceeds 1200 miles in length,—a barrier sufficient to arrest the general progress of settlement, for very many years to come, in a course due west of the Mississippi.

The only direction which remains for extensive free soil settlement in and near the United States is northwards, partially along the immediate banks of the Missouri, about the head-

* Exploration and Surveys for a Railroad Route from the Mississippi River to the Pacific Ocean, page 40.

waters of the Mississippi, and towards the valleys of the Red River and the Assinniboine and the main Saskatchewan. The popular impression that immense areas of land, available for the purposes of agriculture, lie between the Missouri and the Rocky Mountain chain, has, as before stated, been completely refuted by the explorations and surveys for the Pacific Railroad. The now well ascertained aridity of climate and its natural consequence, sterility of soil, both combine to confirm the title of "The Great American Desert," given by the early explorers of the eastern flank of the Rocky Mountains to that extensive region of country. This important fact cannot fail to exercise a powerful influence upon the occupation of British territory north of the 49th parallel of latitude, and on the sources from which that occupation will flow. [From a paper by the author on the Great North-West, published in 1858.]

VIII.

SEASONS AT RED RIVER.

ON THE PROGRESS OF THE SEASONS AND STATE OF THE WEATHER AT RED RIVER SETTLEMENT, FROM 1ST JUNE, 1855, TO 31ST MAY, 1856.

1855. June 5th was the coldest day in the month. Thermometer, 7 a.m., 58; 2 p.m., 63; 9 p.m., 56. The 14th was the hottest day. Thermometer, 7 a.m., 72; 2 p.m., 88; 9 p.m., 71. Three inches of rain fell on the 17th, one on the 19th and six on the 25th.

July 2nd was the coldest. Thermometer, 7 a.m., 56; 2 p.m., 78; 9 p.m., 68; light rain. The 25th was the hottest day. 7 a.m., 87; 2 p.m., 92; 9 p.m., 82. 7th, rain $3\frac{3}{8}$ inches; 10th, rain $\frac{3}{4}$ inches. Thunderstorm on the 17th, rain 3 inches. 26th, 1 inch rain; 29th, 3 inches rain; 30th, 2 inches; total $14\frac{5}{8}$ inches. Wheat out of the ear. On the 12th hay-cutting commenced. Tabani and mosquitoes very numerous and troublesome.

August: — Coldest day, 29th. Thermometer, 7 a.m., 44; 1 p.m., 68; 9 p.m., 56. The hottest day was the 5th. 7 a.m., 67; 2 p.m., 86; 9 p.m., 76. On the 8th, 5 inches of rain fell; 11th, 5¼ inches fell; 14th, 2 inches; 27th, ¼ inch: total, 12½ inches. Barley harvest commenced about the 1st; wheat harvest on the 15th. Slight frost on the 30th.

September:—The coldest day was the 30th. Thermometer average + 48. The hottest day was the 5th; thermometer, 7 a.m., 70; 2 p.m., 80; 9 p.m., 70. Total of rain during the month, 6½ inches. Finished storing wheat on the 8th. A few leaves falling. 26th, grey geese flying to the south.

October: — The warmest day was the 1st. Thermometer, 7 a.m., 56; 2 p.m., 70; 9 p.m., 58. Some snow fell on the 4th. Taking up potatoes on the 8th. White geese flying to the south, and continued to do so up to the 20th, and a few flocks later than that: all the larger kind of ducks leave about the same time. The deciduous trees are bare of leaves, except the oak, and some of the hardier kinds.

November:—The 2nd was the warmest day. Thermometer, 7 a.m., 32; 2 p.m., 38; 9 p.m., 36; 2½ inches rain fell on the 3rd; 5 inches of snow fell on the 11th; 12th, river covered over with ice. The coldest day of the month was the 21st, thermometer, 7 a.m., —12; 2 p.m., +8; 9 p.m. +6. Warm weather from the 21st to the end of the month. Seven inches of snow fell during the month. Flocks of snow birds have made their appearance from the north, and all the summer birds are gone.

December:—The warmest day was the 6th. Thermometer, 7 a.m., +22; 2 p.m., +26; 9 p.m., +30. The coldest day was the 24th; thermometer, 7 a.m., —48; 2 p.m., —30; 9 p.m., —40. We had six days of very cold weather, including the 23rd and 28th. The wind blew from the north during three days before the severe cold began; during its continuance there was very little wind, and for two of the coldest days it was at the south. Eight inches of snow fell.

1856, January:—Tbe warmest day was the 17th. Thermometer, 7 a.m., +10; 2 p.m., +22; 9 p.m., +16. The coldest was the 7th; thermometer, 7 a.m., —36; 2 p.m.—28; 9 p.m., —36. Five inches of snow fell. The average cold for this month has not been great; very little wind.

February:—Coldest day the 2nd. Thermometer, 7 a.m., —36; 2 p.m., —20; 9 p.m., —34. The warmest day was the 20th; thermometer, 7 a.m., +26; 2 p.m., +35; 9 p. m., +24. 6 inches of snow fell. After the 12th, spirits of wine in the glass stood with few exceptions above zero, and the weather has been pleasant.

March:—The coldest day was the 8th; 7 a.m., —32; 2 p.m., 24; 9 p.m., —26. The warmest day was on the 22nd. Thermometer, 7 a.m., +28; 2 p.m., +38; 9 p.m., +34. The thermometer fell during the night a few degrees below zero; but on the whole the weather was pleasant; 6½ inches of snow fell. Much of the snow melted during the month. Barking crows made their appearance about the 20th.

April:—Geese made their appearance on the 2nd, and the snow birds left us for the north. The 12th was the coldest day this month. Thermometer, 7 a.m., +16; 2 p.m., +30; 9 p.m., +24. Warmest day, 23rd: thermometer, 7 a.m., +46; 2 p.m., +66; 9 p.m., +44. About six inches of snow, and five of rain fell. On the 16th the rain began to throw off its winter coat; clear of ice on the 20th. Sturgeon taken in the river in great numbers; the snow all away. Wild fowl to be seen in every direction on the 29th, and sowing wheat commenced.

May:—The coldest day, 11th. Thermometer, 7 a.m., +34; 2 p.m., +43; 9 p.m., +30. The warmest day was the 18th, 7 a.m., +75; 2 p.m., +84; 9 p.m., +56; four inches rain fell on the 26th. On the 4th whip-poor-will began his serenades. The wheat sown on the 29th has germinated, and given a green appearance to the field. On the 9th wild geese abundant in the plains; maple in leaf, gooseberry bushes the same; finished sowing wheat on the 10th.

1856. Wheat sown in the beginning of May, was in the ear on the 13th July, and ripe on the 20th August. The wheat sown on the 29th April was ripe on the 14th August. The hottest day this last summer was the 20th of July. Barley harvest commenced in July; finished cutting wheat on the 28th August; slight frost on the 30th of the same month; potatoes taken up first week of October.

6th September:—Flocks of grey geese flying to the south. *Prunus Americana* ripe and very plentiful in the first part of this month, or rather before this month. Flocks of passenger pigeons are in from the north, and leave from the 20th to the last of the month. On the night of the 7th whip-poor-will gave us his parting song. *Coregonus lucidus* enter the river to spawn. The *Coregonus albus* in Lake Winnipeg commences spawning about the 19th of October, and ends about the 1st November.

This Register was kept by Mr. Donald Gunn, of the Lower Settlement, Red River. For the details of the Register see the Red River Report for 1857.

IX.

Extracts from a Journal kept at "Fort à la Corne," on the Main Saskatchewan, Lat. 55°·30, *Long.* 104°·25.

1856.

April 1.—No frost last night, but thick mist this morning. The weather has been warm although cloudy.

2.—Hard frost last night, but mild during the day.

4.—Slight frost last night, day very mild. Snow dissolved a great deal during the day. Water making its appearance on edge of river.

7.—Froze hard last night, and has been cold most of the day.

8.—Do. do. no thaw during the day. River rising very much, and boat frozen in.

9.—Ice made a start previous to moving.

17.—Warm weather, ice drifting down river.

April 19.—Weather fine. Annual Goose Dance of McLeod took place to-day.

21.—Rain with N. W. wind.

23.—Had good fall of snow during night. Continued snowing without intermission the whole day. Nets set for first time. One sturgeon, ten suckers, and one gold-eye caught.

25.—Hard frost last night.

26.—Weather fine, considerable quantity of ice in river, but melting fast.

May 1.—Weather warm. Change perceived on trees, they are getting a little *green*.

2.—Working in garden; put down peas, onions, radish, and a few greens. Net produced two sturgeon.

6.—Weather warm.

10.—Storm of snow and rain during last night, with a strong north wind, which continued at intervals during the day.

12.—Planted north field with potatoes, and ploughed south field.

13.—Cloudy, rain, with N. W. wind. Planted potatoes in south garden.

14.—Cold north wind. Sowed four beds of Swedish turnips.

21.—Thunder and lightning most of last night. Rain poured down in torrents. River rose considerably to-day.

30.—Saskatchewan Brigade arrived this afternoon. Started same evening.

June 1.—Clear and beautiful to-day.

Sept. 16.—Raining all day, wind east.

17.—Clear but rather cold. Slight frost last night. Wind N. E. (light).

20.—Mild and warm during day. Slight frost last night.

Oct. 2.—Raining all morning; wind W. Cleared up in the afternoon. Men in morning cleared all the *potato* stalks out of north garden, and in afternoon commenced again the potatoes in south garden.

Oct. 13.—S. W. wind. Fall boats started this morning for Carlton.

17.—Fine weather, men employed in garden.

18.—Do. do. putting dung in garden.

22.—Very hard frost over night.

23.—Severe frost last night.

26.—Snowed during night, but thawed as it fell. Blowing very hard.

Nov. 11.—River full of ice.

16.—Weather fine. One cow calved.

31.—Slight fall of snow last night, but day remarkably fine.

Dec. 1.—Weather fine, not in the least cold. Have had no cold weather as yet, compared to last year.

2.—Weather colder than of late.

5.—Slight fall of snow during night.

11.—Very cold.

12.—Cold, very severe.

31.—Snowing most of the day.

1857.

Jan. 2.—Cold, and snowing at intervals.

3.—Very cold.

Mar. 29.—Hard frost last night.

30.—Very warm, snow melting about the fort.

31.—Raining during the night. Slight rain during the day.

April 2.—North wind and cold. No thaw these three days back.

3.—North wind and very cold.

4.—North wind.

5.—Weather milder, a slight thaw.

6.—South wind, thawing a great deal.

8.—Hard frost last night, cold all day. North-west wind accompanied with snow, which continued most of the day.

April 9.—Snowed last night. Cold during day. Water appearing on edges of river.

At this time last year ice started in river. What a difference this year! We cannot go anywhere at present without snow-shoes, our cattle are nearly starved, they cannot go about as the snow is so hard.

10.—Weather still cold, wind variable.

11.—Storm of snow and wind.

12.—Hard frost last night. Cold all day. No thaw.

13.—Blowing hard, accompanied with snow. Day fine and snow dissolving. Turned very stormy in afternoon. North wind with snow.

14.—Still cold. North wind. No thaw.

15.—Southerly wind, but still cold. River still rising at edges. Little or no thaw during day.

16.—Weather clear, but still cold. Little or no thaw. Notwithstanding the late cold weather the ice went off this day.

17.—Weather same. Very little ice drifting down river.

18.—Weather still cold. North wind.

19.—Fine during day. Sun shining bright. Snow melted a good deal.

20.—Day fine, but weather turned cold towards evening. Had a slight fall of snow last night.

21.—Beautiful day. Snow dissolving fast. Little or no ice drifting.

22.—Cloudy and variable, very little thaw.

23.—Strong south wind. Thawing very much.

24.—Snowed without intermission the whole day. Wind variable, and blowing hard.

25.—Beautiful day. Warmest we have had this season.

27.—Cloudy and cold, with slight snow.

28.—Weather fine and warm.

29.—Cold and cloudy. Slight snow.

30.—Beautiful day, but blowing hard.

May 1.—Weather and wind from same quarter. Snow dissolving fast.

3.—Ice drifting all last night, but not much to-day.

5.—Disagreeable day. Snowing without intermission, with a cold north wind. River full of ice.

8.—Stormy northerly wind, and very cold.

12.—Weather warm. Yesterday planted potatoes and onions in south garden, and to-day sowed cabbages in boxes.

15.—Mild, wind south.

18.—Boisterous weather.

20.—Beautiful day. All hands employed planting potatoes. Sowed turnips, carrots, beans, &c. Nets caught three sturgeon and nine suckers.

21.—Very warm. Annual Goose Dance came off.

June 2.—Hard frost last night. Froze my beans, and the hops were affected also.

7.—Rained hard all last night, and continued without intermission all day.

9.—Fine weather, river stiil rising.

15.—Very warm and clear this afternoon.

30.—Beautiful day. Bull dogs so numerous that horses had to be put in stable, and grass cut for them. Starvation is staring the people in the face. Have caught no sturgeon for some time back. Our nets produced nothing to-day.

1858.

April 20.—Warm and clear, south wind.

21.—Ice drifting in river. Large quantity of ice on banks.

22.—Cold north wind.

24.—Slight fall of snow in morning. Rain towards sunset. Still cold, wind south-west.

25.—Warm and fine to-day.

27.—South wind. Warmest day this spring.

April 28.—North wind. Cold and blowing hard.

May 1.—South wind. Warm. Sky overcast with smoke. Large fire close to fort. Clearing up north garden.

7.—Set four men to dig potatoe ground in south garden. Caught one sturgeon, first this spring.

11.—Cold north wind. Cut the potatoes for planting.

12.—Planted potatoes in south field and commenced to dig the north field for sowing. Sowed beetroot, radish, and lettuce.

15.—South wind. Weather cold. Planted north garden with potatoes.

17.—Still cold. Slight fall of snow in night.

18.—Wind from north, and cold. Think we are going to have a second winter.

19.—Continues cold. Wind north.

20.—Weather improving, wind south-west.

21.—Warm and mild. South wind.

22.—Warm and fine.

23.—Warm in morning. Thunder and rain towards sunset.

24.—Warm. Wind south. Clearing up garden. River muddy and water rising fast.

June 1.—Wind south, and weather warm.

8.—Wind north, and appearance of cold. Think we are going to have a cold summer. Garden herbs slow in making their appearance above ground.

15.—Weather continues warm.

July 1.—Boisterous weather. Wind north.

10.—Very warm to-day. Bull dogs so numerous, horses and cattle had to be kept in stable all day. Man hoeing south garden.

21.—A very fine day.

X.

PRICES OF FURS AT FORT GARRY IN JANUARY, 1860:—

Badgers	1*s.* 0*d.*
Beavers (large)	3*s.* 6*d.*
Bears (black)	£1 10*s.* to £2 0*s.* 0*d.*
Fishes	10*s.* 0*d.*
Foxes (silver)	£6 0*s.* 0*d.*
Do. (cross)	15*s.* 0*s.* to £1 0*s.* 0*d.*
Do. (red)	4*s.* 0*d.*
Lynx (large)	4*s.* 0*d.*
Martens	8*s.* to 10*s.* 0*d.*
Minks	3*s.* 0*d.*
Otters	10*s.* to 12*s.* 0*d.*
Racoon	2*s.* 0*d.*
Buffalo skins (dressed)	8*s.* to 10*s.* 0*d.*
Moose skins (dressed)	12*s.* to 15*s.* 0*d.*
Wolves	3*s.* 6*d.*
Rats	1*d.*

XI.

REMARKS ON THE PIGEON RIVER ROUTE, WITH AN ESTIMATE OF THE COST OF OPENING IT FOR COMMERCIAL TRAFFIC. BY J. A. DICKINSON, C. E.*

SIR,

* * * * * * *

On looking at the profile of this route (see profile of the Pigeon River Route, plate of sections), one is struck by its similarity to the profile of an artificial canal; for with the exception of the Pigeon River and the Nameaukan, this route consists of a close chain of lakes, only requiring locks to form a canal sufficient for boats or steamers as large as will ever be needed for carrying the traffic of the country. It will be shown hereafter how the Pigeon River can be avoided, and the Nameaukan presents no difficulties that cannot be easily overcome.

* Reports of Progress, &c. Assinniboine and Saskatchewan Exploring Expedition. Chap. xxi. page 201. Blue Book — presented Aug. 1860.

The profile of the Kaministiquia and Rivière la Seine route shows that out of a length of 240 miles, there is only about ninety-five miles of lake navigation; the remaining 145 miles consists of rivers of various velocities. Now the advantages of lakes over rivers for the purposes of navigation are many and important: 1st, the absence of currents in lakes, which in rivers impede the upward navigation more than they assist the downward. 2ndly, lakes are not subject to the great and sudden changes of level which occur in most rivers. 3rdly, a high flood in a lake would be of no consequence, whereas in a river it might be dangerous and full of difficulties. 4thly, long and straight courses may be obtained on lakes, and the steering of craft is consequently easy; but on rivers, owing to their windings and eddies, steering is troublesome and difficult.

These are some of the reasons for my preferring the route by the lakes to the Kaministiquia and Rivière la Seine route; the others being that it is shorter by several miles, that the portages on it are better, and lastly, that it passes through a country bearing superior timber. This for a long period of years was the route traveled by the old voyageurs, and was only given up, I understand, on account of the length of the Grand Portage, which was supposed to have had a bad effect on the spirits of the men, occurring as it did at the commencement of their journey. I may remark, in confirmation of this opinion, that our guide, who was with us on both routes, and who had made some forty journeys between Lake Superior and Red River, said he much preferred the Pigeon River route to any other, and how could there be a person better qualified for being a judge than he?

The lakes on the Pigeon River route are all deep, free from shoals and rocks, wide, and yet not so wide as to be affected by winds. The works necessary for the improvements of either of these routes are of course of the same character, but for the same reasons that make the Pigeon River route preferable to the other even now, the works on it could be more easily executed; they would cost less and be more permanent. A dam placed across a river is always liable to be more or less damaged by freshets, and most of those on the Rivièrè la Seine route would, moreover, I think, flood a great portion of the country; whereas dams built at the head of the streams issuing from the lakes, would not have to bear any extraordinary force arising from a sudden rush of water, and would therefore be more stable. The portages on the Pigeon River route being better than those on the other, will, of course, cost less for improvements; and if hereafter locks were to be made where the portages are now, the lakes will afford a much surer and more abundant supply of water for them than the rivers. The means that might be adopted for making this Pigeon River route sufficient for any purposes that are likely to be required at present or for some time to come, are the following:—

The repairing, or perhaps remaking, of the old North-West Company's road from Point des Meurons, near Fort William, to Arrow Lake. This road is only about forty-five miles long, and has been reported favourably on by a member of the Red River expedition, who examined it in the year 1857. The difficulties that there would be in improving the portion of the route between Lake Superior and the Height of Land, in any way compel the necessity of this road being made use of. Arrow Lake will then be the commencement of the water communication. Boats capable of carrying five tons, such as are at present used by the Hudson's Bay Company, to be employed on the lakes as far as Rainy Lake. In order to avoid the unloading and re-loading of the boats at the portages, iron tramways might be laid down at these places, by which the loaded boat might be carried from one lake to another, and thus save a great deal of time and trouble. Some of the present portages might be got rid of with little expense, by clearing out the passage between two lakes, and placing a dam so that the water in the lower lake would rise to the level of the upper one, and so deepen the connecting channel and get rid of the rapid or fall in it. This could be done in several cases. On Rainy Lake a steamer might be placed, one of the ordinary lake steamers, which could go as far as Fort Frances on Rainy River. Here, on account of the falls, a transhipment must be made to another steamer, which can ply the whole way between the falls and the north-west corner of the Lake of the Wood. From this point a road to be made across to the Red River Settlement. This road would not be more than 100 miles long, the greater part of it, seventy miles at least, would, I know from my own knowledge (see my Report, No. VII.), be made without any difficulty.

The following is a rough estimate of the cost of these several works:—

	£
Road from Point des Meurons (Fort William) to Arrow Lake	1,000
Tramways over the portages	5,000
Dams	2,000
Road from the Lake of the Woods to Red River	1,500
Depôts	500
Engineering and contingencies	1,500
	11,500

I think it is quite safe to say that the expenses for making the improvements I have suggested would be considerably under 12,000*l.* sterling.

The length of the proposed route, and the time it will take to perform the journey, will be—

	Length.	Hours.
Carriage Roads	145	36
Tramways	3	6
Boat navigation	156	39
Steam navigation	186	18
Total	490	99

Allowing a rest of eight hours in the twenty-four, the journey may easily be accomplished in less than six days. Mr. Dawson has stated that it will take but "three days, as near as may be," to accomplish the journey by the other route; but it appears he takes no thought of the necessary delays at the portages, or of the rest necessary both for the travelers and the boatmen, and he proposes to run at night on unlighted and intricate waters. The journey might no doubt be made in five days, if necessary, by the route I propose; but allowing for accidents and delays, I think it better to calculate on its taking six.

In conclusion, I would say, that before works of any magnitude are undertaken for the opening out of a communication between Lake Superior and Red River, it would be advisable that a more accurate survey than has as yet been taken should be made of the whole country.

Yours truly,

James Austen Dickinson.

H. Y. Hind, Esq.
&c. &c.

XII.

*Table of the Portages, Décharges, Rapids, Lakes, Lake Straits, and Navigable Channels on the Pigeon River Route (the old North-West Company's Route) from Lake Superior to Rainy Lake, showing their Lengths and Distance from Lake Superior.**

Portages.	Length in St. miles.	Décharges.	Length in St. miles.	Rapids.	Length in St. miles.	Lakes.	Length in St. miles.	Lake Straits and Navigable Channels.	Length in St. miles.	Distance from Lake Superior. Statute miles.	Remarks.
No. 1. Grand Portage	8·16									0·00	
								No. 1. Pigeon R.	1·50	8·16	
2. Partridge Portage	0·25									9·66	
								2. Pigeon R.	4·25	9·91	Strong current. Canoes poled up part of the way.
		No. 1. Décharge	0·25							14·16	
								3. Pigeon R.	1·75	14·41	
		2. Décharge	0·35							16·16	
								4. Pigeon R.	3·13	16·51	
3. Fowl Portage	1·13									19·64	
						No. 1. Fowl Lake	4·55			20·77	
4. Mosse Portage	0·41									25·32	
						2. Mosse Lake	4·25			25·73	
5. Great Cherry Por.	0·48									29·98	
						3. Lake	0·25			30·46	
6. Mud Portage	0·15									30·71	
						4. Lake	0·30			30·86	

* The distances are from the International Boundary Survey—made according to the 7th article of the Treaty of Ghent.

Table of Portage, &c.—(continued).

Portages.	Length in St. miles.	Décharge.	Length in St. miles.	Rapids.	Length in St. miles.	Lakes.	Length in St. miles.	Lake Straits and Navigable Channels.	Length in St. miles	Distance from Lake Superior. Statute miles.	Remarks.
No. 7. Lesser Cherry P.	0·13									31·16	
						No. 5. Mountain L.	7·81			31·29	
8. Watap Portage	0·30									39·10	
						6. Watap Lake.	3·70			39·40	
9. Great New Port.	1·46									43·10	
						7. Rose Lake	3·00			44·56	
10. Portage	0·01									47·56	4 feet fall.
						8. Mud Lake	2·62			47·57	
11. Portage	0·21									50·19	
						9. South Lake	2·84			50·40	
12. Height of Land P.	0·26									53·24	
						10. Gun Flint L.E.	2·77			53·50	
		No. 3. Décharge	0·01							56·27	4 feet fall.
						11. Gun Flint L.W.	8·92			56·28	
				No. 1. Rapid	0·01					65·20	2 ft. fall. North canoes let down by line.
								No. 5. Lake Strait	0·25	65·21	
13. Little Rock Port.	0·02									65·46	
								6. Lake Strait	0·75	65·48	
				2. Rapid	0·01					66·23	
								7. Lake Strait	0·25	66·24	

XIII.

*Table of the Portages, Décharges, Rapids, Lakes, Lake Straits and Navigable Channels on the Pigeon River Route (the old North-West Company's Route) from Lake Superior to Rainy Lake, showing their Lengths and Distance from Lake Superior.**

Portages.		Décharges.		Rapids.		Lakes.		Lake Straits and Navigable Channels.		Distance from Lake Superior.	Remarks.
	Length in St. miles.		Length in St. miles.		Length in St. miles.		Length in St. miles.		Length in St. miles.	Statute miles.	
No. 1. Grand Portage	8·16									0·00	
								No. 1. Pigeon R.	1.50	8·16	
2. Partridge Portage.	0·25									9·66	
								2. Pigeon R.	4·25	9·91	Strong current. Canoes poled up part of the way.
		No. 1. Décharge	0·25							14·16	
								3. Pigeon R.	1·75	14·41	
		2. Décharge	0·35							16·16	
								4. Pigeon R.	3·13	16·51	
3. Fowl Portage	1·13									19·64	
						No. 1. Fowl Lake.	4·55			20·77	
4. Moose Portage	0·41									25·32	
						2. Moose Lake.	4·25			25·73	
5. Great Cherry Por.	0·48									29·98	
						3. Lake	0·25			30·46	
6. Mud Portage	0·15									30·71	
						4. Lake	0·30			30.86	
7. Lesser Cherry P.	0·13									31·16	
						5. Mountain L.	7·81			31·29	

* The distances are from the International Boundary Survey—made according to the 7th article of the Treaty of Ghent.

Table of Portages, &c.—(continued).

Portages.	Length in St. miles.	Décharge.	Length in St. Miles.	Rapids.	Length in St. miles.	Lakes.	Length in St. miles.	Lake Straits and Navigable Channels.	Length in St. miles.	Distance from Lake Superior. Statute miles.	Remarks.
No. 8. Watap Portage	0·30									39·10	
						No. 6. Watap Lake	3·70			39·40	
9. Great New Por.	1·46									43·10	
						7. Rose Lake.	3·00			44·56	
10. Portage	0·01									47·56	4 feet fall.
						8. Mud Lake.	2·62			47·57	
11. Portage	0·21									50·19	
						9. South Lake	2·84			50·40	
12. Height of Land P.	0·26									53·24	
						10. Gun-Flint L.E.	2·77			53·50	
		No. 3. Décharge	0·01							56·27	4 feet fall.
						11. Gun-Flint L.W.	8·92			56·28	
				No. 1. Rapid	0.01					65·20	2 ft. fall. North canoes let down by line
								No. 5. Lake Str.	0·25	65·21	
13. Little Rock Por.	0·02									65·46	
								6. Lake Str.	0·75	65·48	
				2. Rapid	0·01					66·23	
								7. Lake Str.	0·25	66·24	
14. Mill-Fall Portage	0·06									66·49	
								8. Lake Str.	0·25	66·55	
15. Island Portage	0·29									66·80	
						12. Lake	1·35			67·09	
										68·44	3 ft. fall, 2 ft. deep. North canoes let down by line.
				3. Rapid	0·02	13. Lake	0.85			68·46	

Table of Portages, &c.—(continued).

Portages.	Length in St. miles.	Décharge.	Length in St. miles.	Rapids.	Length in St. miles.	Lakes.	Length in St. miles.	Lake Straits and Navigable Channels	Length in St. miles.	Distance from Lake Superior. Statute miles.	Remarks.
				No. 4. Rapid	0·07	...	...	...	0.75	69·31	5 ft. fall, 2 chs. wide. North canoes let down by line.
								No. 9. Lake Str.	.	69·38	
				5. Rapid	0·05	...	...	...	0·50	70·13	4 ft. fall, 25 ft. wide. North canoes let down by line
								10. Lake Str.	.	70·18	
				6. Rapid	0·04	...	...	...	0·12	70·68	3 ft. fall. North canoes let down by line.
								11. Lake Str.	.	70·72	
				7. Rapid	0·01	...	...	...	0·20	70·84	1 ft. fall. Run by loaded North canoes.
								12. Lake Str.	.	70·85	
				8. Rapid	0·01	...	...	...	0·45	71·05	1 ft. fall. Run by loaded North canoes.
								13. Lake Str.	.	71·06	
				9. Rapid	0·02	...	...	...	.	71·51	2½ feet fall.
						No. 14. Lake .	5·35	...	0·70	71·53	
		No. 4. Decharge	0·05	...	...	...	...	...	.	76·88	5 feet fall.
								14. Lake Str.		76·93	
No. 16. Portage . .	0·01	...	...	...	...	...	...	...	.	77·63	
						15. L. Seiganagh	10·07	...	.	77·64	
17. Portage . .	0·01	...	...	...	...	...	...	...	.	87·71	
						16. Swamp Lake	0·82	...	.	87·72	
18. Swamp Portage	0·24	...	...	...	...	...	...	...	.	88·54	

Table of Portages, &c.—(continued).

Portages.	Length in St. miles.	Décharges.	Length in St. miles.	Rapids.	Length in St. miles.	Lakes.	Length in St. miles.	Lake Straits and Navigable Channels.	Length in St. miles.	Distance from Lake Superior. Statute miles.	Remarks.
						No. 17. Cypress Lake	5·35			88·78	
No. 19. Portage	0·02									94·13	
						18. Knife Lake.	10·70		0·20	94·15	
20. Portage	0·04									104·85	
								No. 15. Lake Str.		104·89	
21. Portage	0·09									105·09	
						19. Lake	0·60			105·18	
22. Carp Portage	0·15									105·78	
						20. Birch Lake.	4·00			105·93	
23. Portage	0·11									109·03	
						21. Basswood L.	16·00			110·04	
24. Portage	0·10									126·04	
								16. Lake Str.	0·25	126·14	
25. Fir Portage	0·20									126·39	
								17. Lake Str.	0·22	126·59	
				No. 10. Rapid	0·04					126·81	3 feet fall.
								18. Lake Str.	0·35	126·85	
				11. Rapid	0·02					127·20	2 feet fall.
								19. Lake Str.	1·50	127·22	
26. Portage	0·07									128·72	
						22. Lake	0·65			128·79	
27. Portage	0·09									129·44	
						23. Crooked L.	16·80			129·53	
28. Curtain-Fall Por.	0·10									146·33	
				12. Rapid	0·03					146·43	3 feet fall.
						24. Iron Lake.	4·50			146·46	

Table of Portages, &c.—(continued).

Portages.	Length in St. miles.	Décharges.	Length in St. miles.	Rapids.	Length in St. miles.	Lakes.	Length in St. miles.	Lake Straits and Navigable Channels.	Length in St. miles.	Distances from Lake Superior. Statute miles.	Remarks.
No. 29. Bottle Portage	0·25									150·96	
						No. 25. Nequaquon L.	22·10			151·21	
30. Portage	0·12									173·31	
						26. Lake	5·60			173·43	
31. Portage	0·15									179·03	
								20. Lake Str.	0·50	179·18	
32. Portage	0·04									179·68	
								21. Loon's Narrows.	7·60	179·72	
						27. Sand Point L.	9·67			187·32	
						28. Nameukan L.	5·20			196·99	
33. Nu Portage	0·08									202·19	
								22. Lakelet.	0·25	202·27	
34. Portage	0·14									202·52	
								23. Lake Str.	5·20	202·66	
Total	15·33	Total	0·66	Total.	0·33	Total.	160·62	Total	30·92	207·86	Statute miles to Rainy Lake.

SYNOPSIS OF THE FOREGOING TABLE OF THE PIGEON RIVER ROUTE.

	Statute miles.	
Land Carriage	15·33	16·32 stat. miles Interrupted Navigation.
Décharges	0·66	
Rapids	0·33	
Lakes	160·62	191·64 stat. miles Navigable Water.
Lake Straits	30·92	
Aggregate distance	207·86	from Lake Superior to Rainy Lake.

Distance from Lake Superior to Rainy Lake, *viâ* the Kaministiquia Route, = 263·34 statute miles.

XIV.

*List of Photographs of Red River; the Churches of Selkirk Settlement; the Houses and Stores of the Settlers; Indian Tents and Graves; the Prairie; the Forts and Stores of the Hon. Hudson's Bay Company; the Native Races; and the Expedition returning to Crow Wing in Dog Carioles.**

THE RED RIVER.

View of Red River, from the Stone Fort.

View of Red River, from St. Andrew's Church, four miles above the Stone Fort.

Red River; Middle Settlement, eight miles below Fort Garry.

Freighter's Boat on the banks of Red River, seven miles below Fort Garry.

Bishop's Court (the residence of the Bishop of Rupert's Land,) on the banks of Red River.

(These Photographs exhibit the general character of the river.)

CHURCHES OF SELKIRK SETTLEMENT.

Cathedral of St. Boniface (Roman Catholic) and Nunnery, on the banks of Red River, opposite Fort Garry.

St. John's Church, two miles below Fort Garry. (Ch. of Eng.)

Presbyterian Church and Parsonage, seven miles below Fort Garry.

St. Paul's Church, Parsonage, and School House, eight and a half miles below Fort Garry. (Ch. of Eng.)

St. Andrew's Church (Rapids Church), sixteen miles below Fort Garry. (Ch. of Eng.)

St. Andrew's Parsonage.

* These Photographs can be procured from Mr. J. Hogarth, No. 5, Haymarket, London. Price two guineas.

HOUSES AND STORES OF THE SETTLERS.

Residence of Mr. Bannatyne, near Fort Garry.

Mr. McDermot's Store, near Fort Garry.

Farm-houses and Windmills, Middle Settlement.

INDIAN TENTS AND GRAVES.

Ojibway Tents, on the banks of Red River, near the Middle Settlement.

Birch-bark Tents, west bank of Red River, Middle Settlement.

Indian Graves, covered with split sticks.

Indian Graves, covered with birch bark.

THE PRAIRIE.

The Prairie, on the banks of Red River, looking south.

FORTS AND STORES OF THE HONOURABLE HUDSON'S BAY COMPANY.

Fort Garry; at the confluence of Red River and the Assinniboine.

Hon. Hudson's Bay Company's Officers' Quarters: Lower or Stone Fort.

Fur Store: interior of Lower or Stone Fort.

NATIVE RACES.

John McKay: a Cree half-breed.

Letitia: a Cree half-breed.

Susan: a Swampy-Cree half-breed.

Wigwam: an Ojibway half-breed, Lake Superior.

An Ojibway Squaw with Papoose.

Red River Freighter's Boat.

Dog Carioles; Expedition returning to Crow Wing, by the winter road.

A Blackfoot Warrior's Robe, containing a history of his wars, &c., Buffalo hunts, and showing the number of scalps he has taken from his enemies.

INDEX.

THE END.

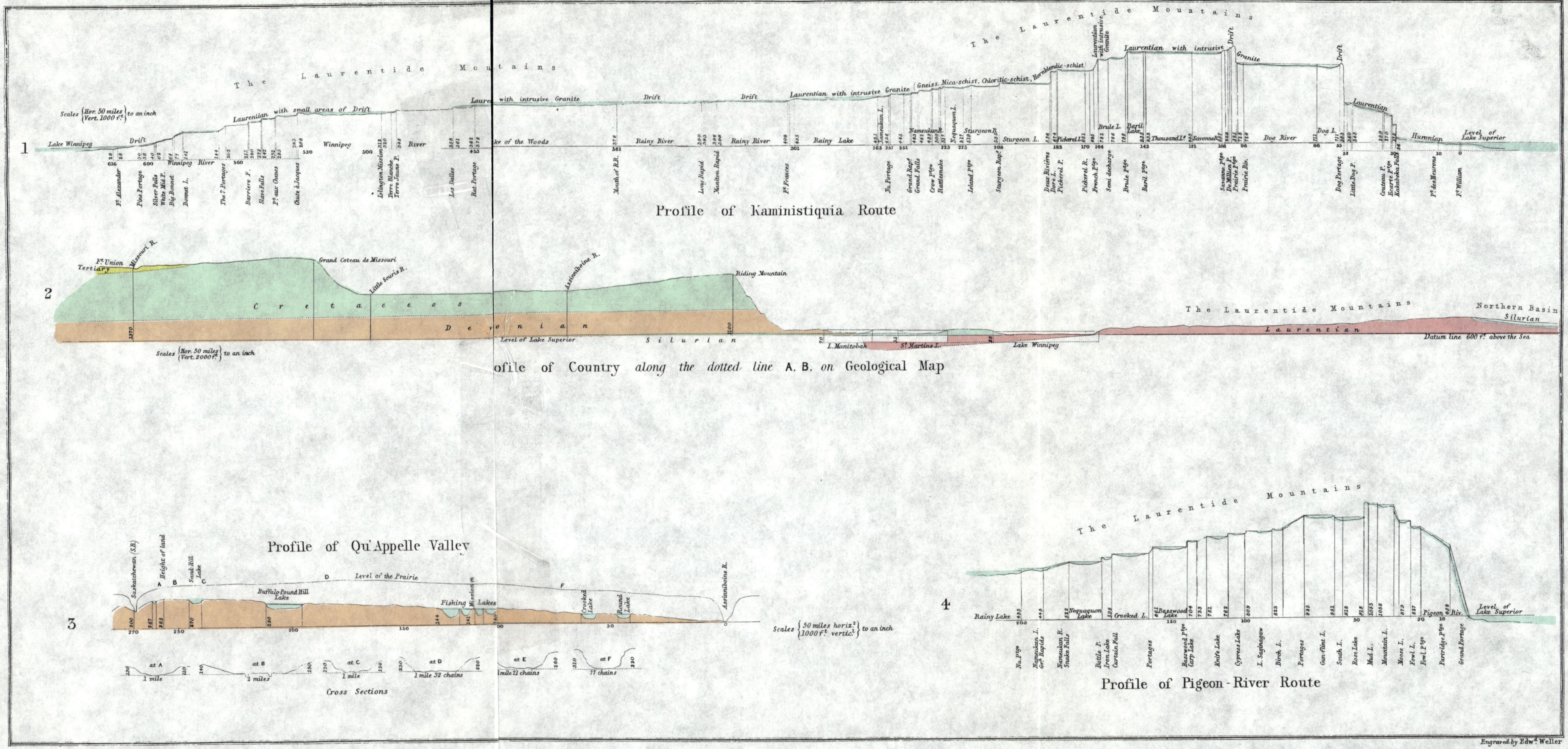

Engraved by Edw.d Weller

London. Longman & Co.

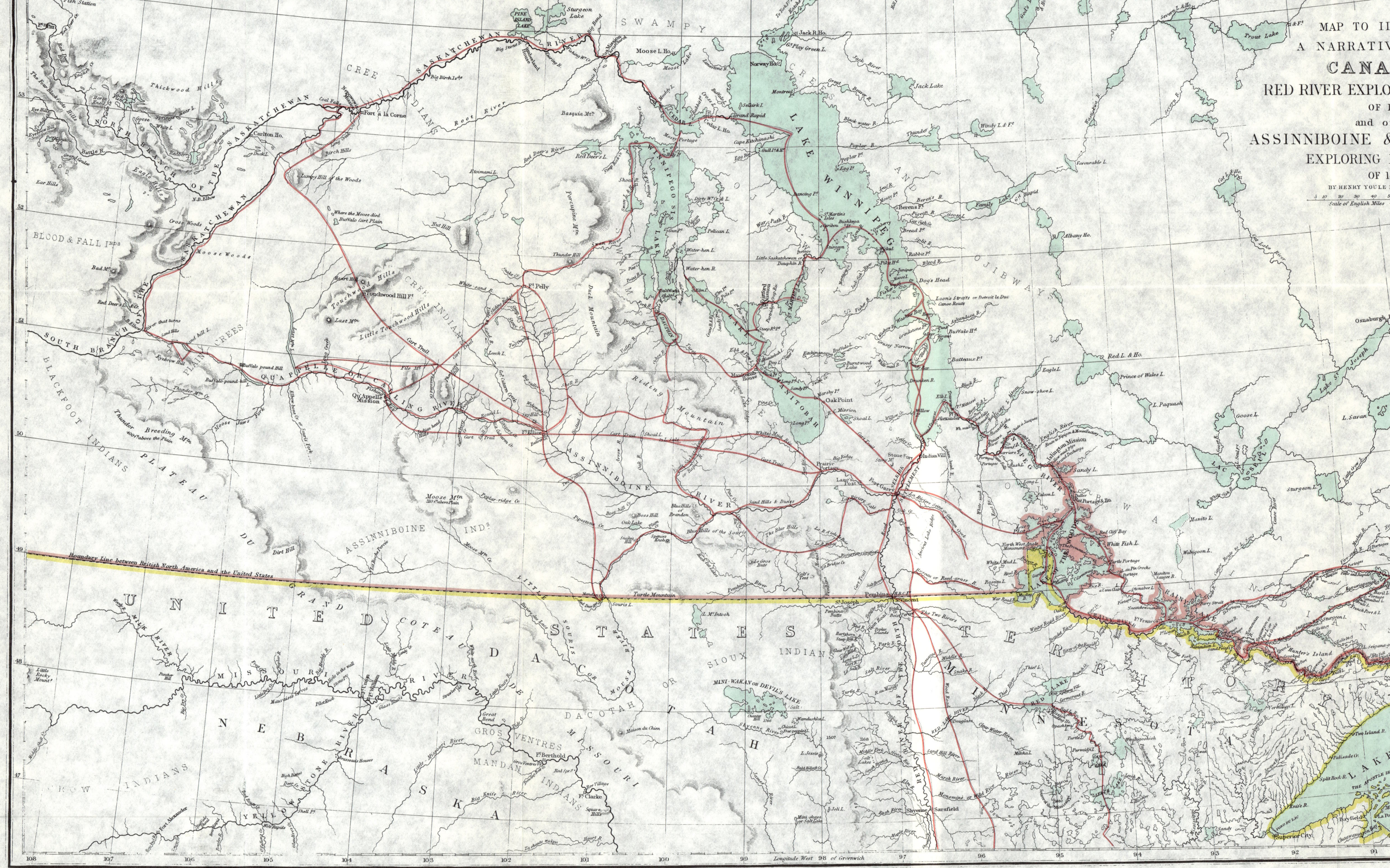

London: Longman & Co.

Note. The Topography south of the Boundary Line and west of Red River is from the Military Map of Nebraska and Dakotah by Lieut. G.K. Warren U.S. Top! Eng^{rs}; East of Red River including the North Shore of Lake Superior from Dr D.D. Owen's Map of Iowa Minnesota and Wisconsin. The North Branch of the Saskatchewan and the Grand Coteau north of the Boundary Line are from Capt. Palliser's Map (Blue Book 1859). The longitudes west of Fort Ellice are from Capt. Palliser's Map.

Map of the Country from LAKE SUPERIOR to the PACIFIC OCEAN, *showing the Western Boundary of Canada & the Eastern Boundary of British Colu*

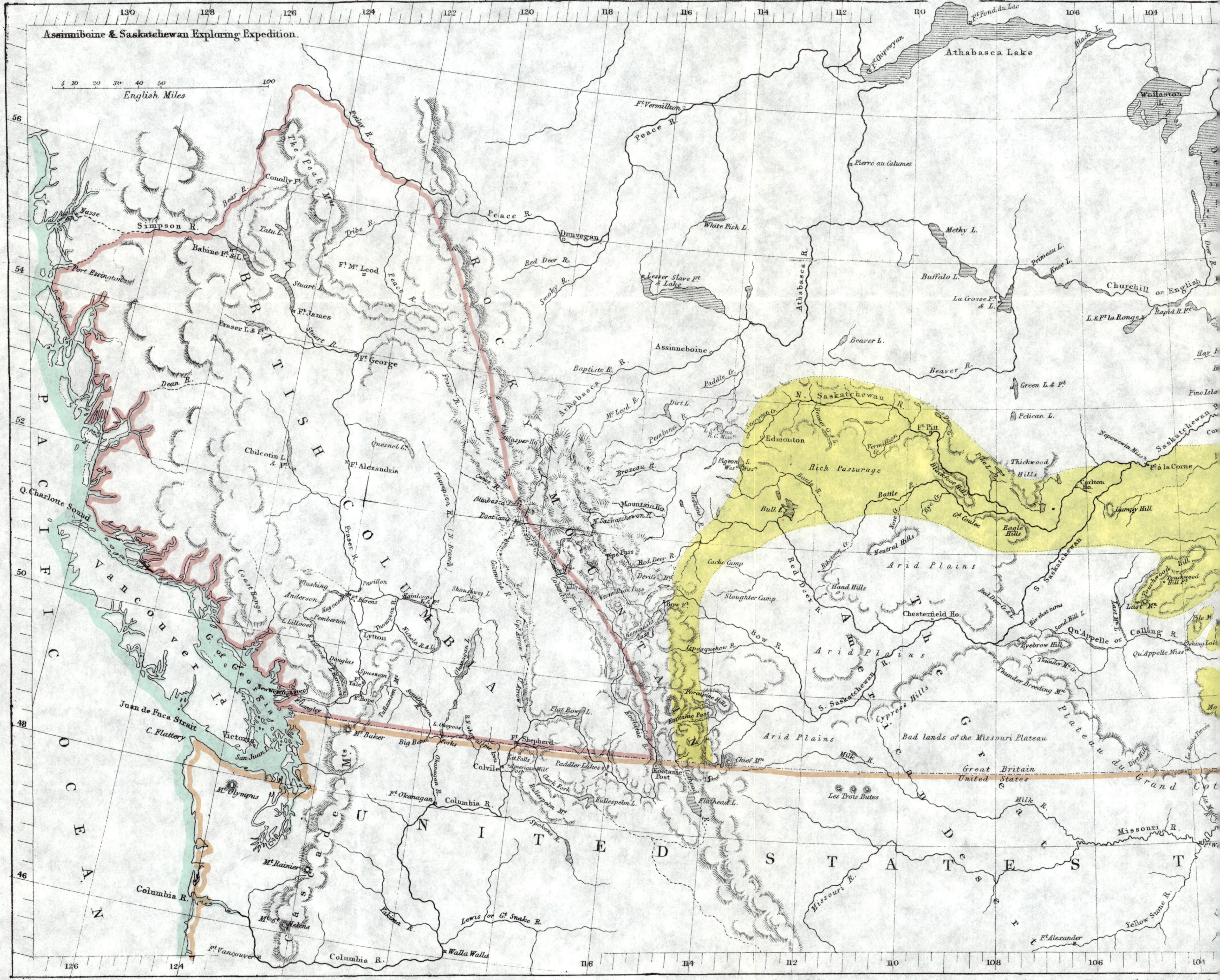

The Fertile Belt is coloured

Authorities:
From Lake of the Woods to Forks of the Saskatchewan .... The Assinniboine & Saskatche
From Forks of the Saskatchewan to the Rocky Mountains .... Captn. Palliser's Expedition.
British Columbia .... Blue Book. Augt. 1860.

Country from LAKE SUPERIOR to the PACIFIC OCEAN, *showing the Western Boundary of Canada & the Eastern Boundary of British Colu*... *so the* Fertile Belt *stretching from the Lake of the Woods to the Rocky Mountains.*

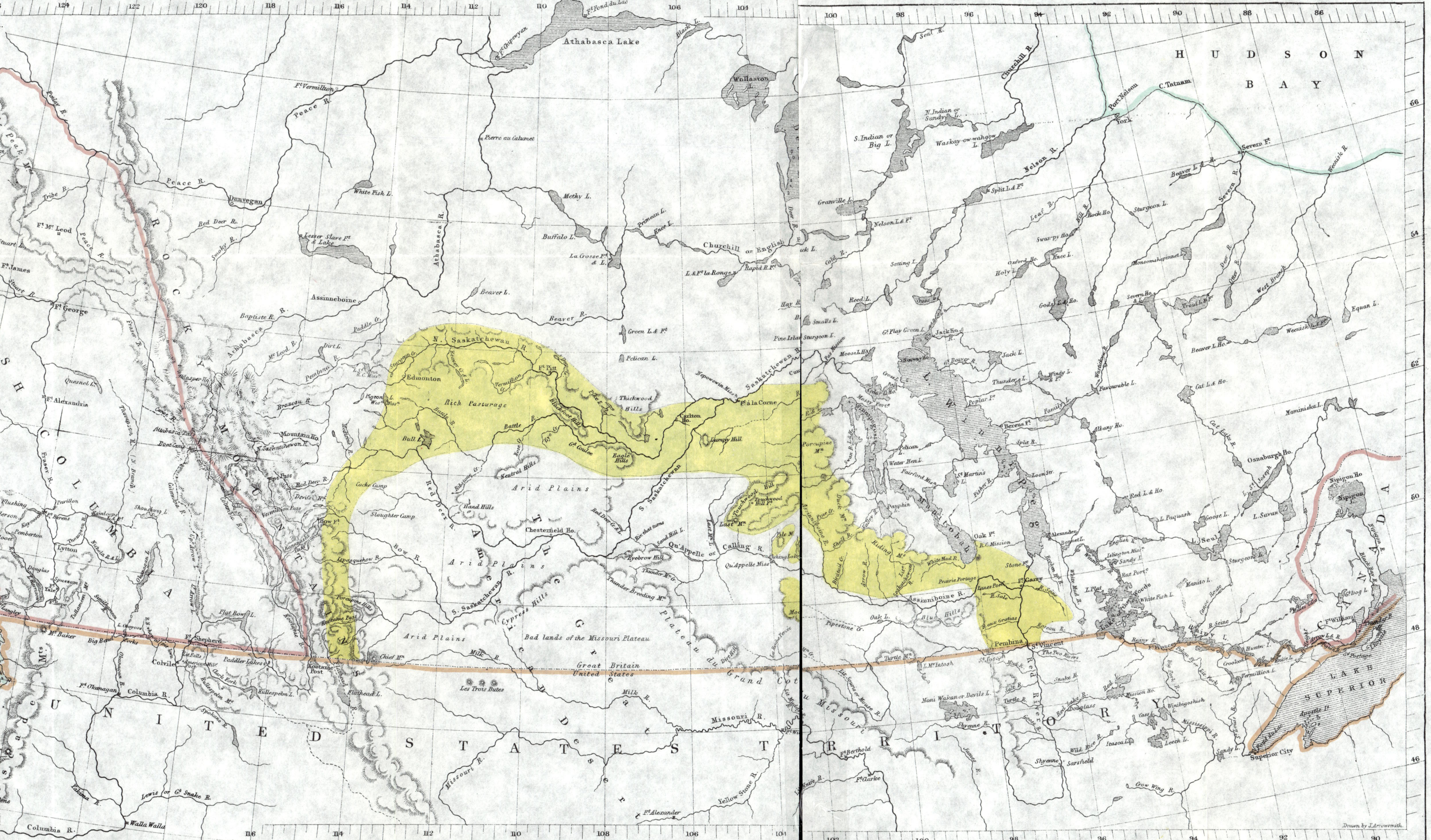

Authorities { From Lake of the Woods to Forks of the Saskatchewan .... The Assiniboine & Saskatche...
From Forks of the Saskatchewan to the Rocky Mountains .... Captn. Palliser's Expedition.
British Columbia .... Blue Book. Augt. 1860.